The Peptides *Volume 1*

THE PEPTIDES

Analysis, Synthesis, Biology

Treatise Editors

E. GROSS AND J. MEIENHOFER

Volume 1

Major Methods of Peptide Bond Formation

The Peptides

Analysis, Synthesis, Biology

VOLUME 1 Major Methods of Peptide Bond Formation

Edited by

ERHARD GROSS

National Institutes of Health
Bethesda, Maryland

JOHANNES MEIENHOFER

Chemical Research Department
Hoffmann-La Roche Inc.
Nutley, New Jersey

ACADEMIC PRESS
New York San Francisco London 1979
A Subsidiary of Harcourt Brace Jovanovich, Publishers

ACADEMIC PRESS, INC.
111 Fifth Avenue, New York, New York 10003

United Kingdom Edition published by
ACADEMIC PRESS, INC. (LONDON) LTD.
24/28 Oval Road, London NW1 7DX

Library of Congress Cataloging in Publication Data

Main entry under title:

The Peptides.

Includes bibliographical references.
CONTENTS: v. 1. Synthesis, part A.
1. Peptides. I. Gross, Erhard. II. Meienhofer, Johannes. [DNLM: 1. Peptides. QU68 P424]
QP552.P4P47 574.1'9245 78-31958
ISBN 0-12-304201-1 (v. 1)

PRINTED IN THE UNITED STATES OF AMERICA

79 80 81 82 9 8 7 6 5 4 3 2 1

Contents

List of Contributors ix
Preface xi
Nomenclature and Abbreviations xiii

Chapter 1 *The Peptide Bond*

Erhard Gross and Johannes Meienhofer

I	Introduction	2
II	Nature of the Peptide Bond	3
III	Cleavage of the Peptide Bond	11
IV	Formation of the Peptide Bond	26
V	Conclusion and Outlook	52
	References	56

Chapter 2 *The Formation of Peptide Bonds: A General Survey*

John H. Jones

I	Introduction	66
II	Coupling between Amino Acid Residues	67
III	*Ab Initio* Assembly of the Backbone	95
IV	Conclusions	97
	References	100

Chapter 3 *Active Esters in Peptide Synthesis*

Miklos Bodanszky

I	The Development of Active Esters	106

II	Classes of Active Esters	109
III	Preparation of Active Esters	128
IV	Catalysis of Active Ester Reactions	138
V	Side Reactions	143
VI	Active Esters and the Strategy of Peptide Synthesis	156
VII	Active Esters in Various Techniques of Peptide Synthesis	165
VIII	Conclusions	174
	References	187

Chapter 4 *The Azide Method in Peptide Synthesis*

Johannes Meienhofer

I	Introduction	198
II	Experimental Procedure	199
III	The Formation of Hydrazides	200
IV	The Generation of Azides	213
V	Azide Segment Condensation	217
VI	Special Applications	228
VII	Opportunities and Constraints	232
	References	233

Chapter 5 *The Carbodiimide Method*

Daniel H. Rich and Jasbir Singh

I	Introduction	242
II	Mechanism of Carboxyl Activation	243
III	Reaction Conditions for Dicyclohexyl-Carbodiimide-Mediated Peptide Bond Formation	252
IV	Reactions of Carbodiimides with Amines	258
	References	260

Chapter 6 *The Mixed Carbonic Anhydride Method of Peptide Synthesis*

Johannes Meienhofer

I	Introduction	264
II	The Experimental Procedure	266
III	Reaction Conditions	267

IV	Coupling Reagents: EEDQ and IIDQ	289
V	Applications	293
VI	Related Methods	304
VII	Opportunities and Constraints	308
	References	309

Chapter 7 *Racemization in Peptide Synthesis*

D. S. Kemp

I	Introduction	317
II	Racemization of Amino Acids, Peptides, and Their Unactivated Derivatives	325
III	Racemization of Activated Acyl Derivatives of Amino Acids and Peptides	336
IV	Control of Racemization during Peptide Synthesis	351
V	Strategies for Minimizing Racemization during Amide-Forming Steps	362
VI	Assays for Chiral Purity	366
	References	378

Author Index 385
Subject Index 407

List of Contributors

Numbers in parentheses indicate the pages on which the authors' contributions begin.

Miklos Bodanszky (105), Department of Chemistry, Case Western Reserve University, Cleveland, Ohio 44106

Erhard Gross (1), National Institutes of Health, Bethesda, Maryland 20014

John H. Jones (65), The Dyson Perrins Laboratory, The University of Oxford, Oxford, England

D. S. Kemp (315), Department of Chemistry, Massachusetts Institute of Technology, Cambridge, Massachusetts 02139

Johannes Meienhofer (1, 197, 263), Chemical Research Department, Hoffmann-La Roche Inc., Nutley, New Jersey 07110

Daniel H. Rich (241), School of Pharmacy, University of Wisconsin, Madison, Wisconsin 53706

Jasbir Singh (241), School of Pharmacy, University of Wisconsin, Madison, Wisconsin 53706

Preface

"The Peptides" is an open-ended treatise providing comprehensive and critical reviews of important developments in three major areas of peptide research: analysis, synthesis, and biology. Volumes on analysis and synthesis are primarily addressed to methodology, whereas those on biology are devoted to individual bioactive peptides or groups of related peptides with emphasis on structure–function relationships. Concise discussions will present information on chemical and physical properties, conformation, and physiological function. Adequate coverage will be extended to questions related to biosynthesis, pharmacology, and clinically relevant aspects.

Unprecedented recent developments in the peptide field and those anticipated in the future necessitated planning this work with flexibility, permitting new topic discussion in each of the three major areas as the need arises. In the mid-sixties, E. Schröder and K. Lübke[1] covered peptide synthesis and biology in a two-volume monograph, but over the past twelve years the literature on peptides has grown at such a rate that it is no longer possible for one or a few authors to review the entire field adequately. "The Peptides" will therefore include contributions from many authors, each one an authority on the topic covered.

"The Peptides" will monitor new developments in the field and critically evaluate their advantages and limitations. In each volume, closely related topics will be comprehensively presented.

"The Peptides" is intended as a reference work for the specialist, as a guide for the student and novice in the field, and as an overview for scientists from a broad range of disciplines concerned with peptides. Prominent among these specialties are analytical, organic, medicinal, and physical chemistry; endocrinology; immunology; and microbiology; as well as numerous facets of pharmacology, physiology, and the clinical sciences. Scientists in these and related areas of research will find this work useful.

[1] E. Schröder and K. Lübke, "The Peptides." Academic Press, New York, 1965 and 1966.

The first volume in this treatise describes methods of peptide synthesis in solution. In an introductory chapter E. Gross and J. Meienhofer discuss the peptide bond. Chapter 2 by J. H. Jones presents a general survey of peptide coupling methods. The four most commonly used procedures of peptide synthesis are discussed in Chapters 3 through 6. Active esters in peptide synthesis are comprehensively described by M. Bodanszky, the azide and mixed anhydride methods are surveyed by J. Meienhofer, the carbodiimide method by D. H. Rich and J. Singh. The concluding chapter by D. S. Kemp reviews racemization, one of the ever-present problems in peptide synthesis.

Subsequent volumes will be devoted to the solid-phase method and to other special procedures and strategies in peptide synthesis, for example, the liquid-phase technique and the oxidation-reduction and four-component condensation methods. Another volume will describe functional group protection.

We wish to thank our authors for their efforts in preparing these chapters and for their patience and cooperation throughout the editorial process. We should also like to express our appreciation to the staff of Academic Press for the efficient production of this book.

Erhard Gross
Johannes Meienhofer

Nomenclature and Abbreviations

In several chapters of this volume the peptide-size nomenclature proposed by M. Bodanszky (1977) at the Fifth American Peptide Symposium has been used. Greek prefixes are retained for di- to decapeptides only. For peptides with more than 10 amino acid residues arabic numerals are used, for example, "15-peptide," "31-peptide," or corticotropin-(1–24), β-LPH-(61–91).

Peptide *segment* and *segment condensation* have been used instead of *fragment* or *fragment condensation.* Amino acids are of the L configuration unless otherwise indicated. No hyphen is used in D-amino acid symbols, e.g., H-Gly-DPhe-Ala-OH.

Abbreviations are kept as short as possible, preferably three letters (four or more letters only if unavoidable). Many abbreviations are indeed those recommended by the IUPAC–IUB Commission on Biochemical Nomenclature (1967, 1970, 1972, 1975). However, the system of complex multiletter symbols including Greek letters and subscript numerals (IUPAC Commission, 1972) will not be used, since these symbols are impractical for typesetting and generally of infrequent use. Short ad hoc symbols instead will be employed and always identified with each citation.

Symbols used in this volume are listed below (except those of the common amino acids).

Abbreviations

AII	angiotensin II
AA	amino acid
AA–tRNA	aminoacyl transfer RNA
Abu	α-aminobutyric acid
Ac	acetyl
Acm	acetamidomethyl
AcOH, HOAc	acetic acid
Ac_2O	acetic anhydride

ACTH	corticotropin
Adc	adamantyloxycarbonyl
AE	active ester method
*a*Hyp	*allo*-4-hydroxyproline
Aib	α-aminoisobutyric acid
*a*Ile	*allo*-isoleucine
AZ	azide method
β-Ala	β-alanine
Bic	5-benzisoxazolylmethyleneoxycarbonyl
Boc	*tert*-butyloxycarbonyl
Bpoc	2-(*p*-biphenylyl)propyl(2)oxycarbonyl
BuOH	butanol
Bz	benzoyl
Bzl	benzyl
CCD	countercurrent distribution
CD	circular dichroic spectroscopy
2Cz	2-chlorobenzyloxycarbonyl
4Cz	4-chlorobenzyloxycarbonyl
DCC	dicyclohexylcarbodiimide (method)
DCU	*N*,*N'*-dicyclohexylurea
Ddz	α,α-dimethyl-3,5-dimethyloxybenzyloxycarbonyl
Dha	dehydroalanine
Dhb	dehydrobutyrine
DME	1,2-dimethoxyethane
DMF	dimethylformamide
DMAc	dimethylacetamide
DMSO	dimethyl sulfoxide
DNA	deoxyribonucleic acid
Dobz	*p*-dihydroxyboronatobenzyloxycarbonyl
2,4-Dnp	2,4-dinitrophenyl
2,6-Dnp	2,6-dinitrophenyl
Dpm	diphenylmethyl (benzhydryl)
EEDQ	*N*-ethyloxycarbonyl-2-ethyloxy-1,2-dihydroquinoline
EK	enkephalin
EtMorph	*N*-ethylmorpholine
Et_3N	triethylamine
EtOAc	ethyl acetate
EtOCOCl	ethyl chlorocarbonate
EtOH	ethanol

Fmc	9-fluorenylmethyloxycarbonyl
fMet–tRNA	formylmethionyl–tRNA
For	formyl
GnRH	gonadoliberin
GTP	guanosine triphosphate
Hal	halogen
HMP	hexamethylphosphoramide
HOAc, AcOH	acetic acid
HOBt	1-hydroxybenzotriazole
HONB	*N*-hydroxy-5-norbornene-2,3-dicarboximide
HOSu	*N*-hydroxysuccinimide
HPLC	high performance liquid chromatography
Hyp	4-hydroxyproline
IBc	isobornyloxycarbonyl
*i*BuOCOCl	isobutyl chlorocarbonate
IIDQ	*N*-isobutyloxycarbonyl-2-isobutyloxy-1,2-dihydroquinoline
im	imidazole
ir	infrared spectroscopy
β-LPH	β-lipotropin
MA	mixed anhydride method
Me	methyl
MeAla	*N*-methylalanine
MeCN	acetonitrile
MeIle	*N*-methylisoleucine
MeLeu	*N*-methylleucine
MeMorph	*N*-methylmorpholine
MeOH	methanol
MePhe	*N*-methylphenylalanine
Met(O)	methionine sulfoxide
Met-ol	methioninol
Moz	4-methyloxybenzyloxycarbonyl
mRNA	messenger ribonucleic acid
β-MSH	β-melanotropin
NBSu	*N*-bromosuccinimide
NCA	*N*-carboxy anhydride
nmr	nuclear magnetic resonance spectroscopy

NO_2z	*p*-nitrobenzyloxycarbonyl
Nps	*o*-nitrophenylsulfenyl
OBt	benzotriazolyl ester
OBzl	benzyl ester
OEt	ethyl ester
OMe	methyl ester
ONb	*p*-nitrobenzyl ester
ONm	*m*-nitrophenyl ester
ONo	*o*-nitrophenyl ester
ONp	*p*-nitrophenyl ester
OPcp	pentachlorophenyl ester
OPfp	pentafluorophenyl ester
OPh	phenyl ester
OPic	picolyl ester
OQ	quinolyl ester
OQc	5-chloroquinolyl(8) ester
OQ(*n*)	quinolyl(*n*) ester ($n = 3,6,8$)
OR	ester (general)
ORD	optical rotatory dispersion spectroscopy
Orn	ornithine
OSu	succinimidyl ester
O*t*Bu	*tert*-butyl ester
2,4,5-OTcp	2,4,5-trichlorophenyl ester
2,4,6-OTcp	2,4,6-trichlorophenyl ester
PCP	pentachlorophenol
pGlu	pyroglutamic acid
Ph	phenyl
Phg	*C*-phenylglycine
Pht	phthaloyl
Pip	pipecolic acid
pmr	proton magnetic resonance spectroscopy
*i*Pr	isopropyl
$i\mathrm{Pr}_2\mathrm{NEt}$	diisopropylethylamine
Δ^3Pro	3,4-dihydroproline
Prot	protecting group (general)
Pyr	pyridine
REMA	repetitive excess mixed anhydride method
RNase	ribonuclease

Sar	sarcosine, *N*-methylglycine
SNp	*p*-nitrothiophenyl ester
SPh	thiophenyl ester
SPy	2-pyridyl thiol ester
*t*Bu	*tert*-butyl
Tcc	2,2,2-trichloroethyloxycarbonyl
TFA	trifluoroacetic acid
TFA_2O	trifluoroacetic anhydride
Tfa	trifluoroacetyl
THF	tetrahydrofuran
Thz	thiazolidine-2-carboxylic acid
Tos	tosyl, *p*-toluenesulfonyl
TOS	*p*-toluenesulfonic acid
tRNA	transfer ribonucleic acid
Trt	trityl, triphenylmethyl
USP	United States Pharmacopeia
Z	benzyloxycarbonyl

REFERENCES

Bodanszky, M. (1977). *In* "Peptides, Proceedings of the Fifth American Peptide Symposium" (M. Goodman and J. Meienhofer, eds.), pp. 1–13. Wiley, New York.
IUPAC–IUB Commission on Biochemical Nomenclature (1967). *J. Biol. Chem.* **242**, 555–557.
IUPAC Commission (1970). *J. Biol. Chem.* **245**, 6489–6497.
IUPAC Commission (1972). *J. Biol. Chem.* **247**, 979–982.
IUPAC Commission (1975). *J. Biol. Chem.* **250**, 3215–3216.

Chapter 1

The Peptide Bond

ERHARD GROSS and JOHANNES MEIENHOFER

I. Introduction 2
II. Nature of the Peptide Bond 3
A. Conformational Analysis in Solution 5
B. X-Ray Crystal Structure Analysis 8
C. Is the Peptide Bond Planar? 10
III. Cleavage of the Peptide Bond 11
A. Hydrolysis and Solvolysis 11
1. Hydrolysis 11
2. Liquid Hydrogen Fluoride Cleavage 15
B. Enzymatic Cleavage of the Peptide Bond 16
1. Limited Proteolysis in the Physiological Environment 16
2. Cascade Processes 16
3. Protein and Peptide Structures Related to β-Lipotropin 19
4. Proteolysis and Structural Elucidation 19
C. Nonenzymatic Cleavage of the Peptide Bond 23
IV. Formation of the Peptide Bond 26
A. Biosynthesis 26
B. Chemical Synthesis 29
1. Basic Features 29
2. Kinetics 31
3. Catalysis 37
4. Racemization 41
C. Strategy of Peptide Chain Assembly 42
1. Basic Scheme 42
2. Strategies and Tactics of Functional Group Protection 43
3. Solution Synthesis 45
4. Solid Phase Synthesis 46
D. Opportunities and Constraints 49
V. Conclusion and Outlook 52
References 56

THE PEPTIDES, VOL. 1

ISBN 0-12-304201-1

I. INTRODUCTION

Den Namen Polypeptide habe ich vorgeschlagen für die Produkte, die durch amidartige Verkettung von Aminosäuren entstehen, und deren einfachster Vertreter das Derivat des Glykocolls, das sogenannte Glycylglycin, $NH_2CH_2CO{-}NHCH_2COOH$, ist. Nach der Anzahl der in ihnen enthaltenen Aminosäuren sollen sie als Di-, Tri-, Tetra-Peptide usw. unterschieden werden. Diese Bezeichnung ist einerseits der Nomenklatur der Kohlehydrate nachgebildet, andererseits ist darin das alte Wort Pepton verwertet, denn ich habe von Anfang an erwartet, und ich bin durch alle nachfolgenden Beobachtungen in dieser Überzeugung bestärkt worden, dass diese künstlichen Produkte den natürlichen Peptonen sehr nahe verwandt sind, mit anderen Worten, dass diese Peptone im wesentlichen ein bisher untrennbares Gemisch von Polypeptiden sind.*

Thus, Emil Fischer defined in 1906 the peptide bond as the amide-like linkage between amino acids and gave the reasons why he chose the term *peptide*. What was called peptones in Fischer's days are the mixtures of peptides formed on the action of proteases on proteins. Much progress has been made and Fischer's seemingly optimistic outlook on the separation of peptide mixtures was warranted in retrospect.

Four years earlier F. Hofmeister (1902) had written reviewingly in an effort to order the facts known about Eiweisskörper (proteins!) at the turn of the century. He remarked as follows:

Man darf daher auf Grund der gegebenen Sachlage die Eiweissstoffe der Hauptsache nach als durch Kondensation von α-Aminosäuren entstanden ansehen, wobei die Verknüpfung durch die Gruppe —CO—NH—CH= als die regelmässig wiederkehrende anzusehen ist.†

Peptides and proteins exhibit the largest structural and functional variation of all classes of biologically active macromolecules. They are of prime importance in the regulation and maintenance of all biological processes. The essential structural features of peptide and protein molecules are chains of amino acids linked to one another by amide bonds (**1**). This structure appears to be relatively straightforward, yet it may be viewed in several

* "I have proposed the name polypeptide for the products formed on linking amino acids via amide type bonds. A derivative of glycocoll, the so-called glycyl-glycine, $NH_2CH_2CO{-}NHCH_2COOH$, is the simplest representative of this class of compounds. On the basis of the number of amino acids found in peptides, they shall be distinguished as di-, tri-, tetra-peptides, etc. The new term is patterned on the nomenclature for carbohydrates, on the one hand, on the other the earlier word pepton finds recognition, as I expected from the beginning, and all subsequent observations support this view, that these artificial products are very closely related to the natural peptones. To put it in different words, these peptones are mixtures of polypeptides for which adequate separation procedures have yet to be developed." (E. Fischer, 1906.)

† "Based on the facts presented, proteins may in essence be considered products formed by the condensation of amino acids. Linkage via the group —CO—NH—CH= must be viewed as regularly recurring." (F. Hofmeister, 1902.)

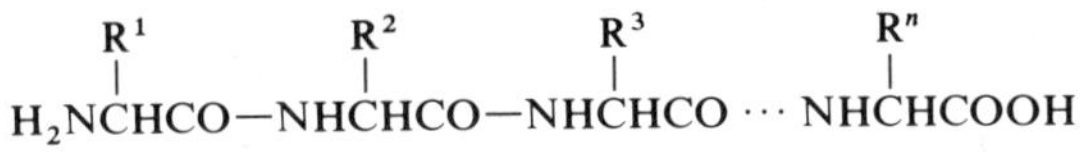

(1)

different perspectives depending on the scientific specialization of the reader. To physicochemists concerned with the conformation of peptides the important structural features of **1** are its bond lengths and angles as shown in Fig. 1. In studies on the spatial electron distributions of peptides, which ultimately determine their chemical and conformational properties, contour maps are obtained from which the peptide bond appears as shown in Fig. 2. For organic chemists interested in special features of the side-chain substituent R, structure **1** assumes proportions similar to those shown in **2**. The

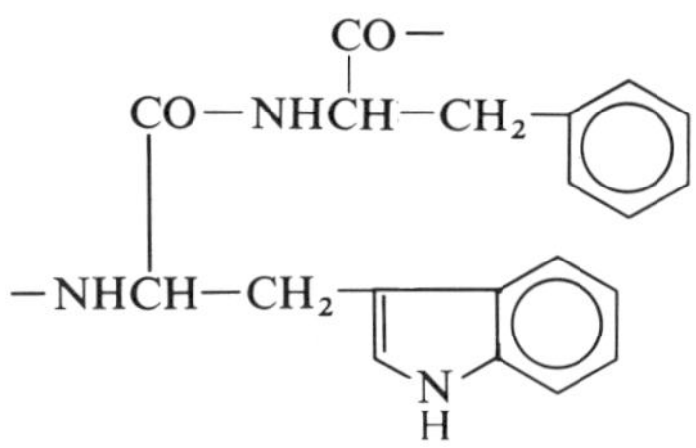

(2)

chemist concerned with peptide synthesis, however, will view structure **1** as an assembly of amino acid building units (**3**) and be quite concerned with the

$$H{-}[NHCHRCO]_n{-}OH$$

(3)

number, *n*, of residues in a target structure. The physicochemical nature of the peptide bond in terms of configuration and preferred conformational states will be discussed briefly in Section II of this chapter. Section III will treat analytical aspects which involve the degradation of the peptide bond by solvolytic, enzymatic, and selective chemical procedures. Section IV will deal with acylation reactions, the tactics of peptide synthesis (protective group selection), the strategy of peptide chain assembly, the present state of the methodology, and the prognosis for the chemical synthesis of proteins.

II. NATURE OF THE PEPTIDE BOND

One of the most fascinating aspects of the unique nature of the peptide bond is the ability of nascent biosynthetic peptide chains to fold spontaneously into the complex three-dimensional structures which are character-

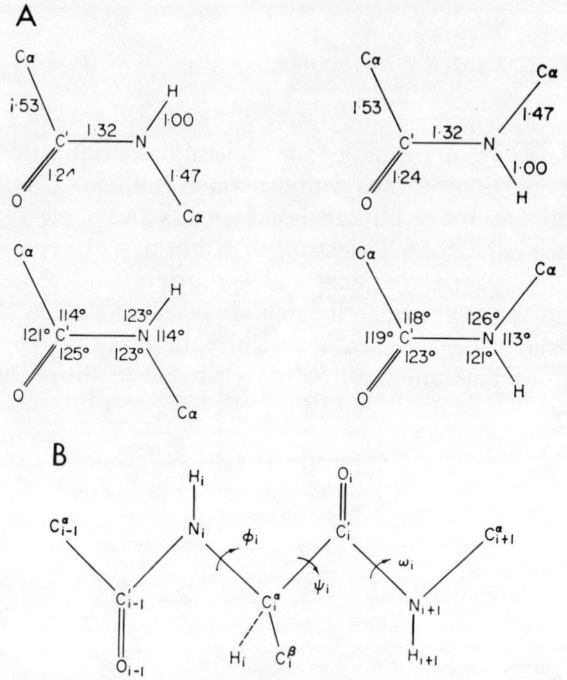

Figure 1. (A) Dimensions of standard trans (left) and cis (right) peptide units (Corey and Pauling, 1953; Ramachandran and Sasisekharan, 1968). Bond angles are in angstroms. (B) Definition of torsion angles ψ_i, ϕ_i, and ω_i in a polypeptide chain. (From Winkler and Dunitz, 1971; reproduced with permission of Academic Press.)

istic of proteins. However, the fact that even some very small linear peptides may assume preferred conformations in solution has only recently been recognized. A striking example is the COOH-terminal tripeptide of oxytocin, Pro—Leu—Gly—NH_2, for which a preferred solution conformation was proposed (Walter *et al.*, 1972) based on 300 MHz pmr studies in dimethyl sulfoxide. Forming a 10-membered, β-turn the peptide is stabilized by a hydrogen bond between the *trans*-carboxamide proton and the C=O of proline, a conformation which was subsequently confirmed by X-ray crystal structure analysis (Reed and Johnson, 1973) which also showed that the prolyl–leucyl peptide bond is significantly *nonplanar*.

The first nmr-derived solution conformation of a peptide hormone was that of oxytocin (Urry and Walter, 1971). A considerable number of peptide conformations have by now been determined either by spectroscopic techniques in solution or by X-ray crystal structure analysis (Marsh and Donohue, 1967; Karle, 1975; Benedetti, 1977). A few prominent examples are discussed below.

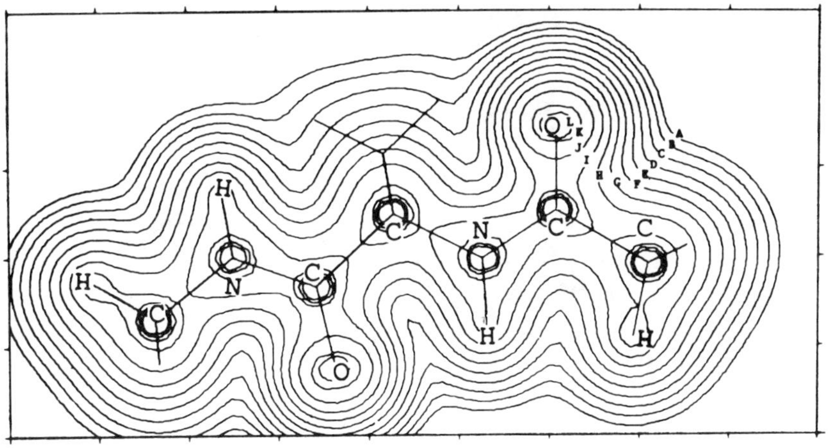

Figure 2. Contours of constant electron density in the plane of the peptide groups in *N*-acetyl-*N*′-methylalanine ($\phi = 180°$, $\psi = 180°$) showing the total molecular density as derived from *ab initio* molecular orbital calculations. (From Hagler and Lapiccirella, 1976; reproduced with permission of John Wiley and Sons, Inc.)

A. Conformational Analysis in Solution

Of the spectroscopic methods useful for the study of peptide conformations (nmr, ORD, CD, and ir), high resolution nmr and ^{13}C-nmr have emerged as the most powerful ones (for reviews, see Bovey, 1972; Bovey *et al.*, 1972; Wyssbrod and Gibbons, 1974; Hruby, 1974; Ovchinnikov and Ivanov, 1975; Ivanov, 1977). In 1971, Urry and Walter proposed a solution conformation for oxytocin (Fig. 3A) which revealed the presence of a type II β-turn involving the sequence -Tyr-Ile-Gln-Asn- within the 20-membered cyclic moiety of the hormone. Thus, the chain is folded back into an antiparallel pleated sheat conformation with the disulfide bridge closing the ring and stabilizing the structure. The backbone NH of asparagine-5 is hydrogen-bonded to the C=O of tyrosine-2 and provides additional intramolecular stabilization. The $CONH_2$-terminal tail moiety forms a second β-turn (type I), comprising residues -Cys-Pro-Leu-Gly-, which folds the tail over one side of the ring and is stabilized by another hydrogen bond between the NH of the leucine-8 and the side chain C=O of the asparagine-5 residues. All peptide bonds are of trans configuration, including the Cys-Pro bond. Similar but not identical conformations have been deduced for lysine–vasopressin, arginine–vasopressin, and arginine–vasotocin (Walter *et al.*, 1974). Most likely the overall backbone structure with the two β-turns involving residues 2–5 and 6–9 is a common conformational feature for all nine neurohypophyseal peptides found in nature. However, the charged

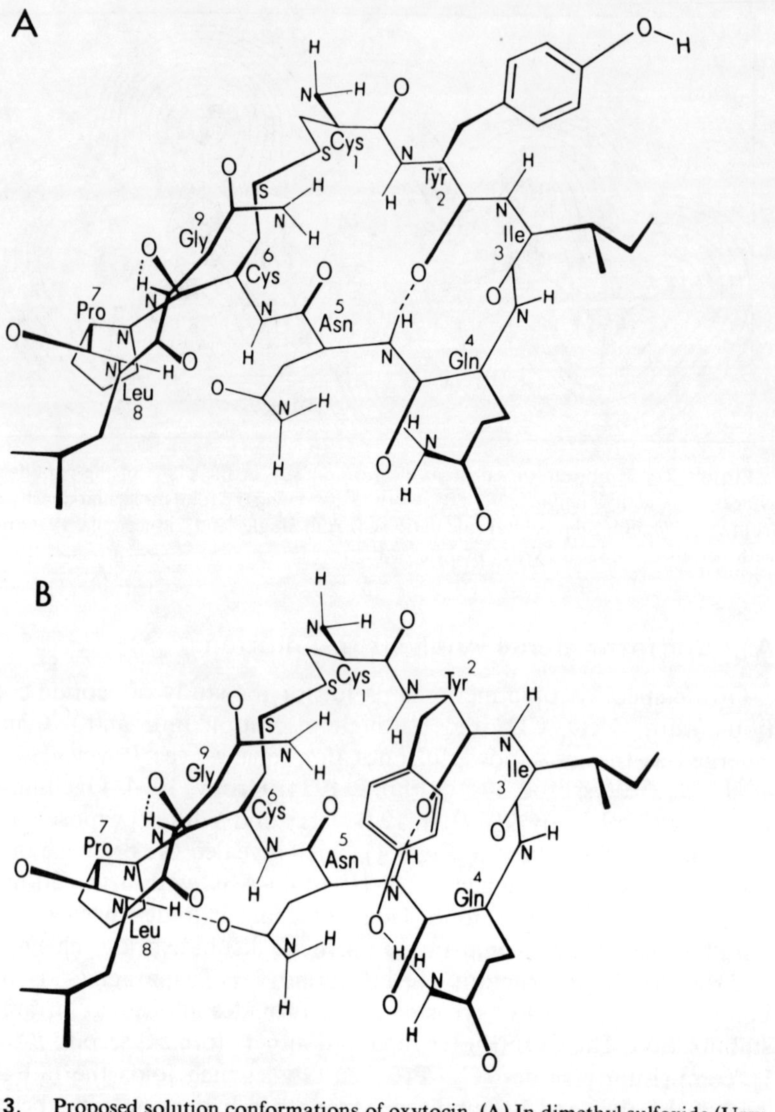

Figure 3. Proposed solution conformations of oxytocin. (A) In dimethyl sulfoxide (Urry and Walter, 1971); (B) at receptor sites (Walter, 1977).

tail portion of lysine–vasopressin seems to possess a *slightly* larger conformational freedom, and those of arginine–vasopressin and arginine–vasotocin *considerably* larger conformational freedom, compared to oxytocin.

During its interaction with receptors, oxytocin is proposed to assume a somewhat different, so-called "cooperative," conformation in which the tyrosine side chain is folded over the 20-membered ring of oxytocin (Fig. 3B). In this model, the tyrosine hydroxyl group acting cooperatively with the asparagine carboxamide group appears as the predominant active element initiating the oxytocin response (Walter, 1977).

In aqueous medium conformational averaging increases significantly and the hydrogen bonds may be loosened but the gross structure is presumed to be conserved. It must be borne in mind that these proposed conformations should be viewed as time averaged, preferred structures of the peptide backbone.

The proposed oxytocin conformation has led to the development of rational guidelines for the design of hormone analogues with selectively enhanced biological activities, more favorable activity ratios, or competitive inhibitory activity (Walter, 1977). Residues in the corner positions 3, 4, 7, and 8 of the three-dimensional structure (Fig. 3B), which are not primarily involved in the intramolecular stabilization of the peptide backbone appear to contain the elements for oxytocic receptor recognition, the *binding element*. The hydroxyl group of the tyrosine side chain, folded over the 20-membered ring of oxytocin, acting cooperatively with the asparagine carboxamide group, is considered to be the *active element*. While modifications of the active element influence potency and/or agonist–antagonist characteristics, substitutions in the corner positions should provide analogues with selectively modified activity profiles. Indeed, the amino acid substitutions which distinguish the naturally occurring neurohypophyseal hormones (see Acher *et al.*, 1973a,b) are confined to position 3, 4, and 8. Using this type of rationale [4-L-threonine, 7-glycine]oxytocin was prepared (Stahl and Walter, 1977) which exhibited an oxytocic to antidiuretic activity ratio of 135,000:1, compared to a ratio of 200:1 for native oxytocin.* Another superselective analogue of arginine–vasopressin, designed along similar guidelines, i.e., [1-deamino,-2-phenylalanine,-7-(L-3,4-dehydroproline)]arginine–vasopressin, exhibited an antidiuretic to pressor ratio of 31,000:1 and a specific antidiuretic activity of 13,000 ± 1,250 USP units/mg, i.e., 26-fold that of the native hormone, thus being the most potent antidiuretic analogue yet prepared (Smith and Walter, 1978).

The significance of this new level of sophistication in analogue design on a predictable basis goes beyond the preparation of oxytocin or vasopressin analogues with exceptionally high potencies and/or specificity for a given activity. The development of conformation–activity relationships and their application to analogue synthesis of other peptide hormones could probably save some of the enormous expenditure of effort that was required for the

* The same analogue was simultaneously prepared by Manning and collaborators (Lowbridge *et al.*, 1977) following a different rationale.

synthesis of hundreds of analogues of neurohypophyseal hormones during the earlier, more empirical structure–activity approach to analogue design.

B. X-Ray Crystal Structure Analysis

Structure studies with small peptides (of up to 12 residues) by the direct method of determination from X-ray intensities have established the following facts (Karle, 1975): (i) that similar conformations do occur in different molecules (for example proline-containing linear peptides may form U-shaped bends to look very similar to cyclic hexaglycyl), (ii) that a particular molecule can have a number of different conformations in the same unit cell of a crystal (for example, cyclohexaglycyl; Karle, 1972), (iii) that five different kinds of hydrogen bonds can be formed, i.e., 3 → 1, 4 → 1: trans I, trans II, and cis, and 5 → 1, and (iv) that cis conformations for peptide units are fairly common and appear to occur only for peptide units with N-substitution, although N-substituted peptide units do not necessarily assume the cis conformation.

For the X-ray structure elucidation of larger peptides heavy atom substitution is required. Some of the most interesting advances have been the determinations of the crystal structures of the pancreatic hormones insulin (Adams *et al.*, 1969; Blundell *et al.*, 1972) and glucagon (Sasaki *et al.*, 1975). The insulin structure shown in Fig. 4 depicts the hexameric form (A) which corresponds to the aggregated 36,000 mol. wt. species observed in neutral solution. The crystal structure provided a clear molecular interpretation of the long known monomer (C), to dimer (B), to hexamer (A) association of insulin. The most important aspect of the insulin monomer structure for its biological activity is the integrity of the overall molecular geometry which stabilizes a hydrophobic receptor binding area comprising 5 residues each from the A- and B-chains (Pullen *et al.*, 1976). No single amino acid residue contributes directly to the bioactivity of the hormone. The relative importance of sequence regions or single amino acid residues, which can vary extensively, depends on their respective contributions to the stability of the compact three-dimensional structure. In addition to the receptor binding area, several other functional surface regions have been identified, e.g., the monomer–monomer and dimer–dimer binding regions and an antigenic region, each comprising certain residues from both the A- and B-chains. The knowledge of the three-dimensional structure of insulin provided great impetus to the establishment of more sophisticated structure–activity relationships (Blundell *et al.*, 1972; Brandenburg *et al.*, 1973) and led to predictive analogue design (Pullen *et al.*, 1976; Friesen *et al.*, 1977). The elucidation of the crystal structure of insulin has been an outstanding

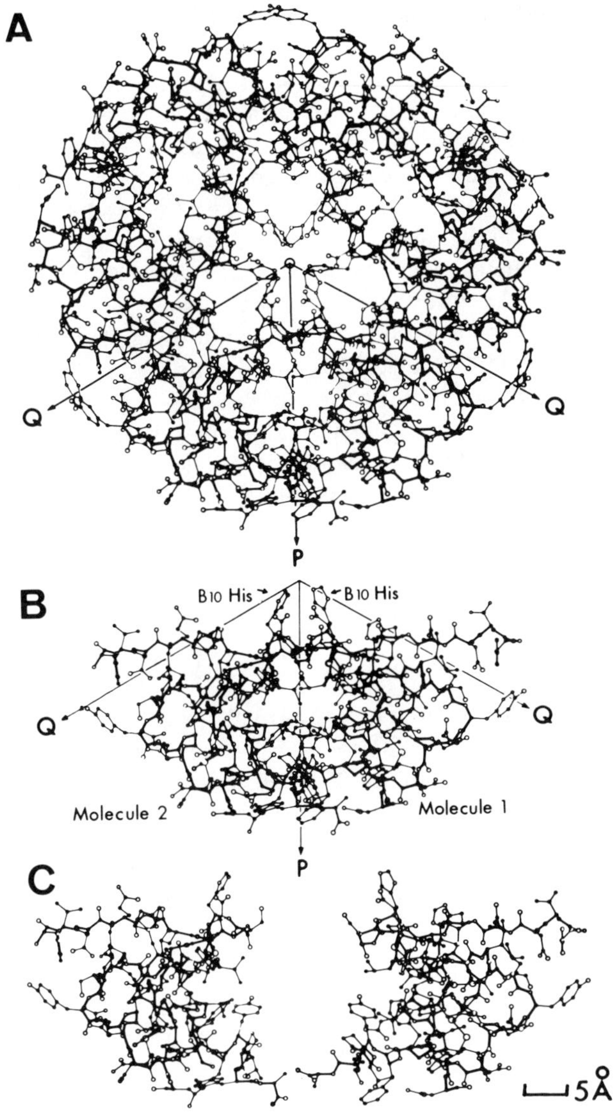

Figure 4. The crystal structure of bovine zinc insulin at 1.9 Å resolution. Two monomeric molecules (C) aggregate in the presence of zinc to form stable dimers (B) which on further aggregation form hexamers (A) in threefold symmetry which constitute a unit cell in the rhombohedral crystal. (Modified from Blundell *et al.*, 1972.)

landmark in peptide hormone research and it is hoped that those of other hormones will follow in due time.

The question whether the X-ray crystal structures and the solution conformations of peptides differ, and if so to which degree, has recently been studied by Bovey (1974) who concluded "It appears that there are more instances of agreement than of disagreement between X-ray and nmr conformations of cyclic polypeptides. There is generally good agreement in the presence of strong stabilizing factors such as proline units, bulky side chains and/or complexed metal ions." As an example, the crystal structure for the COOH-terminal tetrapeptide H—Cys—Pro—Leu—Gly—NH_2 of oxytocin (Rudko *et al.*, 1971) agrees with the β-turn solution conformation proposed by Urry and Walter (1971) for this part of the molecule. For some complex larger structures agreement also seems to exist. The compact three-dimensional crystal structure of insulin appears to be conserved in solution (Mercola *et al.*, 1972) and during its receptor interaction (Pullen *et al.*, 1975). Identity of its crystal structure with its solution conformation and conformation in the DNA receptor complex has also been indicated for the peptide antibiotic actinomycin D (Lackner, 1975; see Meienhofer and Atherton, 1977).

C. Is the Peptide Bond Planar?

One of the most fundamental concepts in all investigations of polypeptide conformations over the past 20 years has been that of the planar trans peptide group (Corey and Pauling, 1953; Ramachandran and Sasisekharan, 1968). In the light of details about peptide conformations that are observed by modern experimental techniques and refined X-ray crystal structure determinations in proteins it has become increasingly more evident in recent years that peptide units may deviate significantly from planarity (Scheiner and Kern, 1977; Wüthrich *et al.*, 1974; Pullman and Pullman, 1974; Ramakrishnan and Balasubramanian, 1972).

For example, in the crystal structure of bovine pancreatic trypsin inhibitor four peptide units (residues 13–16) surrounding the lysine-15 active site residue are particularly nonplanar with observed torsion angles of $\omega = -9°$, $+17°$, $-9°$, $-17°$ (Deisenhofer and Steigemann, 1975; see also Huber *et al.*, 1975). For synthetic cyclotetraglycine nonplanarity of all four bonds has been proposed from nmr evidence (Wüthrich *et al.*, 1974), and in another cyclic tetrapeptide, the antimetabolite derivative dihydrochlamidocin (structure **4**), crystal structure analysis (Flippen and Karle, 1976) revealed nonplanarity of all (trans) peptide units: ω angles deviated by 14° to 24° from the ideal value of 180°. However, even more striking are observations of nonplanar peptide units in small linear peptides in the absence of severe

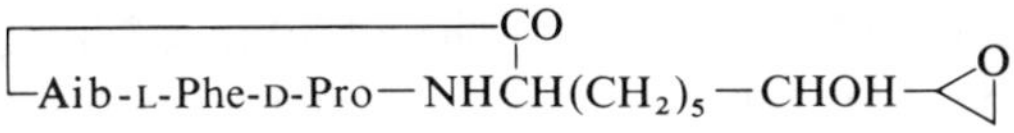

Aib = α-aminoisobutyric acid

(4)

constraints such as those imposed upon cyclic tetrapeptides or larger protein structures. For example, a torsion angle ω of 10.2° has been observed in the crystal structure of glycyl-L-alanine·HCl (Naganathan and Venkatesan, 1972), and in that of glycyl-L-leucine the observed twist amounts to $\omega = -11.4°$ (Pattabhi *et al.*, 1974). The observed nonplanarity of the prolyl–leucyl bond in the crystal structure of Pro-Leu-Gly-NH_2 (Reed and Johnson, 1973) has been mentioned before.

Distortions of individual peptide units from planar form (called "pyramidalizations") have been examined by both semiempirical (e.g., Momany *et al.*, 1970; Winkler and Dunitz, 1971; Ramakrishnan and Balasubramanian, 1972; Ramachandran *et al.*, 1973) and *ab initio* techniques (e.g., Renugopalakrishnan and Rein, 1976). These and recent calculations by Pullman and Pullman (1974) show that relatively little energy is required for significant deviations from planarity to occur, e.g., about 1 kcal/mol up to a torsion of 12°–13° and ca. 2 kcal/mol for a twist of about 20°.

In conclusion, it appears that indications are accumulating that the most important characteristic might not be the planarity of the peptide unit, but rather the *flexibility of the peptide bond.*

III. CLEAVAGE OF THE PEPTIDE BOND

Peptides may be degraded into constituent amino acids by acid or alkaline hydrolysis (for review, see Hill, 1965). Enzymatic degradation by proteolytic enzymes may proceed in stepwise fashion from either terminus (by exopeptidases) or through cleavage of specific single bonds in the interior (by endopeptidases), the specificity being dependent on configuration, amino acid sequence, and conformation (see Hirs and Timasheff, 1972). For selective chemical cleavage of peptides a formidable variety of procedures have been developed (for review, see Spande *et al.*, 1970), with a few capable of cleaving at the α-carbon (e.g., via dehydroalanine residues).

A. Hydrolysis and Solvolysis

1. *Hydrolysis*

Peptides can be readily hydrolyzed to give α-amino acids which usually account quantitatively for the weight of the peptide treated. Mechanistically, the hydrolysis of simple amides has much in common with ester hydrolysis.

Under alkaline conditions nucleophilic addition of hydroxyl ion to the amide carbonyl occurs, followed by the elimination of ammonia (R′ = H) or an amine (Scheme 1). Alkaline hydrolysis is infrequently used in peptide studies in contrast to acidic hydrolysis which is very important.

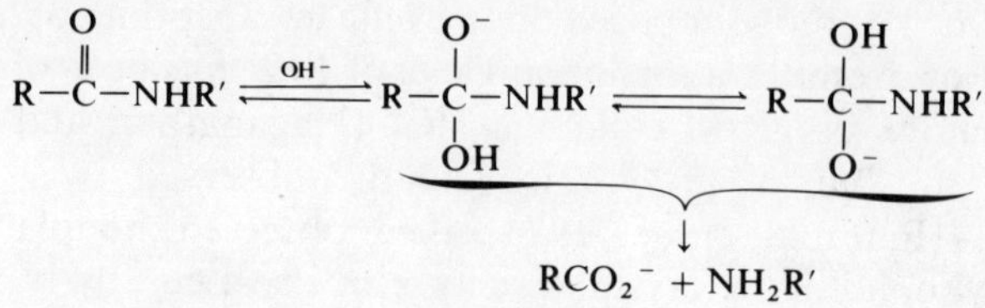

Scheme 1

Under acidic conditions protonation of the amide presumably precedes addition of water (Scheme 2). Energy surface calculations indicated that

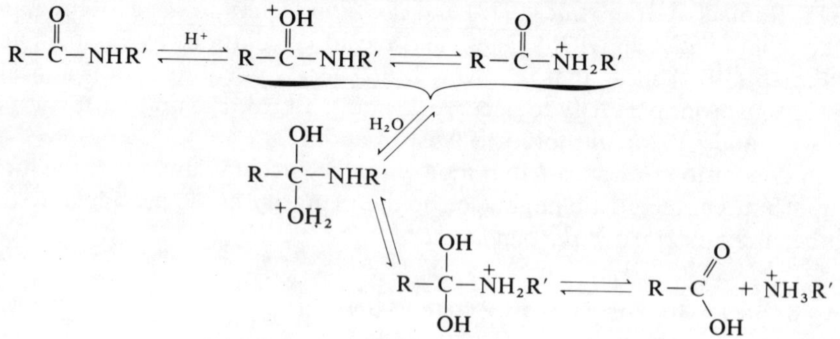

Scheme 2

N-protonation of the peptide unit is favored if the NH_2^+ group is able to rotate, otherwise, O-protonation is favored (Moffat, 1973; see there for other references). The rate of hydrolysis of peptides increases with acid concentration over the range of 3 *M* to 10 *M* (HCl) which operates by assisting leaving group departure (Scheme 3), thereby increasing the electrophilicity

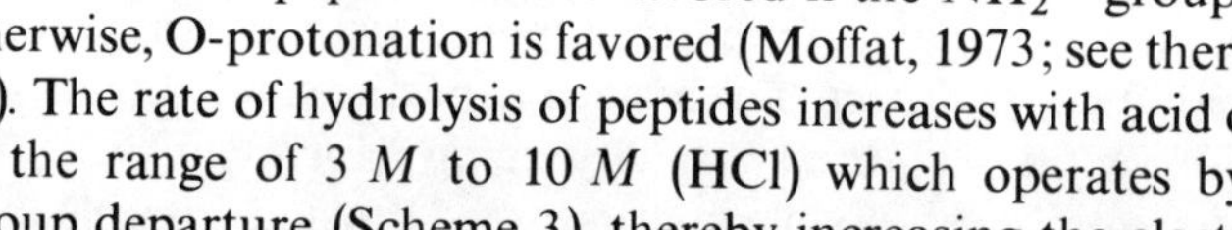

Scheme 3

of the acylating agent (Satchell, 1963). Most peptides are completely hydrolyzed by 16- to 20-hour treatments with constant boiling HCl (6.5 *N*) at 105°–110°C. The importance of complete hydrolysis for the sensitivity and precision of amino acid analysis has been authoritatively reviewed by Moore (1972).

In spite of the extensive use of acid hydrolysis of peptides, quantitative information on the influence of temperature, pressure, acid concentration, and kind of acid used is fragmentary (Hill, 1965). Electrostatic effects decrease the rate of acid hydrolysis of a peptide bond adjacent to the α-ammonium group on both energy and entropy grounds (see Table I), resulting in slow dipeptide hydrolysis (Fig. 5) (Long and Lillycrop, 1963). Steric effects (Muramatu *et al.*, 1963) limit the approach of H^+ ions to the peptide bond and cause the well-known resistance to acid hydrolysis of Ile-Ile or peptides with NH_2-terminal valine or leucine (see also Table I). Peptide bonds formed by aspartic acid or by the amino groups of serine and threonine are very sensitive to acid hydrolysis (see Hill, 1965, for a review and earlier references).

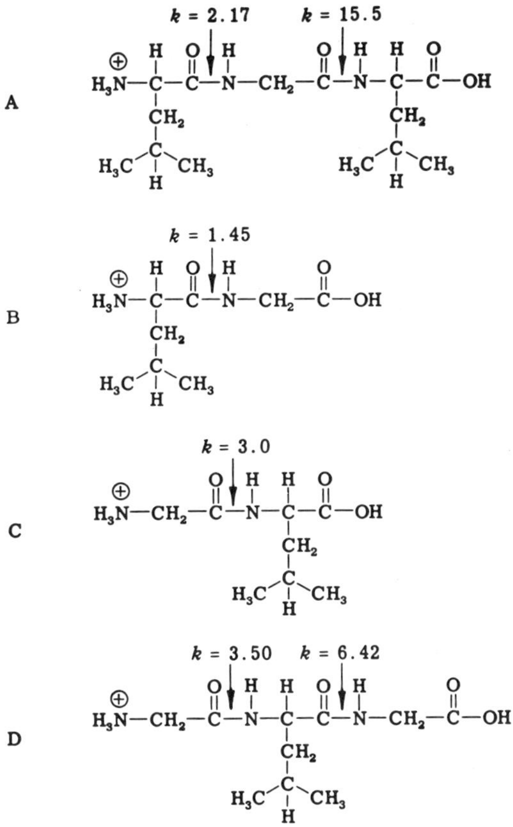

Figure 5. The mean first-order rate constants ($\times 10^3$ min^{-1}) for the hydrolysis of leucyl–glycyl–leucine (A), leucyl–glycine (B), glycyl–leucine (C), and glycyl–leucyl–glycine (D) in 2 *N* HCl at 94.15°C. (From Hill, 1965; reproduced with permission of Academic Press.)

Table I Heat and Entropy Values for the Hydrolysis of Peptide Bonds in Leucylglycylleucine and Glycylleucylglycine[a]

	Peptide					
	Leu-Gly	Leu-Gly-Leu	Gly-Leu-Gly	Gly-Leu	Leu-Gly-Leu	Gly-Leu-Gly
Bond hydrolyzed	Leu-Gly	Leu-Gly	Leu-Gly	Gly-Leu	Gly-Leu	Gly-Leu
$\Delta H^{\ddagger}$ (kcal)	21.6	22.3	20.2	19.0	20.2	19.9
$\Delta S^{\ddagger}$ (cal/mol deg)	−23.4	−20.1	−23.2	−27.1	−21.8	−25.4

[a] From Long and Truscott (1963). Reprinted with permission of the Faraday Society.

2. Liquid Hydrogen Fluoride Cleavage

The discovery of the enormous solvation power of liquid hydrogen fluoride for amino acids, peptides, and proteins (Katz, 1954a,b), and good stability of insulin and corticotropin in this solvent, were followed by observations of loss of biological activity, e.g., of bovine ribonuclease (Gutte and Merrifield, 1971), lysozyme (Koch *et al.*, 1956; Sharp *et al.*, 1973), α-MSH (Lenard and Hess, 1964; Lenard *et al.*, 1964), staphylococcal nuclease (Anfinsen *et al.*, 1967); growth hormone (Li and Yamashiro, 1970), and acyl carrier protein (Marshall *et al.*, 1973). Findings of N → O acyl shifts at serine and threonine residues (Scheme 4) (Shin *et al.*, 1962; Sakakibara *et al.*, 1962),

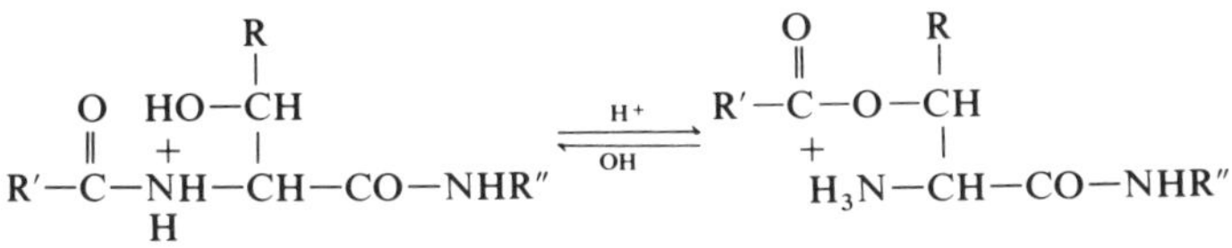

R = H or CH_3
R′, R″ = peptide chain

Scheme 4

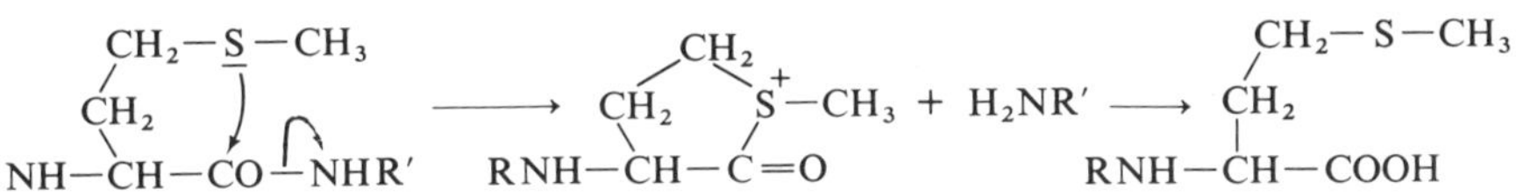

Scheme 5

and specific cleavage of methionyl peptide bonds (Scheme 5) (Lenard *et al.*, 1964), made it very clear that liquid HF was by no means an innocuous or safe solvent/reagent for peptides and proteins. In fact, a procedure for the specific chemical cleavage of peptide chains at serine and threonine residues has been based on the N → O peptidyl shift in liquid HF (Iwai and Ando, 1967).

Notwithstanding these adverse indications, wide use is made at this time of liquid HF for protective group cleavage in peptide synthesis, originating from a report of Sakakibara and Shimonishi (1965; see review by Sakakibara, 1971). It is not clear why peptide chemists should have embraced such a destructive reagent as liquid HF for final deprotection of carefully synthesized peptides when a considerable amount is going to be destroyed in the process. We feel strongly that the exploration of protective groups which can be cleaved under exceptionally mild and/or highly specific conditions (Carpino, 1973), e.g., the benzisoxazole-5-methyleneoxycarbonyl (Bic) group

(Kemp *et al.*, 1975a,b; Kemp and Reczek, 1977) should be one of the most important and urgent areas of research in peptide synthesis.

B. Enzymatic Cleavage of the Peptide Bond

In 1876, the term *enzyme* was used for the first time by W. F. Kühne in a discussion of the pancreatic protease trypsin (Kühne, 1876; cf. also Gutfreund, 1976). One hundred years later the proteolytic enzyme trypsin is understood well enough that it is possible to give an account entitled "Structural Basis of the Activation and Action of Trypsin" (Huber and Bode, 1978). What is the structural basis of this activation? What is the basis for the action of a proteolytic enzyme?

1. *Limited Proteolysis in the Physiological Environment*

For trypsin as well as other proteases, chymotrypsin and pepsin to be mentioned here from among those long known, nature has provided the concept of the inactive precursor in the form of the zymogen, e.g., trypsinogen. In the course of the activation, for instance, of bovine trypsinogen, a single peptide bond is cleaved with the release of the H_2N-terminal hexapeptide Val-Asp-Asp-Asp-Asp-Lys (activation peptide). The chain length of the activation peptide released from trypsinogens may differ from species to species (Neurath, 1975). For all trypsinogens, however, the COOH-terminus of the activation peptides is a residue of lysine, followed by isoleucine in the original sequence. The release of the amino group of this isoleucine residue is a key feature in making the molecule a catalytically potent one. Interaction of this amino group with another site, an aspartic acid residue in position 194 in the peptide chain by formation of an internal salt bridge, causes the conformational change adequate for the display of proteolytic activity (Huber and Bode, 1978).

The limited proteolysis (Linderstrøm-Lang and Ottesen, 1949) of trypsinogen is the result of the action of intestinal enterokinase (Maroux *et al.*, 1971). While trypsin itself is capable of activating trypsinogen and while potentially it can cleave all lysyl and arginyl peptide bonds in the protein substrate, enterokinase appears to be highly specific for that first lysyl peptide bond in the NH_2-terminal region of the zymogen, probably for reasons of the presence of a number of acidic residues in the activation peptide (*vide supra*; Maroux *et al.*, 1971). This high level of specificity makes enterokinase resistant to autodigestion while in storage in intestinal cells. Trypsin, on the other hand, must be stored as zymogen in order to forestall autolysis.

2. *Cascade Processes*

Once released from the zymogen state, trypsin begins to perform important proteolytic functions. In a next step of what is termed a cascade process

(Davie and Ratnoff, 1964), trypsin acts as the catalyst for the conversion of zymogens of the pancreas to the respective enzymes (Neurath, 1964). In cascade processes, limited proteolysis, most frequently the cleavage of a single peptide bond in the protein, provides the new catalytically active molecule for the subsequent step; for instance, trypsin, derived from the zymogen by cleavage of one and one peptide bond only, acts on chymotrypsinogen. Again, one peptide bond, and only one, is cleaved upon this action of trypsin. The product obtained is known as π-chymotrypsin, catalytically a very active species among several on the way to α-chymotrypsin, the final product of the activation process. The peptide bond cleaved in π-chymotrypsin is that between residues 15 and 16, position 16 being occupied by a residue of isoleucine (*vide supra* the activation of trypsinogen by enterokinase and the liberation of the amino group of isoleucine in position 16 of the zymogen to become the NH_2-terminus of the enzyme; a coordinating system of position numbers common to both has been adopted for trypsinogen and chymotrypsinogen for reasons of their close phylogenetic relationship). Cleavage of the peptide bond between residues 15 and 16 in the continuous chain of chymotrypsinogen does not result in the release of a fragment inasmuch as residues 1 through 15 are held to the remainder of the molecule by a disulfide bond. π-Chymotrypsin is the proteolytically active molecule for the next step in the cascade. Via the cleavage of again one peptide bond (between residues 13 and 14) and with the release of serylarginine, δ-chymotrypsin is formed. Followed by the autocatalytic cleavage of two additional peptide bonds and the release of another dipeptide (threonylasparagine which was preceded by tyrosine) the final product of the activation, α-chymotrypsin, is obtained (Desnuelle, 1960; Neurath, 1964). α-Chymotrypsin consists of three individual chains linked via two disulfide bridges (Blow *et al.*, 1969). In similar manner, trypsin initiates the activation of other pancreatic proteases, e.g., that of the carboxypeptidases A and B and that of elastase.

The highly selective cleavage of a limited number of peptide bonds in zymogens thus appears to serve the purpose of bringing into proper three-dimensional relationship the side chains of key amino acids for the display of catalytic action (Huber and Bode, 1978; for chymotrypsin, cf. Blow, 1976).

Limited proteolysis by way of its capability to incorporate restricted but lasting molecular alteration, offers the potential to generate and to control biological function. The cleavage of but one peptide bond in a precursor protein may cause the conversion necessary to establish the molecular species with new physiological activity. Changes brought about in the structure of a protein by limited proteolysis are largely irreversible. The free energy of peptide bond cleavage in the aqueous environment is negative. The level of specificity must be uniquely high for proteases to be capable of executing limited proteolysis. Topographically the environment of the peptide bond to be cleaved must be accommodated rather precisely in the active site of the protease. The control mechanisms considered here are thus quite distinctly different from reversible ones, such as those requiring an allosteric effector, phosphorylation, and adenylation.

Examples of physiological control involving limited proteolysis by cleavage of single peptide bonds in individual reactions or in cascades are numerous. Coagulation, the complement systems and kinins, fibrinolysis, cellular aspects of protease action, proteases in reproduction, cell growth and proteases, and last, but not least, tumor-associated proteases are topics well covered in an account of recent date (Reich *et al.*, 1975).

The role of limited proteolysis in generating biological function by the cleavage of one or a few selected peptide bonds is well documented for insulin (the prohormone and a preprohormone are now known) and β-lipotropin. The latter peptide hormone encompasses within its structural domain the sequences of several biologically active peptides and recently it has been shown that it, in turn, is derived from a precursor three times its molecular weight (*vide infra*).

Insulin, Proinsulin, and Preproinsulin. The subtle regulatory features of zymogen activation and the formation of the active species of a hormone were seen throughout the studies that led to the discovery of the precursor molecule of insulin (proinsulin: Steiner *et al.*, 1974). Proinsulin escaped detection for some time due to the high rates with which zymogen conversion and generation of the active hormone occur once the cell structure has been disrupted (Steiner *et al.*, 1969). This appears to be true also for preproinsulin the isolation and characterization of which (Chan *et al.*, 1976) was possible only after the large scale preparation of pancreatic islets and access to messenger RNA (Duguid *et al.*, 1976) for cell-free protein synthesis. Rapid limited proteolysis leads to the cleavage of a single peptide bond and the removal of a peptide of 23 residues from the preprohormone to give proinsulin. The structure and properties of preproinsulin are reminiscent of those of other presecretory proteins (Blobel and Dobberstein, 1975). While considerable variation in chain length has been observed for prepeptides of different origin, the presence of a region with hydrophobic residues is common to all of them. This clustering of hydrophobic side chains within a 10-residue region of the prepeptides may be responsible for their tendency to interact with membranes and to play their role in the discharge of newly synthesized polypeptide chains from within microsomal cisternae (Blobel and Sabatini, 1971; Milstein *et al.*, 1972). Having discussed trypsinogen and the cleavage of a single peptide bond to form active trypsin, it is worthy of mention that a prepeptide is also attached to the zymogen. Pretrypsinogen has been identified (Devillers-Thiery *et al.*, 1975).

3. Protein and Peptide Structures Related to β-Lipotropin

Another aspect of limited proteolysis and a demonstration of the multiplicity of its options for potent biological control began to unfold in 1975/1976. β-Lipotropin structurally known for various species (Li and Chung, 1976a) for some time and for its lipotropic action, was identified as the precursor of a number of biologically active peptides in 1976. Bradbury *et al.* (1976a) characterized β-melanotropin (β-melanocyte stimulating hormone, β-MSH) and corticotropin (adrenocorticotropic hormone, ACTH) among constituent fragments of β-lipotropin that were isolated in homogeneous form from porcine pituitary glands. The C-fragment of β-lipotropin (residues 61–91) in a rapid sequence of events was identified as one of opioid peptides in several laboratories (Li and Chung, 1976b; Bradbury *et al.*, 1976b; Lazarus *et al.*, 1976) and named β-endorphin. Limited proteolysis in its extension to the selective cleavage of a restricted number of peptide bonds within the range of the structural domain of β-lipotropin is thus a part of neurophysiological control. β-Lipotropin itself is, in turn, a structural segment of a larger precursor of molecular weight 31,000 (Roberts and Herbert, 1977a,b; cf. also Mains *et al.*, 1977). Enkephalin, the pentapeptide with opiate activity (Hughes *et al.*, 1975) corresponds to residues 61–65 in the sequence of β-lipotropin. The specificities (for limited proteolysis) of the activating enzymes are the likely determinants for a given peptide fragment to be formed.

4. Proteolysis and Structural Elucidation

The site-specific action of proteases is the prerequisite for the necessary limited cleavage of peptide bonds in substrates for sequence determination. The distinct specificity of the endopeptidase trypsin for the peptide bonds of lysine and arginine was invaluable for the first successful determination of the structure of a protein, that of bovine pancreatic ribonuclease A (Hirs *et al.*, 1956, 1960). The native molecule is a poor substrate for trypsin and it was necessary to break a conformation, likely to be held tightly, by oxidation of the four disulfide bridges present with performic acid (Hirs, 1956).

The exposure of certain regions of bovine pancreatic ribonuclease and their accessibility to limited proteolysis was elegantly demonstrated by Richards and Vithayathil (1959). The microbial protease subtilisin (from *Bacillus subtilis*; Güntelberg and Otteson, 1954, 1976) cleaves one peptide bond between residues 20 and 21 giving rise to two fragments known as S-peptide (residues 1–20 of bovine pancreatic ribonuclease) and S-protein (residues 21–124). While both fragments are void of ribonuclease activity, recombination of the two (ribonuclease S′), however, restores full enzymatic activity.

X-Ray analytical investigations on ribonuclease (Kartha *et al.*, 1967) and ribonuclease S (Wyckoff *et al.*, 1967) confirmed the surface exposure of the amino acid residues in the region of the peptide bond cleaved.

One peptide bond in ribonuclease may be broken without adverse effect on the molecule's catalytic properties. Obviously, there are forces at play capable of maintaining the conformation necessary for enzyme activity. In trypsinogen and in chymotrypsinogen (*vide supra*) the cleavage of single peptide bonds mobilizes intramolecular forces and their interaction results in the conformational change necessary for the display of enzyme activity. The alterations in trypsinogen and chymotrypsinogen are due to the high specificity of activating enzymes and their potential for limited proteolysis. The controlling parameter for the cleavage of one peptide bond in ribonuclease by subtilisin is temperature. To prevail with the cleavage of only the peptide bond between residues 20 and 21, the digestion must be carried out at 0°C. An increase in temperature to 20°C results in multiple peptide bond cleavage as additional sites are exposed due to further weakening the conformation of the substrate ("heat denaturation"; see Scheme 6 for the example of a tryptic fragment of the heterodetic multicyclic peptide subtilin and its fragmentation by thermolysin at different temperatures).

The forces responsible for the interaction of S-peptide and S-protein and the enzymatic properties of ribonuclease S have been investigated extensively inclusive of the use of synthetic S-peptide and analogues of the same. William H. Stein has given an impressive account of the studies conducted at Rockefeller University with native S-peptide and S-protein (Stein, 1965). Klaus Hofmann extended his work with synthetic S-peptide to offering the S-peptide–S-protein system as model for hormone-receptor interaction (Finn and Hofmann, 1973).

Observations similar to those made early with ribonuclease with regard to the resistance of the native molecule to proteolytic degradation were made again and again. This is true, in particular, for low molecular weight assemblies of amino acids of microbial origin which frequently contain structural features that are rarely seen in peptides of mammalian origin. The structurally rare components range from heteroatoms in cyclic systems to distinct configurational patterns which, in turn, give rise to novel types of conformations. For example, the gramicidins A, B, and C (Sarges and Witkop, 1965), which are pentadecapeptides with an alternating pattern of L- and D-configurations assume the conformation of the π_{LD}-helix (Urry, 1971). Apart from the fact that the NH_2-terminus and the COOH-terminus are blocked by the formyl group and by ethanolamine, respectively, and the direct action of exopeptidases was forestalled for these reasons, none of the endopeptidases known at the time accepted this molecule as substrate. The question remains an open and interesting one: are there sources in nature where endo- or exopeptidases are found that accept the L- and D-alternating configurational pattern in substrates? A role for peptidase action has been claimed for phagocyte chemotaxis (Aswanikumar *et al.*, 1976). The pentapeptide formyl—LPhe-DLeu-LPhe-DLeu-LPhe is a good agonist in leukocyte chemotaxos. Is this peptide a substrate for a peptidase that is related to chemotaxis? The structures of the linear gramicidins had to be solved without fragmentation by proceeding successfully with the stepwise degradation technique devised by Per Edman.

Nisin (Gross and Morell, 1971) and subtilin (Gross *et al.*, 1973) are heterodetic pentacyclic peptides with α,β-unsaturated amino acid and lanthionine residues. Lysine residues are among the constituent amino acids in both peptides, the action of trypsin, however, is badly impaired in either case under commonly applied conditions of digestion. This impairment was less aggrevating for nisin, since alternative nonenzymatic fragmentation (see Section III,C) of methionyl peptide bonds by cyanogen bromide treatment was applicable (Gross and Witkop, 1962). The successful tryptic digestion of subtilin (Kiltz and Gross, 1973) required these conditions: enzyme to substrate ratio 1:20, *N*-ethylmorpholine acetate, pH 7.8, 0.01 *M* $CaCl_2$, 7 days, 25°C. The fragments secured still suffered from a lack in residues whose peptide bonds would lend themselves subsequently to direct nonenzymatic cleavage. All along, residues were present with side chains that are recognized by chymotrypsin. However, as with trypsin, subtilin and its tryptic fragments are not ready-made substrates for chymotrypsin.

Reference has earlier been made to the difference in the action of subtilisin on bovine pancreatic ribonuclease at different temperatures (*vide supra*). A metalloendopeptidase was isolated by Endo (1962) from culture filtrates of the thermostable organism *Bacillus thermoproteolyticus*. Initially named "Thermoase" the protease is now known as thermolysin (Matsubara *et al.*, 1965). When the tryptic fragment T_2 of subtilin was treated with thermolysin (0.05 *N* *N*-ethylmorpholine acetate, pH 7.9, 0.01 *M* $CaCl_2$, 5 days, 25°C) cleavage occurred at the phenylalanine residue in position 20 (the specificity of thermolysin is directed at the peptide bond ahead of amino acid residues with hydrophobic side chains) and to some extent, at the leucine residue in position 16 (cf. Scheme 6). Thermolysin treatment at 60°C of the newly derived fragment comprising residues 3 through 19 of subtilin whose peptide bond Ala-15–Leu-16 was intact resulted in the cleavage of the valine peptide bond in position 12 (Nebelin and Gross, 1973). Again, advantage was taken of the heat denaturation of substrate, the thermophilic endopeptidase thermolysin having maintained its proteolytic power at the elevated temperature of 60°C.

For peptides such as nisin and subtilin, with heterodetic ring systems in which sulfide bridges participate, the type of fragmentation described is very

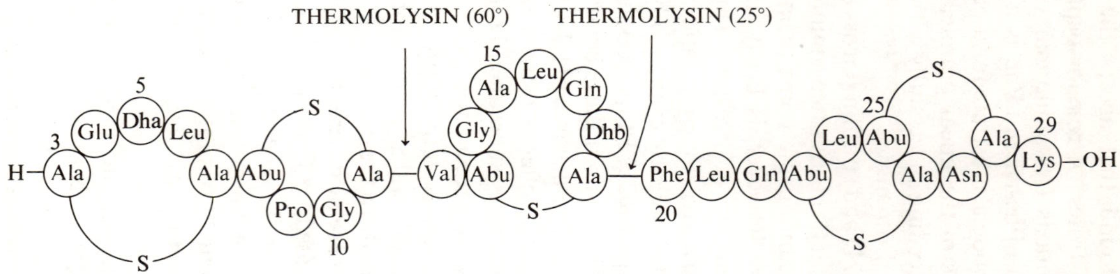

Scheme 6. The structure of the tryptic fragment T_2 of subtilin. The numbering system is that of the parent molecule. Abu, α-aminobutyric acid; Dha, Dehydroalanine; Dhb, dehydrobutyrine; Ala-S-Ala, lanthionine; Aba-S-Ala, β-methyllanthionine.

important. Its successful conduct greatly facilitates the assignment of sulfide bridges. The thermolytic cleavage at the valine residue in position 20 gave the monocyclic fragment (12–19) in which only valine (12) occupied an exocyclic position.

The controlled cleavage of the peptide bond by way of "limited proteolysis" executed by highly specific proteases, is a means of generating biological function and bringing about physiological control. Selective enzymatic cleavage is a necessary tool for structural determination. Chemists and biologists with an interest in the structure and function of the assemblies of amino acids, small and large, have learned to use to advantage the agent enzyme masterfully evolved in nature and to be supplied forever.

C. Nonenzymatic Cleavage of the Peptide Bond

The chemical cleavage of a single peptide bond was vital to the structural elucidation of phalloidin (Wieland and Schön, 1955) one of the toxic peptide constituents of the mushroom *Amanita phalloides.* At the time still considered and termed a "partial hydrolysis," it was the first nonenzymatic cleavage of a peptide bond proceeding according to a mechanistic principle soon to be recognized properly and explored widely.

The peptide bond cleaved in phalloidin is that of an amino acid with unique features in substitution: γ,δ-dihydroxyleucine. Under adequately acidic conditions, such as dilute sulfuric acid, 100°C, 30 min, protonation occurs at both hydroxyl groups. The oxonium ion formed at the γ-carbon atom is the most favored leaving group. With the departure of a molecule of water greatly assisted by the powerful neighboring group effect that comes to bear in a 1,5-type interaction from the electron density of the carbonyl group, formation of the iminolactone (Scheme 7) is initiated. Iminolactones are the subject of ready hydrolysis, here the cleavage of a peptide bond and the release of functional groups which, in general, do offer handles for further degradation and structural investigations.

The neighboring group assisted principle of 1,5–interaction is available in all amino acids with functional groups at the γ-carbon atom. Functions at γ-carbon atoms occur in variety. Those early explored for the purpose of peptide bond cleavage were double bonds in tryptophan (Patchornik *et al.*, 1960) and tyrosine (Schmir and Cohen, 1961). The reagents used were oxidants of the *N*-haloimide type (e.g., *N*-bromosuccinimide). The γ-substituted side chain methyl sulfide function of butyric acid in methionine is the subject of cyanosulfonium salt formation upon reaction with cyanogen bromide. Methylthiocyanate is an excellent leaving group and via homoserine lactone formation methionyl peptide bonds are cleaved in high yield. This cleavage

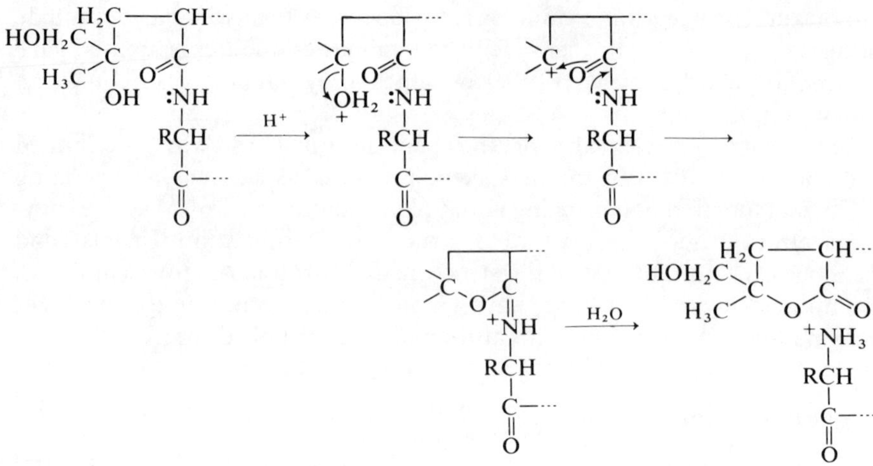

Scheme 7. Cleavage of the γ,δ-dihydroxyleucyl bond in phalloidin (0.2 *N* sulfuric acid, 100°C, 30 min). R = amino acid side chain.

is free from side reactions and widely used in structural work on peptides and proteins (Gross, 1967).

β-Substituted amino acids would have to form β-lactones to bring about cleavage of the peptide bond at their carboxyl groups. The reaction is energetically not favored. Neighboring group participation of preceding carbonyl functions is, however, possible for the β-hydroxyamino acids in $N \rightarrow O$-acyl shifts and for *S*-methylcysteine in cyanogen bromide initiated reactions (see Gross and Morell, 1974). Alternative mechanisms for the scission of the peptide chain at β-substituted amino acids proceed via β-elimination and the formation of dehydroalanine. Under aqueous acidic conditions dehyroalanine is degraded with the formation of amide and pyruvic acid. It was this reaction that led to the discovery of dehydroalanine in nisin after pyruvyllysine, arising from the COOH-terminal lysine, had been isolated and identified while efforts were underway to purify the peptide (Gross and Morell, 1967).

The resistance of nisin to tryptic digestion did not pose a problem. The presence of two residues of methionine invited the use of cyanogen bromide. Execution of the cleavage at 0°C, to preserve α,β-unsaturated amino acids as best possible, gave fragments adequate for subfragmentation by trypsin, under the forced reaction conditions used to fragment subtilin (*vide supra*), and structural elucidation (Gross and Morell, 1971).

The degradation of a second residue of dehydroalanine in nisin (Gross and Morell, 1968) was essential for the determination of the configurations in the only lanthionine residue present. This lanthionine residue is of the meso form.

Aspartic acid, and as of recent hydroxyaspartic acid (*vide infra*), deserve attention as useful sites for the purpose of nonenzymatic fragmentation. Schultz (1967) has proposed the use of dilute acid at elevated temperature to cleave both peptide bonds of aspartic acid. A mechanistic interpretation of the reaction is to be found in Bernhard Witkop's (1961) first review on nonenzymatic methods for the cleavage of peptide bonds.

Cinnamycin (Benedict *et al.*, 1952) and duramycin (Pridham *et al.*, 1956) are peptides produced by *Streptomyces* strains. Both peptides contain lanthionine and β-methyllanthionine and the question arose whether α,β-unsaturated amino acids are also present in analogy to nisin and subtilin. This is apparently not the case, however, a novel addition product of a nucleophile across the double bond of dehydroalanine, lysinoalanine, has been seen for the first time to occur in nature as constituent amino acid in cinnamycin (H. H. Kiltz and E. Gross, unpublished, 1978) and duramycin (E. Gross and J. H. Brown, unpublished, 1978).

Cinnamycin contains lysine, duramycin arginine; both peptides contain three residues each of phenylalanine and one residue of valine. Trypsin and chymotrypsin are not effective in causing peptide bond cleavage. The same is true for an impressive number of other proteases. Thermolysin at 60°C does cleave peptide bonds in either of the two molecules. By contrast to subtilin, however, the results are not satisfactory.

The amino acid composition of cinnamycin and duramycin showed aspartic acid and hydroxyaspartic acid as candidates of the first choice for an attempt to induce nonenzymatic fragmentation. When duramycin, for instance, was exposed to 0.03 *N* hydrochloric acid for 54 hr at 110°C, three amino acids were released in equimolar quantities (E. Gross and J. H. Brown, unpublished, 1978): hydroxyaspartic acid, aspartic acid, and glycine (Scheme 8). Considering the mechanism proposed for the reaction of aspar-

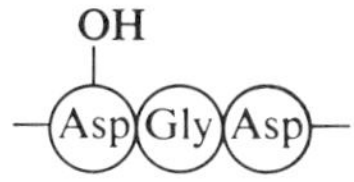

Scheme 8. Partial endocyclic sequence in duramycin. Equimolar quantities of aspartic acid, glycine, and hydroxyaspartic acid are released from the peptide on treatment with 0.03 *N* hydrochloric acid (110°C, 54 hr). The placement of glycine is based on the mechanism proposed for the cleavage of each of both peptide bonds of aspartic and hydroxyaspartic acid under the conditions employed. HO—(Asp), hydroxyaspartic acid.

tic acid in dilute acid at elevated temperature, the glycine residue must be placed between the two aspartic acid residues. Their order, in turn, and the placement of all three residues in endocyclic positions is based on additional structural data (Gross and Brown, 1976).

Nonenzymatic cleavage of the peptide bond is a concept based on chemical reactions possible with amino acids and triggered at functional groups in side chains. Reactions involving the 1,5-type interaction of γ-carbon atoms and the oxygen in the carbonyl group of the same amino acid have been applied successfully, in particular that initiated at methionine by cyanogen bromide.

Enzymatic and nonenzymatic cleavage of the peptide bond are not mutually exclusive. A long record of cases has already been accumulated where one method complemented the other.

The number of amino acids on the list of those found in mammalian tissue is limited, in contrast to those of microbial origin. Novel types of functionalities are to be expected to come from the latter source. There is consolation in the fact that much chemistry remains to be explored for reactions, and the cleavage of their peptide bonds, on amino acids, be they from mammalian or microbial tissue sources.

IV. FORMATION OF THE PEPTIDE BOND

A. Biosynthesis

Prior to the elucidation of the complex ribosomal pathway of protein biosynthesis, *in vivo* peptide bond formation was thought to be effected by comparatively simple enzymatic processes. In efforts to elucidate their mechanisms, thermodynamic aspects were studied and free energy changes associated with condensations of amino acids to peptides and peptides with one another were determined (for review, see Borsook, 1953). As an example, selected free energy and heat of formation data of several amino acids and dipeptides are listed in Table II. Peptide bond formation is an endergonic process with an average estimated ΔF of about 3000 cal/mol for a dipeptide, e.g., 4130 cal/mol for DL-alanylglycine, 3590 cal/mol for glycylglycine, and

Table II Free Energies and Heats of Formation of Some Amino Acids and Peptides[a]

Compound	$\Delta F^\circ_{310.6}$	$\Delta\tilde{H}_{298.1}$	$\Delta\tilde{F}_{98.1}$	$\Delta\tilde{F}_{310.6}$
DL-Alanine	− 86,830	−132,370	− 89,110	− 87,300
Glycine	− 86,920	−122,500	− 89,140	− 87,710
DL-Leucine	− 80,350	−150,900	− 81,760	− 78,870
DL-Alanylglycine	−113,190	−186,410	−114,570	−114,680
Glycylglycine	−115,200	−175,960	−118,060	−115,630
DL-Leucylglycine	−108,225	−203,620	−110,900	−107,065
Water	− 56,200	− 68,320	− 56,690	− 56,200

[a] Data in gram calories. ΔF° refers to the pure substances; $\Delta\tilde{H}$, $\Delta\tilde{F}$ refer to the substance in the standard state in 1 *M* aqueous solution, except in the case of water, where it is mole fraction = 1. Data from Borsook (1953).

3315 cal/mol for DL-leucylglycine (as calculated from the data in Table II). The equilibrium is far in the direction of hydrolysis and the action of a proteolytic enzyme on a simple peptide leads to almost complete scission of the available substrate.

The concept of general peptide biosynthesis by simple groups of enzymes was abandoned* when the genetic code and the central roles of messenger RNA and tRNA were recognized (Crick, 1957). Amino acids are activated by transfer to a specific tRNA each, according to the aminoacylation reaction (Scheme 9). Formation of an enzyme-bound aminoacyladenylate (structure **5**) with liberation of pyrophosphate (step a) is followed by transfer of

$$\mathrm{AA^i + ATP + E^i \rightleftharpoons E^i \cdot AA^i{-}AMP + PP} \qquad \text{(a)}$$

$$\mathrm{E^i \cdot AA{-}AMP + tRNA^i \rightleftharpoons AA^i{-}tRNA^i + AMP + E^i} \qquad \text{(b)}$$

$$\mathrm{AA^i + ATP + tRNA^i \rightleftharpoons AA^i{-}tRNA^i + AMP + PP} \qquad \text{(c)}$$

Scheme 9

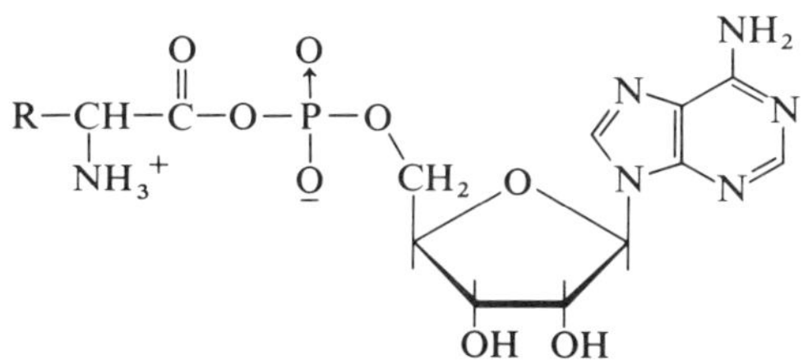

(5)

the amino acid to tRNA with liberation of AMP and regeneration of the free enzyme (step b). The AA—tRNA formed is at an energy level as high as that of ATP which provides for a high group transfer potential. Protein biosynthesis then proceeds through a series of steps including (1) formation of an mRNA—ribosome—initiator—tRNA complex, (2) binding of aminoacyl—tRNA, (3) peptide bond formation, (4) translocation, and (5) termination. These processes are carried out on the ribosome, a 3×10^6 dalton multicomponent particle which constitutes essentially a fully automated machinery (Fig. 6).

Once a linear polypeptide chain is assembled, folding occurs into each protein's exact three-dimensional structure that is essential to execute its biological function. The present state of insight into the molecular mechanisms of protein biosynthesis has been superbly reviewed in a recent multi-authored volume edited by Weissbach and Pestka (1977), to which the reader is referred for details on a wide scope of topics (see also Section III,B).

* However, several small peptides, for example the dipeptides anserine and carnosine and the tripeptide glutathione, originate from enzymatic biosynthesis, in animals and in man, in which the energy is derived from ATP (see Meister, 1965).

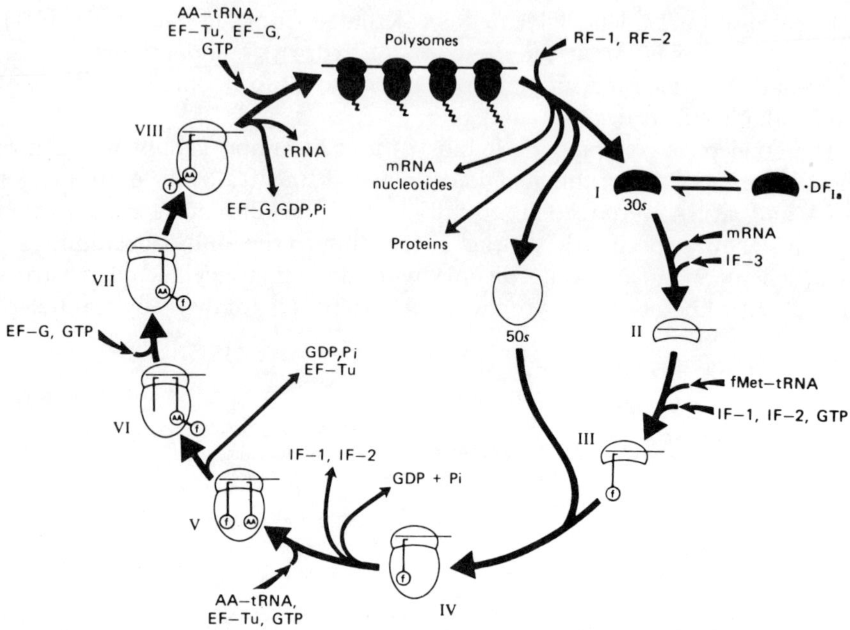

Figure 6. Ribosomal states and the ribosome cycle of protein biosynthesis. The semilunar cap (I) represents the free 30 S subunit. Initiation of protein synthesis involves attachment of mRNA to a 30 S subunit (I) to form complex II; this process requires Mg^{+2} as well as initiation factor IF-3. Subsequent attachment of fMet–tRNA in response to initiation codon AUG to form complex III requires GTP and initiation factor IF-1 and IF-2. Junction of the 50 S subunit to complex III produces complex IV. Enzymatic recognition of internal codons involves factors EF-Tu and GTP. The EF-Tu · GTP · AA-tRNA complex binds to the ribosomes in response to the respective codon to form complex V. Peptide bond formation occurs by transfer of the fMet (peptidyl) group to form fMet–AA (VI); peptidyl transfer requires only ribosomes and K^+. Translocation involves several coordinate processes: release of deacylated $tRNA^{fMet}$ to form state VII, one codon movement of mRNA and ribosome with respect to each other precisely positioning the next codon into position for translocation, and coordinate movement of peptidyl—tRNA (fMet–AA) from the A to P site resulting in state VIII. For every addition of a single amino acid (elongation) the ribosome cycles from state V to VIII. State IV is formally equivalent to VIII. Thus, the elongation epicycle (V to VIII) is superimposed on the overall ribosome cycle and repeats itself on each amino acid addition. The termination sequence (not given in detail) involves a termination codon (UAA). In response to release factors RF-1, RF-2, the completed protein is released. After termination, the ribosome may be dissociated into 30 S and 50 S subunits with release of mRNA. Dissociation factor (DF) bound to 30 S subunits (Ia) may be identical to one of the initiation factors and may be involved in dissociation of 70 S ribosomes following termination. [From S. Pestka (cited in Weissbach and Pestka, 1977, pp. 469 and 475); reproduced with permission of Academic Press.]

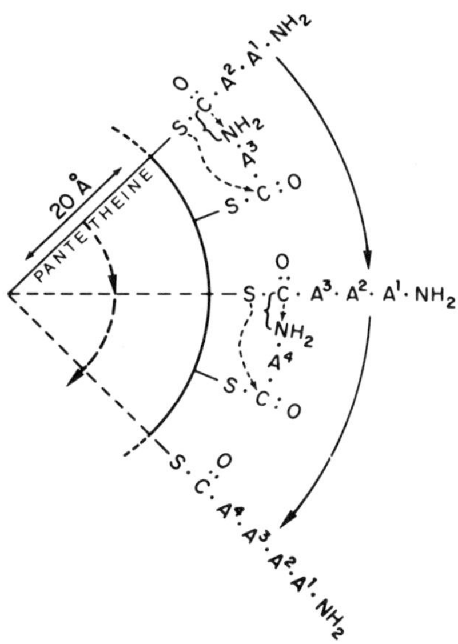

Figure 7. Schematic representation of transpeptidation and transthiolation in gramicidin S and tyrocidin biosynthesis mediated by enzyme-bound pantetheine. Pantetheine is depicted as the central thiol-carrying arm, charged with the nascent chain. It donates a peptidyl group from pantetheine to the amino group of the succeeding amino acid to form the next longer peptide, which sits on the peripheral thiol. The liberated pantetheine—SH performs a translocation by transthiolation of the peptide from peripheral acceptor—SH to its own donor thiol. Successive peptidyl transfers, each involving transpeptidation and transthiolation, are required to form the respective penta- and decapeptidyl chains before release by cyclization. (Modified from Kleinkauf *et al.*, 1971.)

In the event, the elegant studies of Lipmann and collaborators (for review, see Lipmann, 1971) showed that enzyme directed nonribosomal synthesis of peptides occurs widely in microorganisms in the biosynthesis of a multitude of peptide antibiotics. For example, the biosynthesis of gramicidins (Bauer *et al.*, 1972) or tyrocidins is effected by multienzyme complexes through consecutive cycles of pantothenate-assisted peptide condensations which are mediated by thioester activation (see Fig. 7).

B. Chemical Synthesis

1. Basic Features

The principal reaction in the synthesis of peptide chains is acylation of the amino group of an amino acid by the carboxyl group of a second amino acid with formation of an amide bond. The presence of several functional groups in amino acid molecules and the necessity to maintain the integrity of the α-carbon chirality centers during coupling makes peptide synthesis enormously more complicated than simple carboxamide formation. Therefore, a multitude of methods has been developed for peptide synthesis under mild conditions. Without exception these methods are based on the direct electrophylic acylation mechanism (see Satchell, 1974; Jones, 1974) involving activation of the α-carboxyl functionality of NH_2-protected amino acids and peptides. The electrophilicity of the carboxyl carbon of the acylating species RCOX is enhanced by electronegative substituents X (activating groups) which further decrease its already low electron density. This type of activation may be formally subdivided into replacement of (1) the hydrogen or (2) the hydroxyl of the carboxyl group (Wünsch *et al.*, 1974).

1. Carboxyl activation through formal hydrogen replacement (structure **6**, R′=alkyl, aryl, acyl) is typical for active esters, anhydrides, imino anhydrides, and acetals. Hydroxylamine-type esters and anhydrides with acids of sulfur, phosphorus, silicon, and boron are obtained with R′ linked to the carboxyl through the respective heteroatom.

RCOO—R′

(6)

2. Carboxyl activation through formal hydroxyl replacement (structures **7**, **8**), i.e., attachment of the activating group R′X- to the aminoacyl residue through certain heteroatoms (**7**) is typical for (a) azides, diimines, azoles, and acylamines with nitrogen as a link, and (b) thioesters and selenoesters with sulfur and selenium, respectively, as links. Activation by heteroatoms (single or in groups) of type **8** is represented by acid chlorides, bromides, or cyanides.

RCO—XR′ RCO—X

(7) **(8)**

The reactivities of the principal types of acylating agents used in peptide synthesis increase generally in the order

$$\text{RCONR}' < \text{RCOOR}' < \text{RCOOCOR}' < \text{RCOHal}$$

although the actual reactivity within each type depends on the substituent R. As a rule of thumb, the acylating power of RCOX increases with the ability of its leaving group X^- to depart, which in turn is related to the strength of the acid HX (Satchell, 1974). The displacement of X by the incoming amine component may either be of a predominantly synchronous type (Scheme 10) as favored for anhydride or azide reactions (Satchell, 1974) or of a two-step type which proceeds via a carboxyl addition intermediate (shown in Scheme 11) (Bender, 1960) as it is generally believed to occur in ester aminolysis (see

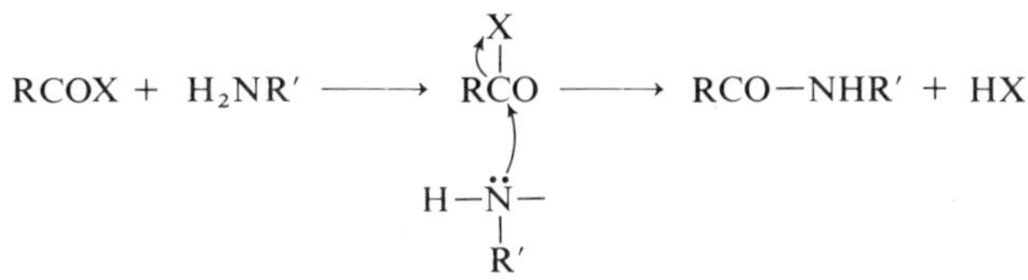

Scheme 10

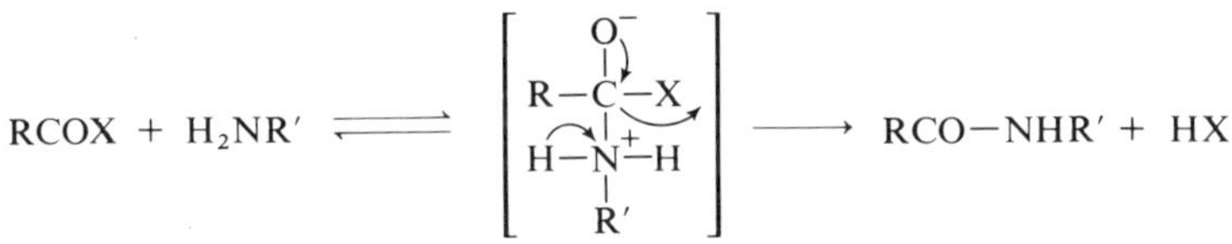

Scheme 11

Bodanszky *et al.*, 1976, pp. 13–17; Jakubke, 1966). The tetrahedral intermediate has never been directly detected or trapped during the course of a displacement reaction but its existence with a lifetime of 10^{-9} to 10^{-10} sec has been deduced from kinetic evidence. A detailed discussion about the significance of the tetrahedral addition intermediate has been presented by Johnson (1967) while Satchell (1974) emphasized arguments against its occurrence.

2. Kinetics

Studies on the kinetics of peptide bond formation have been confined mainly to ester aminolysis. Systematic data and comparison with other methods are lacking. Josef Rudinger (1972) criticized this shortcoming at the 4th American Peptide Symposium and pointed out:

> We know very little about the kinetics of coupling even for simple peptides.... It is perhaps a hope rather than a prediction that peptide chemists, and perhaps others who appreciate the importance of peptide synthesis, will increasingly contribute to the store of mechanistic or even merely quantitative information on which we can draw to make our syntheses more rational.

The situation has been considerably improved since 1972 as apparent from Table III,* but much more remains to be done. However, it should be noted that several recent studies (Kemp *et al.*, 1974a; Kovacs *et al.*, 1975) provide interesting and potentially useful new leads. Before discussing these data,

* Not included in this literature compilation are the extensive data available on aminolysis and hydrolysis of simple *p*-nitrophenyl esters [e.g., solvent isotope effects, entropies of activation, rates of nucleophilic reactions, standard free energy (ΔG) of formation] which have been superbly reviewed by Johnson (1967). Originating from the laboratories of T. C. Bruice, W. P. Jencks, S. L. Johnson, A. R. Day, and others, these data were obtained from reaction kinetics in predominently aqueous media. In contrast, chemical synthesis of peptides is conducted mainly in organic solvents such as dimethylformamide.

Table III Kinetics of Peptide Bond Formation (Literature Survey)[a]

No.	Carboxyl component	Amine component	Solvent	Temperature (°C)	Data	Reference
1	Z-Gly {ONp, SPh, SNp}	H-Ala-OH	Dioxane–H_2O	(18°)	Relative reaction rates	Farrington *et al.* (1955)
2	Z-Phe-OR[b]	Benzylamine	Dioxane–H_2O	(25°)	$t_{1/2}$	Pless and Boissonnas (1963); Pless (1963)
3	Z-AA-OTcp(2,4,5)[c]	Benzylamine	Dioxane	(25°)	$t_{1/2}$	
4	Z-Phe-OTcp(2,4,5)	H-AA′-OMe[d]	Dioxane	(25°)[e]	$t_{1/2}$	
5	Z-AA-ONp[f]	H-Gly-Gly-OH	Dioxane–H_2O (1:4)	(25°)	Rate constants	Khurgin and Dmitrieva (1965)
6	Z-AA-OQ	Benzylamine	Dioxane–H_2O	(25°)	$t_{1/2}$	Jakubke and Voigt (1966a)
7	Z-AA-OQCl	Benzylamine	Dioxane–H_2O	(25°)	$t_{1/2}$	Jakubke and Voigt (1966b)
8	Z-AA-OQ(n)[g]	Benzylamine	Dioxane–H_2O	(25°)	$t_{1/2}$ Mechanism	Jakubke *et al.* (1967)
9	Boc-Leu {-ONo, -ONm, -ONp, -OPcp}	Benzylamine; H-AA′-O*t*Bu[h]; H-AA′-Resin[h]	EtOAc	(25°)	$t_{1/2}$	Bodanszky and Bath (1969); Bodanszky *et al.* (1974)
10	Boc-AA-OH + DCC[i]	Peptide-Resin	DMF		Solid-phase $t_{1/2}$	Gut and Rudinger (1968)
11	Z-Cys(Bzl)-OR[j]	H-Val-OMe	THF	(23°)	Rate constants	Kovacs *et al.* (1970a,b)
12	Z-{Ala, Phe, Ser, Trp}-OR[k]	H-Val-OMe	THF	(23°)	Rate constants	Kovacs *et al.* (1972)
13	Z-{Glu(OMe), Asp(OMe)}-OR[k]	H-Val-OMe	THF	(23°)	Reaction rates	Kovacs *et al.* (1973)
14	Boc-AA {OPfp, ONo, ONp}	H-Gly-Resin			$t_{1/2}$	Kovacs and Penke (1973)
15	Z-Phe {-OBt, -OTcp, -OSu, -ONp}	H-Val-*o*-nitranilide	THF, DMF	(25°)	$t_{1/2}$ HOBt catalysis	König and Geiger (1972, 1973a,b)

16	R-COOPh	Benzylamine	Dioxane Benzene	(40°)	ΔG, ΔH, ΔS Acid catalysis	Muzalewski (1973)
17	Z-/Boc- AA-ONp[l]	H-AA' -OEt[m]/-OtBu	DMF	(30°)	Rate constants	Kemp (1973); Kemp *et al.* (1974a)
18	Z-/Boc- AA-OR[k,n]	H-Val-OMe	THF	(23°)	Predictive factors	Kovacs *et al.* (1975)
19	Z-Gly-R[o]	*N*-Methylbenzylamine *N*-Methyl-*N*-(trimethyl-silyl)benzylamine	$CHCl_3$	(28°)	Rate constants ΔG	Kricheldorf *et al.* (1975)
20	Boc-/Nps- AA OSu/SPy	H-AA-Resin	CH_2Cl_2, DMF toluene	(25°)	$t_{1/2}$, Rate constants	Rudinger and Buetzer (1975)
21	Boc-His-N_3	H-Pip-OMe	DMF		Rate constants Azide ion catalysis	Veber (1975)
22	Z-/Boc- AA -OPfp/-OPcp	Piperidine	CH_3CN C_6H_5Cl	(25°)	IR, NMR Rate constants	Kisfaludy *et al.* (1976a)

[a] The survey was not exhaustive and literature citation is bound to be incomplete. Special abbreviations: Ph, phenyl; Scp, 4-chlorothiophenyl; Ocp, 4-chlorophenyl; 2,4,6-Tbp, 2,4,6-tribromophenyl; OQ, quinolyl (ester).

[b] OR = 27 differently substituted phenyl esters.

[c] AA (in decreasing order of coupling rates) = Gly > Asp(OBzl) > Ala > Tyr(Z) > Cys(Bzl) > Phe > Tyr > Ser > Gln > Pro > Lys(For) > Glu(OMe) > Tyr > Lys(Z) > Ser(tBu) > Val > Ile.

[d] AA' (in decreasing order of coupling rate) = Gly > Phe > Val > Lys(Z) > His > Ala > Arg > Ser > Met.

[e] Solvent influence in order of decreasing coupling rates ($t_{1/2}$, min) = DMF (0.3), EtOAc (2.3), Pyr (3.1), dioxane–H_2O [8:2] (3.3), *N*-methylmorpholine (4.1), dioxane (4.9), $CHCl_3$ (23.0).

[f] AA (in decreasing order of coupling rates) = Ser(Ac) > Asp(OMe) > Gly > Phe > Met > Tyr > Ala > Leu > Pro > Val > Ile.

[g] *n* (in decreasing order of coupling rate) = 8 > 3 > 6 [i.e., quinolyl(8) ester > quinolyl(3) ester > quinolyl(6) ester].

[h] AA' = Gly, Val.

[i] Solid phase synthesis of Boc-Ala-Val-Val-Val-Phe-OResin.

[j] R (in decreasing order of coupling rate) = Pfp > 2,4-Dnp > Su > 2,6-Dnp > Pcp > 2,4,5-Tcp > Np > 2,4,6-Tcp > 2,4,6-Tbp.

[k] R (in decreasing order of coupling rate) = Pfp > Su > Pcp > 2,4,5-Tcp > ONp.

[l] AA (in decreasing order of coupling rates of Z-AA-ONp) = Gly > Ala > Phe > Leu > Pro > Val.

[m] AA' (in decreasing order of coupling rates of H-AA'-OEt) = Gly > Ala > Leu > Val.

[n] AA = Ala, Phe, Trp, Cys(Bzl).

[o] R (in decreasing order of coupling rate) = OSu > Opcp > 2,4,5-OTcp > Imidazolide > ONp > 2,4,6-OTcp > Scp > Ocp.

credit must be given to the very thorough earlier study by Pless and Boissonnas (1963) in collaboration with Stich and Leemann (1963). In examining the reactivities of 27 mostly halogen-substituted phenyl esters (Table III, No. 2), an approximate correlation with the acidity (pK_a) of the corresponding phenol components [or the basicity of the leaving phenoxides (Bruice and Benkovic, 1964, 1966; Gresser and Jencks, 1977)] was observed. Furthermore, the influence of solvents, substituents on the acyl group, and nature of the amino acid residues in the nucleophile on the rate of *N*-benzyloxycarbonyl-L-phenylalanine-2,4,5-trichlorophenyl ester [Z—Phe—OTcp(2,4,5)] aminolysis was determined (see Table III, No. 4). The $t_{1/2}$ values of the reaction of Z—AA—OTcp(2,4,5), of 17 different amino acids, with benzylamine in dioxane were also determined (Table III, No. 3). The activation energy of Z—Phe—OPcp was calculated as 6.75 kcal mol^{-1} under pseudo first-order* and 6.40 kcal mol^{-1} under second-order reaction conditions (Pless and Boissonnas, 1963).

Kemp *et al.* (1974a) reported rate constants for 41 aminolysis reactions of *N*-protected amino acid *p*-nitrophenyl esters with amino acid ethyl or *tert*-butyl esters in dimethylformamide at 30°C (see Table III, No. 17 and Scheme 12). With the exception of reactions involving proline esters as

$$\text{Z—A—ONp} + \text{H—B—OEt} \rightarrow \text{Z—A—B—OEt}$$

$$k_{\text{A-B}} = (k_{\text{A-Gly}})(k_{\text{Gly-B}})\left(\frac{1}{k_{\text{GlyGly}}}\right)$$

Scheme 12

nucleophiles, the observed rate constants fitted the simple second-order rate law given in the equation (Scheme 12). The reaction rates could all be approximated as products of two partial rate factors, which indicate that the steric effect of each amino acid on the coupling rate is exerted independently of the other amino acid. The most important implication of these results is that some 400 rate constants for possible dipeptide formation can be estimated by measuring only 39. More data are needed with other substituted phenyl esters before general conclusions can be drawn.

In a related approach, Kovacs and collaborators (1975) use the additivity principle (Scheme 13) for the prediction of rate constants of peptide bond

$$\log k = \log f_1 + \log f_2 + \cdots \log f_i \quad \text{(a)}$$

$$k = f_1 f_2 \cdots f_i \quad \text{(b)}$$

Scheme 13

formation by active ester aminolysis (see Table III, No. 18). It is postulated that log k is the unweighted sum of contributions from each reactant and

* Using more than one-hundredfold excess amine.

Table IV Predictive Factors[a]

	F_r	F_c	F_c/F_r
Activating ester			
OPfp	15	37	2.5
OTcp(2,4,5)	2.5	0.53	0.21
ONp	1.6	0.22	0.14
OPcp	1.0	1.0	1.0
N-Protecting group			
Boc	0.23	0.94	4.1
Z	1.0	1.0	1.0
Amino acid side chain			
Ala	1.0	1.0	1.0
Phe	7.1	0.70	0.099
Trp	0.89	0.60	0.67
Cys(Bzl)	280	1.3	0.0046
Solvent			
DMSO	23	42	1.8
DMF	10	15	1.5
Dioxane	0.43	0.54	1.3
THF	1.0	1.0	1.0
$CHCl_3$	3.0	0.14	0.047
Amino acid residue			
Gly	105	8.5	0.081
Amino acid side chain			
Ser	29	1.3	0.045
Asp(OMe)	20	0.90	0.045
Asp(OBzl)	22	1.4	0.064
Glu(OMe)	2.5	0.47	0.19

[a] From Kovacs *et al.* (1975).

each component of each reactant and from the solvent (see Scheme 13) where k is the experimental rate constant, f_i is the contribution of the solvent, and f_1, f_2, etc., are contributions of the reactants or their substituents.

Predictive factors F (see Table IV) have been calculated for the relative contributions of (a) N^α-protecting groups, (b) amino acid side chains and residue, (c) active ester groups, and (d) solvents, for both coupling (F_c) and racemization (F_r).* The example in Scheme 14 illustrates the calculation of the coupling rate constant for the reaction of Z—Glu(OMe)—ONp with H—Val—OMe in tetrahydrofuran with the aid of F_c factors listed in Table

* These predictive factors are relative to arbitrary standards, i.e., $k_c = 0.506 \times 10^{-2}\ M^{-1}\ sec^{-1}$ of the coupling reaction of Z—Ala—OPcp with H—Val—OMe in tetrahydrofuran at 23°C is the standard for F_c factors, and $K_r = 0.825 \times 10^{-6}\ M^{-1}\ sec^{-1}$ of the racemization of Z—Ala—OPcp in the presence of triethylamine in tetrahydrofuran at 23°C is the standard for F_r factors (Kovacs *et al.*, 1972).

$$k_c = F_Z F_{\mathrm{Glu(OMe)}} F_{\mathrm{ONp}} (0.506 \times 10^{-2} M^{-1}\mathrm{sec}^{-1})$$

$$k_c = 1.0 \times 0.47 \times 0.22 (0.506 \times 10^{-2} M^{-1}\mathrm{sec}^{-1})$$

$$k_c\ (\mathrm{calc.}) = 0.052 \times 10^{-2} M^{-1}\mathrm{sec}^{-1}$$

$$k_c\ (\mathrm{exptl.}) = 0.045 \times 10^{-2} M^{-1}\mathrm{sec}^{-1}$$

Scheme 14

V (Kovacs *et al.*, 1975). Similar calculations can be made for predicting rates of racemization. One of the obvious conclusions from the data of Table IV is that the active esters of *tert*-butyloxycarbonylamino acids racemize about 2–10 times more slowly than the corresponding benzyloxycarbonyl derivatives.

Table V shows coupling and racemization rate constants of active esters of benzyloxycarbonylamino acids and benzyloxycarbonyl dipeptides, having glycine residues in position 1. Reasonable agreement is seen between experimental and calculated values (Kovacs *et al.*, 1975). For practical coupling reasons chiral integrity of the products is best preserved by maximizing k_c/k_r. For example, in a segment condensation reaction one would select the less racemization-prone tryptophan or alanine COOH-terminal residues for *p*-nitrophenyl or pentachlorophenyl ester activation, rather than the more readily racemizing aspartic acid β-methyl ester or phenylalanine residues.

Compared to the growing amount of kinetic data for active ester coupling,

Table V Experimental and Calculated Racemization and Coupling Rate Constants of Benzyloxycarbonylamino Acid and Benzyloxycarbonylglycyl Dipeptide Active Esters[a]

Compound	$k_r \times 10^6$	$k_c \times 10^2$	k_c/k_r	k_{r2}/k_{r1}	k_{c2}/k_{c1}
Z-Ala-ONp	1.38 (1.32)	0.152 (0.111)	1100	82	8.4
Z-Gly-Ala-ONp	113 (1.39)	1.28 (0.95)	113		
Z-Ala-OPcp	0.82 (0.83)	0.51 (0.506)	6220	140	2.2
Z-Gly-Ala-OPcp	112 (87)	1.1 (4.30)	98.2		
Z-Glu(OMe)-OPcp	2 (2.06)	0.164 (0.238)	820	155	13.8
Z-Gly-Glu(OMe)-OPcp	309 (217)	2.26 (2.02)	73.1		
Z-Glu(OMe)-ONp	3.0 (3.30)	0.045 (0.052)	150	96.7	17.3
Z-Gly-Glu(OMe)-ONp	290 (347)	0.78 (0.445)	26.9		
Z-Asp(OMe)-OPcp	17.6 (16.5)	0.74 (0.455)	420	93	1.0
Z-Gly-Asp(OMe)-OPcp	1640 (1,730)	0.77 (3.87)	4.70		
Z-Trp-OPcp	0.72 (0.734)	0.32 (0.304)	4440	377	14
Z-Gly-Trp-OPcp	272 (77.1)	4.5 (2.58)	165		
Z-Phe-OPcp	3.3 (5.86)	0.29 (0.354)	879	130	7.3
Z-Gly-Phe-OPcp	431 (615)	2.12 (3.01)	49		
Z-Phe-ONp	9.2 (9.37)	0.032 (0.0779)	35	20	9.7
Z-Gly-Phe-ONp	187 (984)	0.31 (0.662)	17		

[a] Data from Kovacs *et al.* (1975). Data in parentheses are calculated rate constants.

those for the carbodiimide, azide, and mixed anhydride procedures are scant or totally lacking. The use of 1-hydroxybenzotriazole catalyzed carbodiimide (DCC—HOBt) coupling continues to increase rapidly but the only relevant kinetics available (see Table III, No. 15) are those on Z—Phe—OBt aminolysis (König and Geiger, 1972, 1973a,b) which are impressive by their clear graphic presentation (Fig. 8). Kinetic information on carbodiimide-mediated peptide coupling would be most desirable, but the complexity of several competing reactions (see Rich and Singh, this volume, Chapter 5) may have discouraged more extensive studies. First attempts to unravel some of the kinetic factors in azide coupling have been made by Veber (1975) (see Table III, No. 21) and a report by Brenner *et al.* (1976) deals with aspects of mixed anhydride coupling.

In conclusion, the results which Kemp *et al.* (1974a) and Kovacs *et al.* (1975) derived from kinetic studies of active ester aminolysis represent most encouraging advances toward a more rational approach to peptide synthesis. Certainly, more data are needed and a unifying treatment of the two approaches would be welcome. The prospects of predicting the rates of peptide segment condensations are very exciting.

3. Catalysis

Catalysis arises from the influence of additional substances (acids or bases) on the reactants. General acid catalysis of ester aminolysis has first been studied by Audrieth and collaborators (Glasoe *et al.*, 1939). Schwyzer *et al.* (1955) observed accelerating effects of acetic acid and other carboxylic acid on the reaction rate of activated esters. Recently, the catalytic effects of

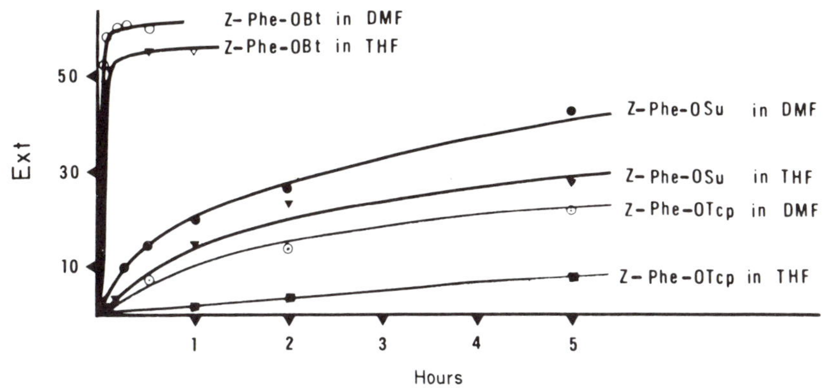

Figure 8. Comparison of reactivities of Z-phenylalanine 1-hydroxybenzotriazole ester (OBt) [1×10^{-2} *M*] with *N*-hydroxysuccinimide (OSu) and 2,4,5-trichlorophenyl ester (OTcp), as measured by reaction with valine-*o*-nitranilide [5×10^{-3} *M*]. (From König and Geiger, 1972; reprinted with permission of Ann Arbor Science Publishers.)

carboxylic acids on the aminolysis of activated esters in dioxane were studied and reaction rates and activation enthalpies and entropies determined (Muzalewski, 1973). General acid catalysis of ester aminolysis by carboxylic acids is most effective in nonpolar solvents such as tetrahydrofuran and negligible in dimethylformamide. In contrast, 1-hydroxybenzotriazole (HOBt, compound **9**) strongly catalyzes the aminolysis of, e.g., *p*-nitrophenyl and 2,4,5-trichlorophenyl esters in polar solvents such as dimethylformamide and dimethylacetamide, presumably via intermediate addition complexes (structure **10**, Scheme 15), see König

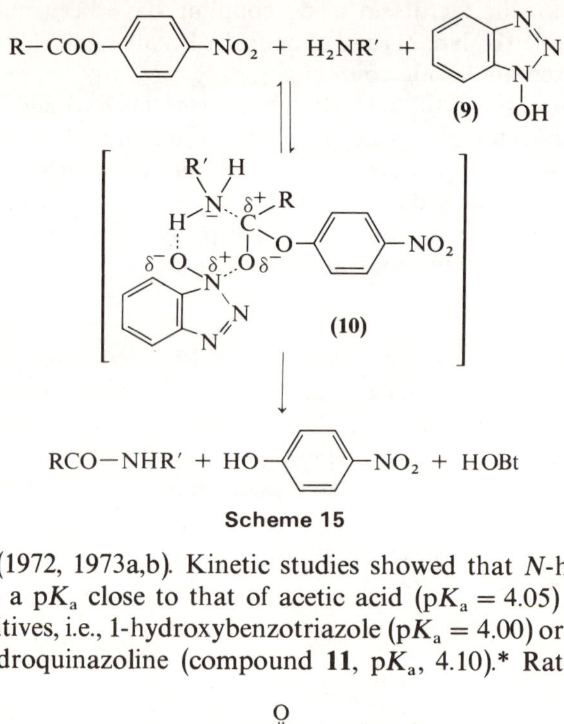

Scheme 15

and Geiger (1972, 1973a,b). Kinetic studies showed that *N*-hydroxy compounds with a pK_a close to that of acetic acid ($pK_a = 4.05$) are the most effective additives, i.e., 1-hydroxybenzotriazole ($pK_a = 4.00$) or 3-hydroxy-4-oxo-3,4-dihydroquinazoline (compound **11**, pK_a, 4.10).* Rate studies and

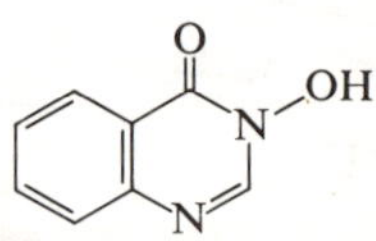

(11)

physical measurements indicate that in the catalyzed process an equilibrium exists between the starting active ester and the intermediate benzotriazole

* However, *N*-hydroxysuccinimide has a weaker catalytic effect although its pK_a (4.05) is identical to that of acetic acid. The effect of 1-hydroxybenzotriazole on *N*-hydroxysuccinimide ester couplings is only marginal.

ester which favors the former ester in acidic medium and the latter in basic medium. Salts of *N*-hydroxy compounds can be used as additives when the amine components are protonated; their free bases are promptly liberated *in situ*.

Intramolecular base catalysis has also been utilized in peptide bond formation by active ester aminolysis. Efforts to minimize racemization have led to investigations of neighboring group effects (anchimeric assistance) which culminated in the development of active esters which are capable of discriminating effectively between aminolysis and racemization.

In the quinolyl(8) esters of *N*-protected amino acids the attack of the amino component at the carbonyl carbon is accelerated by specific nucleophilic catalysis from the quinoline ring nitrogen. For example, the aminolysis of Z—Ala—OQ (compound **12**) is about 8 and 66 times faster than

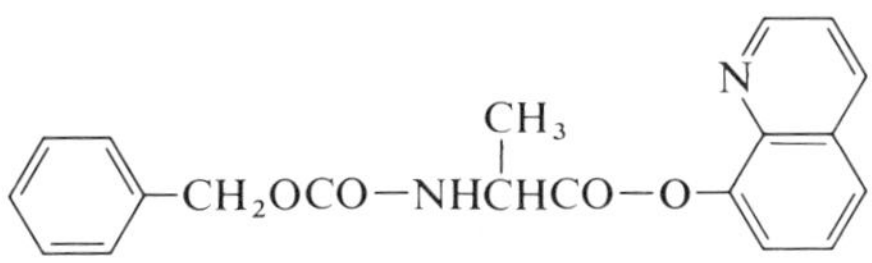

(12)

those of the corresponding quinolyl(3) and quinolyl(6) esters (Jakubke *et al.*, 1967). Without the anchimeric effect in quinolyl(8) esters, the reactivities of the three esters would be expected to be in the order of the acidity of their leaving groups as shown in Scheme 16, but this order is only followed by the quinolyl(3) and quinolyl(6) esters in accordance with a B_{Ac2} reaction mechanism of amide bond formation (Jakubke *et al.*, 1967). The quinolyl(8) esters are distinguished by their ability to form an intramolecular hydrogen bond in the transition state in which the incoming amine is held in close and favorable position (Scheme 17). Racemization via 5(4H)-oxazolone* formation

* Nomenclature according to "Chem. Abstr. 1977 Index Guide," pp. 836G, 837G. Example: Structure **12a** is designated as 2-methyl-4-(phenylmethyl)-5(4H)-oxazolone in *Chem. Abstr.* **83**, 3633CS (1975) *Chemical Substance Index*,

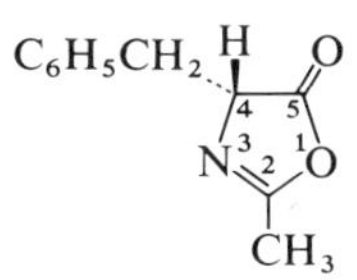

(derived from N^{α}-acetyl-L-phenylalanine)

(12a)

as a reference to Siemion, I. Z., and Baran, G. (1975). *Bull Acad. Pol. Sci., Ser. Sci. Chim.*, **23**, 317–322: "Conformational investigation on azlactones of aromatic *N*-acylamino acids". The term 5(4H)-oxazolone replaces all earlier terms, such as azlactone, 4,5-dihydro-1,3-oxazol-5-one, oxazolinone, 1,3-oxazoline-5-one, 2-oxazoline-5-one, etc., whose use should be discontinued.

Quinolyl(3) ester > quinolyl(6) ester ⋡ quinolyl(8) ester

pK_a: 8.06 8.88 9.89

Scheme 16

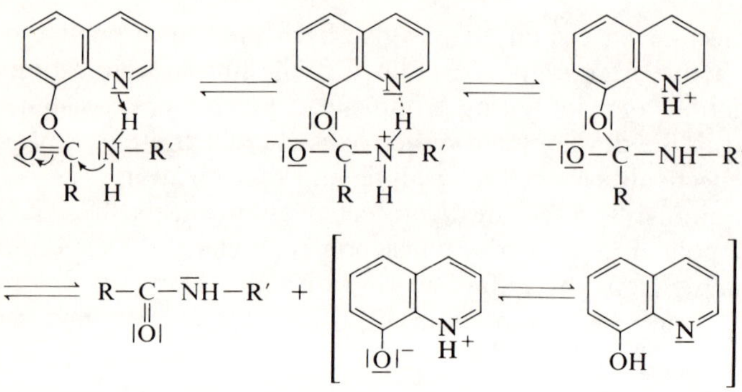

Scheme 17

tion, however, is not assisted because the respective nucleophile, i.e., the carbonyl oxygen atom of the amide grouping, cannot readily attack the weakly activated carboxyl (Scheme 18).

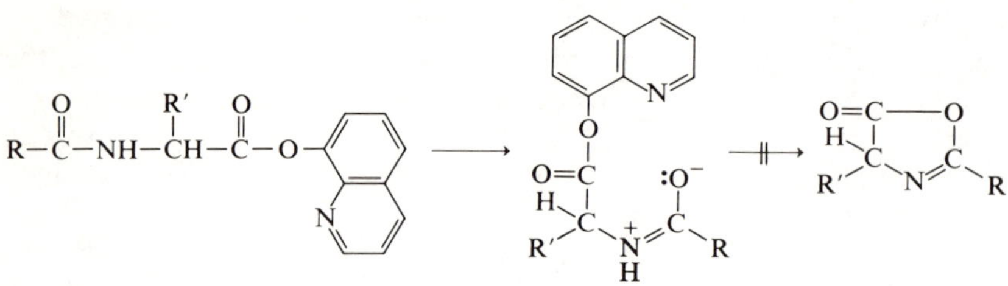

Scheme 18

Intramolecular general base catalysis is similarly responsible for the high reactivity of catechol monoesters (structure **13**) developed by Jones and Young (1967, 1968; see also Young, 1967) and by Kemp and collaborators (structure **14**) (Kemp and Chien, 1967; Kemp, 1970; Kemp *et al.*, 1974b,c,d), for 2-pyridylthiol esters (structure **15**) (Lloyd and Young, 1971), 2-pyridyloxy esters (structure **16**) (Dutta and Morley, 1971) and others; see the review by Bodanszky (this volume, Chapter 3).

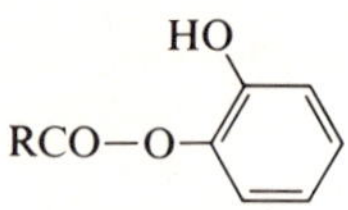

(13)

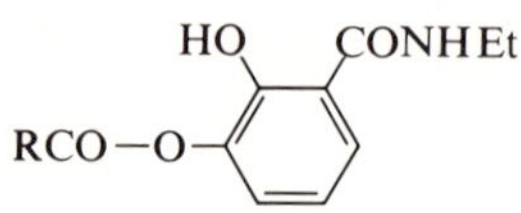

(14)

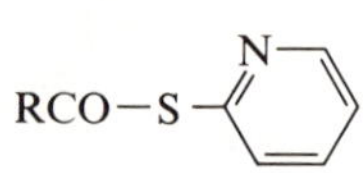

(15)

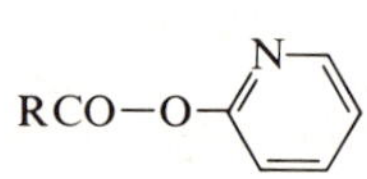

(16)

4. Racemization

Racemization is of major concern in peptide synthesis and has been alluded to several times in the above discussions on kinetics and catalysis. This subject is discussed in detail by Kemp (this volume, Chapter 7; see also Wünsch, 1974, Part I, pp. 32–45). The preservation of chiral integrity during coupling is a cardinal criterion for judging the relative merits of synthetic methods. Fortunately, highly sensitive racemization tests which measure as little as 0.001% (Kemp *et al.*, 1970) to 0.1% (Izumiya *et al.*, 1971) of D-enantiomer have become available in recent years. The carboxyl component is mainly at risk, since both its preparation and its subsequent use in coupling may lead to 5(4H)-oxazolone formation (compound **12**a, Scheme 19) which constitutes the main pathway of peptide racemization (see Kemp,

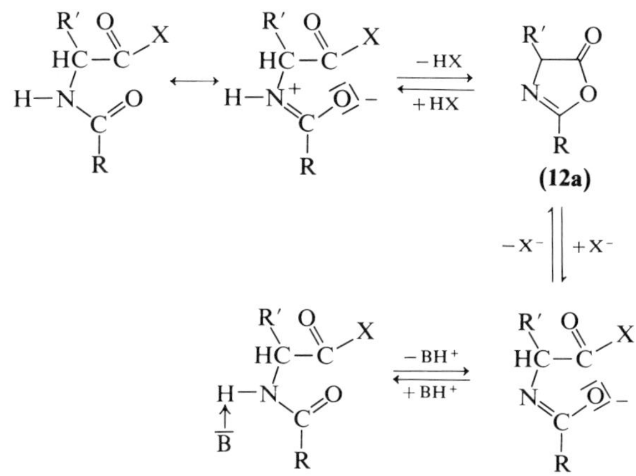

Scheme 19

this volume, Chapter 7). Another significant racemization mechanism involves hydrogen abstraction from the α-carbon followed by enolization (Scheme 20) (Goodman and Glaser, 1970; Kemp and Rebek, 1970).

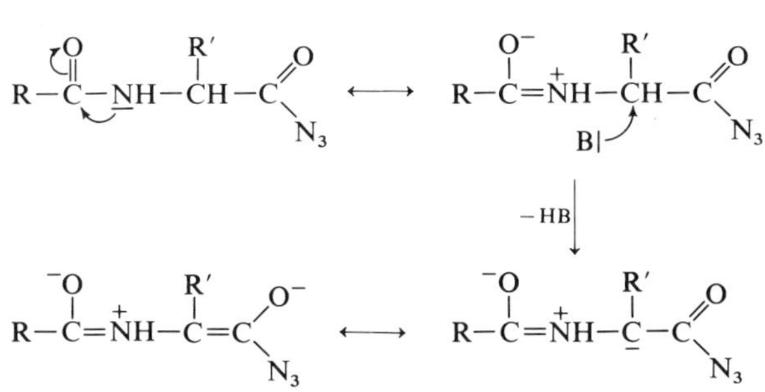

Scheme 20

The very recent proposal of the use of predictive factors which allow calculating the potential risks of racemization for a contemplated peptide coupling (see Tables IV and V) (Kovacs *et al.*, 1975) is a most welcome development and the necessary expansion of the initial data base will be eagerly awaited.

C. Strategy of Peptide Chain Assembly

1. *Basic Scheme*

Condensation of two amino acids (i.e., monoamino–monocarboxylic acids) formally provides peptides with elimination of water (Scheme 21). To obtain a defined product, A–B, the amine function of A and the carboxyl

$$\underset{(A)}{H_3\overset{+}{N}CHR^1COO^-} + \underset{(B)}{H_3\overset{+}{N}CHR^2COO^-} \rightleftharpoons \underset{(\text{mixture of A—A, A—B, B—A, B—B})}{H_3\overset{+}{N}CHR^{1,2}CO\text{—}NHCHR^{1,2}COO^-} + H_2O$$

Scheme 21

function of B must be temporarily blocked. This is the first step of four consecutive operations generally employed in peptide synthesis (Scheme 22). Side chains (R) of certain amino acids contain functionalities which may

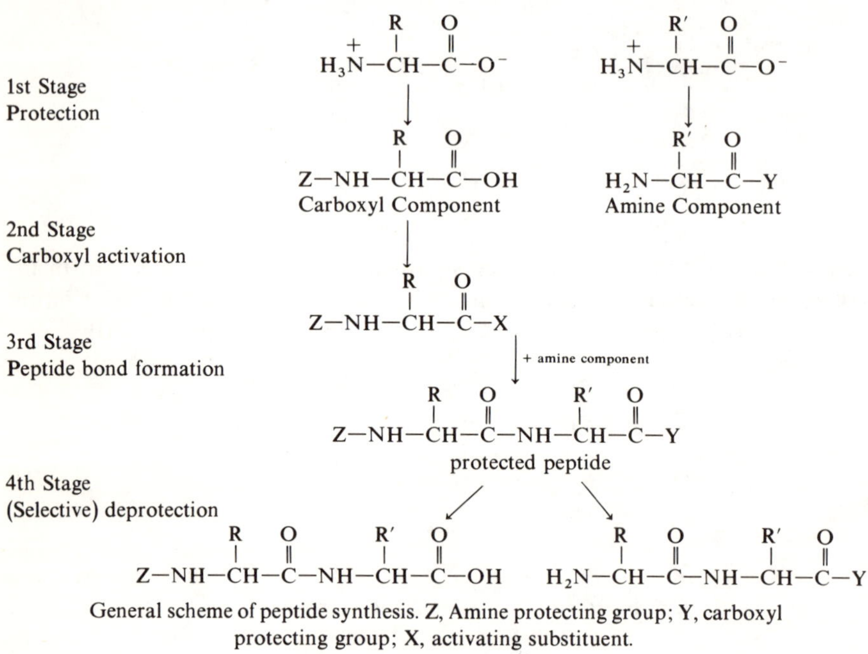

General scheme of peptide synthesis. Z, Amine protecting group; Y, carboxyl protecting group; X, activating substituent.

Scheme 22

need protection that is maintained throughout a synthesis and therefore termed semi-permanent protection.

In the second stage (Scheme 22) the carboxyl group of the N^{α}-protected carboxyl component is activated as discussed in Section IV,A. Following activation, carboxyl components of the anhydride type are used immediately once formed, whereas many kinds of active esters may be isolated and stored as well-defined stable solids. "Coupling reagents" (Klausner and Bodanszky, 1972) conveniently effect selective activation of the carboxyl component in the presence of the amine component.

Peptide bond formation constitutes the third stage. An efficient condensation procedure should provide for rapid peptide bond formation, minimal racemization, few side reactions, ready work-up, and high yield. Because of the stringent requirements only a few methods became widely accepted. They will be discussed in detail in Chapters 2–6.

In the fourth stage protecting groups are removed. Selective cleavage of α-amine or α-carboxyl protection (Scheme 22) affords intermediates for further condensation. Peptide chains are assembled by alternating repetition of stages 2 + 3 and 4. At the end of a synthesis all remaining protecting groups are cleaved. For more details, consult Wünsch (1974), Bodanszky *et al.* (1976), and Schröder and Lübke (1965).

2. *Strategies and Tactics of Functional Group Protection*

The need for protection of certain side chains functionalities, R, (see Scheme 22) of amino acids requires the use of two sets of selectively cleavable blocking groups. While side chain functionalities are frequently blocked through many steps of a synthesis (semipermanent protection), the protection of the α-amino group is removed prior to each coupling step (temporary protection). High selectivity of the respective cleavage reagent is of critical importance for the integrity of the side chain protecting groups. The system presently most popular for selective deprotection employs differential acidolysis. It is used virtually exclusively in solid phase synthesis (Erickson and Merrifield, 1976) and widely in solution techniques (Yajima *et al.*, 1977; Sakakibara, 1977). Its selectivity is relative (kinetic selectivity) and presents increasing problems for the synthesis of larger peptides. The greater the variety of side chain functions the more urgent is the need for absolute selectivity (chemical selectivity) of protecting group cleavage. A well-known ideal system (Schwyzer, 1966; see also Finn and Hofmann, 1976) of complete chemical selectivity involves catalytic hydrogenolysis of the N^{α}-benzyloxycarbonyl group in the presence of *tert*-butyl-derived side-chain protection, which in turn is cleaved by mild acidolysis. This system has recently been applied to sulfur-containing peptides (Kuromizu and Meienhofer, 1974; Felix *et al.*, 1978) by using liquid ammonia as a solvent for catalytic hydrogenolysis. The development of other suitable combinations of protecting groups with chemical selectivity of cleavage (i.e., by different mechanisms) should receive high priority (Veber *et al.*, 1972, 1977; Kemp *et al.*, 1975a,b) in view of the adverse side effects of the widely used HF cleavage of protecting groups in the final stages of peptide synthesis (see Sakakibara, 1971, 1977). High losses of product occur frequently, especially with larger peptides. Specific peptide degradation with liquid HF via N → O-acyl shift (Lenard and Hess, 1964; Lenard, 1969), methionyl peptide bond cleavage (Lenard *et al.*, 1964), and deamination of asparagine and glutamine residues (Robinson *et al.*, 1970) have long been known (see Section III,A2). Acyl carbonium ion formation by glutamic acid γ-carboxyl groups (Feinberg and Merrifield, 1975) may reduce product yields significantly. Development of alternative systems for protecting group cleavage under mild nondegradative conditions in the final stages of peptide synthesis is vital for further progress in this area.

The extent of side chain protection is an important consideration in the planning of a synthesis and depends both on solubility properties of individual peptides and on the selection of the coupling methods. The side chain functionalities of aspartic and glutamic acid, serine, threonine, tyrosine, arginine, and histidine may remain unprotected provided that appropriate coupling procedures (e.g., azide and active ester methods) and compatible chain assembly strategies are used. However, it is mandatory that side chain amine and sulfhydryl groups be blocked throughout a synthesis.

The complete blocking of all functional groups, i.e., "maximum" or "global" protection, provides optimum safety against undesired side reactions and flexibility in the choice of coupling procedures. For example, maximum protection was essential for the eventual success in the difficult solution synthesis of glucagon (Wünsch, 1967). Maximum protection is also necessary in solid phase synthesis because of the use of a large excess of reagents in each acylation step. As a drawback in solution synthesis, many protecting groups presently used can cause larger peptides to become insoluble in organic solvents. Because of this, Hirschmann *et al.* (1969) and Hofmann and collaborators (Beacham *et al.*, 1971) employed "minimum protection" of side chain functionalities in their work on the synthesis of ribonuclease A S-protein and ribonuclease T_1, respectively.

Minimum protection does restrict the choice of coupling methods, e.g., to the use of the azide method in the above studies. However, one of the major benefits of minimum protection vital for both industrial peptide production (Feurer, 1977) and the synthesis of large peptides and proteins, is the greatly facilitated cleavage of the smaller number of protecting groups at the completion of the synthesis to generate the desired free peptide.

D. Opportunities and Constraints

The 25th anniversary of the synthesis of oxytocin (du Vigneaud *et al.*, 1953) which signaled the beginning of modern peptide chemistry, presents a particularly timely vantage point to assess the remarkable progress made in peptide chemistry. A rather respectable number of efficient procedures provided the basis for the synthesis of numerous complex peptides, for example, human insulin (Sieber *et al.*, 1974, 1977) and human big gastrin (Choudhury *et al.*, 1976; Wünsch *et al.*, 1977) by solution methods and fully active bovine pancreatic trypsin inhibitor (Tan and Kaiser, 1976) and ovine β-lipotropin (Yamashiro and Li, 1978) by solid phase synthesis.

While the size of a synthetic peptide bears no relation to its importance, it has often served as a convenient criterion to assess the advances made in the methodology of peptide synthesis. This is mainly so because of (1) the extraordinary increase in the amount of effort and time required for synthesis of ever larger peptides, (2) the severe methodological problems and limitations as the target size increases, and (3) the intention to master the synthesis of proteins (i.e., peptides with over 100 amino acid residues).

All advances notwithstanding, the considerable constraints placed on methodology still impose "an undiminished challenge" (Bodanszky, 1977). The synthesis of proteins by solution methods has not as yet advanced beyond the pioneering level. The remarkable ribonuclease S-protein synthesis by Hirschmann *et al.* (1969) remains the only preparation of a peptide of over 100 residues by solution synthesis, and other efforts in enzyme synthesis (Finn and Hofmann, 1976; Borin *et al.*, 1976; Galpin *et al.*, 1976; Kisfaludy *et al.*, 1976b) have yet to be completed. Constraints in solution synthesis arise mainly from (a) decreasing coupling rates with increasing segment size, resulting in low yields and side-product formation including racemization, (b) low solubility, even in organic solvents of high dissolving power, and (c) difficulties in protecting group removal from large peptides, i.e., incomplete cleavage under mild and nondestructive conditions or partial product decomposition with stronger cleaving agents.

Perusal of the literature indicates that the present state of methodology may permit solution synthesis of peptides with 60–80 residues. This limit will probably be extended by further refinements of existing methods and development of novel approaches such as the "four component segment condensation" (Ugi *et al.*, 1975; Waki and Meienhofer, 1977) or the "amine capture" method (Kemp *et al.*, 1975a,b).

Similarly, attempts of protein synthesis by solid phase procedures have not as yet succeeded to provide products that fully match the physical, chemical, and biologic properties of the natural prototype, but not because of coupling or solubility problems. The constraints of solid phase synthesis encompass (a) formation of microheterogeneous products intractable to complete fractionation by even the most powerful multistage separation procedures presently available, (b) insufficient stability of anchoring and side chain protecting groups, and (c) partial product decomposition upon removal from the solid support [see Sharp *et al.* (1973) for a thorough assessment]. The numerous procedural improvements made since 1973 have been reviewed recently (Merrifield *et al.*, 1977). A synthesis of the 31-peptide β-endorphin (Li *et al.*, 1976), in 30% yield and apparently homogeneous form (see Fig. 10) represents a remarkable example of the refinements made in solid phase methodology.

The present size limit for producing solid phase synthetic peptides that are homogeneous and indistinguishable from the natural material appears to be in the range of 30–40 amino acid residues. However, only a few months ago (Yamashiro and Li, 1978) a synthesis of the 91-peptide ovine β-lipotropin was described in which rigorous purification of the final product provided a material in 0.66% overall yield that was identical with natural β-LPH in 10 different sensitive analytical and biological tests.

It would seem then, that at this time either solution synthesis or solid phase synthesis, or combinations of both, may be utilized to synthesize peptides of up to 15 residues readily. Those of up to 30 or 40 residues will require considerable expertise. For peptides of a size approaching that of proteins, the expenditure of great efforts and patience will be essential.*

While the preparation of increasingly larger natural peptides is necessary, from the medical point of view it is as important, if not more so, that considerable expertise has already been gained in the synthesis of hormone *analogues* which exhibit increased biological potency, modified pharmacologic specificity, longer duration of action and even compatibility with oral administration. Select examples are listed in Table VI. Predictive analogue design has in several instances replaced the time honored trial and error approach. For example, the rationale for various insulin modifications (Friesen *et al.*, 1977) may now be derived from details of its crystal structure (Blundell *et al.*, 1972). Second-generation analogues with most remarkable improvements in biological properties (see Table VI) have been obtained by predictive analogue design, e.g., based on nmr-derived solution conformations of the neurohypophyseal hormones oxytocin (Urry and Walter, 1971; Walter, 1977) and vasopressin (Walter *et al.*, 1974) or on information about sites of metabolic cleavage by proteolytic enzymes (Walter and Hoffman, 1974; Hambrook *et al.*, 1976; Griffiths, 1976; Marks *et al.*, 1976; see also

* The sequence-dependent difficulties arising from unfavorable solubility properties, coupling rates, and racemization rates, remain unpredictable; yet, they may often be decisive for the success or failure in syntheses of large peptides.

Table VI Selected Examples of Peptide Analogues Exhibiting Highly Modified Activities

Native peptide	Analogue	Modified activity	Reference
Met–Enkephalin (Met–EK)	[DAla^2, MePhe^4, Met(O)^5-ol]EK	Potentiation (30,000 ×) Prolongation Orally active	Roemer *et al.* (1977)
Arginine–vasopressin (AVP)	[1-Deamino, DVal^4, DArg^8]vasopressin	Antidiuretic vs. pressor activity shift (125,000:1)	Manning *et al.* (1973)
	[1-Deamino, Phe^2, Δ^3Pro^7]arginine–vasopressin	Antidiuresis potentiation (26 ×)	Smith and Walter (1978)
Angiotensin II (AII)	[Sar^1, Thr(Me)^8]AII	Inhibitor	Khosla *et al.* (1976)
Gonadoliberin (GnRH)	Des-(Gly-NH$_2$^{10})-[DLeu^6]GnRH	Potentiation (~70 ×) Prolongation (~60 ×)	Arimura *et al.* (1974)

Benuck and Marks, 1976) of oxytocin, enkephalin, somatostatin, and other hormones. Future extensions of predictive analogue design to other systems such as endorphin or corticotropin present challenging opportunities of enormous potential for the preparation of clinically useful hormone analogues. However, further refinement in both conformational analysis and synthetic methodology will be required.

V. CONCLUSION AND OUTLOOK

Salient advances have been achieved in the multidisciplinary field of peptide research and development in the 25 years that have passed since the first synthesis of oxytocin was announced by Vincent du Vigneaud and his collaborators (1953). In the same year Bricas and Fromageot (1953) published a comprehensive 125-page review on the contemporary knowledge entitled "Naturally Occurring Peptides" in which they concluded from a survey of 666 literature references that "the number of known naturally occurring peptides is not very great. . . ." (in fact, confirmed structures of only six small peptides could be presented*) and "very little is known about those substances as a group." This illustrates clearly why the historic oxytocin synthesis is commonly considered to be the starting point of modern peptide research. It was indeed a quantum jump in peptide chemistry of dramatic excitatory effects. Impressive advances have since been made. Increasingly refined laboratory techniques and instrumentation provided the tools for successful isolation, structure analysis, and synthesis of hundreds of naturally occurring peptides (not including enzymes). With respect to their biological functions these peptides range from antibiotics to fungal, plant or animal toxins, from circulating hormones to neurotransmitters, hormone release stimulators and inhibitors, from complement to chemotactic factors, from antimetabolites to ionophores, from growth factors to enzyme inhibitors. The study of these peptides, in turn, has stimulated methodological progress. Refined synthetic methods permitted synthesis of numerous analogues for structure–activity studies. Close to 500 analogues of oxytocin and vasopressin and over 100 of such a formidable molecule as insulin have been synthesized. Increasingly rapid and sensitive pharmacological assay techniques have provided the basis for development of ever more meaningful structure–activity relationships. Great progress has been made in understanding three-dimensional structures of both cyclic and linear peptides by X-ray crystal structure analysis or nmr and CD conformational studies in

* These included the dipeptides anserine and carnosine, the tripeptides glutathione and lycomarasmine, the folic acid group of pteroyl oligoglutamates, and the ergot alkaloid group.

solution. These growing insights into the nature of the peptide bond permitted the elaboration of novel rationales for predictive analog design, e.g., for oxytocin, vasopressin, and insulin.

A cardinal question concerns the impact of all the enormous progress in the peptide field on human medicine. Is insulin still the only peptide representative in the pharmaceutical arsenal? A look at Table VII shows that progress has also been made in this area. Quietly and without fanfare, 12 synthetic peptide pharmaceuticals have been developed starting with oxytocin in 1956. Eight companies have had the courage and long-range planning capacity to take upon the unusually difficult task (see Feurer, 1977) to steer multistage peptide syntheses through development and technical production. Perhaps one of the main obstacles for peptides to become widely accepted therapeutics is their failure of being orally effective. In this context the development of orally active Met-enkephalin analogues (see, e.g., Roemer *et al.*, 1977) is encouraging but the recently observed toxicity in man (von Graffenried *et al.*, 1978) also indicates that too many modifications of peptides may precipitate undesired effects.

For the coming years one can confidently predict a continuing search in the discovery of novel biologically important peptides. Only two years ago the entirely new area of opioid peptides began to unfold; certainly at this

Table VII Industrially Produced Peptide Pharmaceuticals[a]

Peptide	No. of amino acids	Trade name	Producer	Introduction
Oxytocin	9	Syntocinon	Sandoz	1956
[Asn^1, Val^5]angiotensin II	8	Hypertensin	CIBA	1959
Lysine–vasopressin	9	Vasopressin	Sandoz	1961
		Diapid (USA)		1970
Corticotropin-(1–24)	24	Synacthen	CIBA	1967
Pentagastrin	5	Peptavlon	ICI	1969
[Deamino]oxytocin	9	Sandopart	Sandoz	1971
[Deamino, DArg]vasopressin [DDAVP]	9	Minirin	Ferring	1973
Thyroliberin	3	TRF	Roche	1974
		Tiregan	Hoechst	
		Thypinone	Abbott	
Salmon calcitonin	32	Calcitonin	Sandoz	1974
		Calcimar	Armour	1975
Gonadoliberin	10	Relisorm L	Abbott	1975
Human calcitonin	32	Cibacalcin	CIBA–Geigy	1977
Human corticotropin	39		Armour	1977

[a] Modified from Fenner (1977).

time we are just beginning to see the tip of the iceberg with respect to brain peptides. The recent development of ultrasensitive high performance liquid chromatographic systems which allow separation and analysis of the hormone content of single individual pituitaries of laboratory animals (rats) opens not only exciting new potentials in peptide isolation, but also provides for meaningful studies of the influences of physiological states, disease, shock, aging, etc., on hormone profiles. Foreseeably, this will stimulate development of yet more sensitive assay procedures and also even more sensitive sequencing instrumentation than the already available 1–10 nanomole capacity. It is similarly predictable that both X-ray crystal structure analysis and conformational analysis in solution will continue to make great strides forward. For example the elucidation of the conformations of somatostatin and β-endorphin will be expected with much anticipation since they might become pivotal for predictive analog design.

Yet there are areas and problems for which it is difficult to predict in which direction future developments will turn. For example the underlying reasons or mechanisms responsible for the extraordinarily high biological potencies and specificities of peptides remain to be poorly understood. Certainly, the receptor concept, i.e., the interaction of a hormone with specific macromolecules which causes activation of the molecule to trigger the expression of physiological activity, has been universally accepted, and increasing information on the properties of isolated receptor proteins is becoming available. Expectations are high for receptor structures, once determined, to provide great insight into the mechanisms of action of the respective peptide ligands. However, it is difficult to be certain about this prospect for several reasons, e.g., knowledge of the X-ray crystal structure of many enzymes has not yet contributed in a major way to a better understanding of the mechanism of enzyme action, and the known receptor of the peptide antibiotic actinomycin D, which happens to be double stranded DNA, and the rather detailed models proposed for their interaction from X-ray studies have failed to clarify the actual mechanism of action of actinomycin D.

It is clear that the mechanism of peptide hormone action differs from that of enzymes, i.e., hormones do not generally seem to possess an "active center" comprising a few vital amino acid residues. Instead several discrete sequences of adjacent amino acids seem to be responsible for different components of the total biological activity spectrum, as for instance in corticotropin (ACTH). The term "sychnologic" has been proposed for this type of molecular organization (Schwyzer, 1973), as opposed to a "rhegnylogic" organization in which the information for biological activity resides in separate residues of the sequence which are brought into close proximity by conformational adaptation (folding).

In the cases of insulin (Blundell *et al.*, 1972) and the oxytocin–vasopressin group of hormones (Walter, 1977) as well as the antibiotic actinomycin D (Lackner, 1975; for review, see Meienhofer and Atherton, 1977), the major factor responsible for their biological specificity and potency appears to be the overall three-dimensional molecular architecture.* Extensive alterations result in total loss of activity, while various modifications of the bioactivity may be obtained by subtle changes. Some of these may have quite surprising effects, e.g., the analogue [1-deamino, 2-phenylalanine, 7-(L-3,4-dehydroproline)] arginine–vasopressin (Smith and Walter, 1978) exhibited 26 times the antidiuretic activity of arginine vasopressin, i.e., its already high activity at the nanomolar level was further increased to the 100 picomole level by the introduction of a double bond in the 7-proline residue.† Thus, nature may be substantially improved upon; and it appears that synthetic analogues, obtained by replacements with amino acids not derived from protein, may exhibit striking biological activity.

In chemical synthesis we can be adamant about the inadequacy of our present approaches to make the synthesis of homogeneous proteins a routine matter. Beyond that it seems uncertain just how to approach the challenge with any chance of success. Two mutually complementary possibilities exist: (i) continuing improvements in the existing arsenal of methods by patiently effecting small changes, expansions, additions, time savings, etc., or (ii) by principally new approaches in which the actual peptide bond formation becomes a concentration-independent intramolecular rearrangement, such as the four component condensation (Ugi *et al.*, 1975) or the amine-capture strategy (Kemp *et al.*, 1975a,b). Only time will tell which of these or other approaches might be successful. Just to make this area yet more exciting, the recent announcement (Itakura *et al.*, 1977) of the synthesis of somatostatin by expression in an *E. coli* culture of a gene that was chemically synthesized and processed by recombinant DNA methods, adds an entire new dimension to the peptide field.

Thus, we may conclude by quoting from Ralph Hirschmann's (1972) outlook at the end of the Third American Peptide Symposium: "The future looks bright, not because we are so clever, but because the building blocks which are at our disposal permit the synthesis of compounds with seemingly unlimited versatility of chemical, physical, and biological properties". As we continue in our research, we may differ in our goals and motivations, but Vincent du Vigneaud may have summed it up in the most unifying way when he once remarked at a staff meeting, "We have to do it, because we owe it to mankind."

* Alternative terms are "geometrical shape" or "overall conformation."

† The other two changes in the 1- and 2-positions are known to exert much lower potentiation. However, they also contributed to the dramatic overall potentiation.

REFERENCES

Adams, M. J., Blundell, T. L., Dodson, E. J., Dodson, G. G., Vijayan, M., Baker, E. N., Harding, M. M., Hodgkin, D. C., Rimmer, B., and Sheat, S. (1969). *Nature* (*London*) **224**, 491–495.

Anfinsen, C. B., Ontjes, D., Ohno, M., Corley, L., and Eastlake, A. (1967). *Proc. Natl. Acad. Sci. U.S.A.* **58**, 1806–1811.

Acher, R., Chauvet, J., and Chauvet, M. T. (1973a). *Eur. J. Biochem.* **40**, 585–589.

Acher, R., Chauvet, J., and Chauvet, M. T. (1973b). *Nature* (*London*), *New Biol.* **244**, 124–126.

Arimura, A., Vilchez-Martinez, J. A., Coy, D. H., Coy, E. J., Hirotsu, Y., and Schally, A. V. (1974). *Endocrinology* **95**, 1174–1177 (1974).

Aswanikumar, S., Schiffmann, E., Corcoran, B. A., and Wahl, S. M. (1976). *Proc. Natl. Acad. Sci. U.S.A.* **73**, 2439–2442.

Atherton, E., and Sheppard, R. C. (1977). *In* "Peptides, Proceedings of the Fifth American Peptide Symposium" (M. Goodman and J. Meienhofer, eds.), pp. 503–505. Wiley, New York.

Atherton, E., Caviezel, M., Over, H., and Sheppard, R. C. (1977). *J. Chem. Soc., Chem. Commun.* pp. 819–821.

Bauer, K., Roskoski, R., Jr., Kleinkauf, H., and Lipmann, F. (1972). *Biochemistry* **11**, 3266–3271.

Beacham, J., Dupuis, G., Finn, F. M., Storey, H. T., Yanaihara, C., Yanaihara, N., and Hofmann, K. (1971). *J. Am. Chem. Soc.* **93**, 5526–5539.

Bender, M. L. (1960). *Chem. Rev.* **60**, 53–113.

Benedetti, E. (1977). *In* "Peptides, Proceedings of the Fifth American Peptide Symposium" (M. Goodman and J. Meienhofer, eds.), pp. 257–273. Wiley, New York.

Benedict, R. G., Dvonch, W., Shotwell, O. L., Pridham, T. G., and Lindenfelser, L. A. (1952). *Antibiot. Chemother.* (*Washington. D.C.*) **2**, 591–594.

Benuck, M., and Marks, N. (1976). *Life Sci.* **19**, 1271–1276.

Blake, J., and Li, C. H. (1976). *J. Chem. Soc., Chem. Commun.* pp. 504–505.

Blobel, G., and Dobberstein, B. (1975). *J. Cell Biol.* **67**, 835–851 and 852–862.

Blobel, G., and Sabatini, D. D. (1971). *In* "Biomembranes" (L. A. Manson, ed.), Vol. 2, pp. 193–195. Plenum, New York.

Blow, D. M. (1976). *Acc. Chem. Res.* **9**, 145–152.

Blow, D. M., Birktoft, J. J., and Hartley, B. S. (1969). *Nature* (*London*) **221**, 337–340.

Blundell, T. L., Dodson, G. G., Hodgkin, D. C., and Mercola, D. A. (1972). *Adv. Protein Chem.* **26**, 279–402.

Bodanszky, M. (1971). *In* "Prebiotic and Biochemical Evolution" (A. P. Kimball and J. Oró, eds.), pp. 217–222. North-Holland Publ., Amsterdam.

Bodanszky, M. (1977). *In* "Peptides, Proceedings of the Fifth American Peptide Symposium" (M. Goodman and J. Meienhofer, eds.), pp. 1–13. Wiley, New York.

Bodanszky, M., and Bath, R. J. (1969). *J. Chem. Soc., D, Chem. Commun.* pp. 1259–1260.

Bodanszky, M., and Natarajan, S. (1975). *J. Org. Chem.* **40**, 2495–2499.

Bodanszky, M., Fink, M. L., Funk, K. W., Kondo, M., Lin, C. Y., and Bodanszky, A. (1974). *J. Am. Chem. Soc.* **96**, 2234–2240.

Bodanszky, M., Klausner, Y. S., and Ondetti, M. A. (1976). "Peptide Synthesis," 2nd ed. Wiley, New York.

Bodanszky, M., Fink, M. L., Klausner, Y. S., Natarajan, S., Tatemoto, K., Yiotakis, A. E., and Bodanszky, A. (1977). *J. Org. Chem.* **42**, 149–152.

Borin, G., Filippi, B., Moroder, L., and Marchiori, F. (1976). *In* "Peptides 1976" (A. Loffet, ed.), pp. 233–238. Editions de l'Université de Bruxelles, Belgium.

Borsook, H. (1953). *Adv. Protein Chem.* **8**, 127–174.

Bovey, F. A. (1972). *In* "Chemistry and Biology of Peptides" (J. Meienhofer, ed.), pp. 3–28. Ann Arbor Sci. Publ., Ann Arbor, Michigan.

Bovey, F. A. (1974). *In* "Peptides, Polypeptides and Proteins" (E. R. Blout *et al.*, eds.), pp. 248–265. Wiley, New York.

Bovey, F. A., Brewster, A. I., Patel, D. J., Tonelli, E. A., and Torchia, D. A. (1972). *Acc. Chem. Res.* **5**, 193–200 (1972).

Bradbury, A. F., Smyth, D. G., and Snell, C. R. (1976a). *In* "Polypeptide Hormones: Molecular and Cellular Aspects," pp. 61–75. Elsevier, Amsterdam.

Bradbury, A. F., Smyth, D. G., and Snell, C. R. (1976b). *Biochem. Biophys. Res. Commun.* **69**, 950–956.

Brandenburg, D., Busse, W.-D., Gattner, H.-G., Zahn, H., Wollmer, A., Gliemann, J., and Puls, W. (1973). *In* "Peptides 1972" (H. Hanson and H. D. Jakubke, eds.), pp. 270–283. North-Holland Elsevier, New York.

Brenner, M., Giger, U., Nyfeler, R., Schenk, P., and Tschopp, A. (1976). *In* "Peptides 1976" (A. Loffet, ed.), pp. 65–70. Editions de l'Université de Bruxelles, Belgium.

Bricas, E., and Fromageot, C. (1953). *Adv. Protein Chem.* **8**, 1–125.

Bruice, T. C., and Benkovic, S. J. (1964). *J. Am. Chem. Soc.* **86**, 418–426.

Bruice, T. C., and Benkovic, S. J. (1966). "Bioorganic Mechanisms," Vol. 1. Benjamin, New York.

Carpino, L. A. (1973). *Acc. Chem. Res.* **6**, 191–198.

Chan, S. J., Keim, P., and Steiner, D. F. (1976). *Proc. Natl. Acad. Sci. U.S.A.* **73**, 1964–1968.

Choudhury, A. M., Kenner, G. W., Moore, S., Ramage, R., Richards, P. M., and Thorpe, W. D. (1976). *In* "Peptides 1976" (A. Loffet, ed.), pp. 257–261. L'Université de Bruxelles, Belgium.

Corey, R. B., and Pauling, L. (1953). *Proc. R. Soc. London, Ser. B* **141**, 10–20.

Crick, F. H. C. (1957). *Biochem. Soc. Symp.* **14**, 25–26.

Davie, E. W., and Ratnoff, O. D. (1964). *Science* **145**, 1310–1312.

Deisenhofer, J., and Steigemann, W. (1975). *Acta Crystallogr., Sect. B* **31**, 238–250.

Desnuelle, P. (1960). *In* "The Enzymes" (P. D. Boyer, H. Lardy, and K. Myrbäck, eds.), 2nd rev. ed., vol. 4, pp. 93–118. Academic Press, New York.

Devillers-Thiery, A., Kindt, T., Scheele, G., and Blobel, G. (1975). *Proc. Natl. Acad. Sci. U.S.A.* **72**, 5016–5020.

Duguid, J. R., Steiner, D. F., and Chick, W. L. (1976). *Proc. Natl. Acad. Sci. U.S.A.* **73**, 3539–3543.

Dutta, A. S., and Morley, J. S. (1971). *J. Chem. Soc. C*, pp. 2896–2902.

du Vigneaud, V., Ressler, C., Swan, J. M., Roberts, C. W., Katsoyannis, P. G., and Gordon, S. (1953). *J. Am. Chem. Soc.* **75**, 4879–4880.

Endo, S. (1962). *J. Ferment. Technol.* (*Japanese*) **40**, 346–353.

Erickson, B. W., and Merrifield, R. B. (1973). *J. Am. Chem. Soc.* **95**, 3750–3756 and 3757–3763.

Erickson, B. W., and Merrifield, R. B. (1976). *In* "The Proteins" (H. Neurath and R. L. Hill, eds.), 3rd ed., Vol. 2, pp. 255–527. Academic Press, New York.

Farrington, J. A., Kenner, G. W., and Turner, J. M. (1955). *Chem. Ind.* (*London*) pp. 601–602.

Feinberg, R. S., and Merrifield, R. B. (1975). *J. Am. Chem. Soc.* **97**, 3485–3496.

Felix, A. M., Jimenez, M. H., Mowles, T., and Meienhofer, J. (1978). *Int. J. Pept. Protein Res.* **11**, 329–339.

Feurer, M. (1977). *In* "Peptides, Proceedings of the Fifth American Peptide Symposium" (M. Goodman and J. Meienhofer, eds.), pp. 448–457. Wiley, New York.

Finn, F. M., and Hofmann, K. (1973). *Acc. Chem. Res.* **6**, 169–176.

Finn, F. M., and Hofmann, K. (1976). *In* "The Proteins" (H. Neurath and R. L. Hill, eds.), 3rd ed., Vol. 2, pp. 105–253. Academic Press, New York.

Fischer, E. (1906). *Ber. Dtsch. Chem. Ges.* **39**, 530–610 (cf. p. 551).

Flippen, J. L., and Karle, I. L. (1976). *Biopolymers* **15**, 1081–1092.

Friesen, H. J., Brandenburg, D., Diaconescu, C., Gattner, H.-G., Naithani, V. K., Nowak, J.,

Zahn, H., Dockerill, S., Wood, S. P., and Blundell, T. L. (1977). *In* "Peptides, Proceedings of the Fifth American Peptide Symposium" (M. Goodman and J. Meienhofer, eds.), pp. 136–140. Wiley, New York.

Galpin, I. J., Handa, B. K., Hudson, D., Jackson, A. G., Kenner, G. W., Ohlsen, S. R., Ramage, R., Singh, B., and Tyson, R. G. (1976). *In* "Peptides 1976" (A. Loffet, ed.), pp. 247–256. Editions de l'Université de Bruxelles, Belgium.

Geiger, R., Jäger, G., König, W., and Volk, A. (1969). *Z. Naturforsch., Teil B* **24**, 999–1004.

Glasoe, P. K., Kleinberg, J., and Audrieth, L. F. (1939). *J. Am. Chem. Soc.* **61**, 2387–2389.

Goodman, M., and Glaser, C. (1970). *In* "Peptides: Chemistry and Biochemistry" (B. Weinstein and S. Lande, eds.), pp. 267–335. Dekker, New York.

Gresser, M. J., and Jencks, W. P. (1977). *J. Am. Chem. Soc.* **99**, 6970–6980.

Griffiths, E. C. (1976). *Horm. Res.* **7**, 179–191.

Gross, E. (1967). *In* "Methods in Enzymology" (C. H. W. Hirs, ed.), Vol. 11, pp. 238–255. Academic Press, New York.

Gross, E., and Brown, J. H. (1976). *In* "Peptides 1976" (A. Loffet, ed.), pp. 183–190. Editions de l'Université de Bruxelles, Bruxelles, Belgium.

Gross, E., and Morell, J. L. (1967). *J. Am. Chem. Soc.* **89**, 2791–2793.

Gross, E., and Morell, J. L. (1968). *FEBS Lett.* **26**, 61–64.

Gross, E., and Morell, J. L. (1971). *J. Am. Chem. Soc.* **93**, 4634–4635.

Gross, E., and Morell, J. L. (1974). *Biochem. Biophys. Res. Commun.* **59**, 1145–1150.

Gross, E., and Witkop, B. (1962). *J. Biol. Chem.* **237**, 1856–1860.

Gross, E., Kiltz, H. H., and Nebelin, E. (1973). *Hoppe Seyler's Z. Physiol. Chem.* **354**, 810–812.

Gruber, K. A., Stein, S., Brink, L., Radhakrishnan, A., and Udenfriend, S. (1976). *Proc. Natl. Acad. Sci. U.S.A.* **73**, 1314–1318.

Güntelberg, A. V., and Ottesen, M. (1954). *C. R. Trav. Lab. Carlesberg, Ser. Chim.* **29**, 36–48.

Gut, V., and Rudinger, J. (1968). *In* "Peptides 1968" (E. Bricas, ed.), pp. 185–186. North-Holland Publ., Amsterdam.

Gutfreund, H. (1976). *FEBS Lett.* **62**, Suppl., E1–E12.

Gutte, B., and Merrifield, R. B. (1969). *J. Am. Chem. Soc.* **91**, 501–502.

Gutte, B., and Merrifield, R. B. (1971). *J. Biol. Chem.* **246**, 1922–1941.

Hagenmaier, H., and Frank, H. (1972). *Hoppe-Seyler's Z. Physiol. Chem.* **353**, 1973–1976.

Hagler, A. T., and Lapiccirella, A. (1976). *Biopolymers* **15**, 1167–1200.

Hambrook, J. M., Morgan, B. A., Rance, M. J., and Smith, C. F. C. (1976). *Nature (London)* **262**, 782–783.

Hill, R. L. (1965). *Adv. Protein Chem.* **20**, 37–107.

Hirs, C. H. W. (1956). *J. Biol. Chem.* **219**, 611–621.

Hirs, C. H. W., Moore, S., and Stein, W. H. (1956). *J. Biol. Chem.* **219**, 623–642.

Hirs, C. H. W., Moore, S., and Stein, W. H. (1960). *J. Biol. Chem.* **235**, 633–647.

Hirschmann, R. (1972). *In* "Chemistry and Biology of Peptides" (J. Meienhofer, ed.), pp. 721–728. Ann Arbor Sci. Publ., Ann Arbor, Michigan.

Hirschmann, R., Nutt, R. F., Veber, D. F., Vitali, R. A., Varga, S. L., Jacob, T. A., Holly, F. W., and Denkewalter, R. G. (1969). *J. Am. Chem. Soc.* **91**, 507–508 (see also pp. 502–507).

Hodges, R. S., and Merrifield, R. B. (1975). *Anal. Biochem.* **65**, 241–272.

Hofmeister, F. (1902). *Ergeb. Physiol., Biol. Chem. Exp. Pharmakol.* **1**, 759–802.

Hruby, V. J. (1974). *In* "Chemistry and Biochemistry of Amino Acids, Peptides and Proteins" (B. Weinstein, ed.), Vol. 3, pp. 1–188. Dekker, New York.

Huber, R., and Bode, W. (1978). *Acc. Chem. Res.* **11**, 114–122.

Huber, R., Bode, W., Kukla, D., Kohl, U., and Ryan, C. A. (1975). *Biophys. Struct. Mech.* **1**, 189–201.

Hughes, J., Smith, T. W., Kosterlitz, H. W., Fothergill, L. A., Morgan, B. A., and Morris, H. R. (1975). *Nature (London)* **258**, 577–579.

Itakura, K., Hirose, T., Crea, R., Riggs, A. D., Heyneker, H. L., Bolivar, F., and Boyer, H. W. (1977). *Science* **198**, 1056–1063.

Ivanov, V. T. (1977). *In* "Peptides, Proceedings of the Fifth American Peptide Symposium" (M. Goodman and J. Meienhofer, eds.), pp. 307–321. Wiley, New York.

Ivanov, V. T., Mikhaleva, I. I., Volpina, O. M., Myagkova, M. A., and Deigin, V. I. (1976). *In* "Peptides 1976" (A. Loffet, ed.), pp. 219–231. Editions de l'Université de Bruxelles, Belgium.

Iwai, K., and Ando, T. (1967). *In* "Methods in Enzymology" (C. H. W. Hirs, ed.), Vol. 11, pp. 263–282. Academic Press, New York.

Izumiya, N., Muraoka, M., and Aoyagi, H. (1971). *Bull. Chem. Soc. Jpn.* **44**, 3391–3395.

Jakubke, H.-D. (1966). *Z. Chem.* **6**, 52–67.

Jakubke, H.-D., and Voigt, A. (1966a). *Chem. Ber.* **99**, 2419–2429.

Jakubke, H.-D., and Voigt, A. (1966b). *Chem. Ber.* **99**, 2944–2954.

Jakubke, H.-D., Voigt, A., and Burckhardt, S. (1967). *Chem. Ber.* **106**, 2367–2372.

Johnson, S. L. (1967). *Adv. Phys. Org. Chem.* **5**, 237–330.

Jones, J. H. (1974). *Chem. Ind. (London)* pp. 723–727.

Jones, J. H., and Young, G. T. (1967). *Chem. Commun.* pp. 35–36.

Jones, J. H., and Young, G. T. (1968). *J. Chem. Soc. C* pp. 436–441.

Karle, I. L. (1972). *In* "Chemistry and Biology of Peptides" (J. Meienhofer, ed.), pp. 117–122. Ann Arbor Sci. Publ., Ann Arbor, Michigan.

Karle, I. L. (1975). *In* "Peptides: Chemistry, Structure and Biology" (R. Walter and J. Meienhofer, eds.), pp. 61–84. Ann Arbor Sci. Publ., Ann Arbor, Michigan.

Kartha, G., Bello, J., and Harker, D. (1967). *Nature (London)* **213**, 862–864.

Katz, J. J. (1954a). *Nature (London)* **173**, 265.

Katz, J. J. (1954b). *Arch. Biochem. Biophys.* **51**, 293–305.

Kemp, D. S. (1970). *In* "Peptides: Chemistry and Biochemistry" (B. Weinstein and S. Lande, eds.), pp. 33–44. Dekker, New York.

Kemp, D. S. (1973). *In* "Peptides 1971" (H. Nesvadba, ed.), pp. 1–19. North-Holland Publ., Amsterdam.

Kemp, D. S., and Chien, S. W. (1967). *J. Am. Chem. Soc.* **89**, 2743–2745 and 2745–2746.

Kemp, D. S., and Rebek, J. Jr. (1970). *J. Am. Chem. Soc.* **92**, 5792–5793.

Kemp, D. S., and Reczek, J. (1977). *Tetrahedron Lett.* pp. 1031–1034.

Kemp, D. S., Wang, S. W., Busby, G., III, and Hugel, G. (1970). *J. Am. Chem. Soc.* **92**, 1043–1055.

Kemp, D. S., Hsia Choong, S. L., and Pekaar, J. (1974a). *J. Org. Chem.* **39**, 3841–3847.

Kemp, D. S., Wang, S.-W., Mollan, R. C., Hsia, S. L., and Confalone, P. N. (1974b). *Tetrahedron* **30**, 3677–3688.

Kemp, D. S., Wang, S.-W., Rebek, J., Jr., Mollan, R. C., Banquer, C., and Subramanyam, G. (1974c) *Tetrahedron* **30**, 3955–3967.

Kemp, D. S., Wrobel, S. J., Jr., Wang, S.-W., Bernstein, Z., and Rebek, J., Jr. (1974d). *Tetrahedron* **30**, 3969–3980.

Kemp, D. S., Roberts, D., Hoyng, C., Grattan, J., Vellaccio, F., and Reczek, J. (1975a). *In* "Peptides: Chemistry, Structure, and Biology" (R. Walter and J. Meienhofer, eds.), pp. 295–305. Ann Arbor Sci. Publ., Ann Arbor, Michigan.

Kemp, D. S., Roberts, D., Hoyng, C. (1975b). *Tetrahedron Lett.* pp. 4625–4628 and 4629–4632.

Khosla, M. C., Munoz-Ramirez, H., Hall, M. M., Smeby, R. R., Khairallah, P. A., Bumpus, F. M., and Peach, M. J. (1976). *J. Med. Chem.* **19**, 244–250.

Khurgin, Yu. I., and Dmitrieva, M. G. (1965). *Tetrahedron* **21**, 2305–2312.
Kiltz, H. H., and Gross, E. (1973). *Hoppe Seyler's Z. Physiol. Chem.* **354**, 802–804.
Kisfaludy, L., Löw, M., Argay, G., Czugler, M., Kömives, T., Sohar, P., and Darvas, F. (1976a). *In* "Peptides 1976" (A. Loffet, ed.), pp. 55–59. Editions de l'Université de Bruxelles, Belgium.
Kisfaludy, L., Löw, M., Schön, I., Szirtes, T., Nyéki, O., Vezer, Cs., Keresztessy, P., Korenczki, F., and Szirmai, M. (1976b). *In* "Peptides 1976" (A. Loffet, ed.), pp. 239–245. Editions de l'Université de Bruxelles, Belgium.
Klausner, Y. S., and Bodanszky, M. (1972). *Synthesis* pp. 453–463.
Kleinkauf, H., Roskoski, R., Jr., and Lipmann, F. (1971). *Proc. Natl. Acad. Sci. U.S.A.* **68**, 2069–2072.
Koch, A. L., Lamont, W. A., and Katz, J. J. (1956). *Arch. Biochem. Biophys.* **63**, 106–117.
König, W., and Geiger, R. (1972). *In* "Chemistry and Biology of Peptides" (J. Meienhofer, ed.), pp. 343–350. Ann Arbor Sci. Publ., Ann Arbor, Michigan.
König, W., and Geiger, R. (1973a). *In* "Peptides 1972" (H. Hanson and H. D. Jakubke, eds.), pp. 158–161. North-Holland Publ., Amsterdam.
König, W., and Geiger, R. (1973b). *Chem. Ber.* **106**, 3626–3635.
Kovacs, J., Mayers, G. L., Johnson, R. H., Cover, R. E., and Ghutak, U. R. (1970a). *J. Chem. Soc., D, Chem. Commun.* pp. 53–54.
Kovacs, J., Mayers, G. L., Johnson, R. H., Cover, R. E., and Ghatak, U. R. (1970b). *J. Org. Chem.* **35**, 1810–1815.
Kovacs, J., Davis, E. J., Johnson, R. H., Cortegiano, H., and Roberts, J. (1972). *In* "Chemistry and Biology of Peptides" (J. Meienhofer, ed.), pp. 359–364. Ann Arbor Sci. Publ., Ann Arbor, Michigan.
Kovacs, J., Cover, R. E., Johnson, R. H., Kalas, T. J., Mayers, G. L., and Roberts, J. E. (1973). *J. Org. Chem.* **38**, 2518–2521.
Kovacs, J., Cover, R., Jham, G., Hsieh, Y., and Kalas, T. (1975). *In* "Peptides: Chemistry, Structure, and Biology" (R. Walter and J. Meienhofer, eds.), pp. 317–324. Ann Arbor Sci. Publ., Ann Arbor, Michigan.
Kovacs, K., and Penke, B. (1973). *In* "Peptides 1972" (H. Hanson and H. D. Jakubke, eds.), pp. 187–188. North-Holland Publ., Amsterdam.
Krail, G., Brandenburg, D., and Zahn, H. (1975). *Makromol. Chem., Suppl.* **1**, 7–22.
Krieger, D. E., Erickson, B. W., and Merrifield, R. B. (1976). *Proc. Natl. Acad. Sci. U.S.A.* **73**, 3160–3164.
Kricheldorf, H. R., Stengele, E., and Regel, W. (1975). *Justus Liebig's Ann. Chem.* **1975**, 1379–1386.
Kühne, W. F. (1876). *Verh. Naturhist.-Med. Ver. Heidelberg* **1**, 1–10.
Kuromizu, K., and Meienhofer, J. (1974). *J. Am. Chem. Soc.* **96**, 4978–4981.
Lackner, H. (1975). *Angew. Chem., Int. Ed. Engl.* **14**, 375–386.
Lazarus, L. H., Ling, N., and Guillemin, R. (1976). *Proc. Natl. Acad. Sci. U.S.A.* **73**, 2156–2159.
Lemaire, S., Yamashiro, D., Behrens, C., and Li, C. H. (1977). *J. Am. Chem. Soc.* **99**, 1577–1580.
Lenard, J. (1969). *Chem. Rev.* **69**, 625–638.
Lenard, J., and Hess, G. P. (1964). *J. Biol. Chem.* **239**, 3275–3281.
Lenard, J., Schally, A. V., and Hess, G. P. (1964). *Biochem. Biophys. Res. Commun.* **14**, 498–502.
Li, C. H., and Chung, D. (1976a). *Nature (London)* **260**, 622–624.
Li, C. H., and Chung, D. (1976b). *Proc. Natl. Acad. Sci. U.S.A.* **73**, 1145–1148.
Li, C. H., and Yamashiro, D. (1970). *J. Am. Chem. Soc.* **92**, 7608–7609.
Li, C. H., Lemaire, S., Yamashiro, D., and Doneen, B. A. (1976). *Biochem. Biophys. Res. Commun.* **71**, 19–25.
Linderstrøm-Lang, K. U., and Ottesen, M. (1949). *C. R. Trav. Lab. Carlsberg* **26**, 403–410.
Lipmann, F. (1971). *Science* **173**, 875–884.

Lloyd, K., and Young, G. T. (1971). *J. Chem. Soc. C* pp. 2890–2896.

Long, D. A., and Lillycrop, J. E. (1963). *Trans. Faraday Soc.* **59**, 907–917.

Long, D. A., and Truscott, T. G. (1963). *Trans. Faraday Soc.* **59**, 918–923.

Lowbridge, J., Manning, M., Hladar, J., and Sawyer, W. H. (1977). *J. Med. Chem.* **20**, 120–123.

Mains, R. E., Eipper, B. A., and Ling, N. (1977). *Proc. Natl. Acad. Sci. U.S.A.* **74**, 3014–3018.

Manning, M., Balaspiri, L., Acosta, M., and Sawyer, W. H. (1973). *J. Med. Chem.* **16**, 975–978.

Marks, N., Stern, F., and Benuck, M. (1976). *Nature* (*London*) **261**, 511–512.

Maroux, S., Baratti, J., and Desnuelle, P. (1971). *J. Biol. Chem.* **246**, 5031–5039.

Marsh, R. E., and Donohue, J. (1967). *Adv. Protein Chem.* **22**, 235–256.

Marshall, G. R., Hancock, W. S., Prescott, D. J., Nulty, W. L., Weintraub, J., and Vagelos, P. R. (1973). *In* "Peptides 1971" (H. Nesvadba, ed.), pp. 185–194. North-Holland Publ., Amsterdam.

Matsubara, H., Singer, A., Sasaki, R., and Jukes, T. H. (1965). *Biochem. Biophys. Res. Commun.* **21**, 242–247.

Meienhofer, J. (1973). *In* "Hormonal Proteins and Peptides" (C. H. Li, ed.), Vol. 2, pp. 45–267. Academic Press, New York.

Meienhofer, J., and Atherton, E. (1977). *In* "Structure-Activity Relationships Among the Semisynthetic Actinomycins" (D. Perlman, ed.), pp. 427–529. Academic Press, New York.

Meister, A. (1965). "Biochemistry of the Amino Acids," 2nd ed., Vol. 1, pp. 452–456. Academic Press, New York.

Mercola, D. A., Morris, J. W. S., and Arguilla, E. R. (1972). *Biochemistry* **11**, 3860–3874.

Merrifield, R. B. (1963). *J. Am. Chem. Soc.* **85**, 2149–2154.

Merrifield, R. B. (1969). *Adv. Enzymol.* **32**, 221–296.

Merrifield, R. B., Barany, G., Cosand, W. L., Engelhard, M., and Mojsov, S. (1977). *In* "Peptides, Proceedings of the Fifth American Peptide Symposium" (M. Goodman and J. Meienhofer, eds.), pp. 488–502. Wiley, New York.

Milstein, C., Brownlee, G. G., Harrison, T. M., and Mathews, M. B. (1972). *Nature* (*London*), *New Biol.* **239**, 117–120.

Mitchell, A. R., Erickson, B. W., Ryabtsev, M. N., Hodges, R. S., and Merrifield, R. B. (1976). *J. Am. Chem. Soc.* **98**, 7357–7362.

Moffat, J. B. (1973). *J. Theor. Biol.* **40**, 247–258.

Momany, F. A., McGuire, R. F., Yan, J. F., and Scheraga, H. A. (1970). *J. Phys. Chem.* **74**, 2424–2438.

Moore, S. (1972). *In* "Chemistry and Biology of Peptides" (J. Meienhofer, ed.), pp. 629–653. Ann Arbor Sci. Publ., Ann Arbor, Michigan.

Muramatu, M., Hirohata, R., Kanda, Y., Shibuya, S., Fujii, S., Nagamatsu, A., and Ono, T. (1963). *Hoppe-Seyler's Z. Physiol. Chem.* **332**, 256–262.

Muzalewski, F. (1973). *In* "Peptides 1971" (H. Nesvadba, ed.), pp. 39–44. North-Holland Publ., Amsterdam.

Naganathan, P. S., and Venkatesan, K. (1972). *Acta Crystallogr., Sect. B* **28**, 552–556.

Natarajan, S., and Bodanszky, M. (1976). *J. Org. Chem.* **41**, 1269–1272.

Nebelin, E., and Gross, E. (1973). *Hoppe-Seyler's Z. Physiol. Chem.* **354**, 807–809.

Neurath, H. (1964). *Fed. Proc., Fed. Am. Soc. Exp. Biol.* **23**, 1–12.

Neurath, H. (1975). *In* "Proteases and Biological Control, Cold Spring Harbor Conferences on Cell Proliferation" (E. Reich, D. B. Rifkin, and E. Shaw, eds.), Vol. 2, pp. 51–64. Cold Spring Harbor Lab., Cold Spring Harbor, New York.

Ovchinnikov, Yu. A., and Ivanov, V. T. (1975). *Tetrahedron* **31**, 2177–2209.

Patchornik, A., Lawson, W. B., Gross, E., and Witkop, B. (1960). *J. Am. Chem. Soc.* **82**, 5923–5927.

Pattabhi, V., Venkatesan, K., and Hall, S. R. (1974). *J. Chem. Soc., Perkin Trans. 2* pp. 1722–1727.

Pless, J. (1963). *In* "Peptides" (G. T. Young, ed.), pp. 69–72. Pergamon, Oxford.

Pless, J., and Boissonnas, R. A. (1963). *Helv. Chim. Acta* **46**, 1609–1625.

Pridham, T. G., Lindenfelser, L. A., Shotwell, O. L., Stodola, F. H., Benedict, R. G., Foley, C., Jackson, R. W., Zaumeyer, W. J., Preston, W. H., and Mitchell, J. W. (1956). *Phytopathology* **46**, 568–575.

Pullen, R. A., Jenkins, J. A., Tickle, I. J., Wood, S. P., and Blundell, T. L. (1975). *Mol. Cell. Biochem.* **8**, 5–20.

Pullen, R. A., Lindsay, D. G., Wood, S. P., Tickle, I. J., Blundell, T. L., Wollmer, A., Krail, G., Brandenburg, D., Zahn, H., Gliemann, J., and Gammeltoft, S. (1976). *Nature* (*London*) **259**, 369–373.

Pullman, B., and Pullman, A. (1974). *Adv. Protein Chem.* **28**, 347–526 (see, esp., pp. 505–512).

Ramachandran, G. N., and Sasisekharan, V. (1968). *Adv. Protein Chem.* **23**, pp. 283–438.

Ramachandran, G. N., Lakshminarayanan, A. V., and Kolaskar, A. S. (1973). *Biochem. Biophys. Acta* **303**, 8–13.

Ramakrishnan, C., and Balasubramanian, R. (1972). *Int. J. Pept. Protein Res.* **4**, 79–90 and 91–99.

Reed, L. L., and Johnson, P. L. (1973). *J. Am. Chem. Soc.* **95**, 7523–7524.

Reich, E., Rifkin, D. B., and Shaw, E., eds. (1975). "Proteases and Biological Control." Academic Press, New York.

Renugopalakrishnan, V., and Rein, R. (1976). *Biochim. Biophys. Acta* **434**, 164–168.

Richards, F. M., and Vithayathil, P. J. (1959). *J. Biol. Chem.* **234**, 1459–1465.

Roberts, J. L., and Herbert, E. (1977a). *Proc. Natl. Acad. Sci. U.S.A.* **74**, 4826–4830.

Roberts, J. L., and Herbert, E. (1977b). *Proc. Natl. Acad. Sci. U.S.A.* **74**, 5300–5304.

Robinson, A. B., McKerrow, J. H., and Cary, P. (1970). *Proc. Natl. Acad. Sci. U.S.A.* **66**, 753–757.

Roemer, D., Buescher, H. H., Hill, R. C., Pless, J., Bauer, W., Cardinaux, F., Closse, A., Hanser, D., and Huguenin, R. (1977). *Nature* (*London*) **268**, 547–549.

Rudinger, J. (1972). *In* "Chemistry and Biology of Peptides" (J. Meienhofer, ed.), pp. 729–735. Ann Arbor Sci. Publ., Ann Arbor, Michigan.

Rudinger, J., and Buetzer, P. (1975). *In* "Peptides 1974" (Y. Wolman, ed.), pp. 211–219. Wiley, New York.

Rudko, A. D., Lovell, F. M., and Low, B. W. (1971). *Nature* (*London*), *New Biol.* **232**, 18–19.

Sakakibara, S. (1971). *In* "Chemistry and Biochemistry of Amino Acids, Peptides and Proteins" (B. Weinstein, ed.), Vol. 1, pp. 51–85. Dekker, New York.

Sakakibara, S. (1977). *In* "Peptides, Proceedings of the Fifth American Peptide Symposium" (M. Goodman and J. Meienhofer, eds.), pp. 436–447. Wiley, New York.

Sakakibara, S., and Shimonishi, Y. (1965). *Bull. Chem. Soc. Jpn.* **38**, 1412–1413.

Sakakibara, S., Shin, K. H., and Hess, G. P. (1962). *J. Am. Chem. Soc.* **84**, 4921–4928.

Sarges, R., and Witkop, B. (1965). *Biochemistry* **4**, 2491–2494.

Sasaki, K., Dockerill, S., Adamiak, D. A., Tickle, I. J., and Blundell, T. (1975). *Nature* (*London*) **257**, 751–757.

Satchell, D. P. N. (1963). *Q. Rev., Chem. Soc.* **17**, 160–203.

Satchell, D. P. N. (1974). *Chem. Ind.* (*London*) pp. 683–687.

Scheiner, S., and Kern, C. W. (1977). *J. Am. Chem. Soc.* **99**, 7042–7050.

Schmir, G. L., and Cohen, L. A. (1961). *J. Am. Chem. Soc.* **83**, 723–730.

Schröder, E., and Lübke, K. (1965). "The Peptides," Vol. 1, Academic Press, New York.

Schultz, J. (1967). *In* "Methods in Enzymology" (C. H. W. Hirs, ed.), pp. 255–263. Academic Press, New York.

Schwyzer, R. (1966). *Naturwissenschaften* **53**, 189–197.

Schwyzer, R. (1973). *In* "Peptides 1972" (H. Hanson and H. D. Jakubke, eds.), pp. 424–436. North-Holland Publ., Amsterdam.

Schwyzer, R., Feurer, M., Iselin, B., and Kägi, H. (1955). *Helv. Chim. Acta* **38**, 69–79; 80–83; 83–91.

Sharp, J. J., Robinson, A. B., and Kamen, M. D. (1973). *J. Am. Chem. Soc.* **95**, 6097–6108.

Shin, K. H., Sakakibara, S., Schneider, W., and Hess, G. P. (1962). *Biochem. Biophys. Res. Commun.* **8**, 288–293.

Sieber, P., Riniker, B., Brugger, M., Kamber, B., and Rittel, W. (1970). *Helv. Chim. Acta* **53**, 2135–2150.

Sieber, P., Kamber, B., Hartmann, A., Jöhl, A., Riniker, B., and Rittel, W. (1974). *Helv. Chim. Acta* **57**, 2617–2626.

Sieber, P., Kamber, B., Hartmann, A., Jöhl, A., Riniker, B., and Rittel, W. (1977). *Helv. Chim. Acta* **60**, 27–37.

Smith, C. W., and Walter, R. (1978). *Science* **199**, 297–299.

Spande, T. F., Witkop, B., Degani, Y., and Patchornik, A. (1970). *Adv. Protein Chem.* **24**, 97–260.

Stahl, G. L., and Walter, R. (1977). *J. Med. Chem.* **20**, 492–495.

Stein, W. H. (1965). *Isr. J. Med. Sci.* **1**, 1229–1243.

Steiner, D. F., Clark, J. L., Nolan, C., Rubenstein, A. M., Margoliash, E., Aten, B., and Oyer, P. E. (1969). *Recent Prog. Horm. Res.* **25**, 207–272.

Steiner, D. F., Kemmler, W., Tager, H. S., and Peterson, J. D. (1974). *Fed. Proc., Fed. Am. Soc. Exp. Biol.* **33**, 2105–2115.

Stich, K., and Leemann, H. G. (1963). *Helv. Chim. Acta* **46**, 1887–1895.

Tan, N. H., and Kaiser, E. T. (1976). *J. Org. Chem.* **41**, 2787–2793.

Udenfriend, S., and Stein, S. (1977). *In* "Peptides, Proceedings of the Fifth American Peptide Symposium" (M. Goodman and J. Meienhofer, eds.), pp. 14–26. Wiley, New York.

Ugi, I., Arora, A., Burghard, H., Eberle, G., Eckert, H., George, G., Gokel, G., Herlinger, H., von Hinrichs, E., Hoffmann, P., Kleimann, H., Klusacek, H., Lam, H.-L., Marquarding, D., Nah, H.-S., Offermann, K., Rehn, D., Stüber, S., Tamasi, M., Urban, R., Wackerle, L., Zahr, S., and von Zychlinski, H. (1975). *In* "Peptides 1974" (Y. Wolman, ed.), pp. 71–91. Wiley, New York.

Urry, D. W. (1971). *Proc. Natl. Acad. Sci. U.S.A.* **68**, 672–676.

Urry, D. W., and Walter, R. (1971). *Proc. Natl. Acad. Sci. U.S.A.* **68**, 956–958.

Vale, W., Brazeau, P., Grant, G., Nussey, A., Burgus, R., Rivier, J., Ling, N., and Guillemin, R. (1972). *C. R. Hebd. Seances Acad. Sci., Ser. D* **275**, 2913–2916.

Veber, D. F. (1975). *In* "Peptides: Chemistry, Structure and Biology" (R. Walter and J. Meienhofer, eds.), pp. 307–316. Ann Arbor Sci. Publ., Ann Arbor, Michigan.

Veber, D. F., Brady, S. F., and Hirschmann, R. (1972). *In* "Chemistry and Biology of Peptides" (J. Meienhofer, ed.), pp. 315–319. Ann Arbor Sci. Publ., Ann Arbor, Michigan.

Veber, D. F., Paleveda, W. J., Jr., Lee, Y. C., and Hirschmann, R. (1977). *J. Org. Chem.* **42**, 3286–3288.

von Graffenried, B., Del Pozo, E., Roubicek, J., Krebs, E., Pöldinger, W., Burmeister, P., and Kerp, L. (1978). *Nature (London)* **272**, 729–730.

Waki, M., and Meienhofer, J. (1977). *J. Am. Chem. Soc.* **99**, 6075–6082.

Walter, R. (1977). *Fed. Proc., Fed. Am. Soc. Exp. Biol.* **36**, 1872–1878.

Walter, R., and Hoffman, P. L. (1974). *Biochim. Biophys. Acta* **336**, 294–305.

Walter, R., Bernal, I., and Johnson, L. F. (1972). *In* "Chemistry and Biology of Peptides" (J. Meienhofer, ed.), pp. 131–135. Ann Arbor Sci. Publ., Ann Arbor, Michigan.

Walter, R., Ballardin, A., Schwartz, I. L., Gibbons, W. A., and Wyssbrod, H. R. (1974). *Proc. Natl. Acad. Sci. U.S.A.* **71**, 4528–4532.

Weissbach, H., and Pestka, S., eds. (1977). "Molecular Mechanisms of Protein Biosynthesis." Academic Press, New York.

Wieland, T., and Schön, W. (1955). *Justus Liebiys Ann. Chem.* **593**, 157–178.

Wieland, T., Birr, C., and Flor, F. (1971). *Angew. Chem., Int. Ed. Engl.* **10**, 336.

Wieland, T., Birr, C., and Flor, F. (1973). *Justus Liebigs Ann. Chem.* **1973**, 1595–1600 and 1601–1605.

Winkler, F. K., and Dunitz, J. D. (1971). *J. Mol. Biol.* **59**, 169–182.

Wünsch, E. (1967). *Z. Naturforsch., Teil B* **22**, 1269–1276.

Wünsch, E., ed. (1974). "Houben-Weyl, Methoden der organischen Chemie," Vol. 15, Parts I and II. Thieme, Stuttgart.

Wünsch, E., Wendlberger, G., Stelzel, P., Thamm, P., and Jaeger, E. (1974). *In* "Houben-Weyl's Methoden der organischen Chemie" (E. Wünsch, ed.), Vol. 15, Part I. Thieme, Stuttgart.

Wünsch, E., Wendlberger, G., Hallett, A., Jaeger, E., Knof, S., Moroder, L., Scharf, R., Schmidt, I., Thamm, P., and Wilschowitz, L. (1977). *Z. Naturforsch., Teil C* **32**, 495–506.

Wüthrich, K., Grathwohl, C., and Schwyzer, R. (1974). *In* "Peptides, Polypeptides and Proteins" (E. R. Blout, F. A. Bovey, M. Goodman, and N. Lotan, eds.), pp. 300–307. Wiley, New York.

Wyckoff, H. W., Hardman, K. D., Allewell, N. M., Inagami, T., Johnson, L. N., and Richards, F. M. (1967). *J. Biol. Chem.* **242**, 3984–3988.

Wyssbrod, H. R., and Gibbons, W. A. (1974). *In* "Survey of Progress in Chemistry" (A. F. Scott, ed.), Vol. 6, pp. 210–326. Academic Press, New York.

Yajima, H., Ogawa, H., Fujii, N., and Funakoshi, S. (1977). *Chem. Pharm. Bull.* **25**, 740–747.

Yamashiro, D. (1964). *Nature (London)* **201**, 76–77.

Yamashiro, D. (1977). *J. Org. Chem.* **42**, 523–525.

Yamashiro, D., and Li, C. H. (1972). *Int. J. Pept. Protein Res.* **4**, 181–185.

Yamashiro, D., and Li, C. H. (1973a). *J. Am. Chem. Soc.* **95**, 1310–1315.

Yamashiro, D., and Li, C. H. (1973b). *J. Org. Chem.* **38**, 591–592 and 2594–2597.

Yamashiro, D., and Li, C. H. (1974). *Proc. Natl. Acad. Sci. U.S.A.* **71**, 4945–4949.

Yamashiro, D., and Li, C. H. (1978). *J. Am. Chem. Soc.* **100**, 5174–5179.

Yamashiro, D., Blake, J., and Li, C. H. (1976). *Tetrahedron Lett.* pp. 1469–1472.

Young, G. T. (1967). *In* "Peptides" (H. C. Beyerman, A. Van de Linde, and W. Maassen Van den Brink, eds.), pp. 55–66. North-Holland Publ., Amsterdam.

Chapter 2

The Formation of Peptide Bonds: A General Survey

JOHN H. JONES

I. Introduction 66
II. Coupling between Amino Acid Residues 67
A. The Use of Acyl Halides and Pseudohalides 67
1. Acyl Chlorides 67
2. Acyl Cyanides 68
3. Acyl Azides 68
B. The Use of Anhydrides 71
1. Symmetrical Anhydrides 71
2. Mixed Anhydrides with Carboxylic Acids 72
3. Mixed Anhydrides with Carbonic Acids 73
4. Mixed Anhydrides with Acids Derived from Phosphorus 75
5. Mixed Anhydrides with Acids Derived from Other Elements 77
C. The Activation of Carboxyl Groups by Addition Reactions 77
1. The Use of Carbodiimides 79
2. The Use of Isoxazolium Reagents 82
D. Reactions Involving Displacement of Acyl Groups from Heterocyclic Nitrogen 83
E. The Activation of Hydrazides by Oxidation 84
F. The Use of Ring-Opening Reactions of Activated Heterocyclic Systems 85
1. *N*-Carboxyanhydrides 85
2. Oxazolones 87
3. Dihydrooxazinones 88
4. The Use of Lactams and Lactones 88
G. Procedures Involving Acyloxyphosphonium Intermediates 89

THE PEPTIDES, VOL. 1

ISBN 0-12-304201-1

H. The Use of Activated Esters 90
I. Polymeric Reagents 94
J. Coupling between Captive Components 95
III. *Ab Initio* Assembly of the Backbone 95
IV. Conclusions 97
References 100

I. INTRODUCTION

The aim of this chapter is to cover all the main types of procedure for peptide bond formation which have been suggested over the years, putting the various methods in perspective. A wide ranging survey at a general level will be offered, but comprehensive citation will not be attempted. Where possible and appropriate, reference will be made to other reviews and leading references without listing all the original papers or all the variations. Detailed accounts of the techniques which have actually been used and matters such as the racemization question, the special case of solid phase synthesis, and so on, will be dealt with in specific chapters elsewhere in this treatise. Selective (Bodanszky *et al.*, 1976a; Finn and Hofmann, 1976; Jones, 1974; Klausner and Bodanszky, 1972) and exhaustive (Wünsch, 1974) reviews of the topic which cite earlier reviews and monographs have appeared recently. Moreover, annual surveys of advances have appeared in the Specialist Periodical Reports series published by the Chemical Society, London, since 1969.

A very wide variety of reactions which lead to the generation of amide functionalities is available; Beckwith (1971) has given a full account of these reactions in the context of the whole of organic chemistry as opposed to the narrower area with which we are concerned here. However, with very few exceptions (see Section III), only procedures involving the formation and aminolysis of activated amino acid derivatives (Scheme 1) have found application in peptide synthesis. The methods by which coupling along the lines of Scheme 1 can be performed fall into three main groups: (1) those in which a reactive acylating agent is formed from the carboxyl component in a separate step, followed by immediate treatment with the amine component;

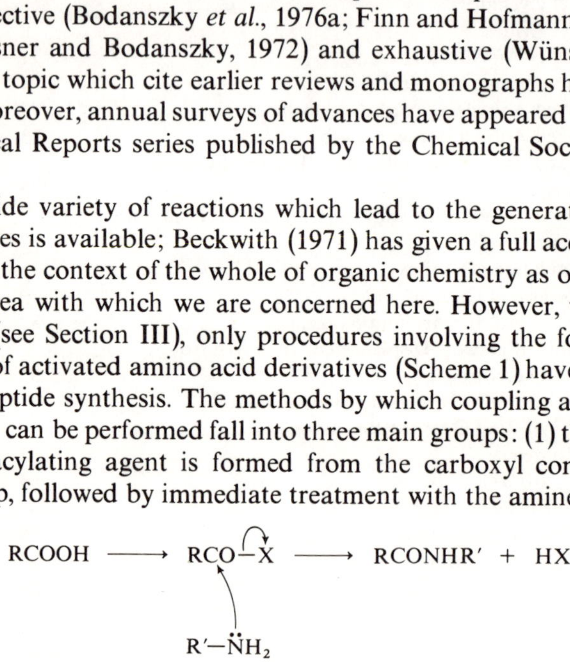

Scheme 1

(2) those in which an isolable acylating agent is formed in a separate step and is purified before aminolysis; and (3) those in which the acylating intermediate is generated in the presence of the amine component by addition of a coupling reagent to a mixture of the amine and carboxyl components ("direct coupling"). We might therefore be moved to use a classification of this kind as a framework for discussion, but the distinctions between the classes are in fact not clearly definable and in any case there is much chemistry which is common to all three types of procedure. An alternative approach to the organization of the material in our syllabus would be to consider techniques in order according to the nature of the leaving group in the aminolysis stage, but this is not very satisfactory either; in several cases of recent invention a cascade of acyl transfer reactions takes place and several species are candidates for consideration as the activated intermediates which suffer aminolysis. We shall therefore discuss the various coupling methods in Section II without the imposition of the awkward restrictions of a rigid system, dealing with methods by groups put together so as to make discussion of their salient features as simple as possible.

II. COUPLING BETWEEN AMINO ACID RESIDUES

It has already been indicated that practically all of the techniques which have ever been used for constructing peptide bonds are of the type outlined in Scheme 1. A great variety of leaving groups and reactions for attaching them to the electrophilic carbon atom of the carboxyl component have been investigated and recommended, although the number of methods actually in regular use is comparatively small because the criteria which an acceptable method must meet are very demanding. Good yields of easily isolated products with freedom from racemization are required, as well as convenience and economy.

A. The Use of Acyl Halides and Pseudohalides

1. Acyl Chlorides

The activation of acylamino acids by conversion to the corresponding acyl chlorides followed by reaction with amino acids or esters under Schotten–Baumann conditions or by treatment with amino esters in organic solvents is in principle the most obvious and simple approach to peptide synthesis [e.g., Scheme 2 (Sheehan and Frank, 1949)] and this kind of method (for a review, see Wünsch, 1974) played an important role in the early days of peptide synthesis. The reagents normally used for acyl chloride

$$\text{Pht—Gly—OH} \xrightarrow{\text{i}} \text{Pht—Gly—Cl} \xrightarrow{\text{ii}} \text{Pht—Gly—Gly—OH}$$

Scheme 2. Conditions: i, PCl_5 in benzene (81%); ii, Gly in MgO—aqueous dioxane (90%).

formation (thionyl chloride, phosphorus pentachloride, etc.) are too vigorous to be compatible with complex or sensitive substrates and, furthermore, simple acylamino acid chlorides cyclize spontaneously to give oxazolones and hence racemic peptides. Benzyloxycarbonylamino acid chlorides are isolable but unstable, decomposing to *N*-carboxyanhydrides on warming (Poduska and Gross, 1961): although they cyclize rapidly to give benzyloxyoxazolones (Jones and Witty, 1977) under basic conditions, optically active peptides can be isolated after reaction with amino esters (see Section II,F,2). The chlorides of trifluoroacetyl-, phthaloyl-, and tosylamino acids are reasonably stable, although the last decompose under Schotten–Baumann conditions (see Rudinger, 1973). Milder alternatives to the classical reagents for acid chloride preparation have been devised, such as compounds **1** (Poduska and Gross, 1961) and **2** (Zaoral and Arnold, 1960),

$$CH_3CCl_2OCH_2CH_3 \qquad (CH_3)_2\overset{+}{N}{=}CHCl\cdot Cl^-$$

(1) **(2)**

but the acid chloride method must nevertheless be regarded as completely obsolete for conventional peptide synthesis, as many more subtle, convenient and efficient methods are available.

2. *Acyl Cyanides*

Suitable protected amino acids can be converted via their bromides to the corresponding acyl cyanides, which react with amino esters to give peptides (e.g., Jones *et al.*, 1965) but the lack of mild methods of generating the acyl cyanides, which may, however, be intermediates in direct coupling by means of diethylphosphoryl cyanide (see Section II,B,3), has precluded development of the method. The fact that the acyl cyanides acylate amine components smoothly is of interest, as acylation is not always the outcome of their reaction with nucleophiles; the action of acqueous acid on benzoyl cyanide, for example, gives an α-keto acid (Oakwood and Weisgerber, 1955) and in their reactions with dichloromethylenetriphenyl phosphorane, aliphatic acyl cyanides behave as if they were ketones (Clement and Soulen, 1974).

3. *Acyl Azides*

The use of acyl azides for peptide bond formation (reviewed by Klausner and Bodanszky, 1974) was introduced by Curtius 75 years ago (1902) but is still an important procedure (see Meienhofer, this volume, Chapter 4, for details). The hydrazides from which the azides are generally made (their production from acyl chlorides is of no interest in the present context) may be obtained by hydrazinolysis of a protected amino acid or peptide alkyl ester, or by selective deprotection of a peptide derivative which has been constructed using a blocked hydrazide for COOH-terminal protection. Conversion to the azide was usually achieved in the early days by adding sodium nitrite to a solution of the hydrazide in a mixture of acetic and aqueous hydrochloric acids at around or just below 0°C (e.g., Boissonnas *et al.*, 1955), but one of the procedures of Honzl and Rudinger (1961), the best of which involves dry acid and an organic nitrite ester in a dry organic solvent at lower temperature, is now generally preferred. The azides are subjected to aminolysis without delay once formed. The essential chemistry of the approach is summarized in Scheme 3.

$$RCO_2R'$$
$$\downarrow \text{i}$$
$$RCONHNH_2 \xrightarrow{\text{iii}} RCON_3 \xrightarrow{\text{iv}} RCONHR''$$
$$\uparrow \text{ii}$$
$$RCONHNH{-}Prot$$

Scheme 3. Conditions: i, NH_2NH_2; ii, deprotection; iii, $NaNO_2$ in aqueous AcOH—HCl or an alkyl nitrite with dry H^+; iv, $R''NH_2$.

Numerous side reactions attending the formation and use of peptide azides have been recognized (see Klausner and Bodanszky, 1974) and the azide method has not been used routinely for building up peptide chains since cleaner methods became available in the 1950's, but it remained completely paramount as a technique for segment condensation until very recently. This was partly due to the way in which the conversions outlined in Scheme 1 lend themselves to the production of activated intermediates from protected peptides but the main consideration was the fact that the coupling of acyl peptides was found to be free from racemization. In fact, for many years it appeared that racemization of acyl peptide azides never took place and confidence in this was one of the cornerstones of peptide synthesis strategy. It is now known that the chiral integrity of acyl peptide azides is not unshakably secure; racemization does occur with excess base in polar solvents, and epimeric peptides have been isolated after segment condensations by the azide method (e.g., Sieber *et al.*, 1970). The danger is greater when a histidine residue with an unprotected imidazole function is in the COOH-terminal position of the carboxyl component; with triethylamine in dimethylformamide even *tert*-butyloxycarbonylhistidine azide is racemized (Veber, 1975). Nevertheless, it does seem that the azide technique is less liable to result in racemization than most methods and with careful control of conditions it can be kept at very low levels. Why this should be is a conundrum which has never been satisfactorily explained, but a possible answer is that the intrinsic reactivity of acyl azides due to the electron withdrawing effect of the leaving group is low, the susceptibility to aminolysis being due to intramolecular general base catalysis by the leaving group. Such activation would be selective for nucleophilic attack by an amine component, which must necessarily lose a proton in the course of the reaction and would not aid racemization by the oxazolone or direct enolization pathways. The transition state envisaged (structures **3a** or **3b** or some-

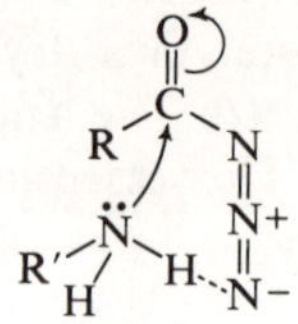

(3a)

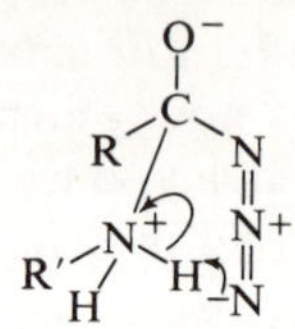

(3b)

thing between the two) is geometrically unconvincing but the proposal (Young, 1967) has the merit of consistency with more firmly based thoughts about the mechanism of aminolysis of active esters which do owe most of their reactivity to intramolecular general base catalysis (see Section II,H). The idea is perhaps supported by the general experience that azide couplings are often very sluggish in the dipolar aprotic solvents which are unavoidable in fragment condensations, despite the fact that simple azides couple rapidly with amino esters in non-polar solvents like ether—a solvent effect in this sense appears to be characteristic (Jones, 1974) of aminolyses assisted in this way. However, it must be admitted immediately that there is no relevant quantitative information available on the reactivity of azides, so the suggestion contained in structures **3a** and **3b** has a very speculative status.

As already intimated, the azide procedure has played a critical part for segment condensation in many outstanding syntheses. Recent examples with full practical detail are given by Kamber *et al.* (1976, Scheme 4), Romovacek *et al.* (1974), Tomatis *et al.* (1977), Brundish and Wade (1976), and Gabriel *et al.* (1977). In addition, the fact that the conversion of hydrazides to azides can be brought about in the presence of an unprotected amino group has made peptide cyclization through azides fairly popular (e.g., Kurath, 1976).

Bpoc–Tyr(*t*Bu)–Gln–Leu–Glu(O*t*Bu)–Asn–Tyr(*t*Bu)–$NHNH_2$ + H–Cys–Asn–O*t*Bu

Trt–Leu–Val–Cys–Gly–OH

↓

Bpoc–Tyr(*t*Bu)–Gln–Leu–Glu(O*t*Bu)–Asn–Tyr(*t*Bu)–Cys–Asn–O*t*Bu

Trt–Leu–Val–Cys–Gly–OH

Scheme 4. Conditions: i, hydrazide/DMF/–10°C/HCl/*t*BuONO/10 min; ii, amine component added with *N*-ethylmorpholine (66%).

B. The Use of Anhydrides

In principle, any acylating agent in which the potential leaving group is the conjugate base corresponding to a reasonably strong acid—i.e., an "anhydride," since such intermediates are formally derived by condensing the carboxyl component with another acidic species and removing the elements of water—can be used for peptide bond formation. In a sense, all acylating agents are anhydrides and, in his classic review of peptide coupling procedures, Albertson (1962) went to the extreme of using the term in this all-embracing way, classifying acyl azides, *O*-acylisoureas, reactive esters, and so on as mixed anhydrides. We shall take the view that this is an unhelpfully general classification, and use the woolly but comfortable criterion that an intermediate calls for treatment under this heading if it is commonly called an anhydride.

1. Symmetrical Anhydrides

Acylamino acid symmetrical anhydrides can be prepared [e.g., Scheme 5 (Wünsch and Wendlberger, 1967)] from the corresponding acylamino acids by use of a variety of reagents including ethoxyacetylene and dicyclohexyl-

$$\text{Z—Phe—OH} \xrightarrow{\text{i}} (\text{Z—Phe})_2\text{O} \xrightarrow{\text{ii}} \text{Z—Phe—Val—O}t\text{Bu}$$

Scheme 5. Conditions: i, DCC/MeCN/ – 5°C (40%); ii, H—Val—O*t*Bu, HCl/Et_3N/MeCN/ – 15°C (83%).

carbodiimide. As we shall see later, symmetrical anhydrides may well be the actual acylating species in procedures involving the direct use of such reagents in some cases. Aminolysis of a symmetrical anhydride (e.g., Scheme 5) is unambiguous but the price for this advantage over some mixed anhydride procedures (see below) is that only half of the acylamino acid taken is incorporated into the product. The symmetrical anhydrides of *tert*-butyloxycarbonyl-, benzyloxycarbonyl- and phthaloylamino acids are generally reasonably stable crystalline substances which are convenient in use (for a table, see Wünsch, 1974). The symmetrical anhydride of benzyloxycarbonylglycine decomposes to **4** on heating (Tadema *et al.*, 1964) or treatment

$$\begin{array}{l}\text{ZNHCH}_2\text{CO}\\ \qquad\quad\;\;|\\ \qquad\quad\text{ZNCH}_2\text{COOH}\end{array}$$

(4)

with tertiary base (Kotake and Saito, 1966), but otherwise these intermediates appear to be reasonably clean acylating agents which could perhaps be used more widely in stepwise synthesis. They are not appropriate for fragment condensation, as it can be anticipated that gross racemization would result from attempts to prepare acylpeptide symmetrical anhydrides.

2. *Mixed Anhydrides with Carboxylic Acids*

A mixed anhydride of benzoylglycine with benzoic acid was presumably involved in the classic synthesis of benzoylglycylglycine (Scheme 6) by Curtius (1881). For an account of the development of this approach in the early days the reader is referred to Schröder and Lübke (1965). There are two main problems with the use of a mixed anhydride of an acylamino acid or peptide with a carboxylic acid in peptide synthesis: first, the anhydride has two electrophilic sites and can therefore undergo aminolysis ambiguously, and second, mixed anhydrides of this sort tend to disproportionate, ultimately with the same result (Scheme 7). However, the selectivity can be

$$PhCOCl + NH_2CH_2CO_2Ag \rightarrow PhCONHCH_2CO_2H + PhCONHCH_2CONHCH_2CO_2H$$

Scheme 6

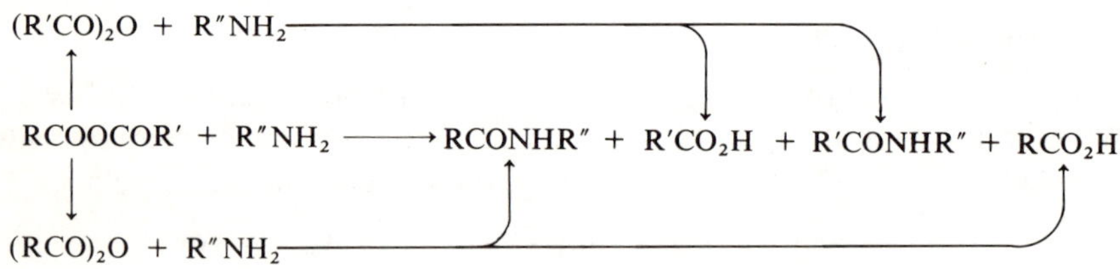

Scheme 7

improved greatly by introducing steric hindrance and inductive depression of electrophilicity so as to direct the attack to the carboxyl component carbonyl, as in the case of mixed pivalic anhydrides [e.g., Scheme 8 (Zaoral,

$$\text{Z—Asn—OH} \xrightarrow{\text{i}} \text{Z—Asn—OCOC(CH}_3)_3 \xrightarrow{\text{ii}} \text{Z—Asn—Cys(Bzl)—OMe}$$

Scheme 8. Conditions: i, $(CH_3)_3CCOCl$/*N*-ethylpiperidine/$CHCl_3$/below 10°C until solution clear; ii, H—Cys(Bzl)—OMe, 5 min later, at −10°C (80%).

1962); for another, more recent example see Schreier and Hermann, 1976], which are the only intermediates in this class to have attained any significant popularity. Mixed pivalic anhydrides appear to be reasonably stable to disproportionation and can be isolated as crystalline materials in some cases (e.g., Leplawy *et al.*, 1960) but they are normally generated and used in separate but rapidly succeeding operations, as in the example of Scheme 8. Pivaloyl derivatives of the amine component are occasionally obtained [in one curious special case in our own work such a product was the major one (Fairweather and Jones, 1972)] but this is not normally a serious problem, and the method is a valuable alternative to the mixed carbonic anhydride method (see below) which, for reasons which are not entirely clear, is rather more popular. In fact in our experience the two procedures are very similar in the advantages they offer and the problems they present: the situations in which one or other method is appropriate are discussed in the next section.

The protected cyclic anhydrides **5** (Le Quesne and Young, 1950) and **6** (John and Young, 1954), derived from glutamic and aspartic acids, which are in a sense mixed carboxylic anhydrides, are well known and easily made

Z—Glu—O (cyclic) **(5)**   Z—Asp—O (cyclic) **(6)**

compounds which are stable to disproportionation, but their aminolysis is more or less indiscriminate so that they are of no general use for peptide bond formation. The special case of the salt **7** of the cyclic anhydride of aspartic acid, which can be obtained by cyclization in the presence of strong acid, is of interest, however, as it has recently been used in a very short preparation of the dipeptide sweetener "aspartame" **(8)** [Scheme 9 (Ariyoshi *et al.*, 1973)].

H_2SO_4 · H—Asp(OH)—OH —i→ *i*PrOSO$_3$H · H—Asp—O (cyclic) **(7)**

—ii→ H—Asp(OH)—Phe—OMe **(8)**

Scheme 9. Conditions: i, H_2SO_4/Ac_2O then *i*PrOH (93%); ii, H—Phe—OMe (61%).

3. Mixed Anhydrides with Carbonic Acids

The most generally successful mixed anhydride method involves the generation and aminolysis of a carboxylic—carbonic anhydride as outlined in Scheme 10 (Boissonas, 1951; Wieland and Bernhard, 1951; Vaughan, 1951b;

RCOOH + R′OCOCl —i→ RCOOCOOR′

—ii→ RCONHR″ + R′OH + CO_2

Scheme 10. Conditions: i, 1 equiv tertiary base/unreactive dry solvent/ca −10°C/a few minutes; ii, R″NH_2.

for details, see Meienhofer, this volume, Chapter 6). Ethyl chloroformate is commonly used, but many prefer isobutyl chloroformate. The fact that one of the carbonyl groups in the activated intermediate is flanked by two oxygen atoms diminishes its reactivity so that nucleophilic attack is directed toward that of the original carboxyl component, and alkoxycarbonyl derivatives of the amine component are not usually formed in more than trace amounts. Disproportionation to symmetrical anhydrides occurs rather easily, however, and low temperatures with minimal activation times are required to avoid this, but providing care is taken, excellent results can be obtained. A typical example is shown in Scheme 11 (Wünsch *et al.*, 1968).

Z—Ser(*t*Bu)—OH $\xrightarrow{\text{i, ii}}$ Z—Ser(*t*Bu)—Gln—Gly—OMe

Scheme 11. Conditions: i, Et_3N/THF/EtOCOCl/−10°C; ii, H—Gln—Gly—OMe, HCl/Et_3N (80%).

Conditions for minimization of racemization have been delineated for model cases (Anderson *et al.*, 1967), but the technique has not found wide application for segment condensation at racemizable residues as less risky procedures are available. The value of the technique lies in its speed and economy. There are some situations where it is especially useful. It often happens, for example, that simple protected peptides in the tri- to hexapeptide range have solubility properties uncomfortably similar to those of dicyclohexylurea, so that the use of dicyclohexylcarbodiimide is very tiresome; a switch to a mixed anhydride procedure in such cases may solve the problem. For backing-off maneuvers in which the amine component is an active ester it is clearly essential for acylation to be as rapid as possible or self condensation will intervene, and the mixed anhydride method (these remarks apply equally to the mixed pivalic anhydride method) is definitely the method of choice here [e.g., Scheme 12 (Fairweather and Jones, 1972)]. The point has

Z—Ala—Gly—OH ⟶ Z—Ala—Gly—Pro—OPcp

Scheme 12. Conditions: i, $Et_3N/CHCl_3/(CH_3)_3CCOCl$/0°C/5 min; ii, Pro—OPcp, HBr/Et_3N (61%).

already been made that segment condensation at a racemizable residue by this technique is dangerous, but if the COOH-terminal residue is not open to racemization it can be useful for the purpose, as in the example of Scheme 13

Z-Gly-Ile-Val-Glu(O*t*Bu)-Gln-Cys(Bzl)-Cys(Bzl)-Ala-Gly-OH

\+ H-Val-Cys(Bzl)-Ser-Leu-Tyr-Gln-Leu-Glu-Asn-Tyr-Cys(Bzl)-Asn-OH

↓

Z-Gly-Ile-Val-Glu(O*t*Bu)-Gln-Cys(Bzl)-Cys(Bzl)-Ala-Gly-Val-Cys(Bzl)-Ser-Leu-Tyr-Gln-Leu-Gly-Asn-Tyr-Cys(Bzl)-Asn-OH

Scheme 13. Activation by EtOCOCl treatment in HMP—THF (90–95% yield).

(Zahn *et al.*, 1965), which proceeded less satisfactorily by other techniques. The commonest application, however, is for the activation of protected amino acids, as in the use of mixed carbonic anhydrides in rapid repetitive stepwise synthesis, the "REMA" approach, which has been developed largely by Beyerman and exemplified most recently by him with syntheses of secretin (van Zon and Beyerman, 1976) and luliberin (Beyerman *et al.*, 1976).

The classical mixed carbonic anhydride procedure using chloroformates [and the recent modification employing an alkoxycarbonylammonium fluoroborate to make the anhydride (Pankstelis and Kim, 1974)] necessarily involves separate activation and aminolysis steps, but mixed ethyl carbonic anhydrides can also be generated by the reaction [Scheme 14 (Belleau and

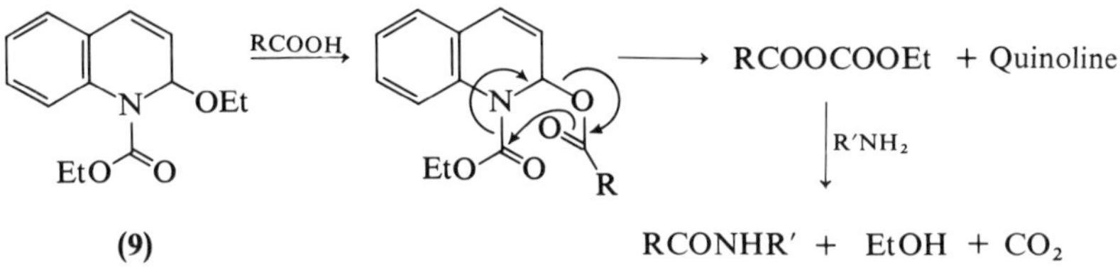

(9)

Scheme 14. Conditions: $RCOOH/R'NH_2/EEDQ/THF/20°C$. The alternative formulation via an alkoxycarbonylquinolinium species which suffers intermolecular reaction giving a mixed anhydride appears equally plausible.

Malek, 1968)] of carboxylic acids with the pseudo base 1-ethyloxycarbonyl-2-ethyloxy-1,2-dihydroquinoline (**9**, EEDQ) and this can be done in the presence of an amine component. Reagent **9**, which is commercially available but in any case is very easily prepared (Fieser and Fieser, 1969), can therefore be used as a direct coupling reagent as in the example shown in Scheme 15 (Muhlemann *et al.*, 1972). Because in this technique the mixed

Boc—Cys(Bzl)—OH · DCHA + HCl · H—Tyr(Bzl)—Gln—Asn—Cys(Bzl)—OBzl

↓

Boc—Cys(Bzl)—Tyr(Bzl)—Gln—Asn—Cys(Bzl)—OBzl

Scheme 15. Conditions: components mixed in DMF and then EEDQ in THF added (67%).

carbonic anhydride is consumed by aminolysis as soon as it is formed, the opportunity for the intervention of side reactions, including racemization when this needs to be considered, is minimal, and since the coproducts are quinoline, ethanol, and carbon dioxide, workup is simple. The analogous reagent (**10**, IIDQ) can be used in the same fashion (Kiso *et al.*, 1973).

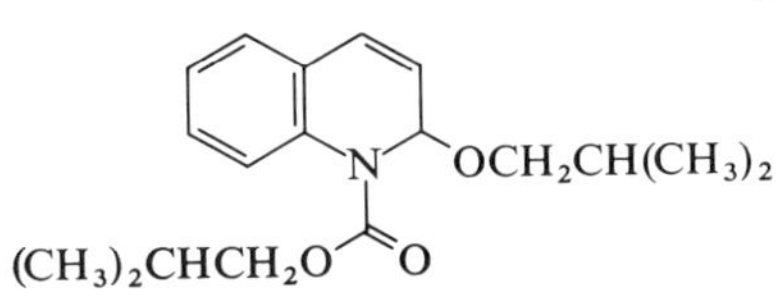

(10)

1-Isobutyloxycarbonyl-2-isobutyloxy-1,2-dihydroquinoline (IIDQ)

4. *Mixed Anhydrides with Acids Derived from Phosphorus*

A large number of coupling procedures involving the formation, either *in situ* or as a separate preliminary to aminolysis, of an activated intermediate by treatment with a phosphorus-based reagent have been devised. A subclass which may proceed via acyloxyphosphonium species is of considerable current interest and promise and therefore merits a separate section later (see Section II,G) but the majority probably involve initial formation of a mixed anhydride by nucleophilic displacement on phosphorus, although the precise chemistry of some (a representative selection of which is given in Table I) is obscure. The diphenylphosphinic mixed anhydride procedure [Scheme 16 (Jackson *et al.*, 1976)] has the advantages over the mixed car-

$$RCOOH \xrightarrow{i} \underset{\textbf{(11)}}{RCO{-}OPOPh_2} \xrightarrow{ii} RCONHR' + Ph_2POOH$$

Scheme 16. Conditions: i, Ph_2POCl/*N*-methylmorpholine/CH_2Cl_2 or EtOAc/−20°C; ii, $R'NH_2$ added 20 min later and temperature allowed to rise to 20°C.

boxylic and carbonic anhydride methods of greater stability to disproportionation and higher regioselectivity in aminolysis of anhydride **11**. Although little has so far been published about this method, the examples

Table I Some Reagents for Activating Carboxyl Components Which Probably Involve the Formation of Mixed Anhydrides with Acids of Phosphorus[a]

Reagent	Formula	Reference
Diphenylphosphoryl chloride	$(PhO)_2POCl$[b]	Cosmatos *et al.* (1961)
Diphenylphosphite-mercuric chloride-pyridine	$(PhO)_2PH/HgCl_2$/pyridine[b,c,d]	Yamazaki and Higashi (1973)
Diphenyl phosphorazidate	$(PhO)_2PON_3$[c]	Hamada *et al.* (1977a)
Diethyl phosphorocyanidate	$(EtO)_2POCN$[c]	Hamada *et al.* (1977b)
Phosphorus oxychloride	$POCl_3$[c]	Wieland and Heinke (1956)
Diphenylphosphinyl chloride	Ph_2POCl[b]	Jackson *et al.* (1976)
Diethyl chlorophosphite	$(EtO)_2PCl$[b,d]	Anderson and Young (1952)
Tetraethyl pyrophosphite	$(EtO)_2POOPO(OEt)_2$[b,c,d]	Anderson *et al.* (1952)
Phosphorus trichloride	PCl_3[c,d]	Goldschmidt and Lautenschlager (1953)
Triphenylphosphite–imidazole	$(PhO)_3P$-imidazole[c,e]	Mitin and Glinskaya (1969)

[a] Procedures involving acyloxyphosphonium intermediates are dealt with in Section II,G. The list is not intended to be comprehensive; the intention is only to indicate the range of types.

[b] Used for activation before addition of the amine component.

[c] Used as a direct coupling reagent.

[d] Can also be used to form a reactive derivative of the amine component before addition of the carboxyl component, but a mixed anhydride between the carboxylic acid and a phosphorus-based acid is involved in this *modus operandi* too.

[e] Presumably involves a phorphorus diester imidazolide as effective reagent; see also Kricheldorf *et al.* (1976).

reported are sufficient to show its promise, and it has been used extensively in the work on the total synthesis of a lysozyme analogue which is in progress at the University of Liverpool (Kenner, 1977). The diethylphosphoryl cyanide and diphenylphosphoryl azide direct coupling reagents, which may possibly involve acyl cyanide or azide formation *in situ*, respectively, but which seem equally likely to proceed *via* mixed or symmetrical anhydrides, have also been shown [e.g., Scheme 17 (Hamada *et al.*, 1977b)]

$$\text{Z—Glu(O}t\text{Bu)—OH} + \text{H—Gly—OEt} \cdot \text{HCl} \longrightarrow \text{Z—Glu(O}t\text{Bu)—Gly—OEt}$$

Scheme 17. Conditions: (*a*) $(PhO)_2PON_3$ and Et_3N added to a mixture of the components in DMF at 0°C, 0°C for 1 hr then 20°C for 24 hr (90%). (*b*) $(EtO)_2POCN$ and Et_3N added to a mixture of the components in DMF at 0°C, 0°C for 4 hr then 20°C for 15 hr (87%).

to be of potential value. The greatest number of examples has been reported for the azide reagent, which gives good results even when hydroxylic side chains are unprotected (Shiori and Yamada, 1974b), and optically pure products can be obtained from couplings of susceptible acyl peptides [e.g., Scheme 18 (Hamada *et al.*, 1977a; see also Ozawa *et al.*, 1977)], although in supersensitive models some racemization was detected (Shiori and Yamada, 1974a).

$$\text{Z—Ser—Asp(O}t\text{Bu)—OH} + \text{H—Ala—Val—Phe—OMe}$$
$$\downarrow$$
$$\text{Z—Ser—Asp(O}t\text{Bu)—Ala—Val—Phe—OMe}$$

Scheme 18. Conditions: $(PhO)_2PON_3/Et_3N/DMF$ (77%, identical with material from a fully stepwise route).

5. *Mixed Anhydrides with Acids Derived from Other Elements*

For the sake of completeness, a selection of mixed anhydride procedures based on acids not derived from carbon or phosphorus is given in Table II; none are of any current significance.

C. The Activation of Carboxyl Groups by Addition Reactions

The nucleophilic addition of carboxylic acids to cumulated double bond or triple bond systems gives adducts which are in general reactive acylating agents because acyl-oxygen fission leads to a carbonyl compound (Scheme 19), and a large number of methods, of both the preactivation and *in situ* activation types, involve chemistry of this kind. The range is indicated in Table III: only the carbodiimide and isoxazolium salt reagents merit detailed individual attention.

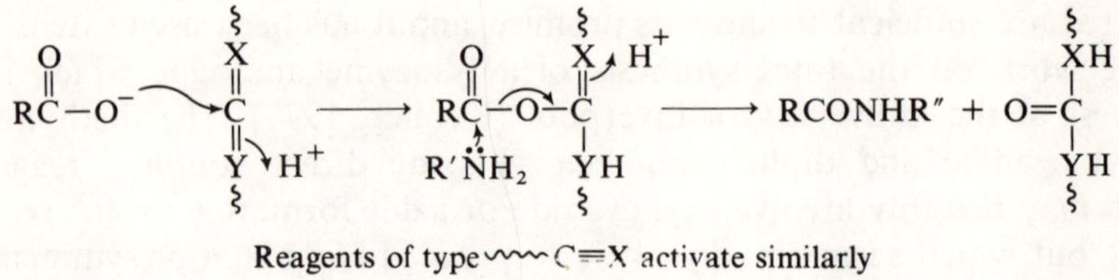

Scheme 19

Table II Some Reagents for Activating Carboxyl Groups by the Formation of Mixed Anhydrides with Acids Not Based on Carbon or Phosphorus

Reagent	Formula	Reference
Dimethylsulfoformimidium complex	SO_3-DMF complex[a,b]	Clayton *et al.* (1957)
p-Toluluesulfonyl chloride	TosCl[a]	Theodoropoulos and Gazopoulos (1962)
Diethyl chloroarsenite	$(EtO)_2AsCl$[a,c]	Vaughan (1951a)
Silicon tetrachloride	$SiCl_4$[a]	Chan and Wong (1971)
Chlorodimethyloxyborane	$(MeO)_2BCl$[a]	Pelter *et al.* (1970)

[a] Used for carboxyl activation before addition of the amine component.

[b] The first formed carboxyl activated intermediate is probably $RCOOSO_3^-$, but may conceivably be $RCOOCH{=}N^+Me_2$ formed by ejection of sulfate instead of dimethylformamide when the carboxyl component attacks the $Me_2N^+{=}CH{-}OSO_3^-$ complex.

[c] Can also be used for conversion of an amine component to a reactive derivative before addition of the carboxyl component but a mixed anhydride is still involved.

Table III Some Reagents for Carboxyl Activation by Addition

Reagent	Structure	Literature
Carbodiimides	$RN{=}C{=}NR'$	See text
Isocyanato esters	$ROCOCHR'N{=}C{=}O$[a]	Goldschmidt (1950)
Ketoketenimines	$RCOCR'{=}C{=}NR''$[b]	See text
Diphenylketene-*p*-tolylimine	$Ph_2C{=}C{=}NPh(p\text{-}Me)$	Stevens and Munk (1958)
Diphenylketene	$Ph_2C{=}C{=}O$	Losse and Demuth (1961)
Ethyloxyacetylene	$EtOC{\equiv}CH$	Arens (1955)
Dimethylamino-*tert*-butylacetylene	$Me_2NC{\equiv}C\text{-}t\text{-}Bu$	Buijle and Viehe (1964)

[a] Acylamino acids add to isocyanates to give carboxylic–carbamic mixed anhydrides which collapse to give, ultimately, peptides so that here the activating reagent is also the amine component (see Babusiaux *et al.*, 1976): gross racemization, however, results from this and the analogous isothiocyanate procedure (Jones and Fairweather, 1969).

[b] Generated by the action of base on an isoxazolium reagent.

1. *The Use of Carbodiimides*

Dicyclohexylcarbodiimide (**12**) has been in use for peptide bond formation for more than twenty years (Sheehan and Hess, 1955), but remains the most important single reagent by an overwhelming margin, especially if we include consideration of solid phase synthesis, where it is used to the near total exclusion of other coupling methods. The reason lies in the great simplicity of the procedure, which involves mere mixing of the amine and carboxyl components with dicyclohexylcarbodiimide in equimolar amounts under appropriate conditions, when *O*-acylisourea generation takes place; this leads to peptide either by direct aminolysis or via a symmetrical anhydride, with the concomitant formation of dicyclohexylurea (Scheme 20).

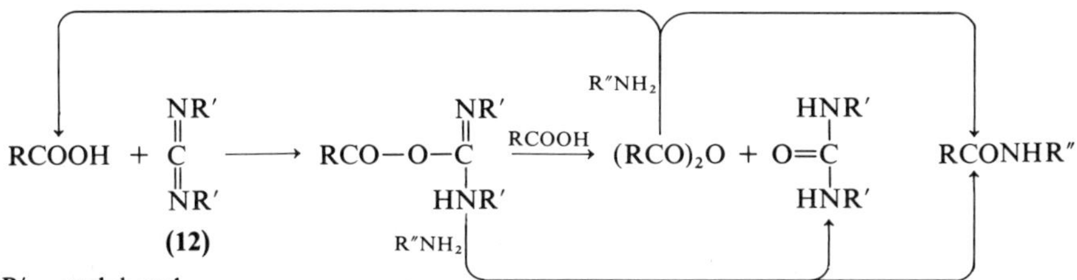

R′ = cyclohexyl

Scheme 20

Since the urea is only sparingly soluble, its separation from the desired product is very straightforward in the case of solution synthesis. It is, on the other hand, soluble enough for removal by extensive washing in solid phase synthesis. A simple example is shown in Scheme 21 (Wünsch and Wendlberger, 1967). Unfortunately the intermediates here are highly reactive and side

$\overset{H^+}{|}$
Z—Arg—OH · Cl^- +

H—Ala—Gln—Asp(O*t*Bu)—OMe → Z—$\overset{H^+}{\text{Arg}}$—Ala—Gln—Asp(O*t*Bu)—OMe · Cl^-

Scheme 21. Conditions: Components mixed in MeCN—DMF and DCC added at −10°C (65%).

reactions can intervene, especially if the amine component is dilatory. Thus extensive racemization is the outcome with susceptible carboxyl components, and collapse of the *O*-acylisourea by intramolecular acyl transfer (Scheme 22) sometimes competes significantly with the desired reaction with

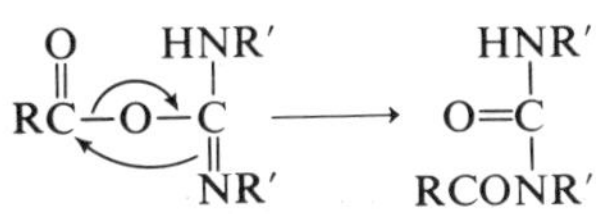

(13)

Scheme 22

external nucleophile; when this happens a much less reactive *N*-acylurea (**13**) is formed, which survives and often gives purification problems as well as reducing the yield. Both these problems can be reduced dramatically by performing the coupling in the presence of a nucleophile which will react very rapidly with the *O*-acylisourea to give an acylating agent which is still reactive enough for aminolysis but which is more discriminating and does not lead to racemization or other side reactions. The first additive to be used in this way was *N*-hydroxysuccinimide [Scheme 23 (Wünsch and Drees,

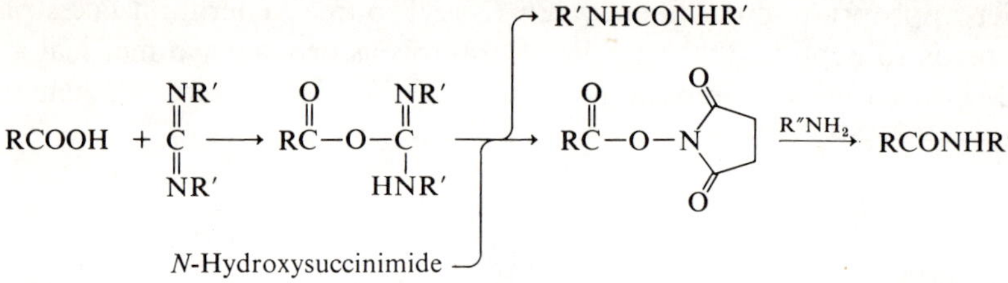

Scheme 23

1966; Weygand *et al.*, 1966)] which suppresses *N*-acylurea formation very effectively and gives a procedure which is sufficiently free from the danger of racemization for it to have been used extensively for acylpeptide coupling [e.g., Scheme 24 (Wünsch and Deimer, 1972)]. Numerous possible additives

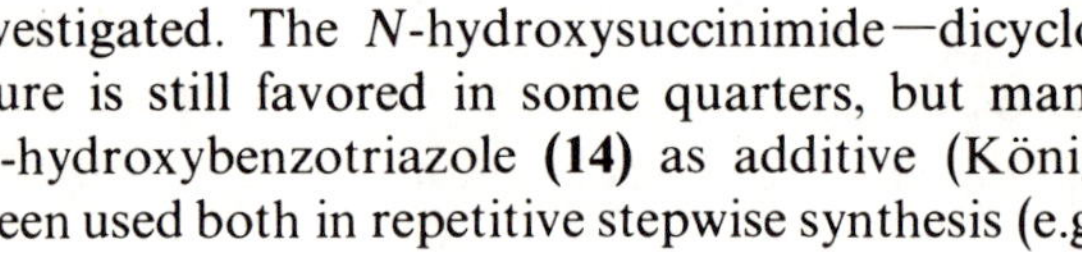

Scheme 24. Conditions: i, *N*-Methylmorpholine/DMF/30 min/−20°C; ii, DCC (1 equiv)/HOSu (1 equiv) added/−20°C/48 hr (76%).

have been investigated. The *N*-hydroxysuccinimide—dicyclohexylcarbodiimide procedure is still favored in some quarters, but many laboratories now prefer 1-hydroxybenzotriazole (**14**) as additive (König and Geiger, 1970); it has been used both in repetitive stepwise synthesis (e.g., Fletcher and

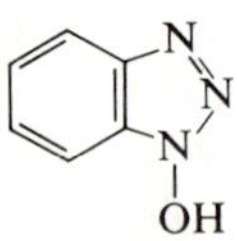

(**14**)

Young, 1972) and for coupling acylpeptides. Scheme 25 (Kamber *et al.*, 1976) is an example of an acylpeptide coupling taken from the recent total synthesis of insulin. Freedom from racemization is not invariable, however, as shown

Bpoc-Tyr(*t*Bu)-Glu-Leu-Glu(O*t*Bu)-Asn-Tyr(*t*Bu)-Cys-Asn-O*t*Bu

Trt-Leu-Val-Cys-Gly-OH + H-Glu(O*t*Bu)-Arg-Gly-Phe-Phe-Tyr(*t*Bu)-Thr(*t*Bu)-Pro-Lys(Boc)-Thr(*t*Bu)-O*t*Bu

↓

Bpoc-Tyr(*t*Bu)-Gln-Leu-Glu(O*t*Bu)-Asn-Tyr(*t*Bu)-Cys-Asn-O*t*Bu

Trt-Leu-Val-Cys-Gly-Glu(O*t*Bu)-Arg-Gly-Phe-Phe-Tyr(*t*Bu)-Thr(*t*Bu)-Pro-Lys(Boc)-Thr(*t*Bu)-O*t*Bu

Scheme 25. Conditions: Components (approx. equiv amounts) mixed with 1 equiv *N*-ethylmorpholine in DMF and HOBt (somewhat more than 1 equiv) with DCC (ca 4 equiv) added. 2 hr/40°C (78.5% after ccd).

by experience (Sieber *et al.*, 1977) at a later stage of that awe inspiring achievement.

Certain side chains can react with dicyclohexylcarbodiimide in an undesirable fashion: thus unprotected imidazole groups give N^{im}-amidino derivatives (Rink and Riniker, 1974) and primary amide side chains are dehydrated (Ressler and Ratzkin, 1961) to a sufficient extent to be a nuisance. The first of these side reactions can be avoided by appropriate protection or negated by subsequent nucleophilic displacement of the amidino group, but the primary amide dehydration reaction is a serious problem which is best met by use of a less vigorously activated intermediate. In solid phase work, for example, it is usual to use an active ester instead of the normal dicyclohexylcarbodiimide routine for the incorporation of asparagine or glutamine residues.

Although dicyclohexylurea is very sparingly soluble, it sometimes separates incompletely from the reaction mixture, and its quantitative separation from the product may be difficult in such cases. The "water-soluble" diimide reagent *N*-ethyl-*N'*-dimethylaminopropyl-carbodiimide **(15)**, which can be

$$CH_3CH_2N{=}C{=}NCH_2CH_2CH_2N(CH_3)_2$$

(15)

used in essentially the same way as dicyclohexylcarbodiimide, enables this problem, and the problem of removing any *N*-acylurea which is formed to be avoided because in this case the co- and by-products have a "handle" by means of which they can be extracted (Sheehan *et al.*, 1961). In principle this seems a very attractive modification, but neither **15** nor any analogously designed reagents have acquired a significant following; it is not clear why this is.

2. The Use of Isoxazolium Reagents

N-Alkylisoxazolium salts react with carboxylates by base-induced ring opening to give ketoketenimines to which the carboxylates add, leading ultimately by intramolecular acyl transfer to enol esters [Scheme 26 (Wood-

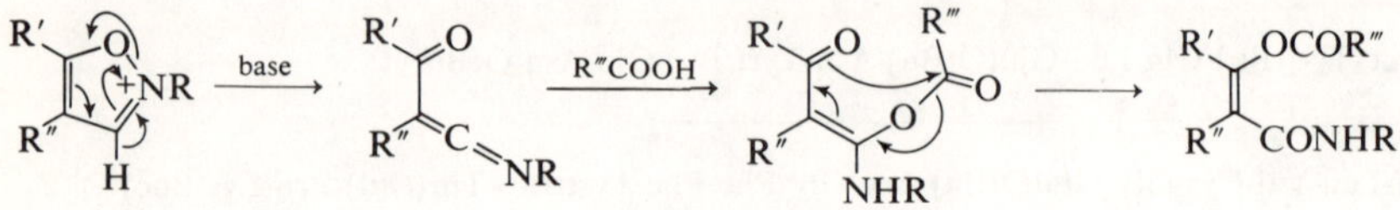

Scheme 26

ward and Olofson, 1966)]. The enol esters thus generated are reactive acylating agents and can therefore be used for peptide bond formation. The best known isoxazolium reagent which has been used in this way [e.g., Scheme 27

Z—Phe—OH → Z—Phe—Arg(Tos)—Trp—Gly—O*t*Bu

Scheme 27. Conditions: i, Et_3N/MeCN/Reagent K/0°C/1 hr; ii, H—Arg(Tos)—Trp—Gly—O*t*Bu (84%).

(Li *et al.*, 1963)] is "Woodward's Reagent K" **(16)** (Woodward *et al.*, 1966). An important feature of its design is the sulfonate group which enables simple separation of the coupling coproduct **(17)** and principal by-product

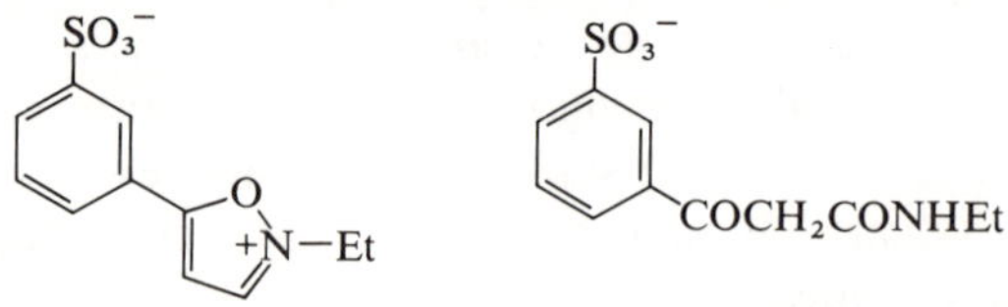

(16) **(17)**

(18) [formed by a further intramolecular acyl transfer within the enol ester **(19)**, Scheme 28] from neutral fully protected peptides. Good yields of very clean products are generally obtained from Reagent K couplings and car-

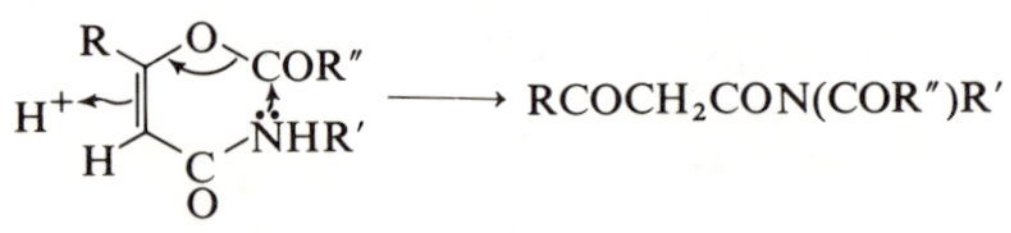

(19) **(18)**

R = *m*-sulfonatophenyl; R′ = Et

Scheme 28

boxyl components which are troublesome with dicyclohexylcarbodiimide (alkoxycarbonylasparagines in particular) can often be added in stepwise syntheses with excellent results, but some racemization occurs in acyl peptide couplings. Because of the side reaction shown in Scheme 28 the enol esters (**19**) from Reagent K are not isolated but are used immediately. Reagent **20** gives enol esters (**21**) which are not susceptible to this reaction and which can be isolated before aminolysis (Woodward and Woodman, 1968),

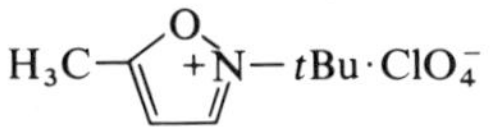

(**20**)

(**21**)

but details of practical applications are lacking. The chemistry outlined in Scheme 26 is exhibited by a range of isoxazolium salts including benzisoxazolium derivatives, and Kemp has used the principle in the design of his carboxyl activating reagent (**22**) which gives (optionally) isolable phenyl esters (**23**) embodying the principle of activation assisted by intramolecular catalysis (cf. Section II,H), as outlined in Scheme 29 (Kemp and Chien,

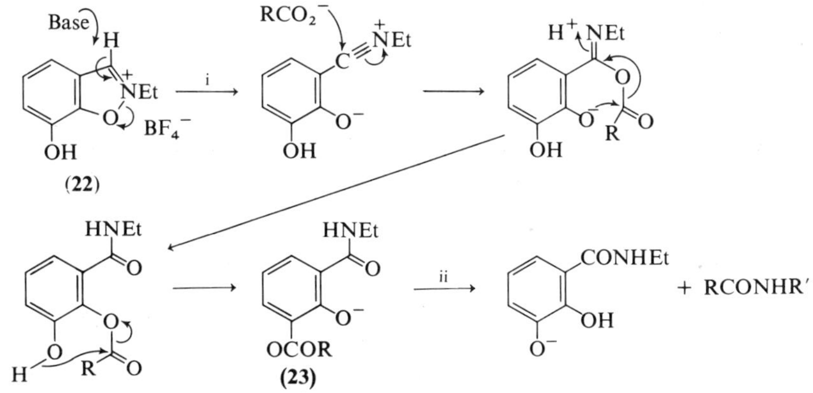

Scheme 29. Conditions: i, $RCOOH/H_2O$ (pH 4.5)—EtOAc/Kemp's reagent; ii, $R'NH_2$/DMF.

1967). These esters form the basis of a practicable method of peptide synthesis (Kemp *et al.*, 1974) but some racemization is found in model systems so it is not suitable for coupling acylpeptides if these terminate in racemizable residues, and the complexity and novelty of the chemistry involved appears to have deterred workers outside Kemp's laboratory from investigating its possible merits over established methods for other cases.

D. Reactions Involving Displacement of Acyl Groups from Heterocyclic Nitrogen

N-Acyl derivatives of imidazoles, pyrazoles, and triazoles are highly reactive acylating agents which are involved in numerous coupling methods, and also in the nucleophilic catalysis of acylation reactions by imidazole. For

Table IV Some Reagents Which Activate Carboxyl Components by Forming *N*-Acyl heterocycles[a]

Reagent	Structure	Literature
Carbonyldiimidazole	Im_2CO[b]	Paul and Anderson (1962)
Carbonyl-di-*s*-triazole	Trz_2CO[b]	Beyerman and Maassen van den Brink (1961)
Thionyldiimidazole	Im_2SO[c]	Wieland and Vogeler (1961)
1-(Diethyloxyphosphino)-1,2,4-triazole	$TrzP(OEt)_2$[c]	Kricheldorf *et al.* (1976)
Triphenylphosphite–imidazole	$ImH/(PhO)_3P$[c]	Mitin and Glinskaya (1969)
Acetylacetone	$CH_3COCH_2COCH_3$[d]	Ried and Schleimer (1958)

[a] Im = 1-imidazolyl; Trz = 1-(1,2,4-triazolyl). See also Wünsch (1974).

[b] Used for activation of a carboxylic acid before addition of the amine component.

[c] Used for direct coupling of amine component–carboxyl component mixtures.

[d] Reacts with acylamino acid hydrazides to give 1-acyl-3,5-dimethylpyrazolides which can be used for peptide synthesis, although they are not very reactive.

example, carbonyldiimidazole has been used for direct coupling as in Scheme 30 (Paul and Anderson, 1962). The actual acylating agent here is probably an *N*-acyl imidazole formed via an anhydride. Table IV gives a

$$\text{Boc—Pro—OH} + \text{Im}_2\text{CO} \xrightarrow{\text{i}} \text{Boc—Pro—Im} \xrightarrow{\text{ii}} \text{Boc—Pro—Phe—OMe}$$

Scheme 30. Conditions: i, THF/1 hr; ii, H—Phe—OMe (72%). Im = 1-imidazolyl.

representative selection of methods of this chemical type. Although the activating reagents here are cheap, the price of using them is high as rather reactive intermediates are involved, with consequent inconvenience (e.g., moisture sensitivity) and liability to side reactions (including racemization when the carboxyl component structure permits), so that these methods are principally of academic interest.

E. The Activation of Hydrazides by Oxidation

The *N*-bromosuccinimide oxidation of hydrazides in the presence of an amine component leads, presumably via an acyldiimine, to a peptide [e.g., Scheme 31 (Wolman *et al.*, 1961)]; in the case of acylamino acid phenylhydrazides (which can also be oxidized with lead tetraacetate) the unstable acylphenyldiimine can be formed separately and isolated (Milne and Kilday, 1965; Milne and Most, 1968). However, some racemization occurs in suscep-

$$\text{Z—Gly—NHNH}_2 + \text{H—Gly—ONb} \cdot \text{HBr} \rightarrow \text{Z—Gly—Gly—ONb}$$

Scheme 31. Conditions: Et_3N (3 equiv)/*N*-bromosuccinimide (2 equiv)/THF/0°C/5 min (86%).

tible cases and the yields are only fair in most of the examples described; these methods, therefore, have attracted little attention.

F. The Use of Ring-Opening Reactions of Activated Heterocyclic Systems

Several of the heterocyclic systems derived from amino acids can be opened by nucleophilic attack giving aminoacyl derivatives of the nucleophile in reactions which form the basis of a number of useful syntheses.

1. *N-Carboxyanhydrides*

The *N*-carboxy or Leuchs' anhydrides **(24)** of α-amino acids are easily obtainable by the action of phosgene on α-amino acids or by thermal decomposition of alkoxycarbonyl-α-amino acid chlorides; they are useful for peptide bond formation in two modes.

First, treatment with small amounts of a nucleophilic initiator in an organic solvent leads to successive ring opening and loss of carbon dioxide giving a new nucleophile which attacks another molecule of *N*-carboxy anhydride and so on, leading to the formation of a homopolyamino acid as outlined in Scheme 32. Other conditions and mechanisms for this exten-

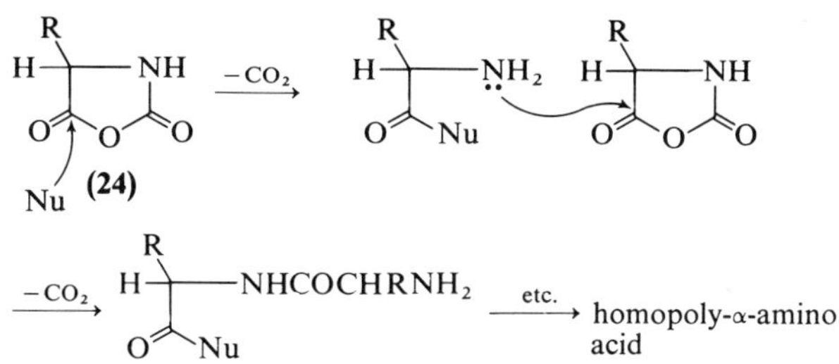

Scheme 32. Conditions: Heat with a trace of nucleophilic initiator.

sively studied polymerization are also known (for reviews, see Katchalski and Sela, 1958; Swarc, 1968), thus making these substances, which have been of great importance as protein models in diverse biological and physicochemical studies (Fasman, 1967) freely available.

Second, providing the conditions are very carefully controlled, *N*-carboxyanhydrides can be used for rapid peptide synthesis in aqueous solution, as exemplified by the dipeptide synthesis shown in Scheme 33 (Hirschmann *et al.*, 1967). Extension to short oligopeptides by repetition is possible [e.g., Scheme 34 (Denkewalter *et al.*, 1966)]. The Merck group which developed the technique used it extensively for the assembly of short fragments in their synthesis of ribonuclease (Hirschmann, 1971). The peptides prepared via *N*-carboxyanhydrides in this way are of high optical

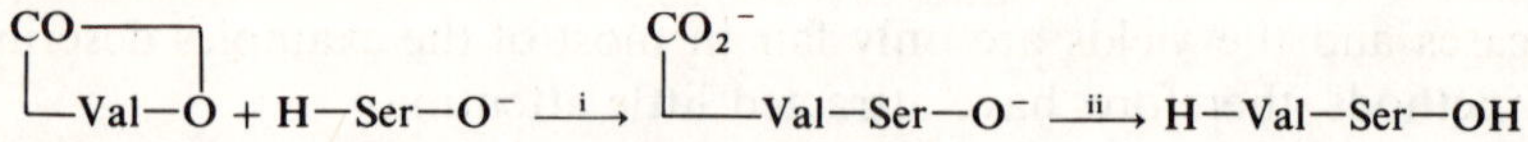

Scheme 33. Conditions: i, 5% excess of the NCA added rapidly to a solution of the amino acid rigorously maintained at pH 10.2 with vigorous mixing at 0°–2°C (complete in less than 2 min); ii, pH adjusted to 4.5 (65% isolated yield).

$$\text{H—Phe—NH}_2 \rightarrow \text{H—Trp—Met—Asp—Phe—NH}_2$$

Scheme 34. Conditions: Treatment with Asp—NCA at pH 10.2, adjustment of pH to 5 and repetition of the cycle with the NCA's of Met and Trp. Synthesis time 1 hr; 30% overall yield of pure material after chromatography.

purity and in principle the great rapidity of the technique and the possibility of using a repetitive régime make it very attractive. There is, however, a snag: the carbamic acids produced by aminolysis of *N*-carboxyanhydrides are unstable and some decarboxylation usually occurs before the aminolysis stage is complete, giving a fresh amino group which reacts with any remaining *N*-carboxyanhydride and results in "over-reaction." Minimization of this and other side reactions requires exceedingly careful control of reaction conditions (and individual optimization for each case) and the progressive accumulation of over-reaction products limits the extent to which repetitive reaction can be taken without prohibitive purification problems. The thio analogues (**25**) have also been used in this manner (Dewey *et al.*, 1971). They

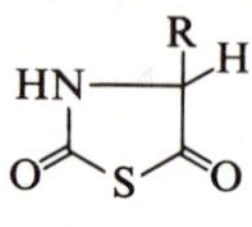

(**25**)

are less prone to yield over-reaction products as the thiocarbamic acids produced by their aminolysis are less fragile. They are preferred in the special cases of glycine and histidine (the *N*-carboxyanhydrides of which are especially subject to troublesome side reactions) but unfortunately they are not secure against racemization. The Merck method has so far not been taken up to a significant extent in other laboratories, presumably because its obvious attractions are offset by the equally clear indications in the published work that the precise execution of every detail of the procedure is so critical that the acquisition of specialist experience and know-how is likely to be a necessary preliminary to the successful use of the method. The modification in which a two-phase system is used [e.g., Scheme 35 (Katakai *et al.*, 1974)] appears to be less critically dependent on control of the reaction conditions and may perhaps recommend itself more widely.

Kricheldorf's discovery (1973; see also Kricheldorf and Fehrle, 1974) that *N*-carboxyanhydrides react with *o*-nitrophenylsulfenyl chloride to give *N*-*o*-

$$H{-}Ala{-}NH_2 \xrightarrow{i} H{-}Glu(OMe){-}Ala{-}NH_2$$
$$\xrightarrow{ii} H{-}Glu(OMe){-}Glu(OMe){-}Ala{-}NH_2$$

Scheme 35. Conditions: i, Glu(OMe)—NCA added to amine component/H_2O/MeCN/Na_2CO_3/−10°C; ii, i repeated (87% overall).

nitrophenylsulfenyl derivatives (**26**) prompted Katakai (1975) to develop the use of these compounds for peptide formation. They are highly reactive and simple to use [e.g., Scheme 36 (Katakai, 1975)]; since the product is N-

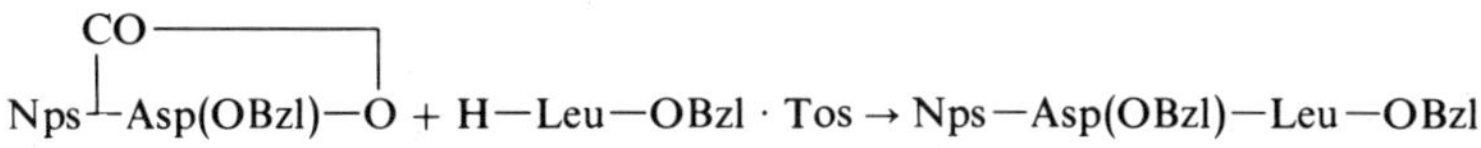

(**26**)

Scheme 36. Conditions: NCA (5% excess) added to a neutralized (Et_3N) solution of the amine componeht in THF or MeCN at room temperature (98%).

protected, over-reaction cannot occur. Racemization is not observed (Katakai, 1975) and the method has been shown, in simple but sufficiently testing examples, to be suitable for repetitive stepwise synthesis (Katakai, 1975; Katakai and Nakayama, 1976, 1977).

2. Oxazolones

Oxazolones derived from simple acylamino acids (e.g., *N*-acetyl, *N*-benzoyl derivatives) and acyl peptides undergo rapid base-catalyzed racemization so that, although they are activated toward aminolysis to give peptides, they are normally quite unsuitable for peptide synthesis. Indeed avoidance of their intervention in coupling is one of the dominant considerations affecting the selection of methods and determination of tactics. When chirality is lacking or there is double substitution at the α-carbon of the COOH-terminal residue of the carboxyl component, however, activation by oxazolone formation is very satisfactory and for simple peptides of α-aminoisobutyric acid (many of the more familiar methods founder on the rocks of the steric problem here) is the method of choice [e.g., Scheme 37 (Leplawy *et al.*, 1960; for further examples, see Wünsch, 1974)].

The cyclodehydration of benzyloxycarbonylamino acids can be achieved by treatment of their unstable acid chlorides with tertiary base. The products were described in detail by Miyoshi (1973) as aziridinones (**27**) but they are in fact (Jones and Witty, 1977) oxazolones (**28**). Miyoshi showed that these

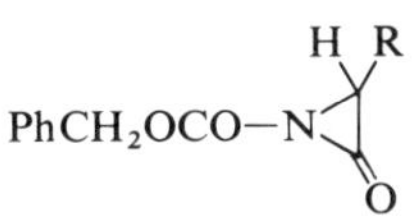

(**27**)

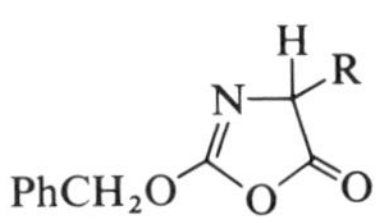

(**28**)

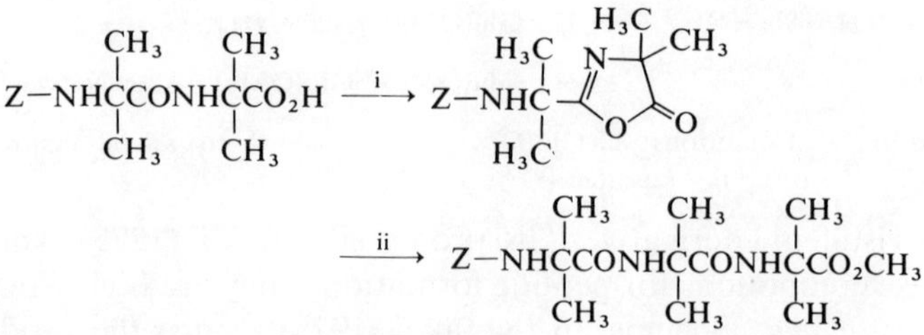

Scheme 37. Conditions: i, Ac_2O/110–120°C/15 min (98%); ii, $H_2NC(CH_3)_2CO_2CH_3$/MeCN/100°C/6 hr (91%).

compounds can be obtained in optically active form and can be used as activated intermediates which couple with amino esters in reasonable yield to give (after recrystallization) optically pure dipeptides. It appears that this type of oxazolone is less optically labile than ordinary oxazolones are so that coupling competes successfully with racemization in this case. As a practicable method of peptide synthesis, however, it has nothing to recommend it as the cyclodehydration is capricious, the oxazolones are unstable (except in the very pure state), the yields are erratic and the danger of racemization seems significant.

3. *Dihydrooxazinones*

Drey *et al.* (1973) have shown that dihydrooxazinone intermediates can be used for the coupling of the sterically hindered 3-amino-3-methylbutanoic acid for which conventional methods fail.

4. *The Use of Lactams and Lactones*

The systems in 1 and 2 above, i.e., *N*-carboxyanhydrides and oxazolones, all involve activation by heterocyclic ring closure between the two α-functions of an amino acid residue. The α-carboxyl group can also be tied back on to a side chain to give an activated species in certain cases. Thus tosyl lactams of type **29** undergo aminolysis with ring opening (e.g., Gut *et al.*, 1968) as do cyclic imidazolides (**30**) of histidine (Sheehan *et al.*, 1959). This type of compound does not appear to have been deliberately used in peptide synthesis, however. The spontaneous resynthesis of the backbone of basic pancreatic trypsin inhibitor after cyanogen bromide cleavage is the result of peptide bond formation by attack at a homoserine lactone (Dyckes *et al.*, 1974). Peptide chains can also be developed from side-chain carboxyl

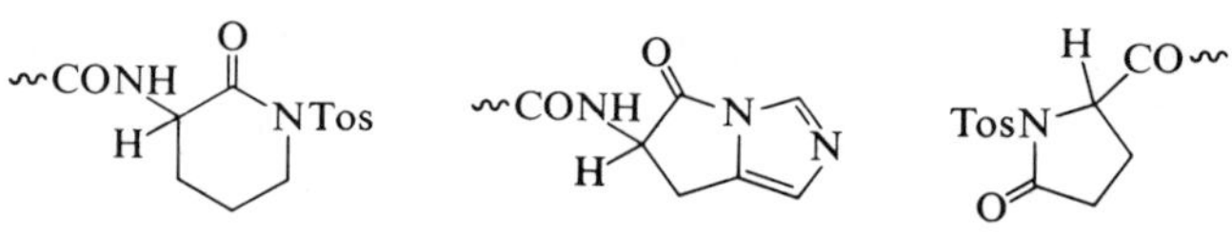

(29) **(30)** **(31)**

groups by aminolysis of tosyl lactams such as **31** (for a review of this approach, see Rudinger, 1973).

G. Procedures Involving Acyloxyphosphonium Intermediates

Acyloxyphosphonium derivatives are very electrophilic at the carbonyl carbon, and such species feature as intermediates in several recently developed procedures. The first of these (Bates *et al.*, 1975) uses as activated phosphorus reagent a bisphosphonium cation which reacts with carboxylates to give first an acyloxyphosphonium species and then a symmetrical anhydride. The dication, as its easily prepared, crystalline and fully characterized di(tetrafluoroborate) (**32**), "Bates' reagent," is used as a direct coupling reagent (Scheme 38), when either or both the acyloxyphosphonium

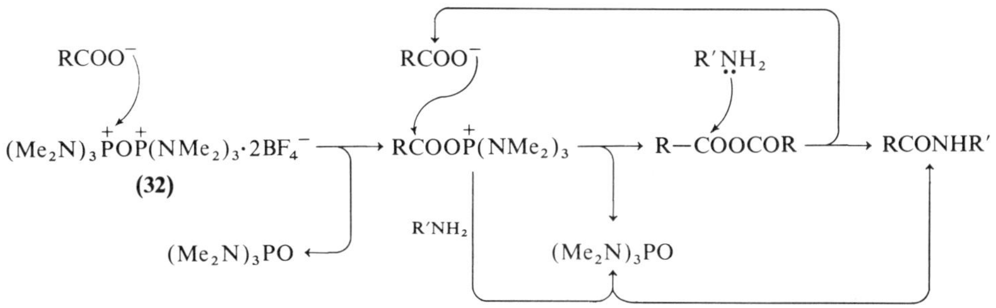

Scheme 38.

cation and the symmetrical anhydride are the actual acylating agents. Excellent yields of easily purified products (the co-product is merely hexamethylphosphoramide) are obtained, as in the example of Scheme 39, which was

Bpoc—Orn(Adc)—Thr(*t*Bu)—Pro—Gly—OH + H—Ser(*t*Bu)—Ala—Asn—Gly—OPh · Tos

↓

Bpoc—Orn(Adc)—Thr(*t*Bu)—Pro—Gly—Ser(*t*Bu)—Ala—Asn—Gly—OPh

Scheme 39. Conditions: 1.5 equiv Bates' reagent/1 equiv of each coupling component/3 equiv iPr_2NH in DMF at 20°C for 16 hr (65%).

troublesome by other methods. Racemization occurs when the structure of the carboxyl component permits. Although this can be minimized by addition of *N*-hydroxysuccinimide or 1-hydroxybenzotriazole, the method is unsuitable for fragment condensation except at glycine or proline. The related reagent (**33**), "BOP reagent," is the outcome of an extensive investigation of possible reagents of this type by Castro's group. Like Bates' reagent it is used as a direct coupling reagent: details of its preparation and application have been reported (Castro *et al.*, 1976).

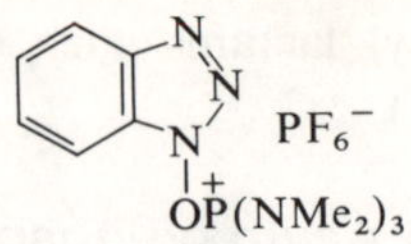

(33)

Acyloxytriphenylphosphonium salts are also reactive acylating agents, and these are presumably involved in the so-called oxidation–reduction condensation (for reviews, see Mukaiyama, 1972, 1976) in one variation of which a carboxylic acid cupric salt is treated with a sulfenamide and triphenylphosphine: the metal ion is required to trap the thiol produced, which otherwise causes side reactions. The chemistry is probably as outlined in Scheme 40; an actual example is given in Scheme 41 (Ueki *et al.*, 1971).

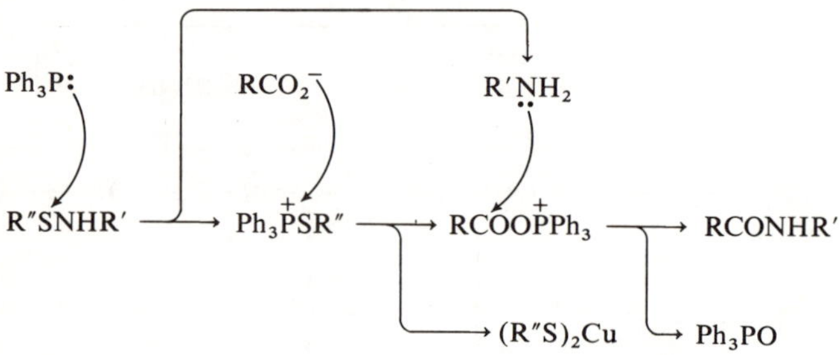

Scheme 40. For the sake of simplicity the symmetrical anhydride which may intervene between the acyloxyphosphonium salt and product is omitted.

$$(\text{Z—Phe—O})_2\text{Cu} + \text{Nps—Gly—OEt} + \text{Ph}_3\text{P} \rightarrow \text{Z—Phe—Gly—OEt}$$

Scheme 41. Conditions: 1:2:4 molar proportions of reactants mixed in CH_2Cl_2 at room temperature and left overnight (97% total yield).

A great deal of work has been invested in the development of oxidation–reduction procedures similar to that illustrated above and some impressive illustrations of the technique have been described (see Mukaiyama, 1976), but, in marked contrast to the case with the Bates' reagent procedure, the final reaction mixtures are inevitably very complex, which is a situation the prudent peptide chemist with any option does not place himself in.

H. The Use of Activated Esters

The use of ester aminolysis (Scheme 42) for peptide synthesis is as old as the subject itself. Emil Fischer (Fischer and Fourneau, 1901; Fischer, 1902) was able to prepare some simple peptides using alkyl esters. It was realized in the early 1950's that the reaction would be facilitated by use of a better

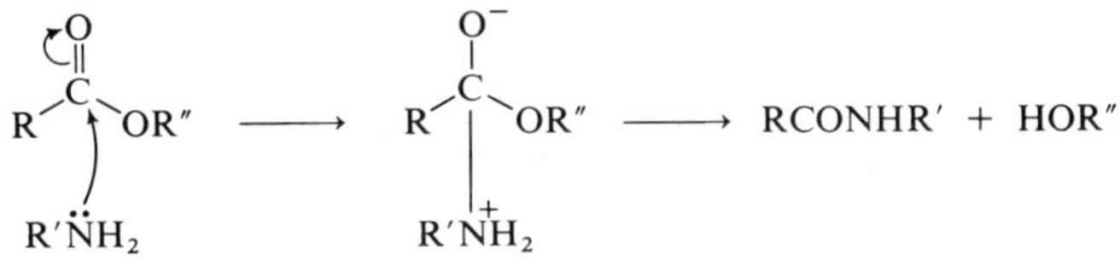

Scheme 42

leaving group. Recent results (Menger and Smith, 1972) reveal that in contrast to the familiar case of ester saponification, aminolysis of phenyl esters in organic solvents proceeds with rate limitation by the collapse of the tetrahedral adduct, not its formation, so the leaving ability of the ester group is probably the most important consideration. Since those early days a very large number of different "active esters" have been investigated (for exhaustive reviews, see Jakubke, 1966; Wünsch, 1974) but we shall confine our remarks and examples to those which are in actual use or which involve a special point of interest (see Bodanszky, this volume, Chapter 2, for details).

Active esters are most commonly prepared by dicyclohexylcarbodiimide-mediated coupling [e.g., Scheme 43 (Bodanszky and du Vigneaud, 1959)]

$$\text{Z—Leu—OH} + \text{HONp} \rightarrow \text{Z—Leu—ONp}$$

Scheme 43. Conditions: 1 equiv DCC added to a mixture of 1 equiv acid and 1.2 equiv phenol in EtOAc at 0°C; temp allowed to rise after 1/2 hr, then 1 hr at 20°C, then urea removed by filtration (86%).

between a protected acid and the ester moiety. Mixed anhydride methods and various specialized reagents can also be used and become important in the occasional case when the active ester proves difficult to separate completely from dicyclohexylurea [e.g., Scheme 44 (Fairweather and Jones,

$$\text{Z—Ala—Gly—Pro—OH} + \text{Cl}_3\text{CCO}_2\text{Pcp} \rightarrow \text{Z—Ala—Gly—Pro—OPcp}$$

Scheme 44. Conditions: 1 equiv of each reactant and 1 equiv Et_3N/EtOAc/20°C/16 hr (69%).

1972)]. The currently popular active esters are generally crystalline, stable materials; because they are at a low level of activation (compared to the activated intermediates discussed so far) they react selectively with the amine component in coupling and side reactions including racemization are generally not a problem, although there is a serious danger of racemization when most active esters are prepared if the carboxyl component is susceptible, especially if exposure to base is involved. The cleanliness of the reaction makes the active ester method, especially when the more reactive types of ester are used, appropriate for repetitive regimes which dispense with isolation of intermediate peptides [e.g., Scheme 45 (Kisfaludy and Nyeki, 1975; see also Bodanszky *et al.*, 1973, 1976b)] as by-products accumulate slowly. Less complete protection than is advisable with more reactive inter-

Boc—Pro—OPfp + H—Phe—ONb · HBr

↓ i

Pro—Phe—ONb · HCl

↓ ii

Z—Asp(OBzl)—Arg(NO_2)—Val—Tyr(Bzl)—Ile—His(Dnp)—Pro—Phe—ONb

Scheme 45. Conditions: i, Et_3N then deprotection; ii, repeated cycles of coupling with appropriate active esters and deprotection (53% based on Boc—Pro—OPfp, in 10 hr).

Boc—Ser(Bzl)—ONp + H—His—Arg(NO_2)—Ile—Ser—OMe · $2CF_3COOH$

↓

Boc—Ser(Bzl)—His—Arg(NO_2)—Ile—Ser—OMe

Scheme 46. Conditions: DMF/*i*Pr_2NEt (66%).

mediates is often sufficient with an active ester [e.g., Scheme 46 (Klausner and Bodanszky, 1977)] although incomplete side chain protection can give rise to problems (see Bodanszky *et al.*, 1977). So-called "salt couplings," in which the amine component has a free COOH-terminal carboxyl group, can be performed by other preactivation techniques (obviously not with direct coupling reagents) but active esters give the best results with this tactic [e.g., Scheme 47 (Bower *et al.*, 1976)]. Some active ester groups can be used for

Boc—Tyr(*t*Bu)—OSu + H—Gly—Gly—OH → Boc—Tyr(*t*Bu)—Gly—Gly—OH

Scheme 47. Conditions: NaOH (1 equiv)/$NaHCO_3$ (2 equiv solid)/aq DMF (74%).

temporary carboxyl blockade (i.e., in backing-off operations, see Scheme 12) and the fact that acylpeptide active esters can be deprotected to give peptide active ester salts has made active esters popular in cyclopeptide synthesis [e.g., Scheme 48 (Ludescher and Schwyzer, 1972)] and *de rigeur* in sequential

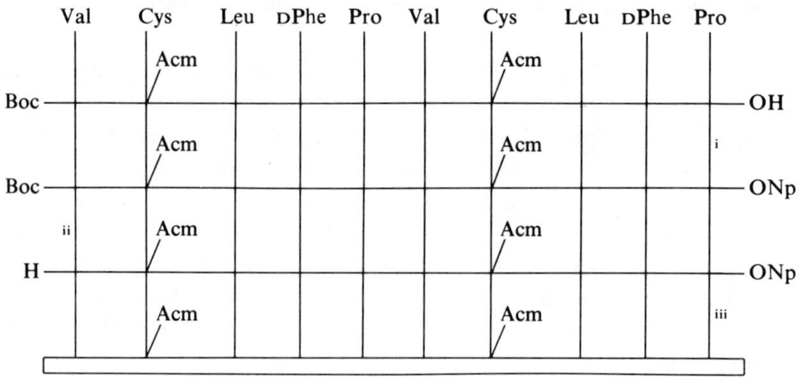

Scheme 48. Conditions: i, $SO(ONp)_2$/pyridine (80%); ii, CF_3CO_2H (92%); iii, pyridine, 55°C (27%).

$$\text{Z—Ala—Gly—Pro—OPcp} \xrightarrow{\text{i, ii}} \text{H—(Ala—Gly—Pro)}_n\text{—OH}$$

Scheme 49. Conditions: i, HBr/AcOH; ii, *N*-methylmorpholine/DMSO.

polypeptide synthesis [e.g., Scheme 49 (Fairweather and Jones, 1972; for a review, see Jones, 1977)].

It is difficult to pronounce firmly upon the relative merits of the various types of active ester as different considerations will apply in different situations. The author's judgment, which is inevitably subjective and prejudiced by selective experience, is as follows on the most popular types:

N-Hydroxysuccinimide esters. These highly reactive esters are the best for most situations when they can be prepared easily, crystallized, and purified (not always the case). They react cleanly and rapidly and isolation of the product is easy as the co-product is very water-soluble. Their storage characteristics are not, however, as good as the *p*-nitrophenyl and pentachlorophenyl esters, and carefully maintained refrigerated dry conditions are required for long-term storage. They are too reactive for use as amine components in backing off procedures, and in any case, amine deprotection of protected *N*-hydroxysuccinimide esters is not usually very satisfactory.

p-Nitrophenyl esters. These esters remain very popular. *p*-Nitrophenol is cheap and its esters are generally crystalline. Quantitative removal of *p*-nitrophenol from coupling products is sometimes difficult. They are widely applicable but better results can often be obtained by switching to a *N*-hydroxysuccinimide or pentachlorophenyl ester.

Pentachlorophenyl esters. These esters are almost invariably crystalline and very easy to prepare. Removal of pentachlorophenol from ether-soluble fully protected peptides is sometimes a nuisance and the *N*-hydroxysuccinimide ester is generally an improvement in such cases. For ether-insoluble peptides, however, the pentachlorophenyl ester method is often very convenient. Amine deprotection by acidolysis or hydrogenolysis is permissible and amino acid and peptide pentachlorophenyl esters can be used in backing off, so that these esters are clearly the ones of choice in sequential polypeptide synthesis.

The esters so far discussed involve no special design features. They are all reactive systems simply by virtue of being *O*-acyl derivatives of moderately strong acids. Some subtlety can be introduced in various ways, however. For example, unreactive esters can sometimes be rendered reactive by a specific reaction not involving the acidic moiety, i.e., a protecting group can be turned into an active ester group (Johnson and Trask, 1968; Cowell and Jones, 1971); see Volume 2 for details (Johnson, 1978). It is also possible to arrange matters so that intramolecular general base catalysis activates what would otherwise be unreactive systems (Jakubke and Voigt, 1966; Jones and Young, 1968). Both principles are illustrated in the sequential polypeptide

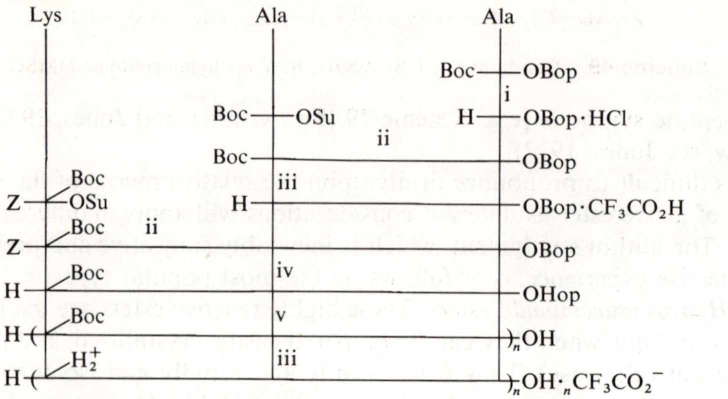

Scheme 50. Conditions: i, HCl/EtOAc; ii, Et_3N/DMF; iii, 90% aq. CF_3COOH; iv, H_2/Pd(C); v, Et_3N/DMSO. Bop = 2-benzyloxyphenyl; Hop = 2-hydroxyphenyl.

synthesis shown in Scheme 50 (Cowell and Jones, 1972). In the early stages the unreactive *o*-benzyloxyphenyl ester is protective of its carboxyl group, but at the required point hydrogenolysis reveals an *o*-hydroxyl group which activates the system by providing intramolecular general base catalysis for aminolysis.

I. Polymeric Reagents

The purpose of the Merrifield solid phase approach is to enable quantitative separation of the peptide under construction from all excess reagents and coproducts after each step by filtration, the peptide being totally insolubilized by attachment to a polymer whereas the reagents and coproducts are in solution. The complementary principle, in which acylating [e.g., Scheme 51 (Kalir *et al.*, 1975)] or activating reagents (e.g., Brown and Williams, 1971; Ito *et al.*, 1975) are polymer-bound has also been explored. Here

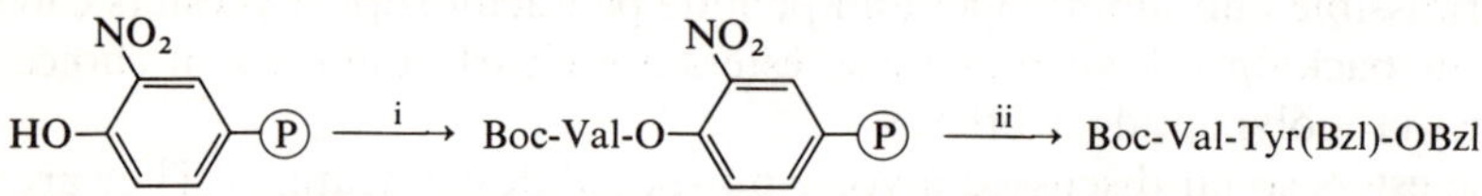

Scheme 51. Conditions: i, Boc—Val—OH/DCC/DMF, then extensive washing; ii, H—Tyr(Bzl)—OBzl/CH_2Cl_2/40 hr/20°C, then filtration (95%). Ⓟ, insoluble polymeric carrier.

it is the excess polymer-bound reagent and co-product of the coupling which are removed by filtration. The peptide is left alone in solution, providing coupling is complete and no co- or by-products are released from the carrier. The chemistry of these methods, a considerable number of which have now been investigated, is the same as the analogous homogenous solution procedures and we shall therefore not discuss them further. The area has been reviewed (Patchornik *et al.*, 1973).

J. Coupling between Captive Components

In principle peptide bond formation between captive components might be very advantageous. In particular it might be possible to achieve the intramolecular coupling with low levels of carboxyl component activation. Kemp (Kemp and Vellaccio, 1975; Kemp *et al.*, 1975) has reported on his progress toward the development of new methods [e.g., Scheme 52 (Kemp *et*

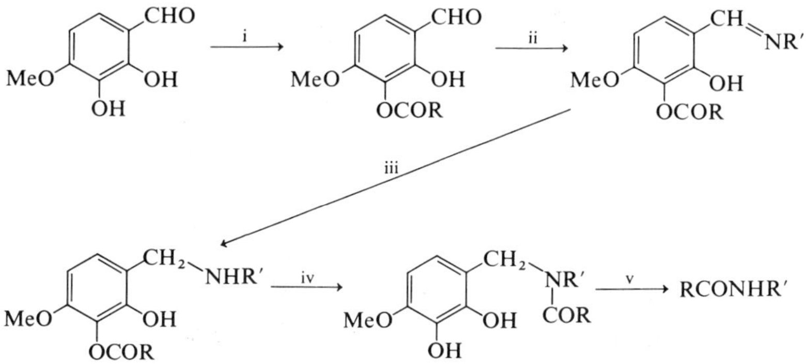

Scheme 52. Conditions: i, DCC/RCOOH; ii, $R'NH_2$/MeCN; iii, pyridine borane/AcOH; iv, unimolecular isomerization; v, HBr/AcOH.

al., 1975)] based on the concept. The same notion is contained in Brenner's aminoacyl insertion method [e.g., Scheme 53 (Brenner *et al.*, 1957)]. It has not proved possible to develop this approach into a practicable method of peptide synthesis although related processes have been used for cyclopeptide (Rothe *et al.*, 1967) synthesis (see also Davies *et al.*, 1973).

III. *Ab Initio* ASSEMBLY OF THE BACKBONE

Acylamino acid amide residues can be assembled by "four component condensations" as outlined in Scheme 54 (for a review, see Ugi, 1971; Ugi *et al.*, 1975). By appropriate choice of the components, i.e., by using a protected amino acid or peptide as carboxyl component, an isocyano ester derived from an amino acid or peptide, a keto component selected according to the α-substituents required, this reaction can be used for generating an amino acid residue already engaged in peptide bonds [e.g., Scheme 55 (Ugi, 1962)]. However, if the creation of chiral α-carbon atoms is involved this requires very precise stereochemical control if the process is to bear comparison with the standard approach, which concentrates on the preservation of existing chirality rather than its controlled creation. Prodigious feats can be accomplished in simple models (e.g., Urban *et al.*, 1976) using the chiral amine **34** as amine component. The organometallic substituent which drives the asym-

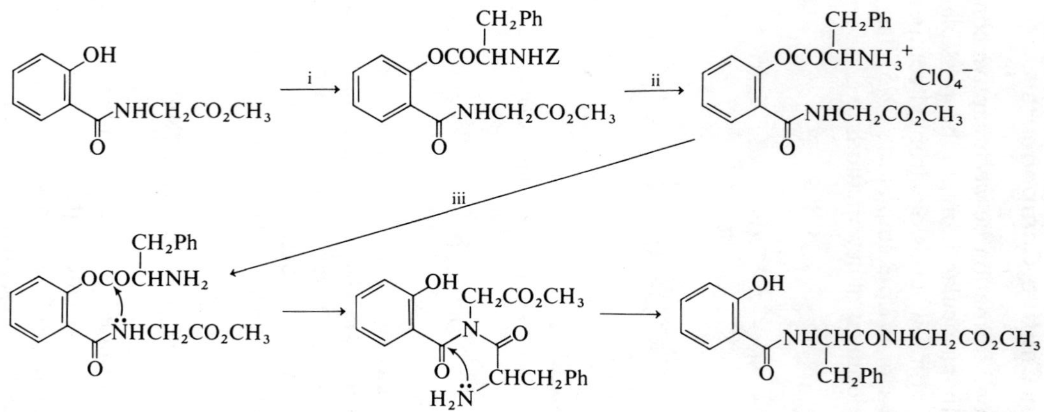

Scheme 53. Conditions: i, Z—DLPhe, phosgene method (81%); ii, H_2/Pd/AcOH/$HClO_4$ (80%); iii, Et_3N/$CHCl_3$ (94%). Coupling of Z—Gly—OH to the product followed by ii and iii gave salicoyl—Gly—DLPhe—Gly—OMe.

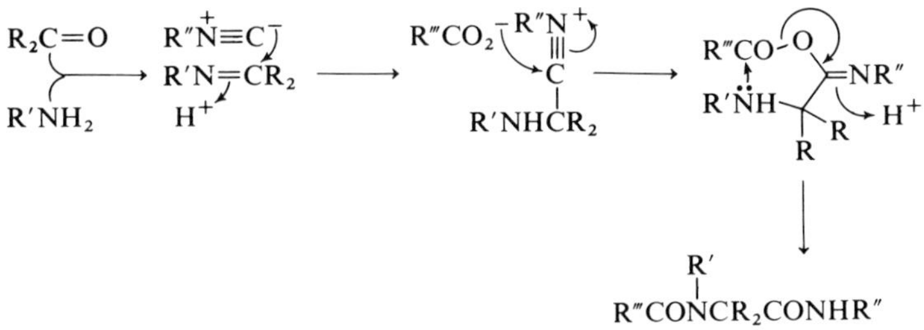

Scheme 54

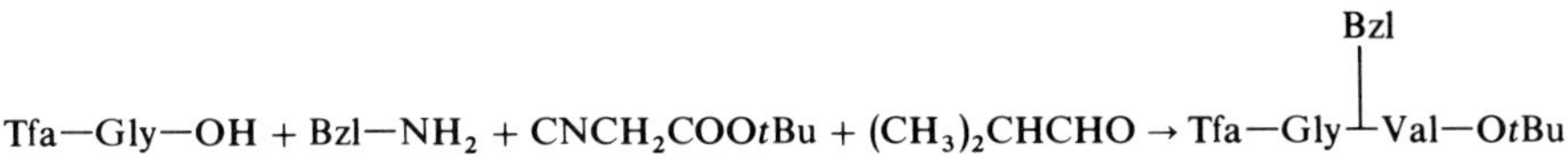

$$\text{Tfa—Gly—OH} + \text{Bzl—NH}_2 + \text{CNCH}_2\text{COO}t\text{Bu} + (\text{CH}_3)_2\text{CHCHO} \rightarrow \text{Tfa—Gly}\overset{\text{Bzl}}{\text{—}}\text{Val—O}t\text{Bu}$$

Scheme 55. Conditions: reactants mixed with cooling in MeOH, then 2 hr at 20°C (87%).

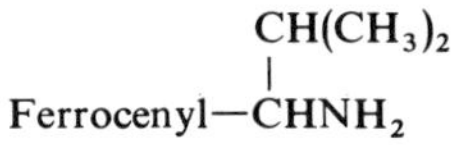

$$\text{Ferrocenyl—}\overset{\text{CH(CH}_3)_2}{\overset{|}{\text{C}}}\text{HNH}_2$$

(**34**: *R* configuration)

metric induction is cleaved from the peptide bond after its formation. The prospects of this approach developing into a practicable general method of peptide synthesis seem slim. For amino acids with symmetrical double substitution at the α-carbon, however, the route is a valuable one (see Maia *et al.*, 1973) as such cases meet severe steric hindrance problems when coupling by conventional means is attempted.

IV. CONCLUSIONS

A few remarks on the relative popularity of the methods surveyed have been made above, but a more precise and objective picture can be obtained by an actual count of applications. This has been done (Table V) for an exhaustive list (Fletcher and Jones, 1976) of peptides, proteins, analogues, and partial sequences which were described in 1975. It is clear that a rather small number of the many methods which have been suggested over the years in fact get used. Furthermore the distribution of popularity shown by Table V is not very different from the situation in 1968 (Jones, 1970) which in its turn was really very similar to that of 1960–1962 (Rydon, 1962). Dicyclohexylcarbodiimide has been the most important single reagent throughout and the only widely adopted new technique since 1960 is the modification using it together with *N*-hydroxysuccinimide or 1-hydroxybenzotriazole. Mixed anhydride and azide procedures have slipped

Table V The Actual Popularity of Methods of Coupling in 1975[a]

Method	Number of papers in which use is described	Total number of couplings by the method
A. Solid phase synthesis: stepwise couplings[b]		
Dicylohexylcarbodiimide alone	35	605
Active ester[c]	11	43[d]
Carbonyldiimidazole	1	19
All other	6	13
	36	680
B. Solid phase synthesis: fragment couplings[e]		
DCC/HOBt or HOSu[f]	3	10
Azide	2	2
All other	1	2
	4	14
C. Solution synthesis: stepwise couplings[b,g]		
Active ester	42	207[h]
DCC/HOBt or HOSu	19	60[i]
DCC alone	21	48
Mixed anhydride	12	54[j]
Azide	11	20
All other	2	2
	65	391
D. Solution synthesis: fragment couplings[e]		
Azide	21	54
DCC/HOBt or HOSu[k]	12	28
Mixed anhydride	5	8
OSu (isolated)	4	6
All other	1	1
	33	97

[a] Compiled by M. J. Witty from a list (Fletcher and Jones, 1976) of syntheses of peptides, proteins, analogues, and partial sequences which were described in 1975, excluding only cases for which insufficient detail was given and syntheses of known or trivial peptides prepared for the sole purpose of exemplifying methodology; the total number of papers was 94.

[b] And fragment couplings at COOH-terminal Gly or Pro.

[c] Mainly for Asn and Gln.

[d] 35 and 8 respectively for ONp and OSu.

[e] Excluding couplings at COOH-terminal Gly or Pro.

[f] 8 and 2, respectively, for HOBt and HOSu as additives.

[g] Including couplings to peptides bound to soluble polymeric supports.

[h] 96, 43, 23, 20, 6, and less than 5 for ONp, OSu, OPcp, OTcp, OPfp, and all others, respectively.

[i] 51 and 9 for HOBt and HOSu, respectively.

[j] 43, 8, and 3 for isobutylcarbonic, ethylcarbonic, and pivalic mixed anhydrides, respectively.

[k] 25 and 3 for HOBt and HOSu, respectively.

back from their position somewhat and active esters have advanced a little but there is no radical newcomer. Why is this? The answer probably involves as much sociology as chemistry. Several methods which appear to have much to offer have fallen on stony ground, and the techniques in use are certainly not ideal or even the best which can be devised. To take the dicyclohexylcarbodiimide reagent as an example, there has never even been a systematic study to see if this is the most satisfactory of its type. Would *a priori* considerations lead to the conclusion that a diimide giving a very sparingly soluble urea would be especially suited for solid phase work? Obviously not; only very recently have other simple dialkyldiimides been seriously examined and they may indeed prove to be better than dicyclohexylcarbodiimide (Sarantakis *et al.*, 1976). The fact is that the dicyclohexylcarbodiimide reagent was well known at the time the solid phase approach was developed. It sufficed, for the moment, so that attention was focussed on the improvement of other more troublesome aspects. By the time these had been sorted out, dicyclohexylcarbodiimide had become part of the regular routine. This kind of story seems to be rather general. All of the coupling methods of enduring popularity give adequate results most of the time, are chemically simple, and are associated with an extensive corporate experience. There has been a marked tendency to stick to well understood and predictable methods of coupling and to confine applications of novel methodology to those aspects of synthetic problems which have proved more critical, particularly purification and selective functional group protection. Ten or fifteen years ago an account such as this would probably have concluded with generalizations along the lines that it was in rapid repetitive synthesis and fragment conjunction that fundamental advances were most sorely needed. This remains true up to a point, but it seems to the present author that these improvements are more likely to flow from better purification techniques than new chemistry. Rapidity of coupling and complete evasion of side reactions are to some extent mutually exclusive characteristics, and the practicability of a rapid repetitive regime is thus limited by the power of the separation techniques which are available. Perhaps sophisticated new fast chromatographic methods such as those described recently by Burgus and Rivier (1976) and Gabriel *et al.* (1977) will be of special significance in this context. Similarly, the demonstration that the separation of an epimeric high molecular weight peptide resulting from racemization at fragment conjunction from the desired isomer can be achieved [e.g., the separation of D-[B_{16}]-insulin and insulin (Sieber *et al.*, 1977)], which might have been thought impossible in 1960, suggests that it may be more profitable to approach the assembly of chirally homogenous peptides from fragments by forcing purification techniques nearer to their limits than by making demands on the coupling methods which may in fact be unrealistic.

REFERENCES

Albertson, N. F. (1962). *Org. React.* **12**, 157–355.
Anderson, G. W., and Young, R. W. (1952). *J. Am. Chem. Soc.* **74**, 5307–5309.
Anderson, G. W., Blodinger, J., and Welcher, A. D. (1952). *J. Am. Chem. Soc.* **74**, 5309–5312.
Anderson, G. W., Zimmerman, J. E., and Callahan, F. M. (1967). *J. Am. Chem. Soc.* **89**, 5012–5017.
Arens, J. F. (1955). *Recl. Trav. Chim. Pays-Bas* **76**, 769–770.
Ariyoshi, Y., Yamatani, T., and Adachi, Y. (1973). *Bull. Chem. Soc. Jpn.* **46**, 2611–2612.
Babusiaux, P., Langeray, R., and Dreux, J. (1976). *Justus Liebigs Ann. Chem.* **1976**, 487–495.
Bates, A. J., Galpin, I. J., Hallett, A., Hudson, D., Kenner, G. W., and Ramage, R. (1975). *Helv. Chim. Acta* **58**, 688–696.
Beckwith, A. J. (1971). *In* "The Chemistry of Amides" (J. Zabicky, ed.), pp. 73–185. Wiley (Interscience), New York.
Belleau, B., and Malek, G. (1968). *J. Am. Chem. Soc.* **90**, 1651–1652.
Beyerman, H. C., and Maassen van den Brink, W. (1961). *Recl. Trav. Chim. Pays-Bas* **80**, 1372–1375.
Beyerman, H. C., Rammeloo, T., Renirie, J. F. M., Syrier, J. L. M., and van Zon, A. (1976). *Recl. Trav. Chim. Pays-Bas* **95**, 143–148.
Bodanszky, M., and du Vigneaud, V. (1959). *J. Am. Chem. Soc.* **81**, 5688–5691.
Bodanszky, M., Funk, K. W., and Fink, M. L. (1973). *J. Org. Chem.* **38**, 3565–3570.
Bodanszky, M., Klausner, Y. S., and Ondetti, M. A. (1976a), "Peptide Synthesis," 2nd ed., Chapter 5. Wiley, New York.
Bodanszky, M., Lin, C. Y., Yiotakis, A. E., Mutt, V., and Said, S. I. (1976b). *Bioorg. Chem.* **5**, 339–350.
Bodanszky, M., Fink, M. L., Klausner, Y. S., Natarajan, S., Tatemoto, K., Yiotakis, A. E., and Bodanszky, A. (1977). *J. Org. Chem.* **42**, 149–152.
Boissonas, R. A. (1951). *Helv. Chim. Acta* **34**, 874–879.
Boissonas, R. A., Guttmann, S., Jaquenoud, P.-A., and Waller, J. P. (1955). *Helv. Chim. Acta* **38**, 1491–1501.
Bower, J. D., Guest, K. P., and Morgan, B. A. (1976). *J. Chem. Soc., Perkin Trans. 1* pp. 2488–2492.
Brenner, M., Zimmermann, J. P., Wehrmüller, J., Quitt, P., Hartmann, A., Schneider, W., and Beglinger, U. (1957). *Helv. Chim. Acta* **40**, 1497–1517.
Brown, J., and Williams, R. E. (1971). *Can. J. Chem.* **49**, 3765–3766.
Brundish, D. E., and Wade, R. (1976). *J. Chem. Soc., Perkin Trans. 1* pp. 2186–2189.
Buijle, R., and Viehe, H. G. (1964). *Angew. Chem., Int. Ed. Engl.* **3**, 582.
Burgus, R., and Rivier, J. (1976). *In* "Peptides 1976" (A. Loffet, ed.), pp. 85–94. Editions de l'Université de Bruxelles, Belgium.
Castro, B., Dormoy, J.-R., Dourtoglon, B., Erin, G., Selve, C., and Ziegler, J.-C. (1976). *Synthesis* pp. 751–752.
Chan, T. H., and Wong, L. T. L. (1971). *J. Org. Chem.* **36**, 850–853.
Clayton, D. W., Farrington, J. A., Kenner, G. W., and Turner, J. M. (1957). *J. Chem. Soc.* pp. 1398–1407.
Clement, B. A., and Soulen, R. L. (1974). *J. Org. Chem.* **39**, 97–98.
Cosmatos, A., Photaki, I., and Zervas, L. (1961). *Chem. Ber.* **94**, 2644–2655.
Cowell, R. D., and Jones, J. H. (1971). *J. Chem. Soc. C* pp. 1082–1090.
Cowell, R. D., and Jones, J. H. (1972). *J. Chem. Soc., Perkin Trans. 1* pp. 1814–1819.
Curtius, T. (1881). *J. Prakt. Chem.* **24**, 239–240.
Curtius, T. (1902). *Ber. Dtsch. Chem. Ges.* **35**, 3226–3228.

Davies, J. S., Hassall, C. H., and Hopkins, K. H. (1973). *J. Chem. Soc., Perkin Trans. 1* pp. 2614–2618.
Denkewalter, R. G., Schwam, H., Strachan, R. G., Beesley, T. E., Veber, D. F., Schoenewaldt, E. F., Barkemeyer, H., Paleveda, W. J., Jr., Jacob, T. A., and Hirschmann, R. (1966). *J. Am. Chem. Soc.* **88**, 3163–3164.
Dewey, R. S., Schoenewaldt, E. F., Joshua, H., Paleveda, W. J., Jr., Schwam, H., Barkemeyer, H., Arison, B. H., Veber, D. F., Strachan, R. G., Milkowski, J., Denkewalter, R. G., and Hirschmann, R. (1971). *J. Org. Chem.* **36**, 49–59.
Drey, C. N. C., Lowbridge, J., and Ridge, R. J. (1973). *J. Chem. Soc., Perkin Trans. 1* pp. 2001–2006.
Dyckes, D. F., Creighton, T., and Sheppard, R. C. (1974). *Nature (London)* **247**, 202–204.
Fairweather, R., and Jones, J. H. (1972). *J. Chem. Soc., Perkin Trans. 1* pp. 1908–1915.
Fasman, G. D. (1967). "Poly-α-Amino Acids." Dekker, New York.
Fieser, M., and Fieser, L. F. (1969), "Reagents for Organic Synthesis," Vol. 2, p. 191. Wiley (Interscience), New York.
Finn, F. M., and Hofmann, K. (1976). *In* "The Proteins" (H. Nemath and R. L. Hill, eds.), 3rd ed., Vol. 2, Chapter 2, pp. 105–253. Academic Press, New York.
Fischer, E. (1902). *Ber. Dtsch. Chem. Ges.* **35**, 1095–1106.
Fischer, E., and Fourneau, E. (1901). *Ber. Dtsch. Chem. Ges.* **34**, 2868–2877.
Fletcher, G. A., and Jones, J. H. (1976). *Amino Acids, Pept. Proteins* 8, Appendices to Chapter 3, 294–309.
Fletcher, G. A., and Young, G. T. (1972). *J. Chem. Soc., Perkin Trans. 1* pp. 1867–1874.
Gabriel, T. F., Jimenez, M. H., Felix, A. M., Michalewsky, J., and Meienhofer, J. (1977). *Int. J. Pept. Protein Res.* **9**, 129–136.
Goldschmidt, S. (1950). *Angew. Chem.* **62**, 538.
Goldschmidt, S., and Lautenschlager, H. (1953). *Justus Liebigs Ann. Chem.* **580**, 68–82.
Gut, V., Rudinger, J., Walter, R., Herling, P. A., and Schwartz, I. L. (1968). *Tetrahedron* **24**, 6351–6363.
Hamada, Y., Shiori, T., and Yamada, S. (1977a). *Chem. Pharm. Bull.* **25**, 221–223.
Hamada, Y., Rishi, S., Shiori, T., and Yamada, S. (1977b). *Chem. Pharm. Bull.* **25**, 224–230.
Hirschmann, R. (1971). *Intra-Sci. Chem. Rep.* **5**, 203–228.
Hirschmann, R., Strachan, R. G., Schwam, H., Schoenewaldt, E. F., Joshua, H., Barkemeyer, B., Veber, D. F., Paleveda, W. J., Jr., Jacob, T. A., Beesley, T. E., and Denkewalter, R. G. (1967). *J. Org. Chem.* **32**, 3415–3425.
Honzl, J., and Rudinger, J. (1961). *Collect. Czech. Chem. Commun.* **26**, 2333–2344.
Ito, H., Takamatsu, N., and Ichikizaki, I. (1975). *Chem. Lett.* pp. 577–578.
Jackson, A. G., Kenner, G. W., Moore, G. A., Ramage, R., and Thorpe, W. D. (1976). *Tetrahedron Lett.* pp. 3627–3630.
Jakubke, H.-D. (1966). *Z. Chem.* **6**, 52–67.
Jakubke, H.-D., and Voigt, A. (1966). *Chem. Ber.* **99**, 2419–2429.
John, W. D., and Young, G. T. (1954). *J. Chem. Soc.* pp. 2870–2873.
Johnson, B. J. (1978). *In* "The Peptides," Vol. 2. Academic Press, New York.
Johnson, B. J., and Trask, E. G. (1968). *J. Org. Chem.* **33**, 4521–4524.
Jones, D. S., Kenner, G. W., and Sheppard, R. C. (1965). *J. Chem. Soc.* pp. 4393–4396.
Jones, J. H. (1970). *Amino Acids, Pept., Proteins* **2**, 143–191.
Jones, J. H. (1974). *Chem. Ind. (London)* pp. 723–727.
Jones, J. H. (1977). *In* "Chemistry and Biochemistry of Amino Acids Peptides and Proteins" (B. Weinstein, ed.), Vol. 4, pp. 29–63. Dekker, New York.
Jones, J. H., and Fairweather, R. (1969). *Makromol. Chem.* **128**, 279–283.
Jones, J. H., and Witty, M. J. (1977). *J. Chem. Soc., Chem. Commun.* pp. 281–282.

Jones, J. H., and Young, G. T. (1968). *J. Chem. Soc. C* pp. 436–441.
Kalir, R., Warshawsky, A., Fridkin, M., and Patchornik, A. (1975). *Eur. J. Biochem.* **59**, 55–61.
Kamber, B., Riniker, B., Sieber, P., and Rittel, W. (1976). *Helv. Chim. Acta* **59**, 2830–2840.
Katakai, R. (1975). *J. Org. Chem.* **40**, 2697–2702.
Katakai, R., and Nakayama, Y. (1976). *Biopolymers* **15**, 747–755.
Katakai, R., and Nakayama, Y. (1977). *J. Chem. Soc., Perkin Trans. 1* pp. 292–294.
Katakai, R., Oya, M., Toda, F., Keikichi, U., and Iwakura, Y. (1974). *J. Org. Chem.* **39**, 180–182.
Katchalski, E., and Sela, M. (1958). *Adv. Protein Chem.* **8**, 243–492.
Kemp, D. S., and Chien, S. W. (1967). *J. Am. Chem. Soc.* **89**, 2743–2745.
Kemp, D. S., and Vellaccio, F., Jr. (1975). *J. Org. Chem.* **40**, 3003–3004 and 3464–3465.
Kemp, D. S., Wrobel, S. J., Jr., Wang, S.-W., Bernstein, Z., and Rebek, J., Jr. (1974). *Tetrahedron* **30**, 3969–3980.
Kemp, D. S., Grattan, J., and Reczek, J. (1975). *J. Org. Chem.* **40**, 3465–3466.
Kenner, G. W. (1977). *Proc. R. Soc. London, Ser. A* **353**, 441–475.
Kisfaludy, L., and Nyéki, O. (1975). *Acta Chim. Acad. Sci. Hung.* **86**, 343–345.
Kiso, Y., Kai, Y., and Yajima, H. (1973). *Chem. Pharm. Bull.* **21**, 2507–2510.
Klausner, Y. S., and Bodanszky, M. (1972). *Synthesis* pp. 453–463.
Klausner, Y. S., and Bodanszky, M. (1974). *Synthesis* pp. 549–559.
Klausner, Y. S., and Bodanszky, M. (1977). *J. Org. Chem.* **42**, 147–149.
König, W., and Geiger, R. (1970). *Chem. Ber.* **103**, 788–798.
Kotake, H., and Saito, T. (1966). *Bull. Chem. Soc. Jpn.* **39**, 853.
Kricheldorf, H. R. (1973). *Angew. Chem. Int. Ed. Engl.* **12**, 73–74.
Kricheldorf, H. R., and Fehrle, M. (1974). *Chem. Ber.* **107**, 3533–3547.
Kricheldorf, H. R., Fehrle, M., and Kaschig, J. (1976). *Angew. Chem., Int. Ed. Engl.* **15**, 305–306.
Kurath, P. (1976). *Helv. Chim. Acta* **59**, 1127–1132.
Leplawy, M. T., Jones, D. S., Kenner, G. W., and Sheppard, R. C. (1960). *Tetrahedron* **11**, 39–51.
Le Quesne, W. J., and Young, G. T. (1950). *J. Chem. Soc.* pp. 1954–1959.
Li, C. H., Gorup, B., Chung, D., and Ramachandran, J. (1963). *J. Org. Chem.* **28**, 178–181.
Losse, G., and Demuth, F. (1961). *Chem. Ber.* **94**, 1762–1766.
Ludescher, U., and Schwyzer, R. (1972). *Helv. Chim. Acta* **55**, 2052–2060.
Maia, H. L., Ridge, B., and Rydon, H. N. (1973). *J. Chem. Soc., Perkin Trans. 1* pp. 98–105.
Menger, F. M., and Smith, J. H. (1972). *J. Am. Chem. Soc.* **94**, 3824–3829.
Milne, H. B., and Kilday, W. (1965). *J. Org. Chem.* **30**, 64–71.
Milne, H. B., and Most, C. F., Jr. (1968). *J. Org. Chem.* **33**, 169–175.
Mitin, Yu. V., and Glinskaya, O. V. (1969). *Tetrahedron Lett.* pp. 5267–5270.
Miyoshi, M. (1973). *Bull. Chem. Soc. Jpn.* **46**, 212–218 and 1489–1496.
Mühlemann, M., Titov, M. I., Schwyzer, R., and Rudinger, J. (1972). *Helv. Chim. Acta* **55**, 2854–2860.
Mukaiyama, T. (1972). *Synth. Commun.* **2**, 243–265.
Mukaiyama, T. (1976). *Angew. Chem., Int. Ed. Engl.* **15**, 94–103.
Oakwood, T. S., and Weisgerber, C. A. (1955). *Org. Synth., Collect.* **3**, 114–115.
Ozawa, K., Shiori, T., and Yamada, S. (1977). *Chem. Pharm. Bull.* **25**, 122–125.
Pankstelis, J. V., and Kim, M. (1974). *J. Org. Chem.* **39**, 1499–1507.
Patchornik, A., Fridkin, M., and Katchalski, E. (1973). *In* "The Chemistry of Polypeptides" (P. G. Katsoyannis, ed.), pp. 315–333. Plenum, New York.
Paul, R., and Anderson, G. W. (1962). *J. Org. Chem.* **27**, 2094–2099.
Pelter, A., Levitt, T. E., and Nelson, P. (1970). *Tetrahedron* **26**, 1539–1544.
Poduska, K., and Gross, H. (1961). *Chem. Ber.* **94**, 527–537.
Ressler, C., and Ratzkin, H. (1961). *J. Org. Chem.* **26**, 3356–3360.
Ried, W., and Schleimer, B. (1958). *Justus Liebigs Ann. Chem.* **619**, 43–46.

Rink, H., and Riniker, B. (1974). *Helv. Chim. Acta* **57**, 831–836.
Romovacek, H., Drabarek, S., Kawasaki, K., Dowd, S. R., Obermeier, R., and Hofmann, K. (1974). *Int. J. Pept. Protein Res.* **6**, 435–445.
Rothe, M., Rothe, I., Toth, T., and Steffen, K.-D. (1967). *In* "Peptides" (H. C. Beyerman, A. van de Linde, and W. Maassen, van den Brink, eds.), pp. 8–14. North-Holland Publ., Amsterdam.
Rudinger, J. (1973). *In* "The Chemistry of Polypeptides" (P. G. Katsoyannis, ed.), pp. 87–123. Plenum, New York.
Rydon, H. N. (1962). *R. Inst. Chem., Lect. Ser.* No. 5.
Sarantakis, D., Teichman, J., Lien, E. L., and Fenichel, R. L. (1976). *Biochem. Biophys. Res. Commun.* **73**, 336–342.
Schreier, E., and Hermann, P. (1976). *J. Prakt. Chem.* **318**, 502–506.
Schröder, E., and Lübke, K. (1965). "The Peptides," Vol. 1. Academic Press, New York.
Sheehan, J. C., and Frank, V. S. (1949). *J. Am. Chem. Soc.* **71**, 1856–1861.
Sheehan, J. C., and Hess, G. P. (1955). *J. Am. Chem. Soc.* **77**, 1067–1068.
Sheehan, J. C., Hasspacher, K., and Yeh, Y. L. (1959). *J. Am. Chem. Soc.* **81**, 6086.
Sheehan, J. C., Cruickshank, P. A., and Boshart, G. L. (1961). *J. Org. Chem.* **26**, 2525–2528.
Shiori, T., and Yamada, S. (1974a). *Chem. Pharm. Bull.* **22**, 849–854.
Shiori, T., and Yamada, S. (1974b). *Chem. Pharm. Bull.* **22**, 859–863.
Sieber, P., Riniker, B., Brugger, M., Kamber, B., and Rittel, W. (1970). *Helv. Chim. Acta* **53**, 2135–2150.
Sieber, P., Kamber, B., Hartmann, A., Jöhl, A., Riniker, B., and Rittel, W. (1977). *Helv. Chim. Acta* **60**, 27–37.
Stevens, C. L., and Munk, M. E. (1958). *J. Am. Chem. Soc.* **80**, 4069–4071.
Swarc, M. (1968), "Carbanions, Living Polymers and Electron Transfer Processes," Chapter 10, pp. 558–619. Wiley (Interscience), New York.
Tadema, G., Harryvan, E., Panneman, H. J., and Arens, J. F. (1964). *Recl. Trav. Chim. Pays-Bas* **83**, 345–363.
Theodoropoulos, D., and Gazopoulos, J. (1962). *J. Org. Chem.* **27**, 2091–2093.
Tomatis, R., Guggi, A., Salvadori, S., and Rocchi, R. (1977). *Int. J. Pept. Protein Res.* **9**, 137–147.
Ueki, M., Maruyama, H., and Mukaiyama, T. (1971). *Bull. Chem. Soc. Jpn.* **44**, 1108.
Ugi, I. (1962). *Angew. Chem., Int. Ed. Engl.* **1**, 8–21.
Ugi, I., ed. (1971). "Isonitrile Chemistry," Academic Press, New York.
Ugi, I., Arora, A., Burghard, H., Eberle, C., Eckert, H., George, G., Gukel, G., Herlinger, H., von Hinrichs, E., Hoffmann, P., Kleinmann, H., Klusacek, H., Larn, H.-L., Marquarding, D., Nah, H.-S., Offermann, K., Rehn, D., Stüber, S., Tamarnasi, M., Urban, R., Wackerle, L., Zahr, S., and von Zychlinski, H. (1975). *In* "Peptides 1974" (Y. Wolman, ed.), pp. 71–91. Wiley, New York.
Urban, R., Eberle, G., Marquarding, D., Rehn, D., Rehn, H., and Ugi, I. (1976). *Angew. Chem., Int. Ed. Engl.* **15**, 627–628.
van Zon, A., and Beyerman, H.-C. (1976). *Helv. Chim. Acta* **56**, 1729–1740.
Vaughan, J. R. (1951a). *J. Am. Chem. Soc.* **73**, 1389.
Vaughan, J. R. (1951b). *J. Am. Chem. Soc.* **73**, 3547.
Veber, D. F. (1975). *In* "Peptides: Chemistry, Structure and Biology" (R. Walter and J. Meienhofer, eds.), pp. 307–316. Ann Arbor Sci. Publ., Ann Arbor, Michigan.
Weygand, F., Hoffmann, D., and Wünsch, E. (1966). *Z. Naturforsch., Teil B* **21**, 426–428.
Wieland, T., and Bernhard, H. (1951). *Justus Liebigs Ann. Chem.* **572**, 190–194.
Wieland, T., and Heinke, B. (1956). *Justus Liebigs Ann. Chem.* **599**, 70–80.
Wieland, T., and Vogeler, K. (1961). *Angew. Chem.* **73**, 435.
Wolman, Y., Gallop, P. M., and Patchornik, A. (1961). *J. Am. Chem. Soc.* **83**, 1263–1264.

Woodward, R. B., and Olofson, R. A. (1966). *Tetrahedron, Suppl.* **7**, 415–440.
Woodward, R. B., and Woodman, D. J. (1968). *J. Am. Chem. Soc.* **90**, 1371–1372.
Woodward, R. B., Olofson, R. A., and Mayer, H. (1966). *Tetrahedron, Suppl.* **8**, 321–346.
Wünsch, E., ed. (1974). "Houben-Weyl's Methoden der organischen Chemie," 4th ed., Vol. 15, Part II. Thieme, Stuttgart.
Wünsch, E., and Deimer, K.-H. (1972). *Hoppe-Seyler's Z. Physiol. Chem.* **353**, 1246–1254.
Wünsch, E., and Drees, F. (1966). *Chem. Ber.* **99**, 110–120.
Wünsch, E., and Wendlberger, G. (1967). *Chem. Ber.* **100**, 160–172.
Wünsch, E., Zwick, A., and Jaeger, E. (1968). *Chem. Ber.* **101**, 336–340.
Yamazaki, N., and Higashi, F. (1973). *Bull. Chem. Soc. Jpn.* **46**, 3821–3824.
Young, G. T. (1967). *In* "Peptides" (H. C. Beyerman, A. van de Linde, and W. Maassen van den Brink, eds.), pp. 55–66. North-Holland Publ., Amsterdam.
Zahn, H., Meienhofer, J., and Schnabel, E. (1965). *Acta Chim. Acad. Sci. Hung.* **44**, 109–128.
Zaoral, M. (1962). *Collect. Czech. Chem. Commun.* **27**, 1273–1277.
Zaoral, M., and Arnold, Z. (1960). *Tetrahedron Lett.* No. 14, pp. 9–12.

Chapter 3

Active Esters in Peptide Synthesis

MIKLOS BODANSZKY

I. The Development of Active Esters 106
II. Classes of Active Esters 109
A. Alkyl Esters 109
B. Aryl Esters 112
C. Enol Esters 118
D. Active Esters Formed as Intermediates in Reactions with Coupling Reagents 119
E. Hydroxylamine Derivatives 123
F. Insoluble Active Esters 125
III. Preparation of Active Esters 128
A. Synthesis through Activated Derivatives of Protected Amino Acids 128
B. Synthesis through Activated Derivatives of the Hydroxyl Components 132
C. Esterification with the Application of Condensing Agents 133
IV. Catalysis of Active Ester Reactions 138
A. Solvent Effects 139
B. Catalysis with Metals 141
C. Acid–Base Catalysis 141
V. Side Reactions 143
A. Side Reactions in the Preparation of Active Esters . . 143
1. The Consequence of Steric Hindrance 143
2. Side Reactions Caused by Hydroxyl Groups in Amino Acid Side Chains 145
3. Side Reactions Caused by the Guanidine Group of Arginine 147
4. Side Reactions Involving the Carboxamide Group in the Side Chains of Asparagine and Glutamine . . 148

THE PEPTIDES, VOL. 1

ISBN 0-12-304201-1

B. Side Reactions during Acylation with Active Esters . .
1. *O*-Acylation 149
2. Chain Terminating *N*-Acylation 149
3. Formation of Aminosuccinimide Derivatives in the Acylation of Aspartic Acid Containing Peptides with Active Esters 150 151
C. Racemization of Active Esters 152
VI. Active Esters and the Strategy of Peptide Synthesis . . . 156
A. Active Esters in Stepwise Chain Lengthening 157
B. Active Esters in Fragment Condensation 159
C. Active Esters in Cyclization 161
D. Active Esters in Polymerization 163
VII. Active Esters in Various Techniques of Peptide Synthesis . 165
A. Synthesis in Solution 165
1. Introduction of Single Amino Acids into Naturally Occurring Peptides 165
2. Introduction of Protecting Groups with Active Esters 166
3. Introduction of Acetyl Groups, Radioactive and Fluorescent Groups and Biotin into Peptide Chains . 167
4. Crosslinking with Active Esters 169
5. *In Situ* Peptide Synthesis 171
B. Active Esters in Solid-Phase Peptide Synthesis . . . 172
VIII. Conclusions 174
References 187

I. THE DEVELOPMENT OF ACTIVE ESTERS

Formation of the peptide bond under mild conditions requires a derivative of the participating carboxyl group with enhanced reactivity:

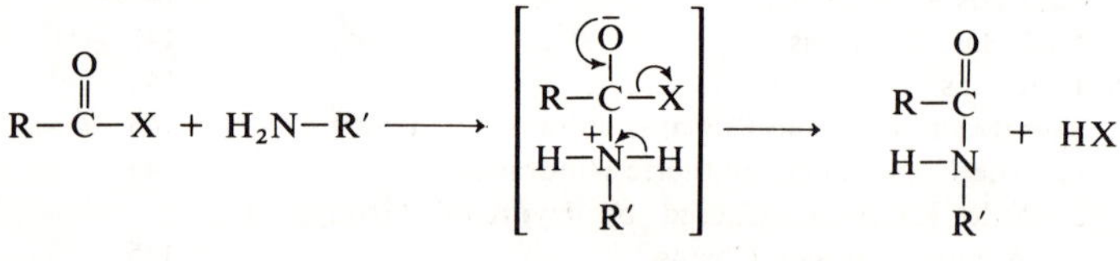

(1)

Simple alkyl esters of carboxylic acids can be reactive enough to acylate amines; e.g., the preparation of nicotinic acid amide from the methyl ester and ammonia proceeds smoothly at room temperature:

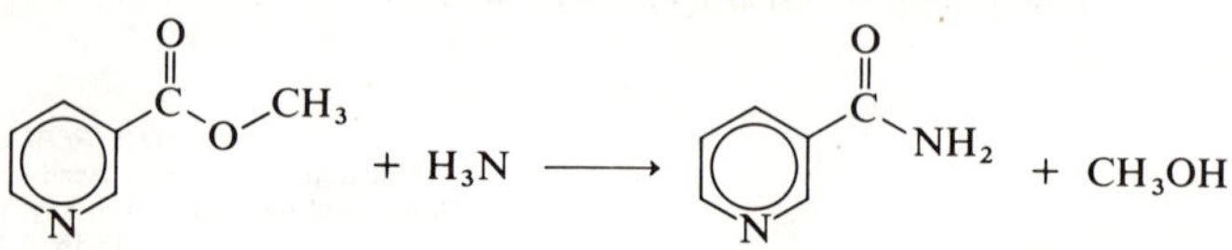

(2)

In this case, the electron-withdrawing effect of the aromatic system enhances the reactivity of the ester group. However, alkyl esters of protected amino acids react too slowly for practical application. In fact, methyl, ethyl, or benzyl esters of protected amino acids or peptides are sufficiently stable toward amines to be used as carboxyl protecting groups. It is obvious why Emil Fischer (1902), in the synthesis of ethyloxycarbonylglycyl-glycyl-leucine ethyl ester from protected diglycine ethyl ester and leucine ethyl ester (reaction 3), could achieve only moderate success. The self-condensation of

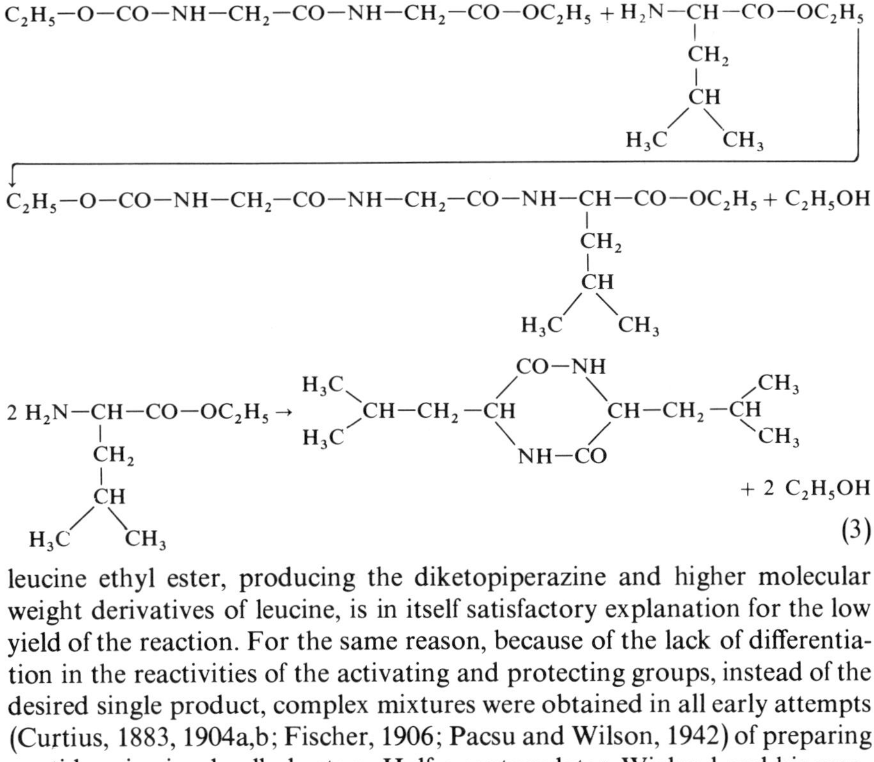

(3)

leucine ethyl ester, producing the diketopiperazine and higher molecular weight derivatives of leucine, is in itself satisfactory explanation for the low yield of the reaction. For the same reason, because of the lack of differentiation in the reactivities of the activating and protecting groups, instead of the desired single product, complex mixtures were obtained in all early attempts (Curtius, 1883, 1904a,b; Fischer, 1906; Pacsu and Wilson, 1942) of preparing peptides via simple alkyl esters. Half a century later, Wieland and his associates (1951) applied thiophenyl esters (compound **1**), and the group led by Schwyzer (Schwyzer *et al.*, 1955a) introduced negatively substituted methyl esters, such as the cyanomethyl esters of protected amino acids (compound **2**). Both types of esters showed greatly increased reactivity in aminolysis, as compared with unsubstituted alkyl esters, and are therefore the forerunners

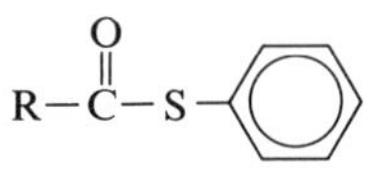

(1)

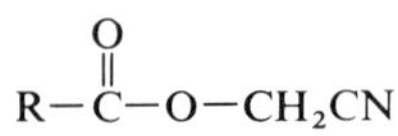

(2)

of the long series of *active esters* that were proposed and often applied for the formation of the peptide bond.

The fundamental studies of Gordon *et al.* (1948, 1949) revealed that the ammonolysis and aminolysis rates of esters are related to the electron-withdrawing properties of the acid or the alcohol component of the ester. Withdrawal of electrons enhances the electrophilic character of the carbonyl carbon, and thereby facilitates the formation of the tetrahedral intermediate (cf. reaction 1) with the nucleophile (the amine). In the experiments of the same authors, vinyl esters and phenyl esters exhibited extreme reactivity toward ammonia. The recognition by the present author (Bodanszky, 1955, 1957a) that the reactivity of thiophenyl esters (Wieland *et al.*, 1951) (compound **1**) is due less to their being derivatives of thiols and more to the fact that they are phenyl esters, prompted investigations that led to the introduction of nitrophenyl esters of protected amino acids (Bodanszky, 1955) in peptide synthesis. The enhancement of the electron-withdrawing effect of the aromatic nucleus by the strong negative effect of one or two nitro groups rendered these compounds practical tools that were applied in the synthesis of many important peptides. Their general usefulness was first demonstrated in a synthesis of oxytocin (Bodanszky and du Vigneaud, 1959a,b).

The practical results achieved with active esters, particularly with *p*-nitrophenyl esters, stimulated further research in this area. Among the numerous contributions proposing aryl esters with different negative substituents, the thorough, comparative studies of Pless and Boissonnas (1963), resulting in the introduction of 2,4,5-trichlorophenyl esters, deserve special mention. A new chapter was opened in the development of active esters by a paper of Nefkens and Tesser (1961) in which they propose the use of the esters of protected amino acids with *N*-hydroxyphthalimide (compound **3**),

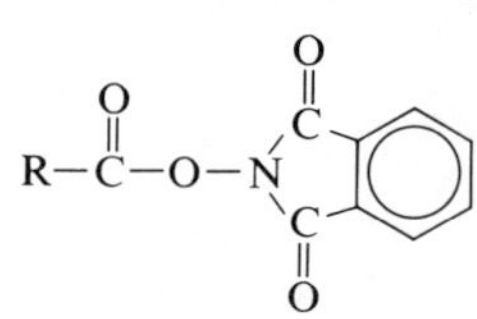

(3)

that could be regarded as mixed anhydrides of the (protected) amino acid and a hydroxamic acid. A long series of *O*-acyl derivatives of substituted hydroxylamine was subsequently recommended for peptide synthesis. Some of these were favorably received and often applied; they will be discussed in Section II.

The reactivity of vinyl esters explains that *enol* esters are powerful *N*-acylating agents. Several coupling reagents (cf. Klausner and Bodanszky, 1972) react with the carboxyl component in the peptide bond-forming reaction, usually by addition, to generate an enol ester as the reactive intermediate. An outstanding example is the reagent developed by Woodward *et al.* (1961) (reaction 4) which activates the carboxyl of protected amino acids or peptides by producing an enol ester.

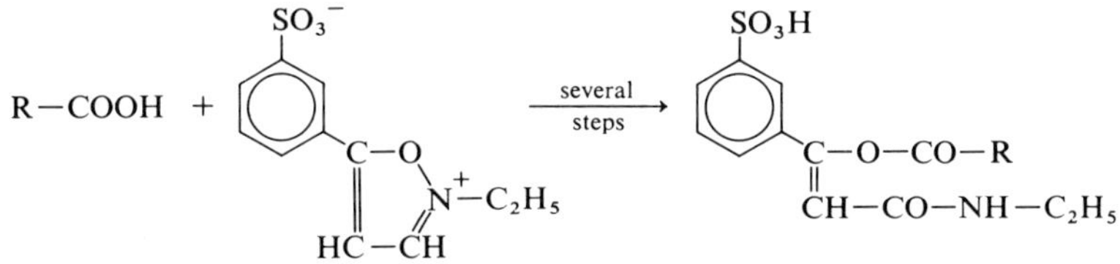

(4)

A very interesting and perhaps not yet fully exploited method in active ester reactions is the application of insoluble active esters (Fridkin *et al.*, 1965; Wieland and Birr, 1966). These permit the use of excess acylating agent without causing any problem in removing and even recovering the unreacted excess.

The advantages of active ester reactions, e.g., in having isolated stable intermediates (commercially) available which may be stored for ready use, can be offset by the relatively moderate rate of the acylation reaction. This circumstance stimulated the development of highly reactive esters on one hand and the acceleration of the aminolysis by the addition of catalysts on the other. At this time, the most promising catalyst is 1-hydroxybenzotriazole, recommended by König and Geiger (1973). Catalysis of the acylation reaction might render active esters more useful in solid-phase synthesis, where they have found only limited application, thus far.

II. CLASSES OF ACTIVE ESTERS

A. Alkyl Esters

The moderate reactivity in aminolysis of methyl and ethyl esters of protected amino acids or peptides seems to preclude their practical application in coupling. Studies with Cu^{2+} and Co^{3+} complexes of glycine esters and ethylenediamine or diethylenetriamine (Yamada *et al.*, 1970; Wu and Busch, 1972) suggest that conditions can be found under which directed synthesis of peptides from alkyl esters might be possible.

With strong nucleophiles, such as hydrazine or hydroxylamine, otherwise almost inert methyl and ethyl esters are sufficiently reactive to form an amide bond. This is indeed one of the commonly used routes to the preparation of peptide hydrazides. Similarly, ammonolysis of lower alkyl esters is often applied in the synthesis of peptide amides (e.g., du Vigneaud *et al.*,

1953). In these reactions, the nucleophiles, hydrazine, or ammonia are applied in excess and in high concentration to compensate for the poor reactivity of the alkyl esters.

Lactones, prepared by intramolecular esterification of β-hydroxyamino acids, because of the strain in the 4-membered ring, are more reactive than simple alkyl esters. Thus, the β-lactone (reaction 5) was prepared from *N*-

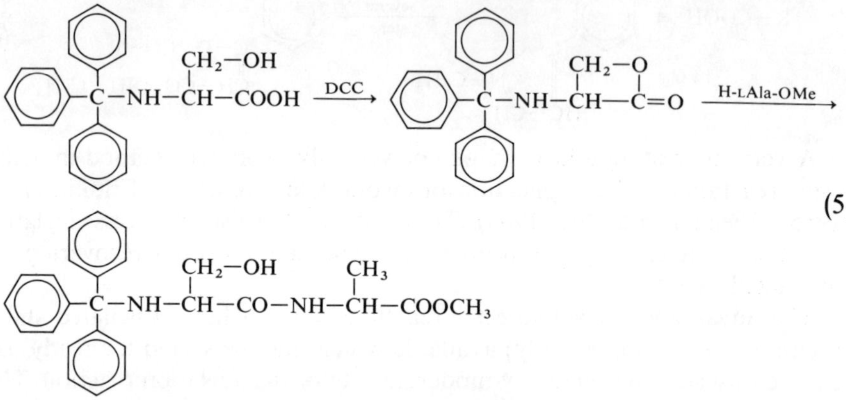

(5)

trityl-L-serine with dicyclohexylcarbodiimide as condensing agent (Sheehan *et al.*, 1959) and applied to the acylation of an amino acid ester. The corresponding benzyloxycarbonyl derivative of L-serine was obtained in excellent yield when the condensation with dicyclohexylcarbodiimide was catalyzed with 1-hydroxybenzotriazole (König and Geiger, 1970a). In γ-lactones, the 5-membered ring is too stable to be readily opened by nucleophilic attack. Nevertheless, the COOH-terminal homoserine lactone moiety formed from methionine-containing peptides during cleavage with CNBr (Gross and Witkop, 1961) can acylate the newly produced amino group of the residue that follows the original methionine, if the reacting groups are held in proximity, e.g., by a disulfide bridge (reaction 6). Such an *intramolecular* acylation with homoserine lactone was observed by Dyckes *et al.* (1974).

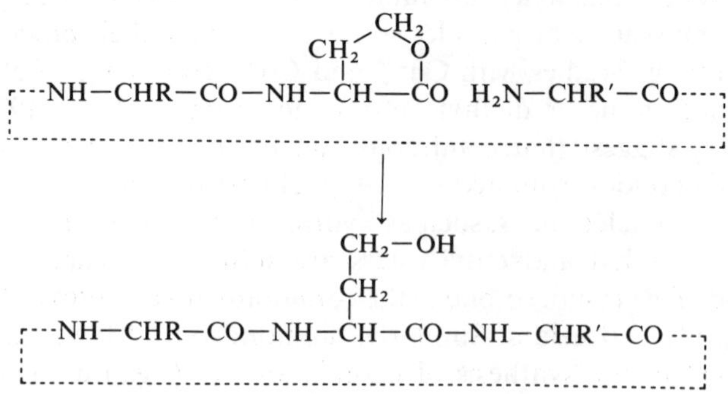

(6)

These examples, however, represent only exceptions. A more general application of alkyl esters requires activation through substitution of the alkyl group with electron-withdrawing substituents. For a strong negative inductive effect, the substituents must be near to the carbonyl group; hence, substituted *methyl* esters are *a priori* the most promising derivatives. Such considerations led Schwyzer and his associates (1955a,b,c) to investigate the applicability of a series of substituted methyl esters in the formation of the peptide bond. For a comparison of the effectiveness of different negative substituents, the yields of the reactions between various esters of hippuric acid and benzylamine (reaction 7) were determined.

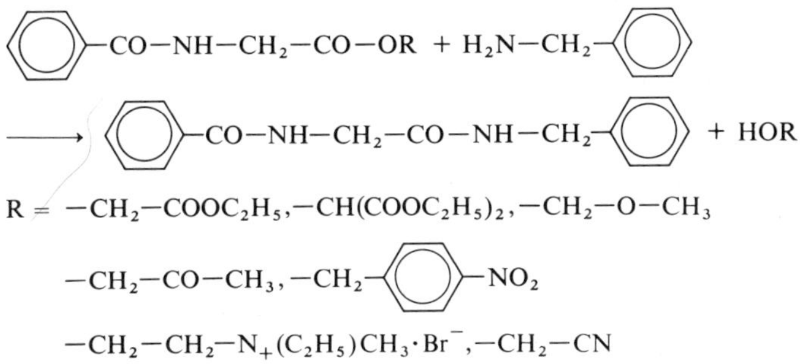

(7)

The cyanomethyl ester ($R = -CH_2-CN$) gave the best results. This ester was readily prepared from the triethylammonium salt of benzoylglycine and chloroacetonitrile:

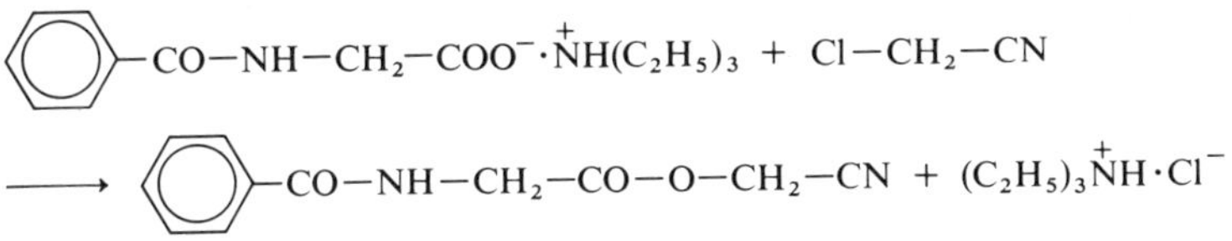

(8)

The same approach was then applied for the synthesis of benzyloxycarbonylamino acid cyanomethyl esters (Schwyzer *et al.*, 1956) as well. These activated esters were used for the preparation of dipeptides (Schwyzer and Iselin, 1955) and of a tripeptide derivative with the NH_2-terminal sequence of oxytocin: *N*-benzyloxycarbonyl-*S*-benzyl-L-cysteinyl-L-tyrosyl-L-isoleucine (Iselin *et al.*, 1955).

Some additional negatively substituted methyl esters, e.g., propargyl esters (compound **4**) (Bodanszky, 1957b) and esters of glycolic acid amide (compound **5**) (Goodman and Steuben, 1959), were briefly investigated, but were not applied in practical peptide synthesis. Of the activated alkyl esters, only the cyanomethyl esters gained popularity, and even this was limited to a brief period.

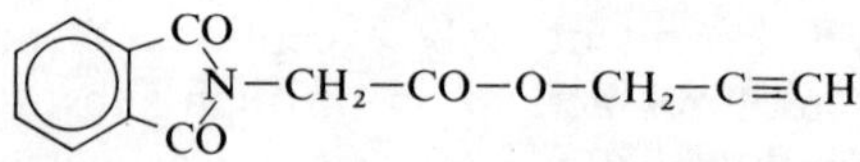

(4)

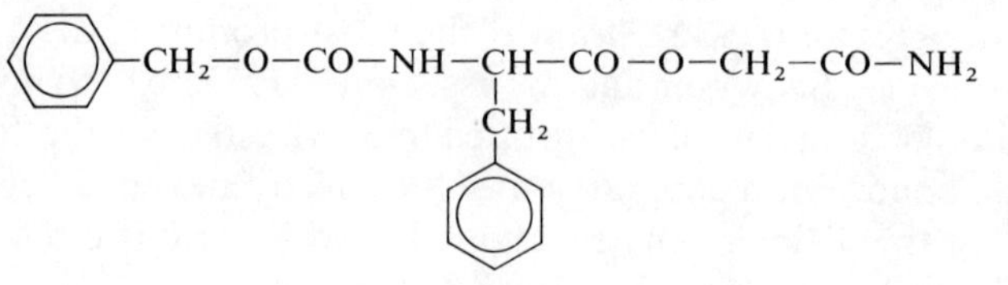

(5)

B. Aryl Esters

A reactive thiol ester grouping was identified in coenzyme A, an important intermediate of many biochemical acylation reactions (Lynen and Reichert, 1951). The structure of acetylcoenzyme A (compound **6**) suggests that the

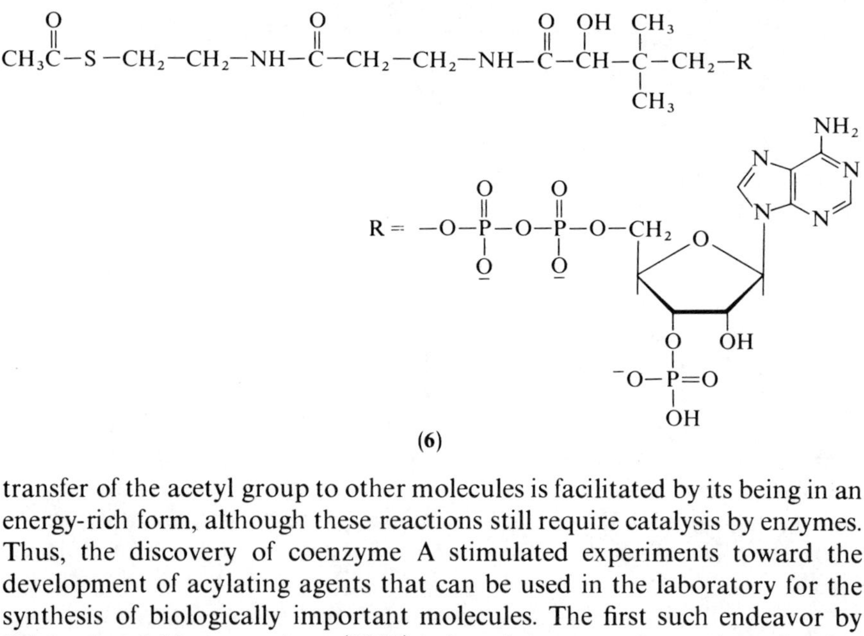

(6)

transfer of the acetyl group to other molecules is facilitated by its being in an energy-rich form, although these reactions still require catalysis by enzymes. Thus, the discovery of coenzyme A stimulated experiments toward the development of acylating agents that can be used in the laboratory for the synthesis of biologically important molecules. The first such endeavor by Wieland and his co-workers (1951) led to the preparation and application (Wieland and Schäfer, 1952; Wieland and Bernhard, 1953) of thiophenyl esters (compound **1**) of protected amino acids, and also to studies of S → N acyl migration in derivatives of cysteamine (Wieland and Bokelmann, 1952). Similar considerations prompted Schwyzer (1953) to investigate the esters of aliphatic thiols with protected amino acids, particularly the *S*-esters of thioglycolic acid (compound **7**), as potential acylating agents in peptide synthesis.

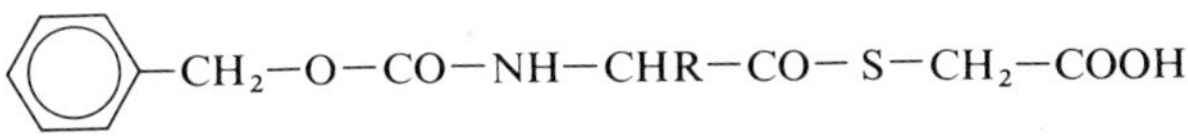

(7)

In the light of the already mentioned studies of Gordon *et al.* (1948), the electron-withdrawing carboxyl group in compound (**7**) activates the methyl group of the alcohol component, thioglycolic acid. Similarly, in thiophenyl esters (compound **1**), the aromatic nucleus itself is responsible for the withdrawal of electrons from the carbon atom of the ester carbonyl. In the enhancement of the electrophilic character of this carbon atom, the role of the sulfur in thiol esters is probably not more than that of a mediator. With its larger and therefore more readily distorted electron shell, it can convey electronic effects from the alcohol component to the carbonyl group more efficiently than oxygen in common esters. The recognition that the principal role in activation is played by the aromatic ring and not by the sulfur atom prompted the present author (Bodanszky, 1955) to examine substituted phenyl esters of protected amino acids, in which negative substituents were chosen to further enhance the electron-withdrawing effect of the phenyl group.* From the four nitrophenyl esters (compounds **8**–**11**) considered, the

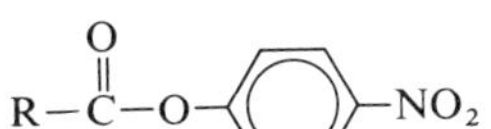

(**8**)

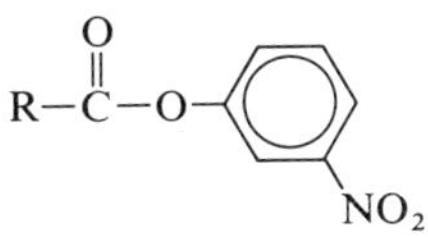

(**9**)

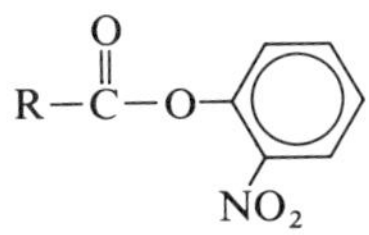

(**10**)

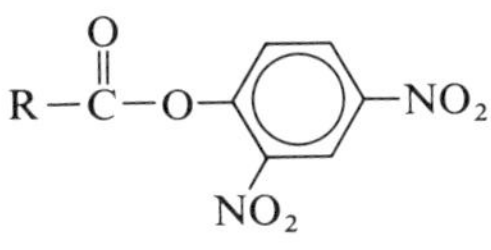

(**11**)

p-nitrophenyl esters were selected for practical application because, in preliminary experiments, no major differences were observed in the acylation rates of the monosubstituted esters (compounds **8**–**10**), and the paraderivatives (compound **10**) could be more readily secured in crystalline form. The

* Without electron-withdrawing substituents, phenyl esters (Wieland and Schäfer, 1951; Wieland and Jaenicke, 1956) are only moderately reactive. This is shown by the fact that their reactivity was overlooked by Karrer and Heynemann (1948), who first prepared phenyl esters of acylamino acids, and also by their usefulness in the *protection* of the carboxyl function (Kenner and Seely, 1972; cf. also Kenner, 1959) during chain lengthening.

2,4-dinitrophenyl esters (compound **11**) seemed to be excessively reactive and hence also too sensitive to hydrolysis. In retrospect, it seems that *o*-nitrophenyl esters would have been a more judicious choice. The advantages of *o*-nitrophenyl esters over the para derivatives became clear only when their relative reactivities were reexamined (Bodanszky and Bath, 1969; Bodanszky *et al.*, 1972a) under the conditions of solid-phase peptide synthesis (Merrifield, 1963). This problem will be discussed in the section dealing with the application of active esters in synthesis carried out on a polymeric support. The new active ester method proved to be a practical procedure and, within a few years, *p*-nitrophenyl esters of protected amino acids became popular tools in peptide synthesis. Their acceptance was due in part to the high yields achieved and the good quality of the products obtained through them (e.g., Bodanszky and du Vigneaud, 1959a,b), and also to their stability that allowed the preparation and storage of active esters on a large scale. Because of this stability, *p*-nitrophenyl esters were the first reactive derivatives of protected amino acids that could be made commercially available.

A combination of the idea of negatively substituted phenyl esters with the thiol ester concept is represented by the *p*-nitrothiophenyl esters (compound **12**) proposed by Farrington *et al.* (1955; cf. also Farrington *et al.*, 1957) and

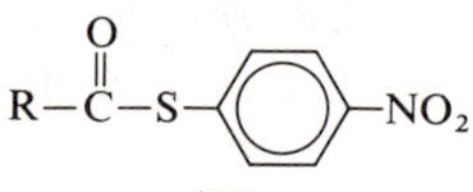

(12)

Kenner *et al.* (1958). These are very reactive acylating agents, but this advantage is somewhat offset by the unpleasant properties of the *p*-nitrothiophenol produced in their aminolysis reactions. Similar disadvantages may have caused limitations in the acceptance of esters of selenophenol (compound **13**), selenonaphthol (compound **14**), and 8-mercaptoquinoline (compound **15**), all introduced by Jakubke (1963, 1964, 1965a,b, 1966), and of the powerful derivatives of 2-mercaptopyridine (compound **16**) recommended by Lloyd and Young (1968, 1971). This assump-

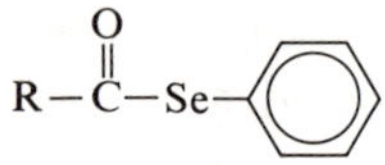

(13)

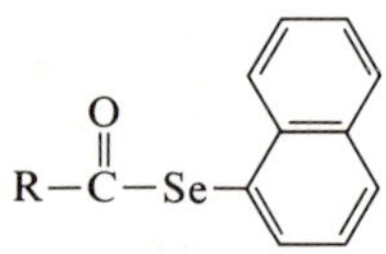

(14)

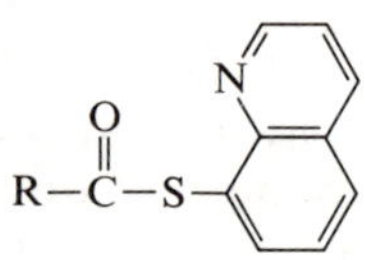

(15)

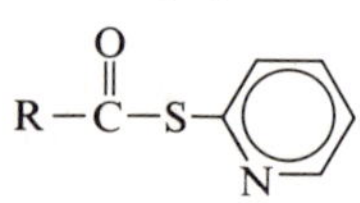

(16)

tion is supported by the practical application of the esters of 2-hydroxypyridine, the oxygen analogues of compound (**16**) (Dutta and Morley, 1971). The probably less reactive 3-hydroxypyridine esters were recommended earlier (Taschner *et al.*, 1965a,b,c).

The usefulness of *p*-nitrophenyl esters stimulated further experiments with aryl esters carrying various negative substituents other than the nitro group. Among the first attempts in this direction are the 2,4,6-trichlorophenyl esters (compound **17**) and the pentachlorophenyl esters (compound **18**) of Kupryszewszki (1961a; cf. also Kupryszewszki and Kaczmarek, 1961). An obvious

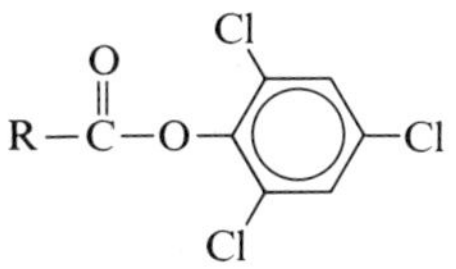

(**17**)

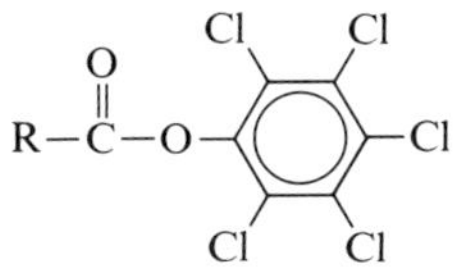

(**18**)

advantage of chlorine as a substituent over the nitro group is its relative resistance to catalytic hydrogenation. An extensive and systematic investigation of substituted aryl esters of protected amino acids was then carried out by Pless and Boissonnas (1963). Their studies proved that the reactivities of aryl esters can be correlated with the acidity of their phenol components: both properties are influenced by the electron-withdrawing or electron-releasing effects of the substituents. Yet, the electronic effects can be canceled out by steric hindrance caused by the substituents, particularly if they are in ortho position to the phenolic hydroxyl. These investigations led to the selection of the esters of 2,4,5-trichlorophenol (compound **19**). At this time, next to nitrophenyl esters, 2,4,5-trichlorophenyl esters are among the most often applied active esters in peptide synthesis.

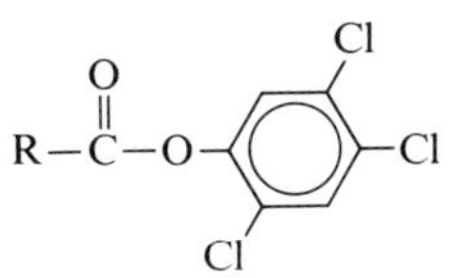

(**19**)

The steric hindrance in pentachlorophenyl esters (compound **18**) is due to *two* substituents in ortho position. In this respect, the 2,6-dichloro-4-nitrophenyl esters proposed by Stewart (1968) are probably not better. A major improvement can be recognized in the pentafluorophenyl esters (compound **20**) of Kisfaludy and his associates (1967, 1970; cf. also Kovacs *et al.*, 1967c). In these compounds, the ortho substituents cause no major steric hindrance, since the van der Waals radius of fluorine is much smaller than that of chlorine, in fact, only slightly larger than the radius of hydrogen.

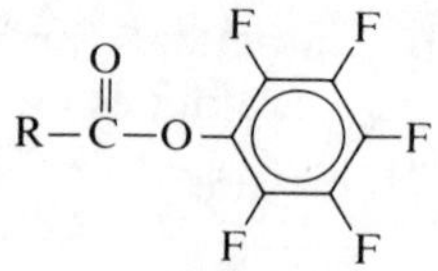

(20)

Since the fluorines provide strong, negative effect without causing bulkiness, the pentafluorophenyl esters are efficient even under crowded conditions, such as those found in solid-phase peptide synthesis. The successful use of these esters in rapid syntheses of oxytocin, angiotensin, etc. (e.g., Kisfaludy *et al.*, 1974) augurs well for their future application. Some concern might be felt on account of aggressive, irritant nature of pentafluorophenol released in the coupling reactions. In case of problems caused by "overactivation" (Brenner, 1967), less extensively substituted fluorophenyl esters, e.g., derivatives of a trifluorophenol, could be considered.

The aromatic esters discussed so far were activated with nitro groups and/or halogens as electron-withdrawing substituents. Similar enhancement of reactivity could be achieved with the cyano group (compound **21**) (Iselin *et al.*, 1957); the propionyl group (compound **22**) (Kaczmarek *et al.*, 1967); a trimethylammonium substituent (compound **23**) (Mitin and Nadezhdina, 1968); the sulfonamide groups (compound **24**) (Kupryszewszki and Muzalewski, 1967) and compound (**25**) (Kupryszewszki *et al.*, 1968); the phenylazo group (compound **26**) (Barth and Losse, 1964); and a combination of two substituents in the *o*-nitro-*p*-sulfophenyl esters (compound **26a**) (Y. S. Klausner and T. H. Meiri, personal communication, 1977). The latter are water-soluble and permitted coupling with di- and tripeptide esters in aqueous solution.

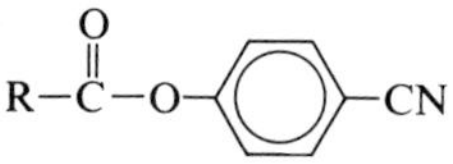

(21)

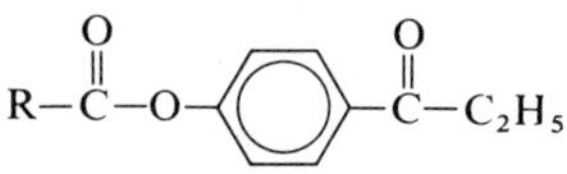

(22)

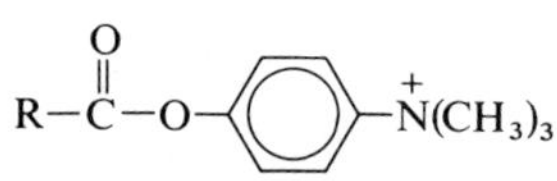

(23)

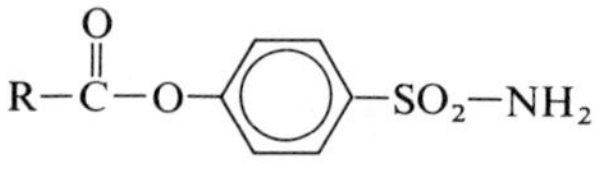

(24)

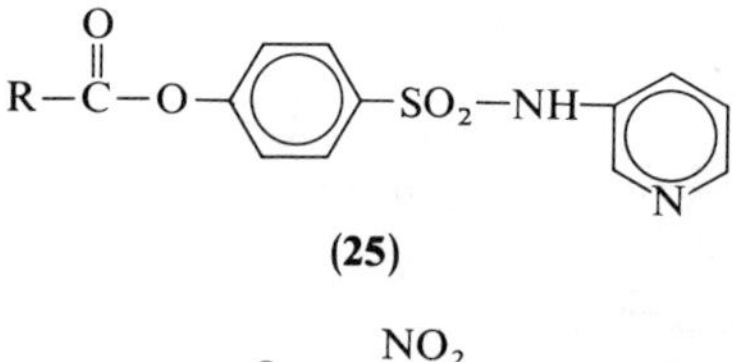

(25)

(26)

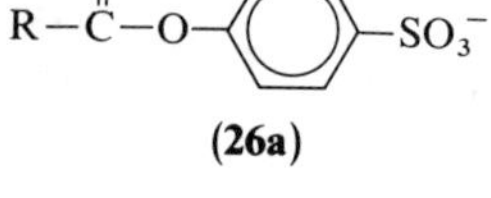

(26a)

Two aryl ester groups deserve special attention because they can be prepared through fairly inert intermediates. This permits the use of the intermediates as temporarily protected amino components in synthesis. Conversion to active esters can then take place at a preselected stage of the synthesis. The catechol esters of Jones and Young (1968b) are obtained through the monobenzyl derivatives (compound **27**). Catalytic hydrogenation or acidolysis converts them (reaction 9) to the active ester form (compound **28**):

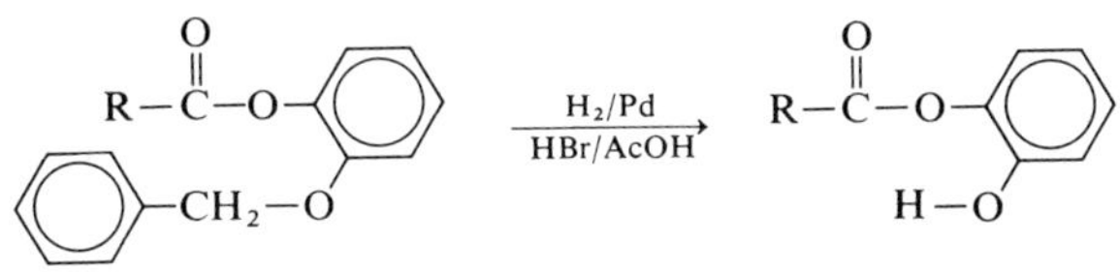

(9)

(27) **(28)**

The second hydroxyl can be temporarily protected by the phenacyl group as well (compound **29**) (Trudelle, 1971):

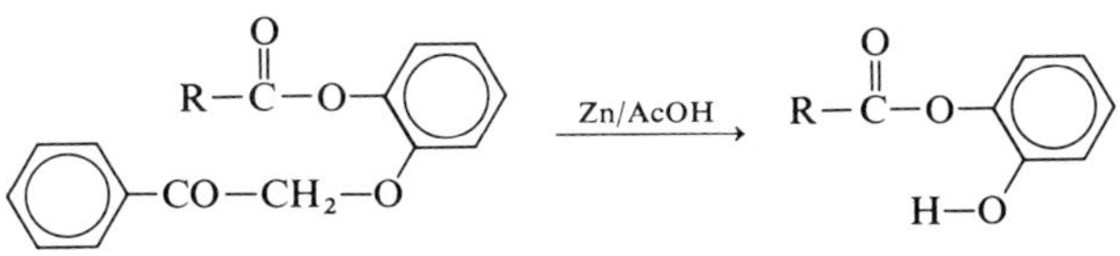

(10)

(29) **(28)**

Aryl esters with a sulfone as activating group were introduced by Iselin *et al.* (1957). Subsequently, Johnson and Jacobs (1968; cf. also Johnson and Ruettinger, 1970) proposed that these esters be prepared via the corresponding sulfide (compound **30**) that is activated by oxidation (reaction 11) to the sulfone (compound **31**):

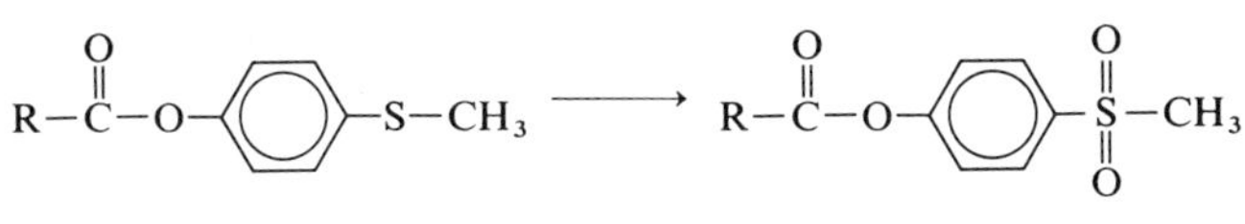

(11)

(30) **(31)**

In this connection it should be mentioned that simple, unsubstituted phenyl esters, once considered as activated derivatives of protected amino acids (Wieland and Schäfer, 1951), were shown, more recently, to be useful as protecting groups of the COOH-terminal carboxyl of peptides (Kenner and Seely, 1972; cf. also Kenner, 1959).

The aromatic systems in the alcohol components need not be limited to benzene, pyridine, or quinoline. This is shown by the esters of 1-phenyl-3-methyl-5-hydroxypyrazole (compound **32**) introduced by Losse and his coworkers (1965).

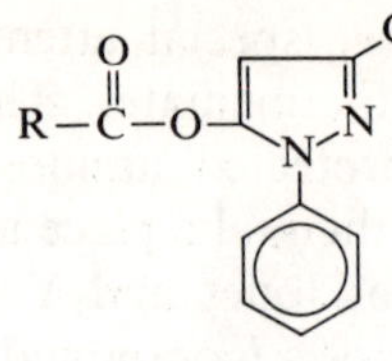

(32)

C. Enol Esters

Vinyl alcohol is the enol form of acetaldehyde. The exceptional reactivity of vinyl esters observed by Gordon *et al.* (1948, 1949) can be readily explained by the electron-withdrawing effect of the vinyl group. The first application of this principle by Weygand and Steglich (1961) demonstrated its potential usefulness in peptide synthesis. A practical preparation of vinyl esters involves the $PdCl_2$-catalyzed transesterification of protected amino acids with vinyl acetate:

$$R{-}COOH + CH_3{-}CO{-}O{-}CH{=}CH_2 \xrightleftharpoons{PdCl_2} R{-}CO{-}O{-}CH{=}CH_2 + CH_3COOH \quad (12)$$

Substituted vinyl esters were studied by Wieland and his associates (1970, 1972). The intermediate tertiary alcohols are inert enough to play the role of protected derivatives. Activation occurs through acid-catalyzed elimination of water (reaction 13):

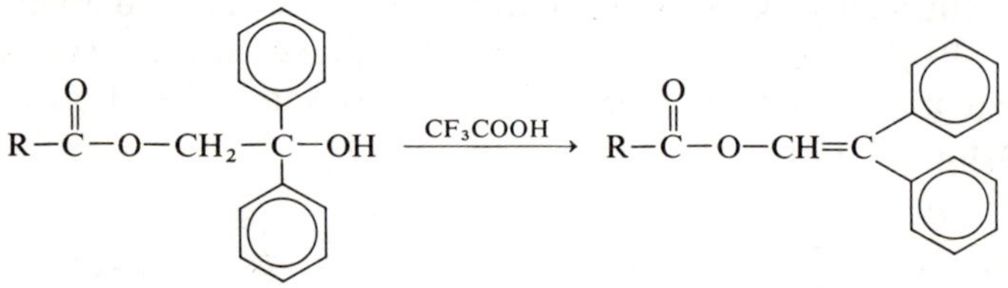

(13)

The well-known protection of carboxyl groups through their phenacyl esters can be activated in an analogous manner (Treiber, 1973). Elimination of water from the cyanohydrin intermediate results in a substituted vinyl ester (reaction 14).

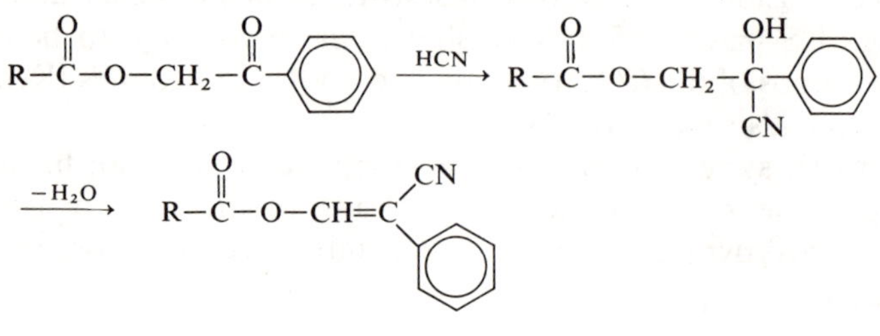

(14)

The esters of 1-phenyl-3-methyl-5-hydroxypyrazole (compound **32**) already mentioned among aryl esters can be considered equally well as enol esters of 1-phenyl-3-methyl-5-pyrazolone (compound **33**). It is probably an

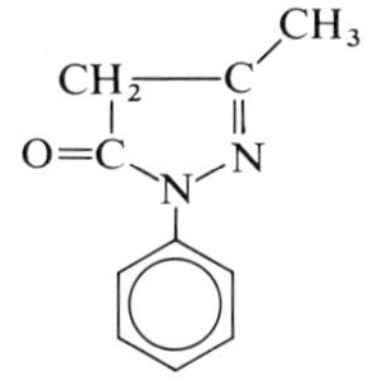

(33)

advantage that the keto form of this compound is not a true ketone but rather an amide. Thus, it is unlikely that the carbonyl derivative eliminated in the acylation reaction would react with the amino group of the still unreacted amino component. Such an undesired interference seems to be likely with enol esters derived from 2,4-pentanedione (compound **34**)

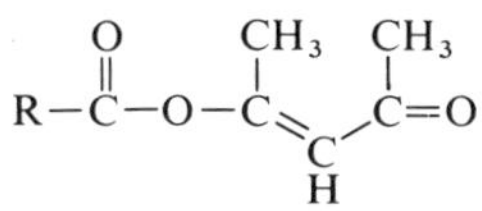

(34)

(Kupryszewski, 1961b). Here the ketone (acetylacetone) liberated in the peptide bond-forming step might condense with amino acid esters or other derivatives of amino acids or peptides with a free terminal amino group to produce an H-bond stabilized enamine (compound **35**) (Dane *et al.*, 1962).

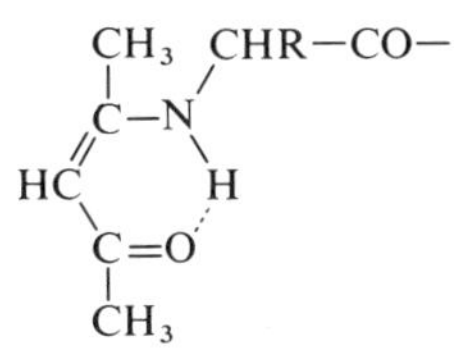

(35)

D. Active Esters Formed as Intermediates in Reactions with Coupling Reagents

Soon after the breakthrough in peptide synthesis resulting from the introduction of mixed anhydrides and also at the time of the emergence of active ester procedures, two condensing agents or coupling reagents were proposed almost simultaneously: ethoxyacetylene (compound **36**) by Arens (1955; cf. also Panneman *et al.*, 1959) and dicyclohexylcarbodiimide (compound **37**),

$CH_3{-}CH_2{-}O{-}C{\equiv}CH$

(36)

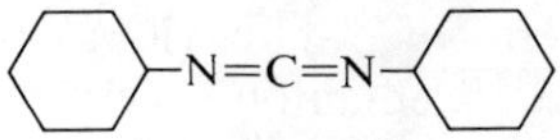

(37)

DCC, by Sheehan and Hess (1955). Addition of protected amino acids to ethoxyacetylene (**36**) generates substituted vinyl esters that readily undergo aminolysis with the liberation of (the enol form of) ethyl acetate:

$$R{-}COOH + HC{\equiv}C{-}O{-}C_2H_5 \longrightarrow R{-}\overset{\overset{O}{\|}}{C}{-}O{-}\overset{\overset{CH_2}{\|}}{C}{-}OC_2H_5$$

$$\xrightarrow{H_2NR'} R{-}\overset{\overset{O}{\|}}{C}{-}NH{-}R' + [CH_2{=}\overset{\overset{O-H}{|}}{C}{-}OC_2H_5] \downarrow CH_3{-}\overset{\overset{O}{\|}}{C}{-}OC_2H_5 \qquad (15)$$

Two alternative routes were proposed (Khorana, 1955) for the reactions of the active ester intermediates derived from DCC (compound **37**). The *O*-acylisourea type addition products of protected amino acids react with the amino component to form the desired amides, but also with the still available carboxyl component to form symmetrical anhydrides, which in turn acylate the amino components to produce the same amides. In either case, elimination of a substituted isourea occurs, that immediately rearranges to *N,N'*-dicyclohexylurea (DCU, compound **38**):

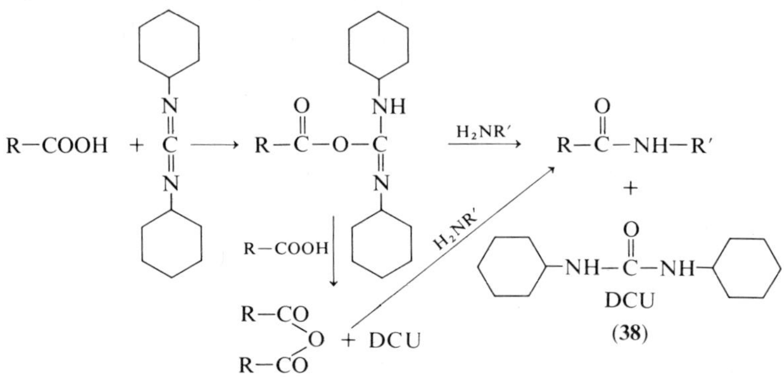

(16)

Beyond doubt, DCC is the most often applied reagent in peptide synthesis. Therefore it is not surprising that it stimulated the development of novel coupling reagents. Some of these form addition products with protected amino acids that show structural similarity with the *O*-acylisoureas: they contain the same partial structure (compound **39**) responsible for the withdrawal of electrons from the carbonyl carbon of the

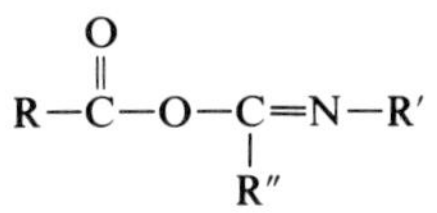

(39)

carboxyl component. The addition products of ketenimines, recommended by Stevens and Munk (1958), are good examples for this analogy:

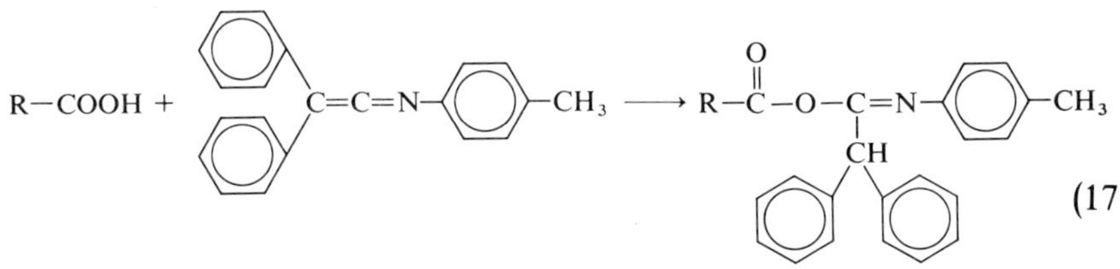

(17)

A similar arrangement and activation can be recognized in azlactones,* e.g., from benzoylamino acids (compound **40**). These too are good acylating agents.

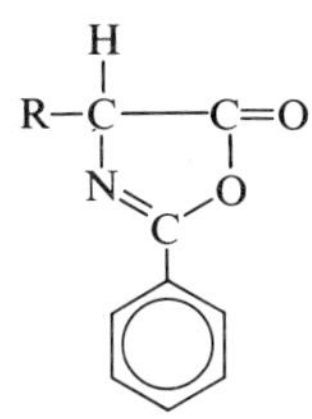

(40)

The addition products of protected amino acids and the ynamines proposed by Buijle and Viehe (1964) are enol esters of amides (reaction 18):

* Nomenclature according to "Chem. Abstr. 1977 Index Guide," pp. 836G, 837G. Example: Structure (**a**) is designated as 2-methyl-4-(phenylmethyl)-5(4H)-oxazolone in *Chem. Abstr.* **83**, 3633CS (1975) *Chemical Substance Index*,

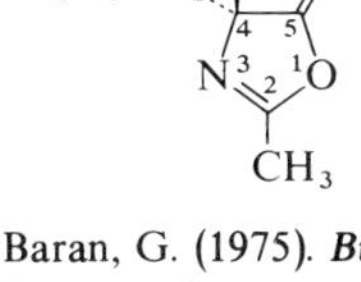

(**a**) (derived from N^{α}-acetyl-L-phenylalanine)

as a reference to Siemion, I. Z., and Baran, G. (1975). *Bull. Acad. Pol. Sci., Ser. Sci. Chim.* **23**, 317–322: "Conformational investigation on azlactones of aromatic *N*-acylamino acids." The term 5(4H)-oxazolone replaces all earlier terms, such as azlactone, 4,5-dihydro-1,3-oxazol-5-one, oxazolinone, 1,3-oxazoline-5-one, 2-oxazoline-5-one, etc., whose use should be discontinued. In this chapter the term azlactone will be used.

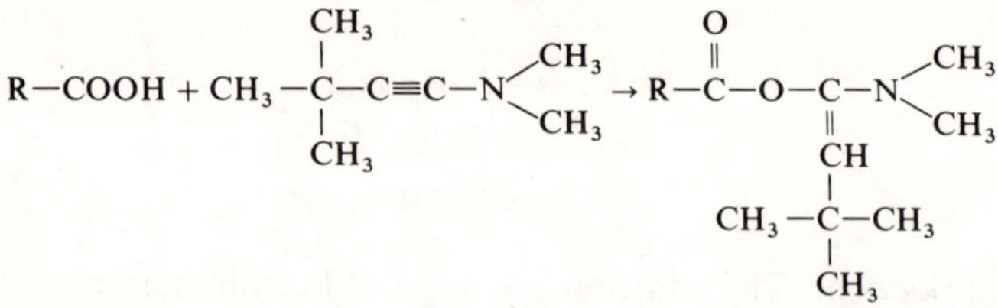

(18)

The same is true for Woodward's reagent K (compound **41**) (Woodward *et al.*, 1961). Here the enol esters of an amide are generated through the addition of the carboxyl components to a ketenimine. The final intermediates, however, that participate in the acylation of the amino group, are enol esters with a substituted acetophenone as the alcohol component (Scheme 1):

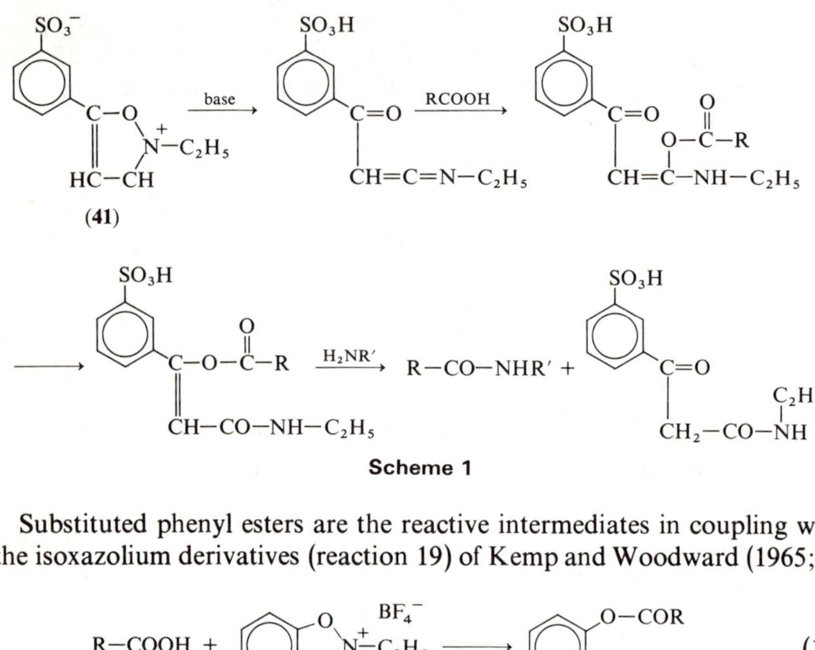

Scheme 1

Substituted phenyl esters are the reactive intermediates in coupling with the isoxazolium derivatives (reaction 19) of Kemp and Woodward (1965; cf.

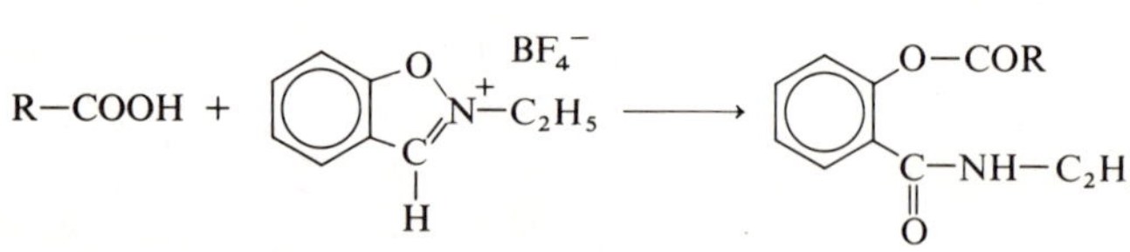

(19)

also Kemp, 1967) and also with the very promising reagent 2-ethyl-7-hydroxybenzo-1,2-oxazolium tetrafluoroborate (Kemp and Chien, 1967), in which the acyl groups migrate from the original point of attachment to the hydroxyl in ortho position:

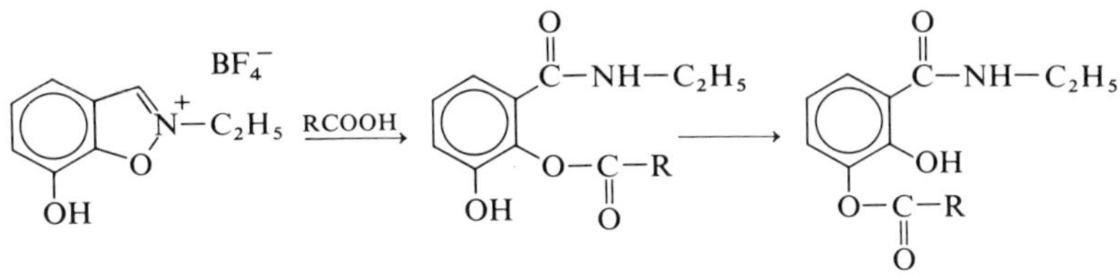

(20)

One of the two insertion methods of Brenner and his co-workers (1955, 1957) should be mentioned here, although it involves no coupling reagent; i.e., the reactive intermediates are esters of the substituted phenol, salicylamide (reaction 21).

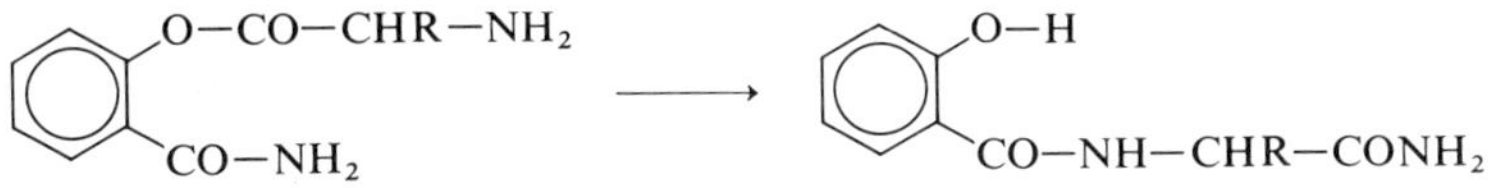

(21)

E. Hydroxylamine Derivatives

The development of active esters took a new turn with the remarkable studies of Nefkens and Tesser (1961) that resulted in the introduction of the potent *O*-acyl derivatives of *N*-hydroxyphthalimide (compound **42**):

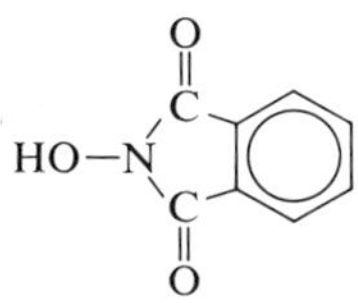

(42)

This was followed by a series of related methods, the most popular of these being the *N*-hydroxysuccinimide ester procedure (Anderson *et al.*, 1963, 1964), which has the advantage of a water-soluble by-product, *N*-hydroxysuccinimide:

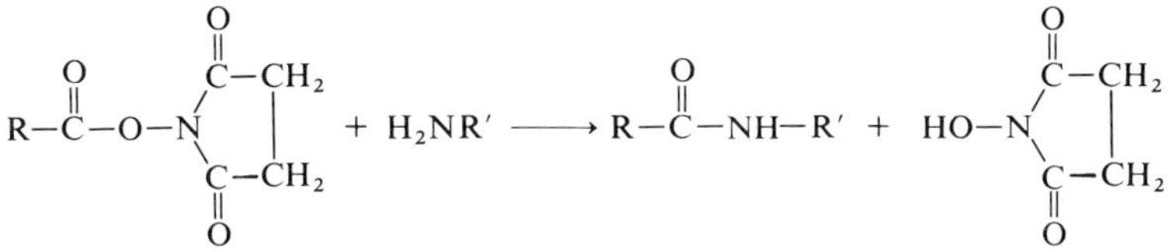

(22)

From other proposed *N*-acylhydroxylamine esters, we mention here the derivatives of *N*-hydroxyglutarimide (compound **43**) (Jeschkeit, 1968), *N*-ethoxycarbonylhydroxylamine (compound **44**) (Jeschkeit, 1969), *N*-*p*-chlorobenzoylhydroxylamine (compound **45**) (Govindachari *et al.*, 1966), *N*-benzoylhydroxylamine (compound **46**) (Taschner *et al.*, 1967; cf. also

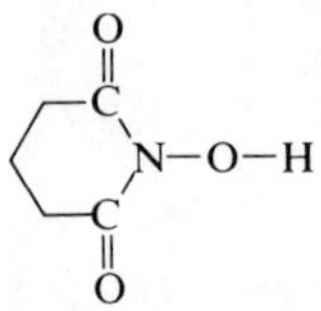

(43)

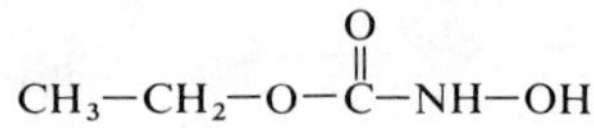

(44)

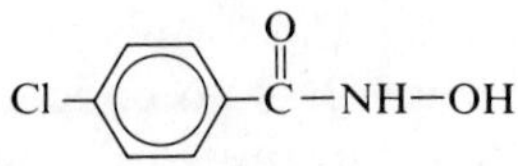

(45)

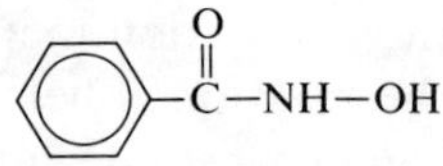

(46)

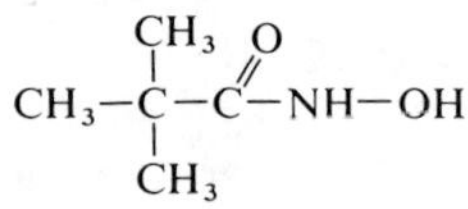

(47)

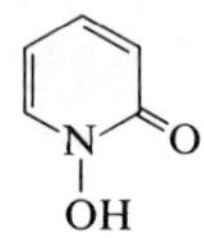

(48)

Lubiewska-Nakonieczna *et al.*, 1970), *N*-trimethylacetylhydroxylamine (compound **47**) (Rajappa *et al.*, 1967), and 1-hydroxy-2-pyridone (compound **48**) (Paquette, 1965; Sarantakis *et al.*, 1968; Taylor *et al.*, 1970):

N-acylhydroxylamines are indeed acidic compounds, called appropriately "hydroxamic acids." Hence their *O*-acyl derivatives should be considered mixed anhydrides rather than esters. Yet the reactivity of this class of compounds is due more to their being derivatives of hydroxylamine than to their mixed anhydride character. The effect of the nitrogen atom in hydroxylamine derivatives becomes operative only when the nucleophile, the amino component, has already approached the ester group. Analogous anchimeric assistance can be recognized in the intramolecularly catalyzed aminolysis of esters of 8-hydroxyquinoline, 2-hydroxy- and 2-mercaptopyridine, and of 1-hydroxybenzotriazole (Scheme 2).

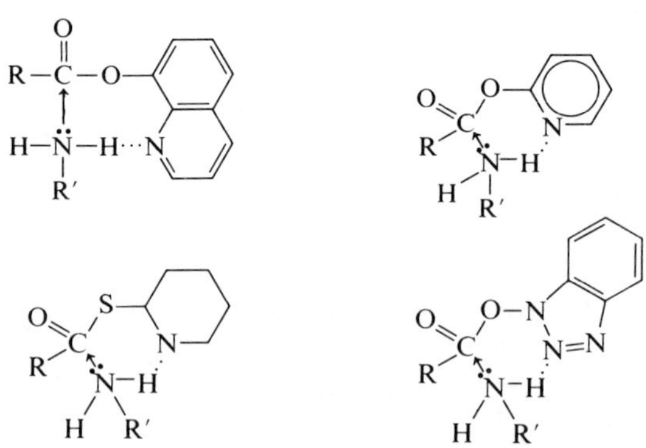

Scheme 2

The importance for reactivity of the neighboring N and O atoms in *O*-acylhydroxylamines is shown by the fact that *O*-acyl derivatives of substituted hydroxylamine without any *N*-acyl group that would produce a hydroxamic acid, are also reactive.* Good examples are esters of *N*-hydroxypiperidine (reaction 23) developed in the laboratory of Young

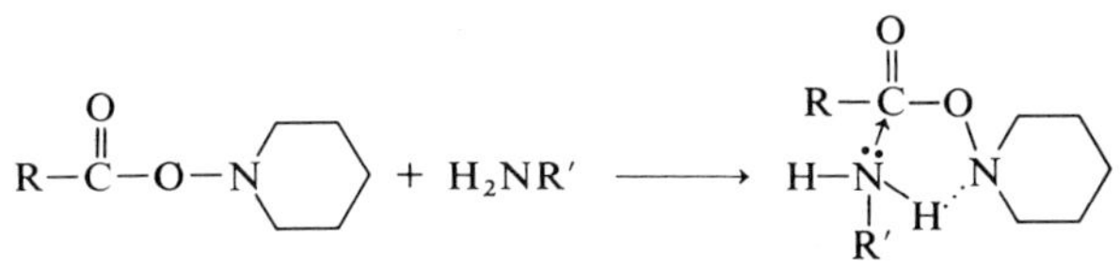

(23)

(Beaumont *et al.*, 1965a,b; Handford *et al.*, 1965) and *O*-acyl oximes [compound **49** (Losse *et al.*, 1964, 1965; Bittner *et al.*, 1965; Fujino and Nishimura, 1969) and compound **50** (Itoh, 1972)].

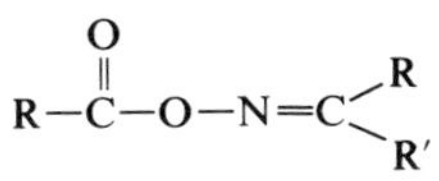

(49)

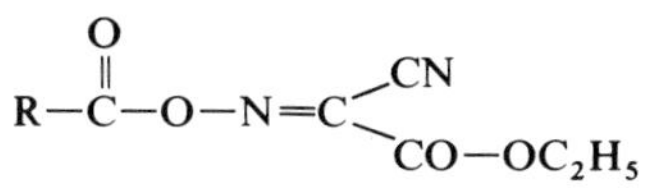

(50)

Important additions to the group of active esters derived from hydroxylamine are the *O*-acyl derivatives of 1-hydroxybenzotriazole (compound **51**) (König and Geiger, 1970a) and of the hydroxydihydrobenzotriazinone (**52**)

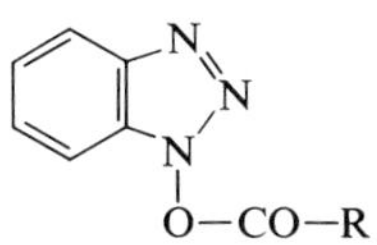

(51)

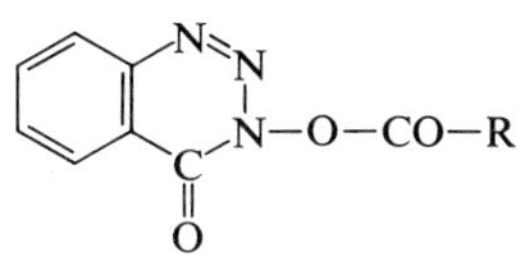

(52)

(König and Geiger, 1970b). The derivatives of compound (**52**) can be considered as mixed anhydrides of protected amino acids with a hydroxamic acid. The main significance of 1-hydroxybenzotriazole lies in its ability to reduce racemization in DCC mediated condensations, and to catalyze the aminolysis of active esters such as *p*-nitrophenyl esters or 2,4,5-trichlorophenyl esters. This aspect of 1-hydroxybenzotriazole esters will be discussed in Sections IV and V.

F. Insoluble Active Esters

The stimulus provided by the introduction of solid-phase peptide synthesis (Merrifield, 1963) is really exceptional. A multitude of publications demonstrates that investigations toward improved solid supports, efficient

* A similar neighboring group effect explains the enhanced reactivity, relative to *p*-nitrophenyl esters, of 4-nitroguaiacyl esters (Bankowski and Drabarek, 1971, 1974).

automated systems, new protecting groups, and activating methods designed for solid phase work, etc., were and are carried out in many laboratories. A most interesting result of the vigorous activity in this area is the development of the "reverse solid-phase synthesis," a procedure in which, not the growing peptide chain, but the activated derivatives of protected amino acids are anchored to an insoluble, polymeric support.

The earliest versions of this approach are the polymeric nitrophenyl esters (compound **53**) proposed by Fridkin *et al.*, (1965, 1966, 1968), and the esters of protected amino acids with a polymer derived from 4,4′-dihydroxydiphenylsulfone (compound **54**) (Wieland and Birr, 1966). In

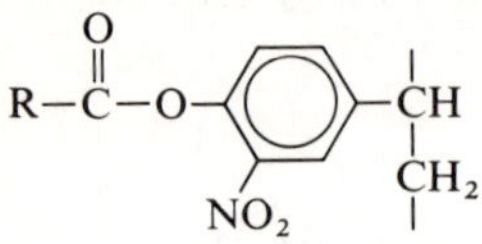

(53)

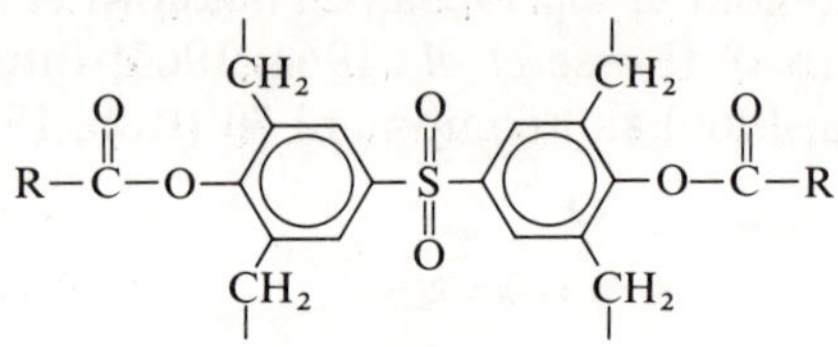

(54)

the latter case, condensation with formaldehyde was applied in the preparation of the insoluble active esters. The nitrophenol-derived polymers were further improved by the insertion of groups that can create some distance between the nitrophenol moiety and the polyethylene chain. Such insoluble esters were prepared by Panse and Laufer (1970; compound **55**) and by Kalir *et al.* (1974; compound **56**). Insoluble active esters could be formed

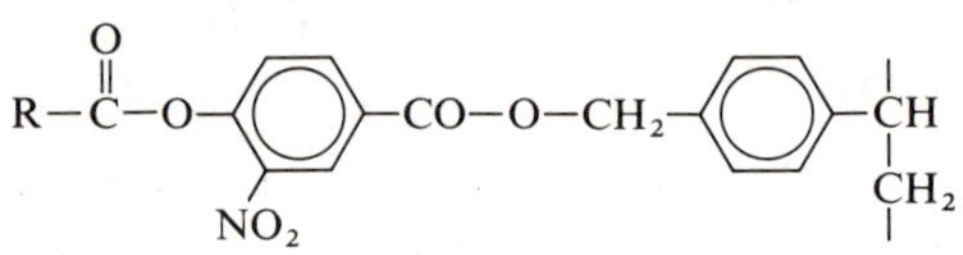

(55)

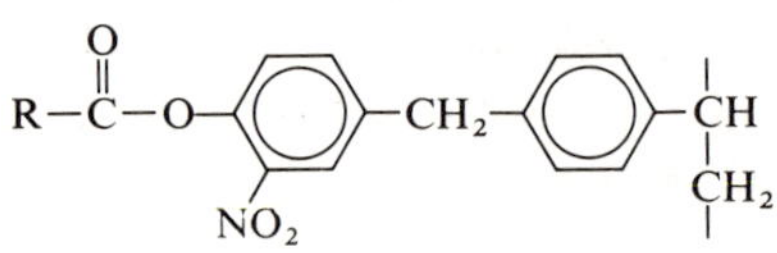

(56)

with polymeric derivatives of hydroxylamine. A copolymer of ethylene and maleic acid anhydride was treated with hydroxylamine, and the resulting *N*-hydroxysuccinimide derivative used for the preparation of active esters (Patchornik *et al.*, 1967; Laufer *et al.*, 1968), as shown in reaction (24).

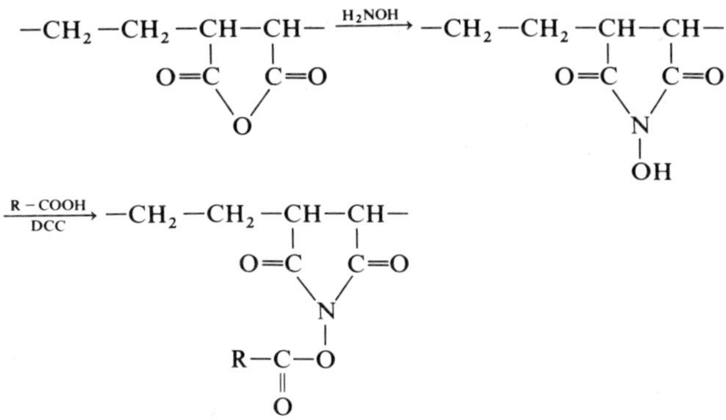

(24)

Hydroxylamine-related insoluble active esters were introduced also by Guarneri *et al.* (1972) (Scheme 3). A promising type of polymeric active ester is based on 1-hydroxybenzotriazole (Fridkin *et al.*, 1975) (Scheme 4).

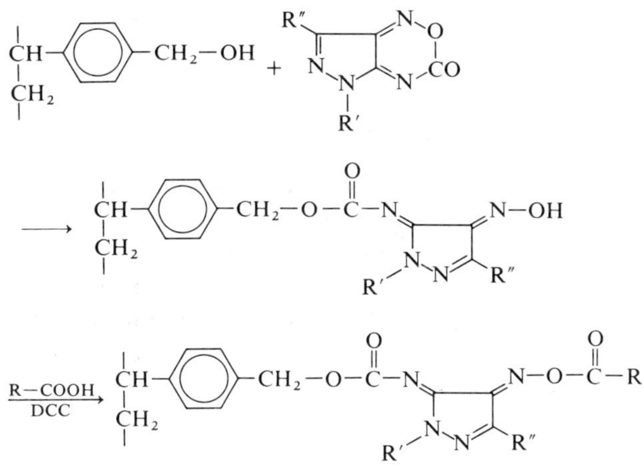

Scheme 3

There are some obvious advantages in the use of insoluble active esters. The excess reagent can be readily removed from the reaction mixture by filtration and can be used in a subsequent acylation. The by-product, the polymeric phenol or hydroxylamine derivative, is also separated in the same easy manner. A mechanization of the procedure is also possible: e.g., the solution of an amino component is allowed to pass through a column of an insoluble active ester and the effluent, after appropriate deblocking (e.g., by photolysis), can be applied to a second column containing the activated derivative of the second (protected) amino acid to be introduced, and so on. Such a process, although feasible, has not been used, so far, in practical

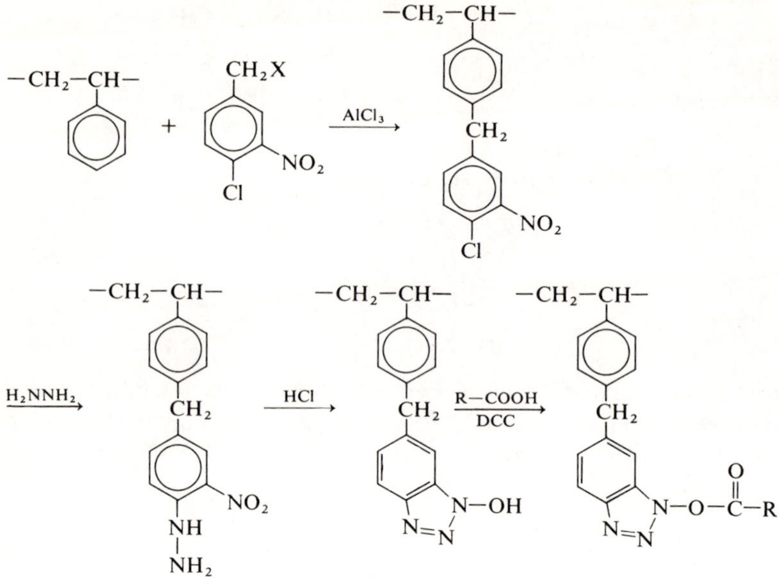

Scheme 4

syntheses. Some of the advantages might be offset, however, by difficulties caused by the insolubility of many protected intermediates in the commonly used organic solvents, including the most generally used one, dimethylformamide. The viscous, often semisolid mixtures, that can form in the coupling reactions, might render it impossible to work in columns and even the otherwise simple separation of the polymer from the product could become cumbersome or even impractical.

III. PREPARATION OF ACTIVE ESTERS

A. Synthesis through Activated Derivatives of Protected Amino Acids

In the earliest examples of the preparation of phenyl esters of acylamino acids (Karrer and Heynemann, 1948), sodium phenolate was allowed to react with acylamino acid chlorides:

(25)

Phthalylamino acid chlorides were used in the synthesis of nitrophenyl esters (Bodanszky, 1955), but the HCl liberated in the reaction limited this approach to amino acids protected with acid-resistant protecting groups:

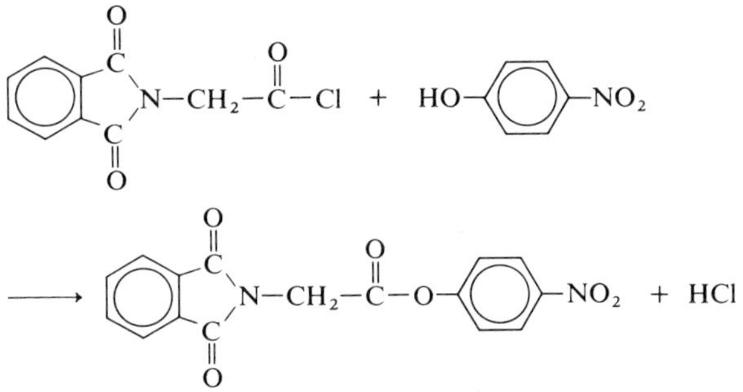

(26)

Preparation of aryl esters of benzyloxycarbonylamino acids via acid chlorides is possible in the presence of an acid-binding agent such as triethylamine (Wieland and Jaenicke, 1956). For a more generally applicable esterification procedure, the carboxyl group must be activated with substituents that produce less aggressive acids. Mixed anhydrides that were useful in the direct preparation of peptides could be adopted (Wieland *et al.*, 1951) also for the esterification of protected amino acids with thiophenol (reaction 27) or with phenol (Wieland and Schäfer, 1951). Mixed anhydride inter-

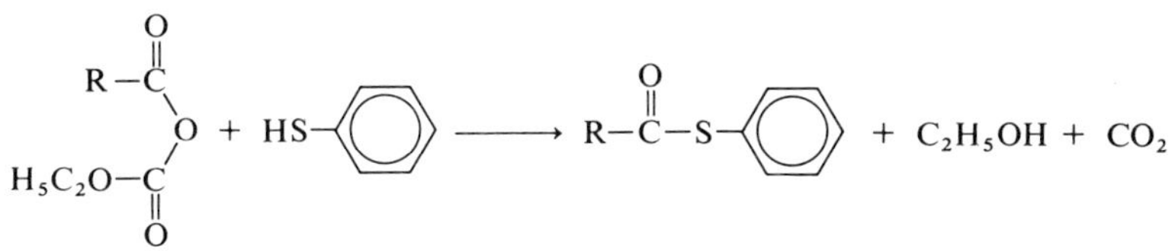

(27)

mediates can be assumed in the $POCl_3$ mediated esterification method (reaction 28) of Wieland and Heinke (1958). The reactions of protected

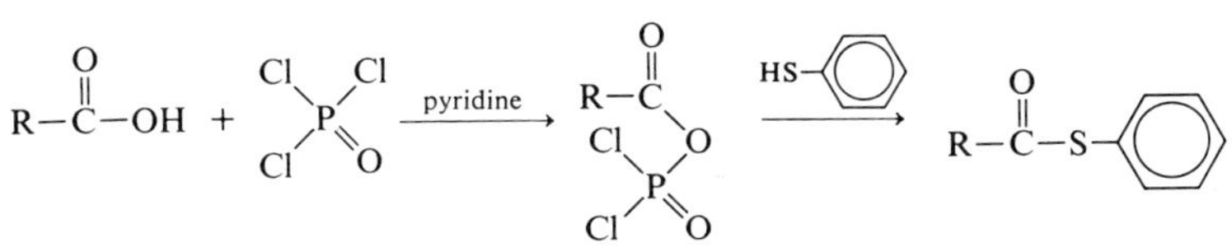

(28)

amino acids with aryl phosphites in pyridine (Farrington *et al.*, 1955; cf. also Mitin and Vlasov, 1968) are seemingly base-catalyzed transesterifications (reaction 29). In fact, both this and the analogous procedure involving aryl

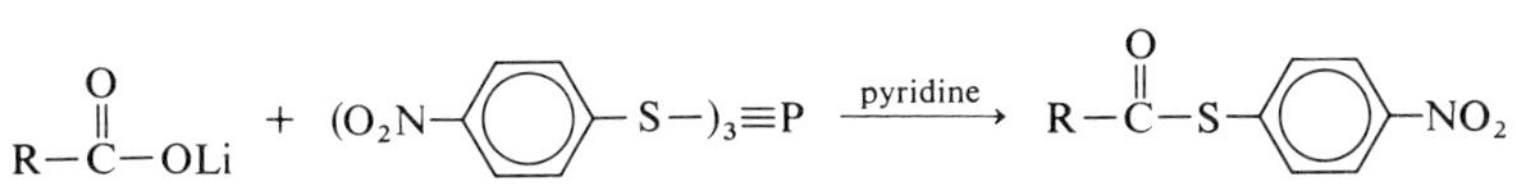

(29)

sulfites (Iselin *et al.*, 1957; Iselin and Schwyzer, 1960) could proceed via mixed anhydrides:

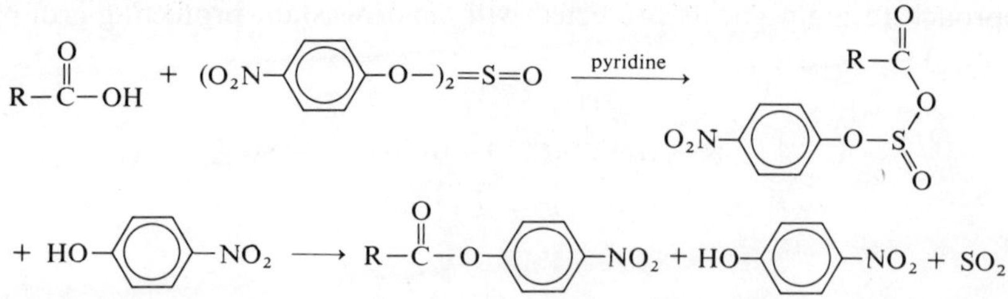

(30)

A mixed anhydride intermediate was indicated in the experiment (Iselin *et al.*, 1957) in which benzyloxycarbonylglycine was allowed to react in pyridine with an equimolar amount of phenyl sulfite in the presence of a fivefold excess of *p*-nitrophenol. The *p*-nitrophenyl ester of protected glycine was the major product of the reaction, while only a minor amount of phenyl ester was formed:

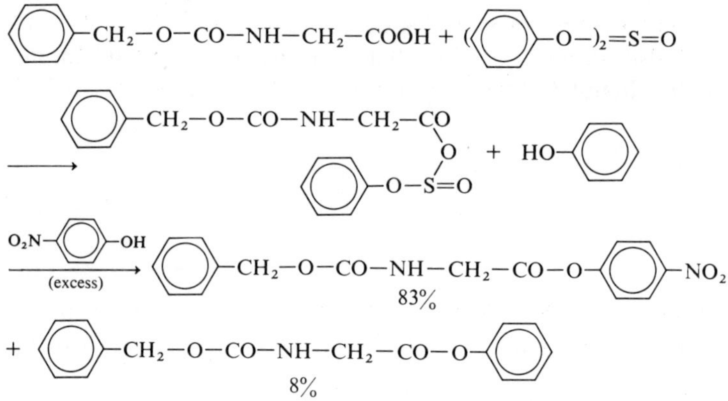

(31)

Furthermore, the phenyl ester of benzyloxycarbonylglycine, when exposed, in pyridine, to a similarly large excess of *p*-nitrophenol, was recovered unchanged; no base-catalyzed transesterification took place.

Similar mixed anhydride pathways might be followed in the esterification of protected amino acids with aryl carbonates (Wieland *et al.*, 1962) (e.g., compound **57**), with *p*-nitrophenyl chloroformate (compound **58**) (Glatt-

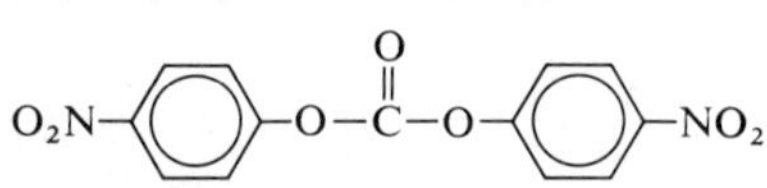

(57)

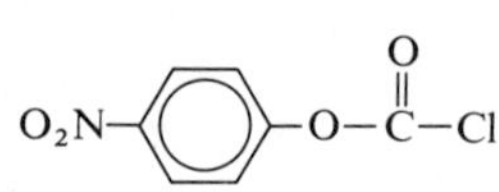

(58)

hard and Matter, 1963) and *o*-phenylene *p*-nitrophenyl phosphite (compound **59**) (Iselin *et al.*, 1957).

The attractive and often practical method of Sakakibara and Inukai (1964, 1965), the reaction of *p*-nitrophenyl trifluoroacetate (compound **60**)

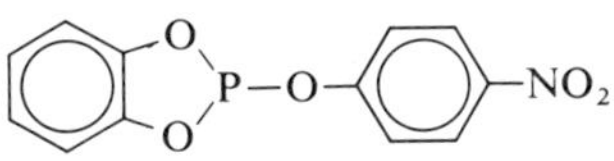

(59)

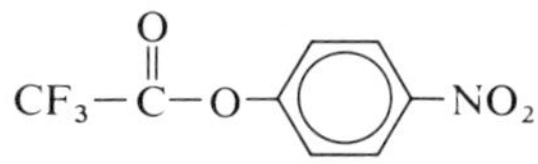

(60)

with protected amino acids in pyridine, could be interpreted as a true transesterification procedure. More recent evidence, however, points to a mixed anhydride mechanism in this case as well. Fletcher *et al.*, (1973) observed* that in the acylation of trifluoroacetate salts of amino components with active esters (with a tertiary amine added to the liberation of the free primary amine), in addition to the desired acylation product, the trifluoroacetyl derivative of the amino component was also formed (reaction 32). This suggests that in the preparation of protected amino acid active

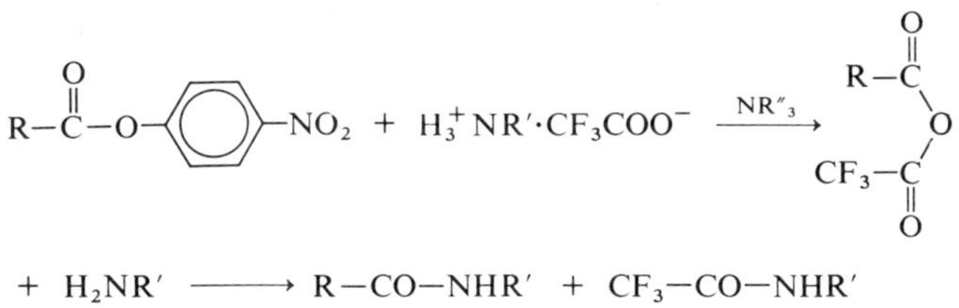

(32)

esters with Sakakibara's reagent (compound **60**), a mixed anhydride intermediate is operative (reaction 33).

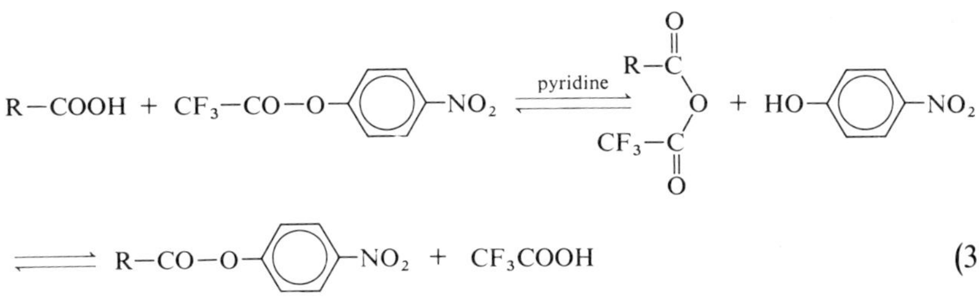

(33)

A different mechanism was proposed for the reaction of acylamino acids and aryl sulfites in pyridine by Yamazaki and his co-workers (1974). These authors, based on the analogy observed with reactive *N*-phosphonium salts of pyridine formed by the oxidation of phosphorus acid esters, suggest that aryl sulfites react according to reaction (34):

* An analogous reaction producing acetyl derivatives of amino components when their acetate salts are allowed to react with active esters was reported by Bodanszky (1969).

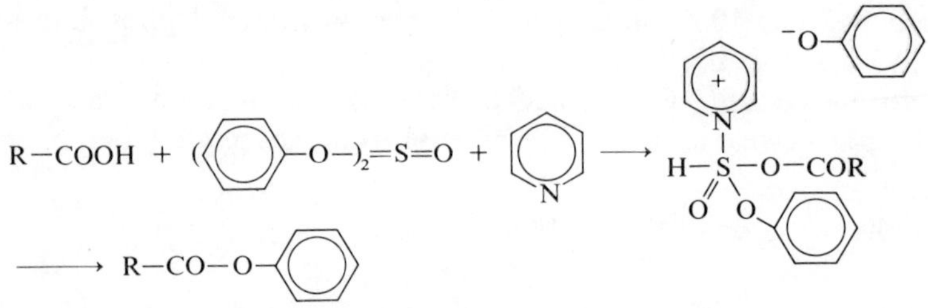

(34)

B. Synthesis through Activated Derivatives of the Hydroxyl Components

This approach is usually not suitable for the preparation of aryl esters; the replacement of the hydroxyl group by halogen leads to rather inert aryl halogenides, although negative substituents on the aromatic nucleus can facilitate the displacement of the halogen by carboxylate. For example, the formation of reactive intermediates, presumably 2,4-dinitrophenyl esters, was observed (Heikens *et al.*, 1954) during dinitrophenylation of amino acids and peptides with 2,4-dinitrofluorobenzene:

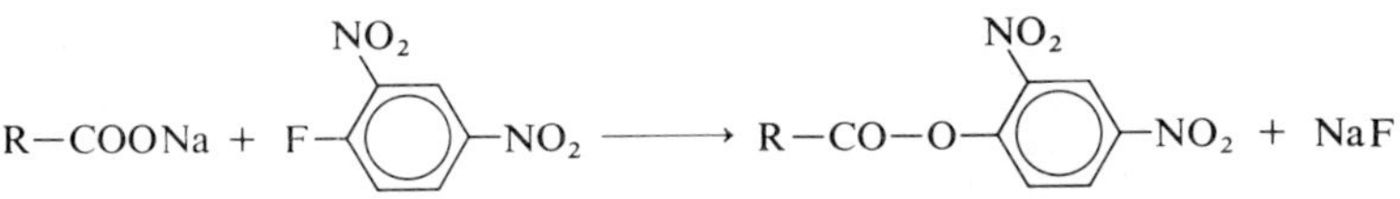

(35)

More practical is the use of reactive alkyl halogenides for the synthesis of activated methyl esters. The cyanomethyl ester of hippuric acid (Schwyzer *et al.*, 1955a) was prepared by the action of chloroacetonitrile on the triethylammonium salt of the acid:

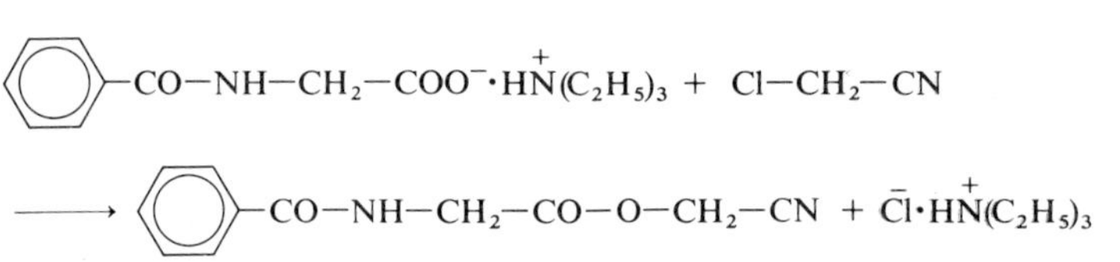

(36)

Subsequently, this procedure was applied for the preparation of cyanomethyl esters of amino acids carrying commonly used amino protecting groups (Schwyzer *et al.*, 1955b).

An interesting approach to benzhydroxamic acid esters of *N*-protected amino acids was designed by Taschner and his co-workers (1967; cf. also Lubiewska-Nakonieczna *et al.*, 1970), wherein the silver salts of the acids are brought into reaction with benzhydroxamic acid chloride. The initially formed enol esters spontaneously rearrange to the *N*-acyloxy derivatives:

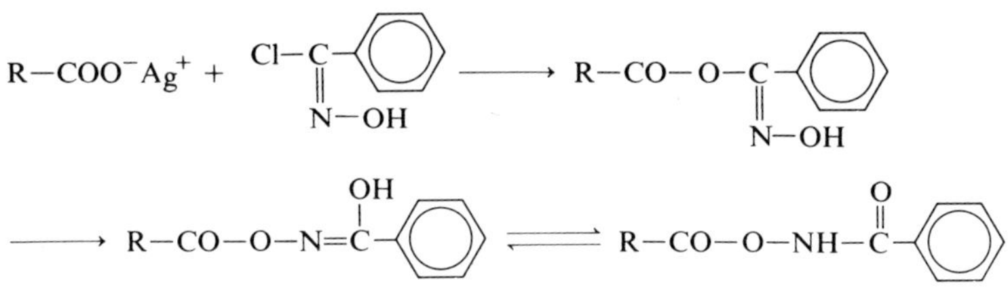

(37)

The possibility of the addition of protected amino acids to an isoxazolium salt, that can be considered an activated derivative of a phenol, is demonstrated by the example of the procedure (reaction 38) of Rajappa and Akerkar (1966). Crystalline active esters were obtained even in cases in which difficulties were encountered with other methods.

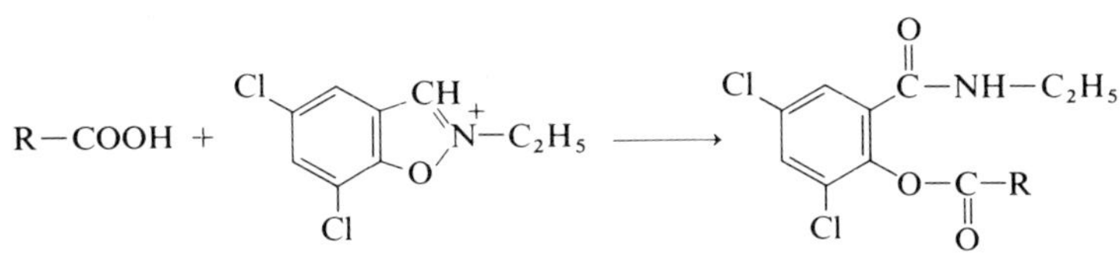

(38)

C. Esterification with the Application of Condensing Agents

The procedures described in this section generally proceed through reactive intermediates of the carboxyl components, the protected amino acids. They are operationally different; the activation of the carboxyl group and its coupling to the alcohol component are carried out in a single step. This can simplify the execution of the reactions considerably. Indeed, the application of dicyclohexylcarbodiimide (DCC; compound **61**) for the preparation of aryl esters was a major improvement in the methodology of peptide synthesis. Carbodiimides were introduced for the formation of the peptide bond by Sheehan and Hess (1955), but their usefulness as condensing agents in esterification was not exploited simultaneously. The reaction between the protected amino acid, DCC and the nucleophile *p*-nitrophenol can follow the two alternative pathways (Scheme 5) pointed out by Khorana (1955). In the symmetrical anhydride route, the regenerated protected amino acid is automatically recycled and the final products in both alternatives are the same: the protected amino acid active ester and dicyclohexylurea (DCU; compound **62**).

The preparation of *p*-nitrophenyl esters with the aid of DCC was first proposed (Elliott and Russel, 1957; Rothe and Kunitz, 1957) without the isolation of the active esters; the reaction mixtures, after the removal of

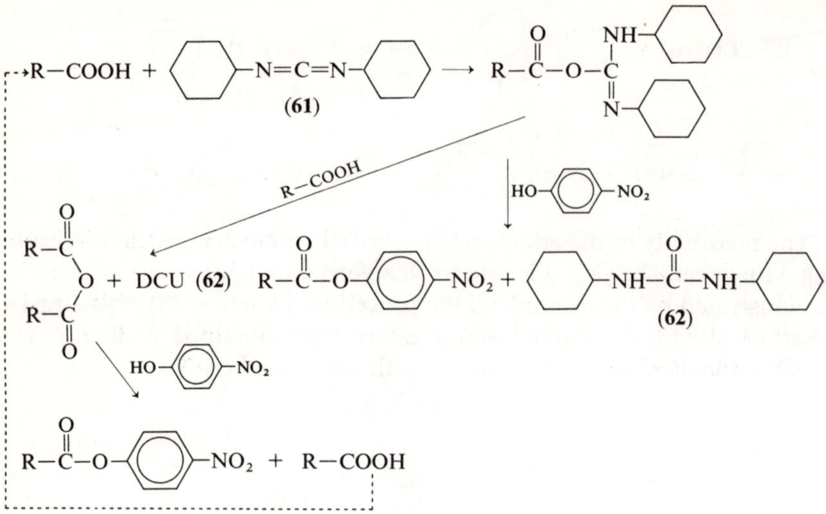

Scheme 5

DCU, were used directly for the acylation of the amino components. This simplification is not without drawbacks; e.g., the active esters have to be prepared separately for each reaction, they cannot be stored, distributed, or analyzed. Isolation of the active esters, prepared by the DCC approach, in homogeneous and mostly in crystalline form, was described by Bodanszky and du Vigneaud (1959a,b, 1962).

An interesting and in some cases advantageous variation of the DCC mediated esterification was invented by Kovacs and his co-workers (1969). They used a complex formed between DCC and the substituted phenol (compound **63**) as an esterification reagent.

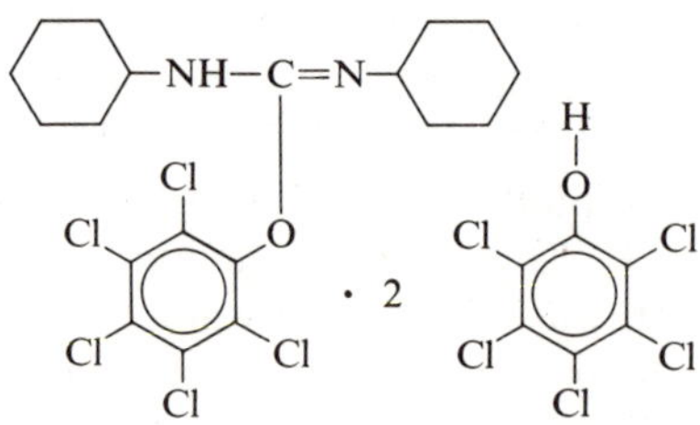

(63)

Since there is no fundamental difference between the synthesis of an amide and that of an ester, several reagents used for the formation of a peptide bond (Klausner and Bodanszky, 1972) were adapted for the preparation of active esters as well. The coupling reagent of Arens (1955; cf. also Panneman *et al.*, 1959), ethoxyacetylene, was proposed for the preparation of *p*-

nitrophenyl esters (Bodanszky and Birkhimer, 1962). In this case, the formation of symmetrical anhydrides (Scheme 6) seems to be quite substantial; in several cases where these anhydrides were insoluble in the solvent used, they separated in crystalline form.

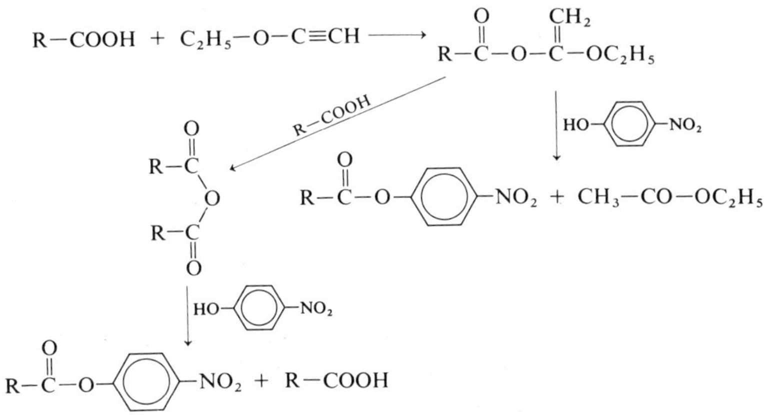

Scheme 6

The elegant method introduced by Staab (1957) for the formation of the peptide bond involves the reaction between protected amino acids or peptides with carbonyldiimidazole (compound **64**) producing an acylimidazole which, in turn, reacts with a nucleophile, the amino component. The same acylimidazoles, however, can be attacked by phenols as nucleophiles producing an aryl ester rather than an amide. This second alternative (Scheme 7) was proposed by Law (1965):

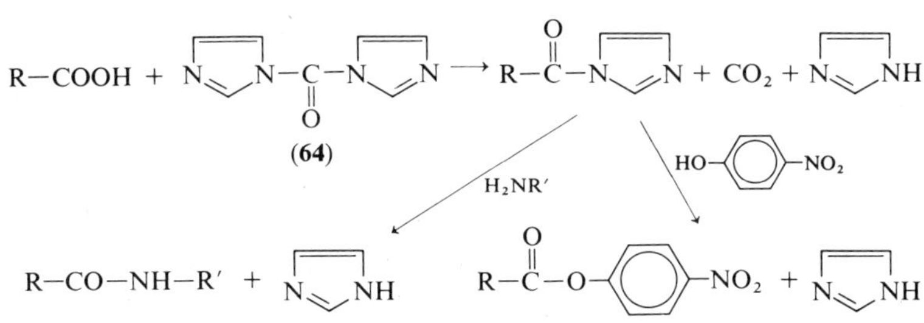

Scheme 7

The use of diphenylketene (Elmore and Smyth, 1963) for the synthesis of aryl esters in a single operation (reaction 39) is mentioned here rather than in section III,A, although it proceeds via a mixed anhydride. The fact that an optically pure sample of benzyloxycarbonylglycyl-D-phenylalanine *p*-nitrophenyl ester was secured by this approach suggests that it might be applicable for fragment condensations via active esters.

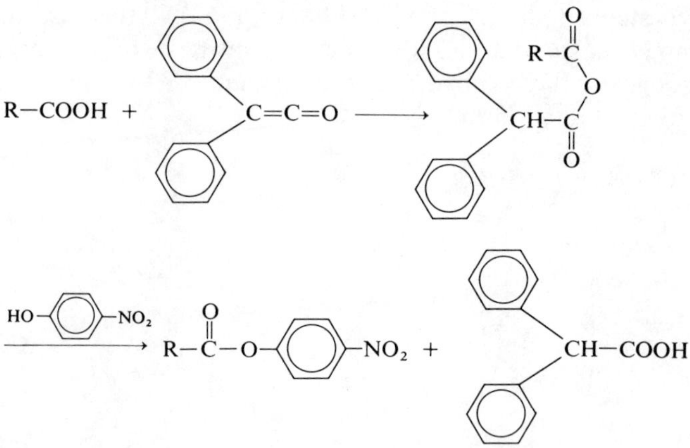

(39)

The carbonates of strongly acidic oximes were found (Itoh, 1974) to be very suitable for the preparation of active esters. The substituent of the reagent used for the suppression of racemization (Itoh, 1973), 2-hydroximino-2-cyanoacetate (compound **65**), was replaced by a series of

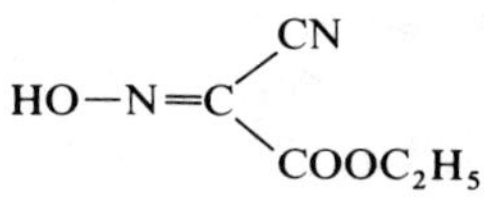

(**65**)

alkyl and aryl groups, and then esterified with various chlorocarbonates (reaction 40):

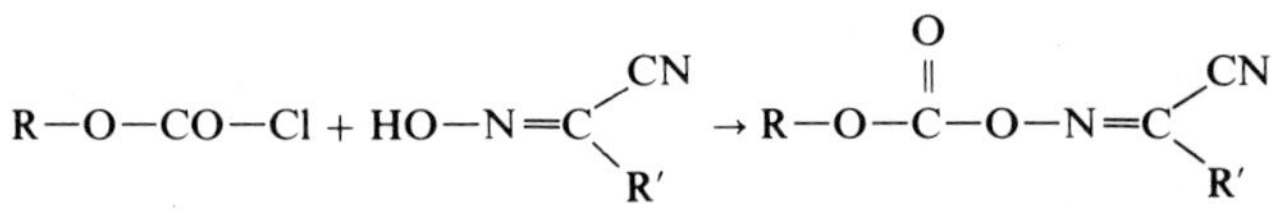

(40)

Carbonates derived from compound (**65**) and arylchloroformates; for example, *p*-nitrophenyl 2-hydroxyimino-2-cyanoacetamide carbonate (compound **66**) were used for the preparation of active esters of protected amino acids.

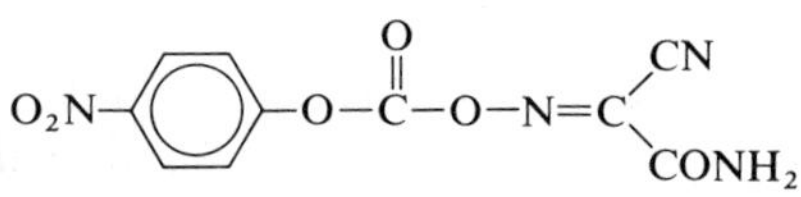

(**66**)

The yields were not uniformly high. Mixed anhydrides were assumed as reactive intermediates in the esterification.

Hexachlorocyclotriphosphatriazine (compound **67**), a coupling reagent proposed by Martinez and Winternitz (1972), was shown by the same

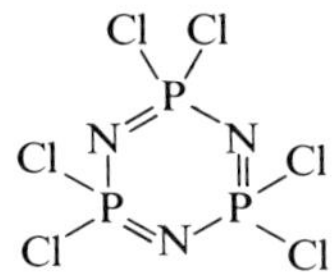

(67)

authors (Martinez and Winternitz, 1975) to be also applicable for the formation of active esters:

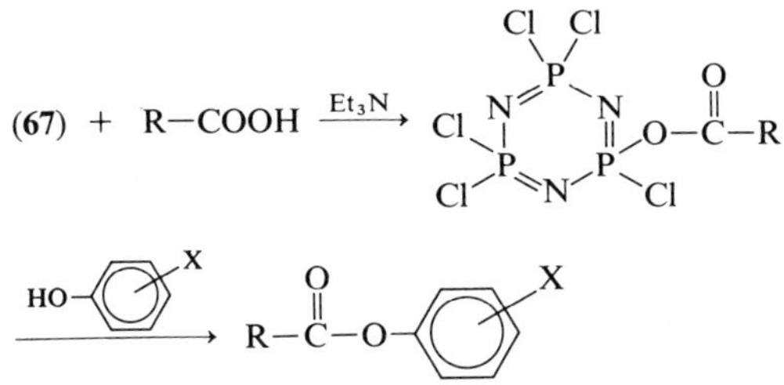

(41)

A series of *p*-nitrophenyl esters, pentachlorophenyl esters, and *o*-nitrophenyl esters were prepared in fair yield.

An example of the very original oxidation–reduction method of peptide synthesis invented by Mukaiyama and co-workers (1968) is shown in reaction (42). Further development of the method (Mukaiyama *et al.*, 1970a) resulted in the application of 2,2′-dipyridyldisulfide (reaction 43). This latter

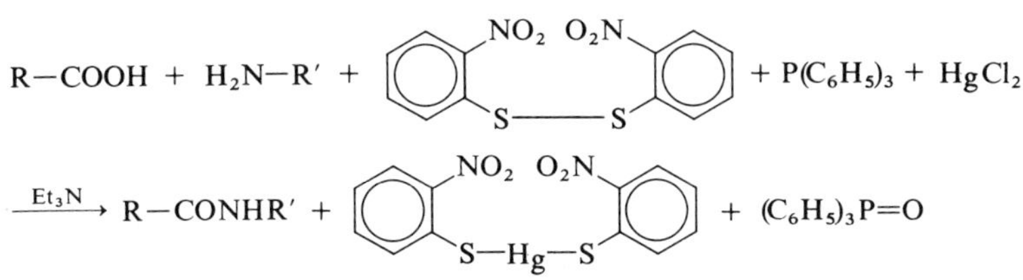

(42)

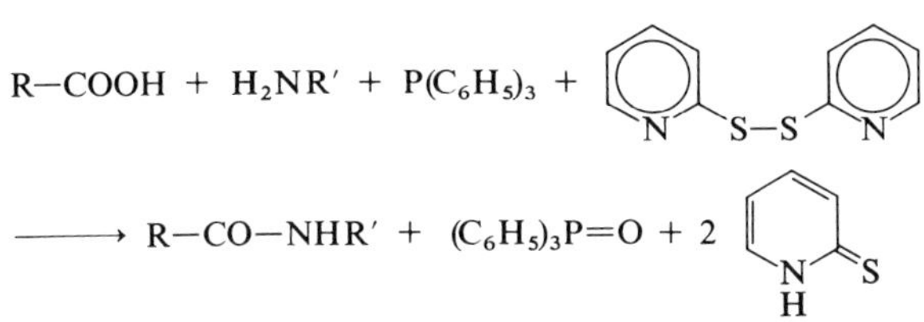

(43)

version of the oxidation—reduction method was applied for the preparation of active esters. Instead of the generally used triphenylphosphine, a phosphine substituted with a sulfonic acid group (compound **68**) was recom-

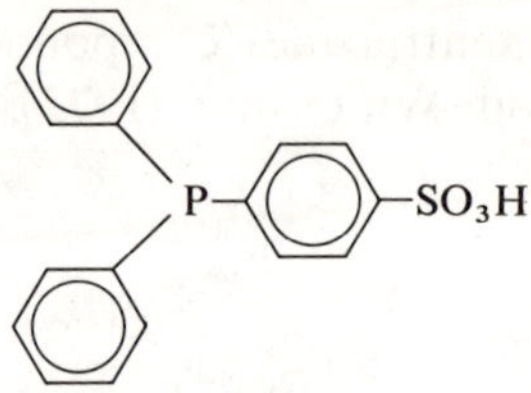

(68)

mended (Mukaiyama *et al.*, 1970a), because the corresponding phosphinoxide by-product is water-soluble and thus readily removed from the reaction mixture.

IV. CATALYSIS OF ACTIVE ESTER REACTIONS

The aminolysis rates of most active esters lag behind those observed with acid chlorides or mixed anhydrides. This is reflected also in the stretching frequencies of the ester carbonyl group in these acylating agents. The carbonyl band of *p*-nitrophenyl esters is at about 1760–1780 cm^{-1}, while the frequency of the CO group of the amino acid in a carboxylic acid mixed anhydride is ca 1830 cm^{-1}. Moderate reaction rates have some advantages. They allow the monitoring of the acylation, and permit better temperature control and more facile adjustments in alkalinity and in the amount of acylating agent used than in the case of a more aggressive reactant. Nevertheless, for the practical execution of the synthesis of a longer peptide chain, reasonably fast reactions are, beyond question, necessary. For example, acylation can take a few hours, but should not require several days. It is understandable, therefore, that many attempts were made to accelerate the reactions of active esters. An increase in reaction rates is particularly important when steric factors impede acylation. Very pronounced steric hindrance was observed (Smith *et al.*, 1952) with derivatives of valine or isoleucine as acylating intermediates. Branching at the β-carbon of these amino acids strongly interferes with the approach of the nucleophile. Superimposed on the steric effect is the electron-release by the branched aliphatic side chain, further diminishing the rate of the desired reaction. The influence on the reaction rates of polymeric supports and of the growing peptide chain itself will be discussed in Section VII.

The peptide bond forming step, with few exceptions, is a bimolecular reaction, and hence its rate is highly dependent on the concentration of the reactants. In the building of long peptide chains, this fact becomes painfully obvious as the time needed for the incorporation of an amino acid residue increases as the chain grows. For practical results, either highly reactive intermediates, such as mixed anhydrides, or active esters with exceptionally high reactivity, e.g., pentafluorophenyl esters, have to be selected. Alternatively, the reactions can be carried out with moderately reactive esters, but under conditions which enhance their reaction rates. The effect of solvents is the first to be considered in this section. The application of weak acids and bases, sometimes combined in bifunctional catalysts, will be discussed subsequently.

A. Solvent Effects

The reaction of hippuric acid cyanomethyl esters (Schwyzer *et al.*, 1955a) with benzylamine is slow; at room temperature in 1 hr, at 0.1 *M* concentration in acetonitrile only a 7% yield could be realized. The yield under the same conditions increased to 12% in dioxane, 14% in chloroform, 26% in benzene, 44% in water-saturated ethyl acetate, 60% in ethanol, and to 74% in a 2:3 mixture of dimethylformamide and water. Similar relationships with respect to the effect of solvents were found in the reaction of benzyloxycarbonyl-L-phenylalanine 2,4,5-trichlorophenyl ester with benzylamine (Pless and Boissonnas, 1963). In a 0.01 *M* solution of the ester with 100% excess of the amine, the half-time of the reaction ($t_{1/2}$) was 23 minutes in chloroform and only 0.3 minutes in dimethylformamide. Neat or aqueous dioxane, *N*-methylmorpholine, pyridine, and ethyl acetate form a group of solvents providing intermediate rates, ethyl acetate ($t_{1/2}$ 2.3 min) being the most practical among them. Dimethylformamide, however, has not only the advantage of the highest reaction rate, but is also a rather general solvent of protected intermediates in peptide synthesis, particularly in the case of compounds of larger molecular weight. In the last-mentioned study, the reaction rates were determined from the changes in the absorption of the unreacted active ester and of the liberated 2,4,5-trichlorophenol (Stich and Leemann, 1963). The spectrophotometric determination of the *p*-nitrophenol formed in the aminolysis or hydrolysis of *p*-nitrophenyl esters is commonly used in kinetic studies.

The tendency recognized in these experiments, namely higher reaction rates in polar and slow reactions in nonpolar solvents, cannot be generalized. Active esters can be quite different in this respect. For example, the esters of protected amino acids and 2-hydroxypyridine (Dutta and Morley, 1971) are extremely potent in dichloromethane; the half reaction time of the 2-pyridyl ester of *tert*-butyloxycarbonyl-L-alanine with glycinamide was almost 100 times less than the $t_{1/2}$ value determined under the same conditions for the corresponding *p*-nitrophenyl ester. The difference between the two kinds of active esters was less pronounced when valine *tert*-butyl ester was used as the amino component, but the 2-pyridyl ester was still about five times faster. In dimethylformamide, however, the relative rate of the two reactions was reversed: the *p*-nitrophenyl ester was 3 times more reactive than the 2-pyridyl ester. The same marked effect on the acylation rates by these two solvents was found also in experiments in which a polymer-bound amino component was used.

The causes of these variations in the solvent effects with different kinds of active esters are not well understood. An interaction between solvent and the electron-withdrawing substituent of the aryl groups of some active esters is indicated in a remarkable difference between *p*-nitrophenyl esters and *o*-nitrophenyl esters (Bodanszky *et al.*, 1974b) of protected amino acids. The aminolysis rates of *p*-nitrophenyl esters are increased by an order of magnitude if the reactions are carried out in ethyl acetate, dioxane or tetrahydrofuran rather than in chloroform and dichloromethane. They increase by a further order of magnitude in dimethylformamide or dimethyl sulfoxide. The reaction rates of *o*-nitrophenyl esters also depend upon the solvent used in the acylation reaction, and the effect of the solvents remains qualitatively the same as in the reactions of their para-isomers. Yet, instead of a factor of about 10, the increase by changing from dichloromethane to ethyl acetate is only by a factor of 2, and the further increase that stems from turning from ethyl acetate to dimethylformamide is by a factor of about 3. The finding that the *o*-nitrophenyl esters tend to assume a quasicyclic conformation (Bodanszky *et al.*, 1974b) allows the tentative conclusion that the effect of solvents is reduced in *o*-nitrophenyl esters because in the single-turn helix (Fig. 1) the intramolecular involvement of the nitro group in a

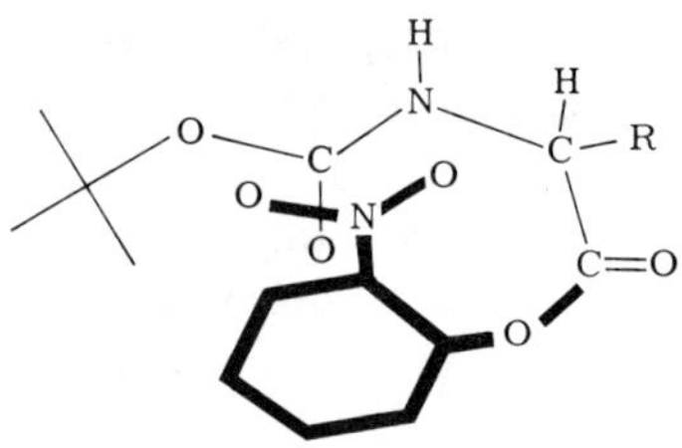

Figure 1. The conformation tentatively proposed for *o*-nitrophenyl esters of Boc amino acids

dipole—dipole (or dipole-induced dipole) interaction more or less prevents intermolecular interactions with solvent molecules. The reduced solubility of *o*-nitrophenyl esters, relative to *p*-nitrophenyl esters, in several organic solvents supports this view. The enhanced ability of *o*-nitrophenyl esters to penetrate crowded environments, such as the matrix of polymers, is in harmony with the assumption of a molecule that carries a smaller solvate shell than the one present in *p*-nitrophenyl esters.

B. Catalysis with Metals

Only rudimentary information is available on this subject. Catalysis of aminolysis of methyl esters with Cu^{2+} and Co^{3+} incorporated in the form of complexes (Yamada *et al.*, 1970; Wu and Busch, 1972) was already mentioned in Section II. More in-depth investigations in this area might bring interesting, perhaps even surprising results.*

C. Acid–Base Catalysis

The experiments to increase the otherwise moderate reactivity of cyanomethyl esters included also the addition of weak acids and bases as catalysts (Schwyzer *et al.*, 1955b). Of the acids, acetic acid was found to be efficient enough to be applied in the synthesis of peptides corresponding to partial sequences of oxytocin (Iselin *et al.*, 1955). The bases used in their studies, pyridine and triethylamine, have found no practical application so far. The effect of weak acids on the aminolysis rates of active esters were further investigated by Taschner and co-workers (1964), who could accelerate the reaction rates of phenyl esters by the addition of formic acid, acetic acid, trimethylacetic acid, or benzoic acid to such an extent that the otherwise moderately active phenyl esters reacted with the rapidity of *p*-nitrophenyl esters. Best results were achieved when acetic acid was added in equimolar amount to the mixture of active ester and amine. The addition of acetic acid had only a minor effect on the reaction rate of a *p*-nitrophenyl ester. An even more detailed study on acid catalysis of active esters was carried out by Nakamizo (1969), who determined the optimal concentrations of several acids and proposed a bifunctional concerted mechanism that involves an 8-membered transition state such as the one shown in Eq. (44). A

(44)

different transition state (compound **69**) was visualized by Muzalewski (1973).

* Catalysis of the reaction between glycyl–glycine and *p*-nitrophenyl acetate with Cu and Zn ions was attempted by Koltun and Gurd (1959).

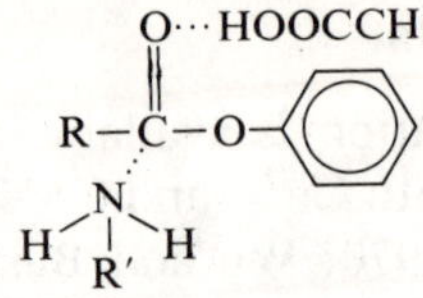

(69)

An impressive catalytic effect of acetic acid on the reactions of the esters of 1-hydroxypiperidine was reported by Jones *et al.* (1967) and Jones and Young (1968a).

A major impetus for the studies on the catalysis of ester aminolysis was provided by Wieland and Vogeler (1964), who proposed imidazole as a catalyst for the preparation of peptides with simple alkyl esters as acylating agents. They were able to acylate the tetraethylammonium salt or *tert*-butyl ester of glycine with the *methyl* ester of benzyloxycarbonyl-DL-alanine in imidazole as solvent at 60°–80°C in 2–3 hr. Both a specific catalytic mechanism (reaction 45) and general base catalysis (reaction 46) seem to be operative.

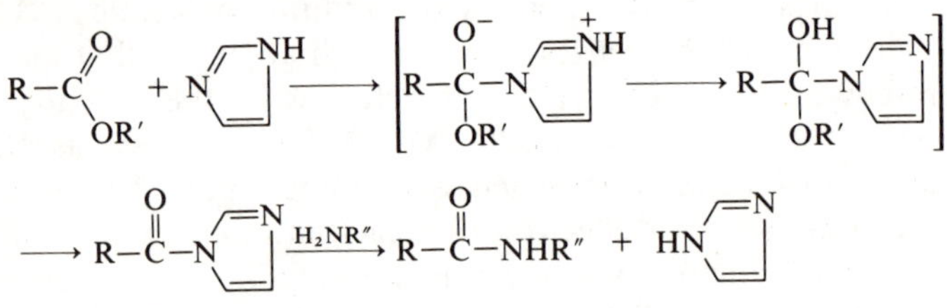

(45)

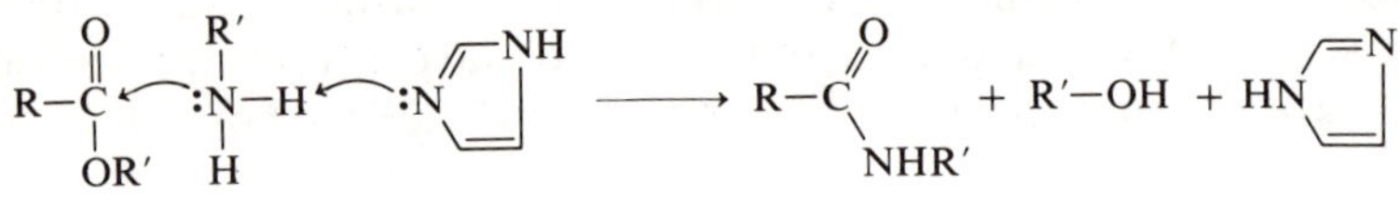

(46)

Acceleration of the aminolysis of *p*-nitrophenyl ester with the addition of imidazole was demonstrated by Mazur (1963), and of the analogous reaction of thiophenyl esters by Wieland and his associates (1963). At the same time, Beyerman and Maassen van den Brink (1963, cf. also Beyerman *et al.*, 1965) proposed pyrazole (compound **70**) and 1,2,4-triazole (compound **71**) as bifunctional catalysts. An additional series of polar compounds was studied,

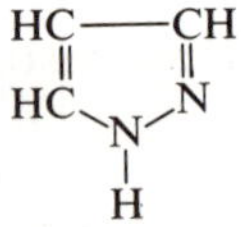

(70)

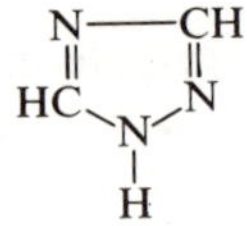

(71)

with respect to catalytic effect in the aminolysis of *p*-thiocresyl esters, by Wieland and Kahle (1966). The catalytic effect of pyrazole, imidazole, and 1,2,4-triazole was overshadowed by the larger catalytic influence of dimethylformamide or dimethyl sulfoxide. Obviously, in these compounds as solvents, the catalysis by the azoles could not be observed.

Catalysis of *p*-nitrophenyl esters and 2,4,5-trichlorophenyl esters by 1-hydroxybenzotriazole (compound **72**) seems to have major significance (König and Geiger, 1972, 1973) because the aminolysis rates are accelerated in dimethylformamide, the solvent most widely used in peptide synthesis.

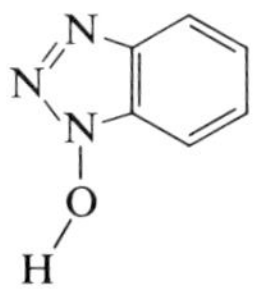

(72)

An intramolecular assistance of aminolysis in phenyl esters by a neighboring methoxy group was recognized by Bankowski and Drabarek (1972).

V. SIDE REACTIONS

A. Side Reactions in the Preparation of Active Esters

1. The Consequence of Steric Hindrance

Most of the side reactions encountered during the preparation of active esters of protected amino acids are connected with functional groups in amino acid side chains. However, in some cases the esterification step can fail because of steric hindrance, even with monoamino monocarboxylic acids. Branching at the β-carbon* in valine and isoleucine leads, in the dicyclohexylcarbodiimide mediated preparation of active esters of these amino acids, to the formation of significant amounts of *N*-acylurea derivatives. The ureides can become the major product if the alcohol component is a poor nucleophile, that cannot well compete with the intramolecular $O \rightarrow N$ migration of the *O*-acylisourea intermediate (Scheme 8). Thus, while no major difficulties were encountered in the preparation of *p*-nitrophenyl esters of benzyloxycarbonyl-L-isoleucine (Bodanszky and du Vigneaud, 1959b), poor results were obtained in the synthesis of *o*-nitrophenyl esters of

* The steric hindrance in amino acids with α-branching, e.g., α-aminoisobutyric acid, is overwhelming. The preparation of active esters of such acids might be impractical.

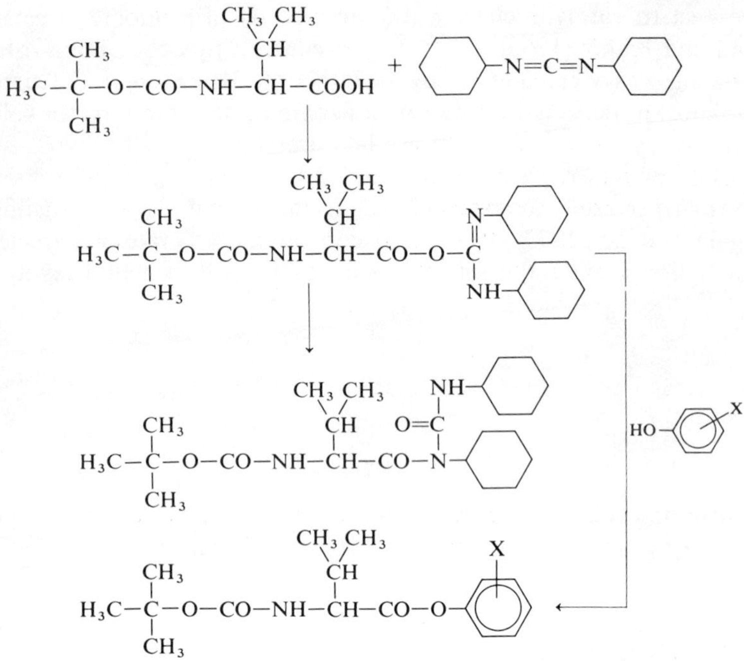

Scheme 8

protected valine and isoleucine (Bodanszky *et al.*, 1973, 1974a) under the conditions recommended (Bodanszky and du Vigneaud, 1962) for the para isomers. The *o*-nitrophenyl esters had to be separated by chromatography from the significant amount of *N*-acylurea derivative formed as a by-product. The reduced nucleophilicity of *o*-nitrophenol due to intramolecular hydrogen bonding (compound **73**) is the most likely cause of this failure. It

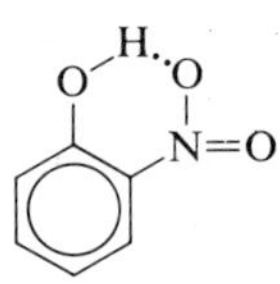

(73)

was expected that pyridine might interfere with the intramolecular hydrogen bonding and also enhance the nucleophilicity of the phenolic oxygen. Indeed, in pyridine as the solvent, the preparation of *o*-nitrophenyl esters, including those of protected valine and isoleucine, proceeded smoothly and with good yield.

2. Side Reactions Caused by Hydroxyl Groups in Amino Acid Side Chains

Early attempts for the esterification of benzyloxycarbonyl-L-threonine with *p*-nitrophenol failed. The report of the successful preparation of the 2,4-dinitrophenyl ester of benzyloxycarbonyl-L-threonine (Rocchi *et al.*, 1963) prompted an examination of the reaction of this protected amino acid with dicyclohexylcarbodiimide (Bodanszky and Ondetti, 1966a). The isolation of *O*-acyl derivatives, such as the dimeric lactone (compound **74**),

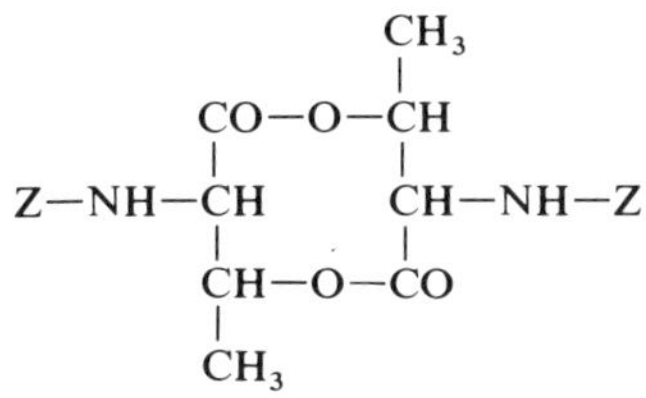

(74)

revealed that the alcoholic hydroxyl in the side chain of threonine, although secondary and hindered, is sufficiently reactive to be a competitor of *p*-nitrophenol in the attack on the *O*-acylisourea intermediate of the esterification reaction. A higher reaction rate of 2,4-dinitrophenol with the reactive intermediate accounts for the more favorable outcome of the reaction. The earlier assumption (Sheehan *et al.*, 1956) that in coupling with DCC only the amino group of threonine requires protection is valid as long as an excellent nucleophile, such as an amino group, is available. With less potent nucleophiles, e.g., *p*-nitrophenol, interference by the side chain hydroxyl becomes a major factor. In alternative procedures, the role of the alcoholic hydroxyl is less pronounced. Esterification of threonine or tyrosine, protected only on their amino groups, could be carried out by the application of di-*p*-nitrophenyl sulfite (Meienhofer *et al.*, 1970), although purification of the crude product by gel-permeation chromatography on Sephadex LH-20 was necessary.

The unhindered primary alcoholic hydroxyl in the side chain of serine is obviously a reactive group that should require protection, at least during the preparation of active esters, when a strongly reactive intermediate is involved, and the desired reaction is the acylation of a phenolic hydroxyl group. Still, the use of 2,4-dinitrophenol, its hydroxyl being a better competitor than that of *p*-nitrophenol, allowed the preparation of the 2,4-dinitrophenyl ester of benzyloxycarbonyl-L-serine (Bodanszky and Ondetti, 1966a). The excellent nucleophile 1-hydroxybenzotriazole produced (in condensation with DCC) the β-lactone of the N^α protected amino acid (König and Geiger, 1970a).*

A crystalline active ester of benzyloxycarbonyl serine was obtained by Rajappa and Akerkar (1966) by reacting the protected amino acid with 5,7-dichloro-*N*-ethylbenzoisoxazolium fluoroborate (reaction 47). Here the

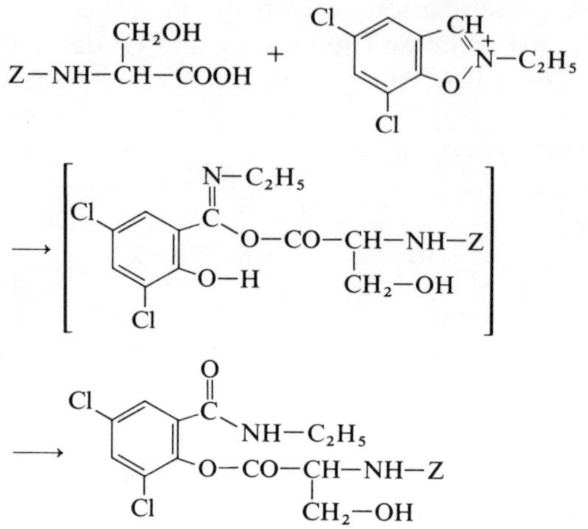

(47)

esterification with the phenolic hydroxyl proceeds through intramolecular displacement, rather than by an intermolecular reaction as in the case of DCC. This might explain the absence of major interference by the alcoholic hydroxyl group.

Hydroxyls are not the only functional groups in amino acid side chains that can interfere with the preparation of active esters. The need for protection of the β- and γ-carboxyl groups in amino-protected derivatives of aspartic and glutamic acid is quite obvious. Also, the activated intermediates, such as the *O*-acylisourea derivatives, when DCC is used as condensing agent, will react with strong nucleophiles such as the ε-amino group in lysine or the sulfhydryl group in cysteine, and to our best knowledge no attempts were made for the preparation of active esters of these amino acids with their side chain functions unprotected. The reactivity of the imidazole in histidine is similar to that of the mercapto group of cysteine. Accordingly, for the synthesis of active esters of histidine both the α-amino and the imidazole NH group have to be blocked. For example, N^α, N^{im}-bisbenzyloxycarbonyl-L-histidine was prepared by Meienhofer (1962). The NH group in the indole

* Active esters of serine and threonine with unprotected hydroxyl groups could be obtained through hydrogenolysis of the *tert*-butyloxycarbonyl-*O*-benzylamino acid 2,4,5-trichlorophenyl esters (M. Bodanszky, J. Henes, and A. Yiotakis, unpublished).

moiety of tryptophan does not interfere with the reactions that yield active esters of this amino acid.

3. Side Reactions Caused by the Guanidine Group of Arginine

The guanidine group of arginine is relatively inert, because it remains protonated except under extremely basic conditions. Still, it is often desirable to avoid the ionic character of protonated guanidines and the concomitant solubility problems. Therefore, derivatives of arginine protected also in their side chains, most frequently with nitro or *p*-toluenesulfonyl (tosyl) groups, are favored in many laboratories. While these protecting groups indeed prevent the participation of the guanidine in intermolecular reactions, they are unable to block intramolecular acylation by the activated carboxyl of the same arginine residue. Thus, the lactone, a piperidone (compound **75**), was formed as the major product in the reaction of N^{α}-benzyloxycarbonyl-nitro-L-arginine with DCC and *p*-nitrophenol (reaction 48) (Bodanszky and Sheehan, 1960; cf. also Gödicke and Losse, 1967). A more detailed study of this side reaction was reported by Paul *et al.* (1961). Under carefully selected conditions and with di-*p*-nitrophenyl sulfite as esterification reagent, it is possible (Meienhofer *et al.*, 1970) to prepare compound (**76**) in acceptable

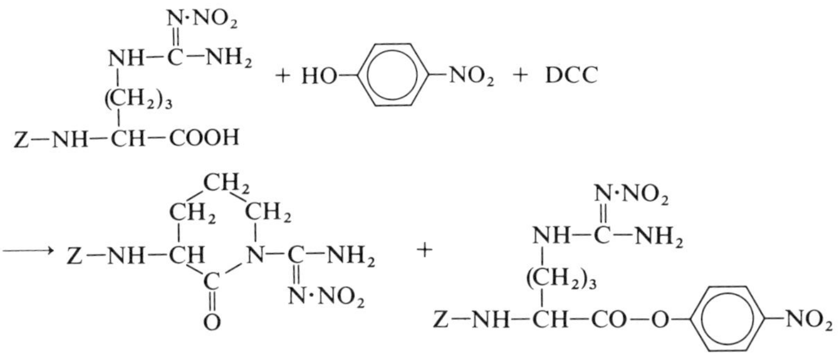

(48)

(**75**) (**76**)

yield and, after chromatography, in pure form. Nevertheless, the tendency for lactam formation persists even in the active ester (**76**). It is not stable in solution and will generate some lactam (compound **75**) during acylation. Protection of the guanidine with the tosyl group also offers only partial protection against intramolecular acylation (Schnabel and Li, 1960). An active ester of arginine tosylated in the guanidine was obtained by Guttmann and his associates (1962). A better protection of the guanidine function is provided by *two* benzyloxycarbonyl (Nicolaides *et al.*, 1960) or by *two* adamantyloxycarbonyl (Jäger and Geiger, 1970) groups. The fully protected guanidine group does not interfere with the synthesis of active esters and the protection is expected to remain intact also during the subsequent coupling reaction.

An alternative against the formation of piperidone derivatives such as compound (**75**) is esterification with hydroxyl components, which are better competitors for the activated carboxyl group than *p*-nitrophenol or the partially protected (e.g., nitro-) guanidine. Both 2,4-dinitrophenol (Bodanszky and Ondetti, 1966a) and pentachlorophenol (Kovacs and Ceprini, 1965) are suitable for this purpose. The pentachlorophenyl ester of benzyloxycarbonyl-nitro-L-arginine can be obtained in crystalline form. The steric hindrance caused by the chlorine substituents in both ortho positions (2 and 6) might be the explanation for the stability of this active ester. The same steric hindrance might, however, reduce its reactivity in intermolecular reactions as well. A crystalline 1-hydroxysuccinimide ester of *tert*-butyloxycarbonylnitro-L-arginine was prepared by Ondetti and his associates (1970).

4. *Side Reactions Involving the Carboxamide Group in the Side Chains of Asparagine and Glutamine*

It is not *a priori* obvious that the rather inert carboxamide groups in asparagine and glutamine should pose problems in the preparation of active esters of these amino acids. Yet, one of the most serious difficulties in peptide synthesis arises during their activation (Gish *et al.*, 1956; Ressler, 1956; Ressler and Ratzkin, 1961). Loss of water from the side chain carboxamide leads to the formation of significant amounts of β-cyanoalanine derivatives in the case of asparagine, and of the homologous nitrile, although to a lesser extent, in the derivatives of glutamine (reaction 49). The condensing reagent,

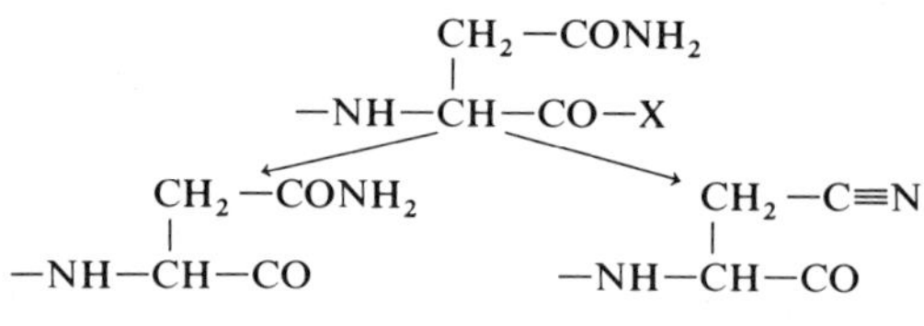

(49)

e.g., DCC, does not react directly with the carboxamide group; once asparagine or glutamine are parts of a peptide chain their carboxamide remains intact in the subsequent chain lengthening procedure. Only vigorous conditions, such as treatment with ethylene chlorophosphite in triethyl phosphite (Ressler and Kashelikar, 1966), can produce dehydration of carboxamides in peptides. Thus, the side reaction under discussion occurs intramolecularly in the reactive intermediate. The same reaction also takes place, to a varying extent, in other procedures of activation and coupling, e.g., in the application of the mixed anhydride method. The use of the carboxyl component and DCC in large excess suppresses nitrile formation (Corley *et al.*, 1972), probably because acylation is fast and the reactive intermediate, at least the part of it that indeed participates in the acylation of the amino component, is short lived. The oxidation–reduction method (Mukaiyama *et al.*, 1968) was claimed to be free from this side reaction, but this claim has still to be substantiated.

Protection of the carboxamide group by a variety of substituted benzyl or benzhydryl groups offers a more reliable solution for this problem, but the introduction of these protecting groups further increases the already arduous task of peptide synthesis.

The use of active esters of N^{α}-protected asparagine is probably the best approach to pure asparaginyl peptides. While dehydration does occur in the DCC mediated ester formation (reaction 50), the two products, the desired active ester (compound **77**) and the active ester of the β-cyanoalanine (compound **78**) are readily separated by fractional crystallization (Bodanszky and

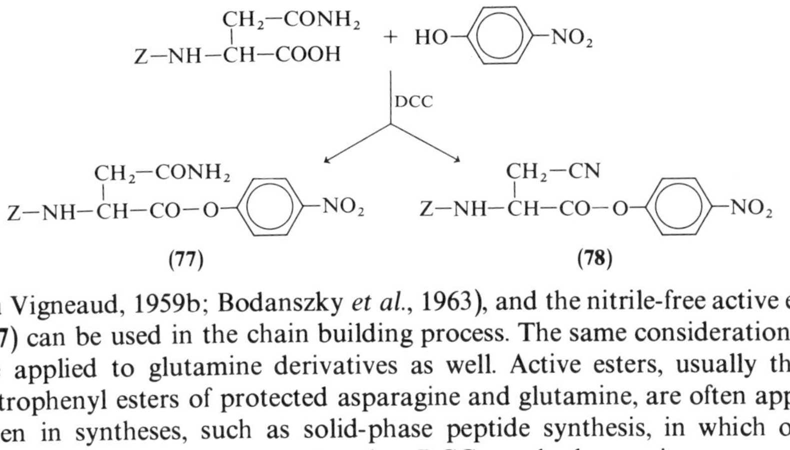

(50)

(**77**) (**78**)

du Vigneaud, 1959b; Bodanszky *et al.*, 1963), and the nitrile-free active ester (**77**) can be used in the chain building process. The same consideration can be applied to glutamine derivatives as well. Active esters, usually the *p*-nitrophenyl esters of protected asparagine and glutamine, are often applied even in syntheses, such as solid-phase peptide synthesis, in which other residues are incorporated by the DCC method or via symmetrical anhydrides.

B. Side Reactions during Acylation with Active Esters

1. *O*-Acylation

The view, held also for a long time by this author, that with moderately active acylating agents, such as azides or active esters, only strongly nucleophilic groups, the ε-amino group in lysine and the sulfhydryl group in cysteine need to be protected, seems to require some revision. The guanidino group of arginine and the carboxamides in the side chains of asparagine and glutamine are indeed inert enough to be left unblocked. The same assumption about the hydroxyl groups of serine, threonine and tyrosine residues was based on the fact that several active esters, e.g., *p*-nitrophenyl esters, can be recrystallized from hot ethanol, but it is not tenable any more; *O*-acylation has been reported again and again in the literature (e.g., Ramachandran and Li, 1963; Zahn *et al.*, 1965; Garner *et al.*, 1971; Geiger *et al.*, 1974).

A reexamination of this problem (Bodanszky *et al.*, 1977a) revealed that the presence of histidine in the amino component can greatly enhance O-acylation (cf. Stewart, 1968). Interestingly, the extent of imidazole catalyzed O-acylation is considerably reduced by the addition of an equimolecular amount of 1-hydroxybenzotriazole. The role of this additive is not easily interpreted since other weak acids, e.g., acetic acid, fail to produce a similar effect, and hydroxybenzotriazole itself, in the absence of imidazole, enhances rather than prevents O-acylation (cf. Klausner and Chorev, 1975). When no such complicating factors are present and a larger excess of acylating agent and tertiary base is avoided, O-acylation does not occur to any significant extent. For example, a protected 11-peptide (compound **79**) could be

$$\text{Boc—Ala—Val—}\underset{}{\overset{\text{2CZ}}{\text{Lys}}}\text{—}\overset{\text{2CZ}}{\text{Lys}}\text{—Tyr—Leu—Asn—}\overset{\text{Bzl}}{\text{Ser}}\text{—Val—Leu—Thr—NH}_2$$

(79)

prepared in very good yield and purity by stepwise chain lengthening with nitrophenyl esters (Bodanszky *et al.*, 1976), although the hydroxyl groups in its threonine and tyrosine side chains were not protected.

2. Chain Terminating N-Acylation

The observation that during chain lengthening with active esters, and with amino components present as acetate salts, some permanently blocked peptides are formed as by-products led to the postulation (Bodanszky, 1969) of mixed anhydride intermediates (reaction 51) which can produce, in addition

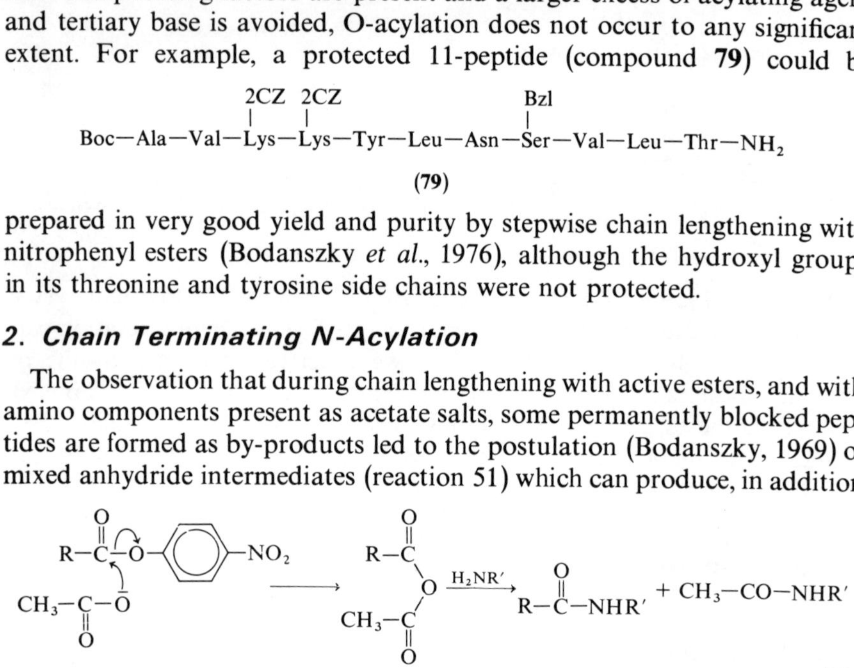

(51)

to the desired peptide, the *N*-acetyl derivative of the amino component as well. Similar side reactions were encountered later by Meienhofer and coworkers (1973) and by Fletcher *et al.* (1973). Mild alkaline treatment, e.g., piperidine in dimethylformamide, can be used to undo the harm caused by unwanted trifluoroacetylation,* but no ready solution can be suggested for the removal of an *N*-acetyl group. Excess base and basic solvents such as pyridine probably enhance the concentration of carboxylate and thereby the rate of nucleophilic displacement of the hydroxyl component from the active ester. Further studies are needed for the identification of the conditions that are conducive to this kind of chain termination.

3. Formation of Aminosuccinimide Derivatives in the Acylation of Aspartic Acid Containing Peptides with Active Esters

Active esters can produce mixed anhydrides, not only with the carboxyls of acetate or trifluoroacetate, but also with the unprotected side chain carboxyl group of an aspartyl residue in the amino component. In the acylation of the tetrapeptide amide Trp—Met—Asp—Phe—NH_2 with active esters of *tert*-butyloxycarbonyl–glycine, ring closure of the aspartyl residue occurred and an aminosuccinimide derivative (compound **80**) formed in considerable

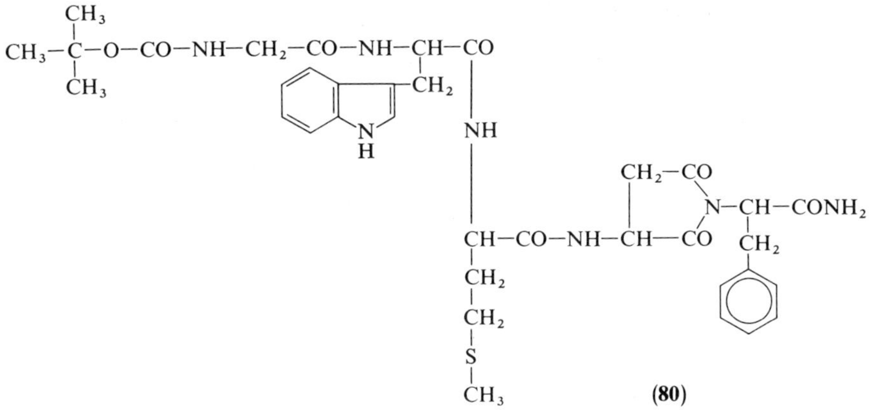

(80)

amounts (Bodanszky and Natarajan, 1975). The attack by the carboxylate of the aspartyl side chain was probably facilitated by the absence of a side

* The trifluoroacetyl group is removable with bases in organic solvents such as dimethylformamide. This raises an additional concern: the formation of failure sequences, that is, peptides in which one or more of the constituent residues are missing. This, at least in principle, could occur when, in stepwise chain lengthening, trifluoroacetyl peptides, present as impurities, are partially deblocked during the subsequent acylation reaction, which is carried out in the presence of a tertiary base. The addition of a small amount of HCl to the acid used for deprotection, e.g., to formic acid (Ohno *et al.*, 1972; Eberle *et al.*, 1975), might eliminate this side reaction, since the deprotected peptides would be hydrochlorides rather than formates. It should be possible to use HCl in trifluoroacetic acid in a similar manner, thereby retaining trifluoroacetic acid as the solvent.

chain in the active ester. Steric hindrance in amino acids other than glycine should limit the extent of this side reaction. The amount of the by-product could be reduced to a minimum by the omission of tertiary base in the coupling; the amphoteric amino component can be fully acylated. Activation of the side chain carboxyl of the aspartyl residue led to by-product formation, through an intermolecular reaction with the still unreacted portion of the amino component (Natarajan and Bodanszky, 1976). So far, no similar side reactions have been reported with glutamic acid side chains. Nevertheless, this experience can be used as an argument against minimal protection. On the other hand, the occurrence of many side reactions is probably related to the degree of activation of the carboxyl component (Brenner, 1967). Thus, the moderately active azides were often used to acylate amino components with unprotected carboxyl groups. Also, the weakly reactive unsubstituted α-phenyl ester of benzyloxycarbonyl-L-glutamic acid was applied by Klieger and Schröder (1963) as the acylating intermediate in the preparation of a series of peptides.

C. Racemization of Active Esters

It can be postulated, *a priori*, that activation of the carboxyl group of (α-amino-protected) amino acids should render them susceptible to base-catalyzed racemization (reaction 52). In most amino acids protected with a

$$\text{Z–NH–}\overset{\text{R}}{\underset{\text{H}}{\text{C}}}\rightarrow\overset{\text{O}}{\overset{\|}{\text{C}}}\rightarrow\text{X} + \text{B} \rightleftharpoons \text{BH}^{+} + \text{Z–NH–}\overset{\text{R}}{\underline{\text{C}}}\text{–}\overset{\text{O}}{\overset{\|}{\text{C}}}\text{–X} \qquad (52)$$

urethane-type protecting group, the tendency of carbanion formation by such direct abstraction of proton from the α-carbon is not significant. Measurable racemization occurs, however, when other acyl groups, such as acetyl

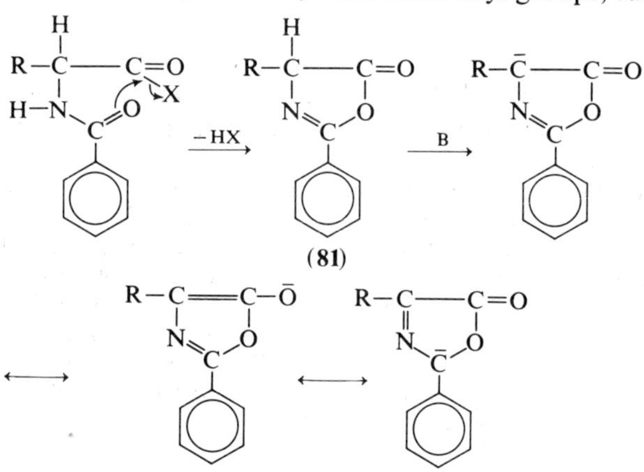

(53)

or, particularly, benzoyl, are attached to the α-amino group of the carboxyl-activated amino acid. Racemization takes place via axazolones (compound **81**), in which proton abstraction leads to a resonance-stabilized carbanion (reaction 53). In benzyloxycarbonylamino acids or *tert*-butyloxycarbonylamino acids, the electron distribution characteristic for urethanes prevents the loss of the NH proton and no azlactone is produced. When forced by extreme activation or high temperature, the reaction takes a different course and *N*-carboxy anhydrides (compound **82**) are formed (reaction 54):

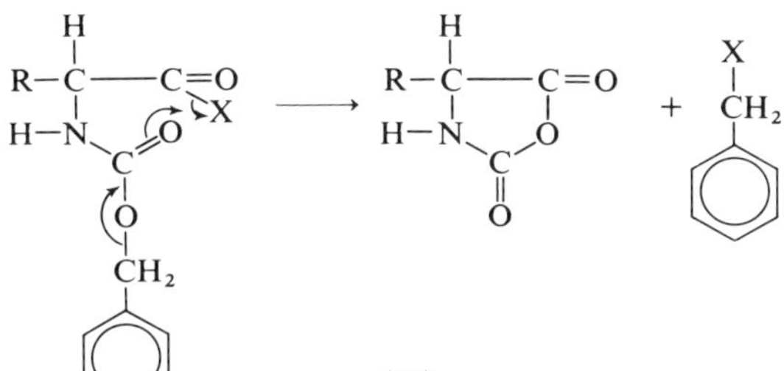

(54)

(82)

The relative chemical stability of active esters of protected amino acids, as compared with acid chlorides, azides, or mixed anhydrides, facilitates studies on the influence of solvents, temperature, protecting groups, the effect of concentration, and nature of tertiary bases. The observation of the changes in the optical rotation of the solutions of active esters was proposed by Bodanszky and Birkhimer (1960) as a method for the study of racemization. The first studies were prompted by an unpublished communication from Stouffer and du Vigneaud on the racemization of *N*-benzyloxycarbonyl-*S*-benzyl-L-cysteine *p*-nitrophenyl ester in the presence of triethylamine. The mechanism of the racemization is not obvious. The formation of racemic *N*-benzyloxycarbonyl-*S*-benzylcysteine thiobenzyl ester (Bodanszky and Bodanszky, 1967) suggests a process that involves the reversible base-catalyzed elimination of benzylmercaptan (reaction 55). An alternative route, isoracemization via a "conducted tour" mechanism (reaction 56), based on studies of the racemization of the pentachlorophenyl ester of the

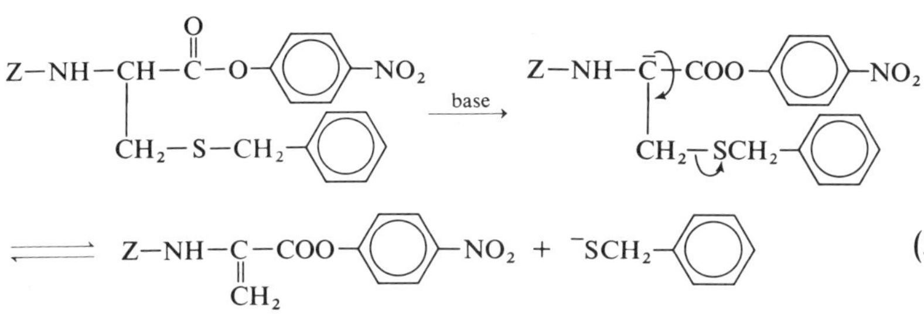

(55)

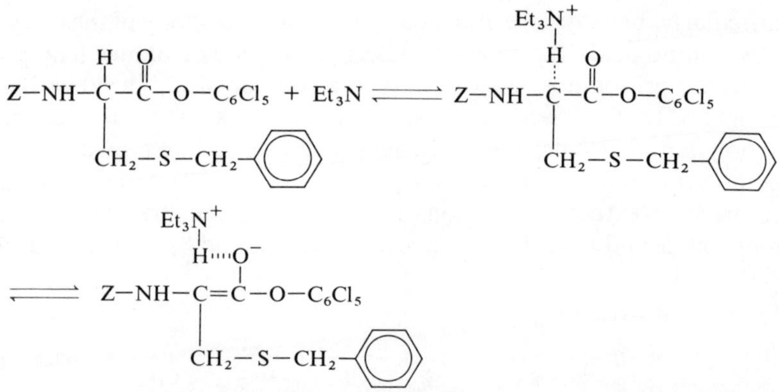

(56)

same protected amino acid, was proposed by Kovacs and associates (1968b, 1971). Derivatives of *S*-benzylcysteine and of the analogous *O*-benzylserine are not the only examples of racemization of active esters. Electron-withdrawing side chain substituents render the active esters of β-cyanoalanine (Liberek, 1963; Liberek and Grzonka, 1964) aspartic acid β-methyl ester and glutamic acid γ-methyl ester (Kovacs *et al.*, 1973) also susceptible to the racemizing effect of bases.

A reassuring result of studies on the influence of different bases (Bodanszky and Bodanszky, 1967) is that racemization of *N*-benzyloxycarbonyl-*S*-benzyl-L-cysteine *p*-nitrophenyl ester is prevented if triethylamine, the tertiary base commonly used for the liberation of amino components from their salts, is replaced by the hindered base, diisopropylethylamine. This base, however, has no significant influence on the rate of racemization of amino acid derivatives which racemize via oxazolones. Yet, the oxazolone route, as discussed earlier, is not operative in amino acids protected by a urethane-type group.

The ready racemization of active esters of protected phenylglycine, even if the protecting group is of the urethane type, is obviously due to the benzylic character of the α-carbon (compound **83**). The very interesting racemization-

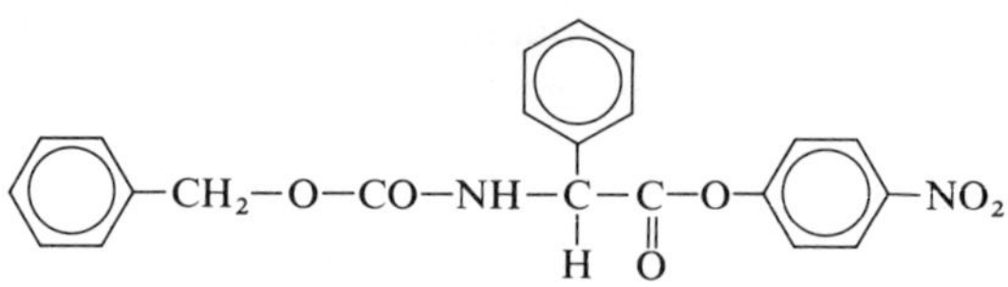

(83)

inhibiting effect of an ortho methoxy substituent in phenyl esters (Bankowski and Drabarek, 1971) deserves further investigations.

The influence of various α-amino protecting groups on the rate of racemization was reported by Anderson *et al.* (1965), who confirmed the previously noted (Bodanszky and Birkhimer, 1960) difference between the benzyloxycarbonyl and phthalyl groups. The latter are less potent in preventing racemization of active esters. An important contribution by Zimmerman and Anderson (1967) is the recognition of the mode of action of additives, such as 1-hydroxysuccinimide, used to reduce racemization in the coupling of peptides with DCC (Weygand *et al.*, 1966; Wünsch and Drees, 1966; Weygand and Ragnarsson, 1966). The formation of 1-hydroxysuccinimide esters as intermediates is the mechanism by which racemization is reduced. In this respect the use of additives is reminiscent of the preparation and use without isolation of *p*-nitrophenyl esters by Rothe and Kunitz (1957) and by Elliott and Russel (1957). Yet, while *p*-nitrophenol, 2,4,5-trichlorophenol, pentachlorophenol, or 8-hydroxyquinoline as additives are not too effective in reducing racemization, derivatives of hydroxylamine, e.g., 1-hydroxysuccinimide, *N*-hydroxyphthalimide and *N*-hydroxypiperidine are much superior to phenols in this respect: no racemate was detected in the Anderson–Callahan (1958) test. This is in harmony with the findings of Jones and Young (1968a) on the chiral stability of esters of 1-hydroxypiperidine. The resistance to base-catalyzed racemization of *N*,*N*-dialkylhydroxylamine esters (Beaumont *et al.*, 1965a) could be due in part to their only moderate intrinsic activation: they become potent acylating agents in the presence of the amino component. The same might be true for the esters of catechol (Jones and Young, 1968b).* On the other hand, the racemization-reducing ability of 1-hydroxybenzotriazole might require a different explanation. Here the intermediate *O*-acyl derivatives (active esters) can rearrange to an *N*-acyl derivative (König and Geiger, 1970a). The significance of this rearrangement on the prevention of racemization may require further studies.

The large number of papers dealing with the racemization of active esters could be interpreted as indication for the seriousness of this problem. In fact, all methods of activation can lead to racemization, but it is most readily studied in active esters, because these are relatively stable and can be observed for periods long enough for physical measurements with commonly used instruments. Moreover, active esters of only a few amino acids are racemized under the influence of tertiary bases and even these are sufficiently inert to the action of hindered bases. The real problem is not the racemization of active esters of protected amino acids, but the preparation and coupling of active esters of protected *peptides*. Numerous attempts were made to evaluate the racemization of various active esters of peptides. In some cases the favorable ratio between the rate of coupling to the rate of racemization

* The reduced tendency for racemization of 4-nitroguaiacyl esters was attributed to hydrogen bonding between the methoxy group and the chiral center of the amino acid (Bankowski and Drabarek, 1972).

(e.g., Kovacs *et al.*, 1967a; Kisfaludy *et al.*, 1967) suggests that these esters could be useful in the coupling of peptides. Unfortunately, these studies shed light only in one, and not necessarily the major part of the problem. The risk of racemization is not particularly great in the coupling reaction, where the amino component competes with the tertiary base for the acylating agent. Also, racemization, during coupling, probably can be avoided simply by keeping tertiary bases absent from the reaction mixture. This can be achieved by the prior isolation of the amino component in its free amine form. What remains problematic is the *preparation of optically pure active esters of protected peptides.* The preparation of esters generally requires the conversion of the COOH-terminal carboxyl group into an intermediate (e.g., an *O*-acylisourea) that is more reactive than the active ester to be prepared and hence also more prone to racemization. Nevertheless, active esters of peptides were prepared and used for coupling. That this approach is not always safe is shown by the formation of a considerable amount of a diastereoisomeric by-product in a coupling reaction that involved the *N*-hydroxysuccinimide ester of a protected peptide (Wünsch, 1975).

The major difficulty in the preparation of optically pure active esters of protected peptides lies in the diverse rates with which different amino acids racemize. The α-amino protecting group also has a major influence on the rate of racemization. In peptides, the penultimate residue plays the role of this protecting group, thus its effect will be different from peptide to peptide. Not less important are the nature of the solvent, that of the hydroxyl component or of the activating reagent, and even the concentration of the reactants. Therefore, no simple generalizations are possible. It might be necessary to establish the optical purity individually for each active ester prepared from peptides. Obvious exemptions are peptides with proline or glycine as COOH-terminal residues. While the former is quite resistant to racemization, glycine has no chirality. The preparation of optically pure active esters by the so-called "backing-off" procedure (Goodman and Stueben, 1959) will be discussed in the next section. An already mentioned method for the preparation of active esters of *N*-benzoylhydroxylamine via the activated form of this hydroxyl component (Taschner *et al.*, 1967; cf. also Lubiewska-Nakonieczna *et al.*, 1970) produced optically pure peptide esters. This kind of approach deserves further attention.

VI. ACTIVE ESTERS AND THE STRATEGY OF PEPTIDE SYNTHESIS

Decision between the alternatives of stepwise chain building versus fragment condensation is one of the primary concerns in the planning of the construction of a peptide chain. From the theoretically possible strategies

(Bodanszky and Ondetti, 1966b), stepwise chain-lengthening starting from the NH_2-terminal residue has to be excluded, until a racemization-free method for the preparation of active esters of peptides is found. The risk of racemization introduces serious limitations into the fragment-condensation strategy as well, but well selected fragments can be linked together via active esters. A by now classical approach is stepwise chain-building starting with the COOH-terminal residue.

A. Active Esters in Stepwise Chain Lengthening

This is the best application of active esters. The acylating agents are derived from protected amino acids. If the α-amino protecting groups offer protection not only against unwanted acylation but against racemization as well, then this crucial problem of peptide synthesis is reduced to the incorporation of a few residues in which the effect of side chains more or less cancels the effect of the protecting group. The case of *N*-benzyloxycarbonyl-*S*-benzyl-L-cysteine has already been discussed (see reactions 55 and 56). In this and similar cases, the use of hindered tertiary bases, particularly diisopropylethylamine (Bodanszky and Bodanszky, 1967), or the acylation of the amino components as free amines rather than a mixture of their salts and a tertiary base, provide the desired remedy. Therefore, the application of active esters is probably the method of choice in the stepwise strategy which was, in fact, originally designed (Bodanszky, 1960) in conjunction with *p*-nitrophenyl esters (Bodanszky, 1955). The practicality of the new approach at that time was first demonstrated in an efficient synthesis of oxytocin (Bodanszky and du Vigneaud, 1959a,b), as shown in Scheme 9. Among later examples, the synthesis of a major part of β-corticotropin (Schwyzer and Sieber, 1966) and the first synthesis of the 27-residue-long chain of porcine secretin (Bodanszky and Williams, 1967; Bodanszky *et al.*, 1967) may be mentioned.

A particular advantage of the stepwise strategy with active esters as acylating intermediates seemed to be the option to build peptides carrying minimal protection. It was thought (cf., e.g., Bodanszky and Klausner, 1973) that only the protecting of the ε-amino group of lysine and the sulfhydryl group of cysteine are mandatory, while other side chain functions in the amino components, such as carboxyl and hydroxyl groups, can be left unprotected. This view had to be revised. Side reactions encountered at more frequent intervals, e.g., the attack by side chain carboxyls or by the hydroxyl group of serine and tyrosine on active esters, were discussed in Section V,B,1. These observations caution against the unlimited application of the principle of minimal protection. As in so many other problems of peptide chemistry, no real generalizations are possible and a case by case determination of the need for protection must be recommended. Still, stepwise chain lengthening

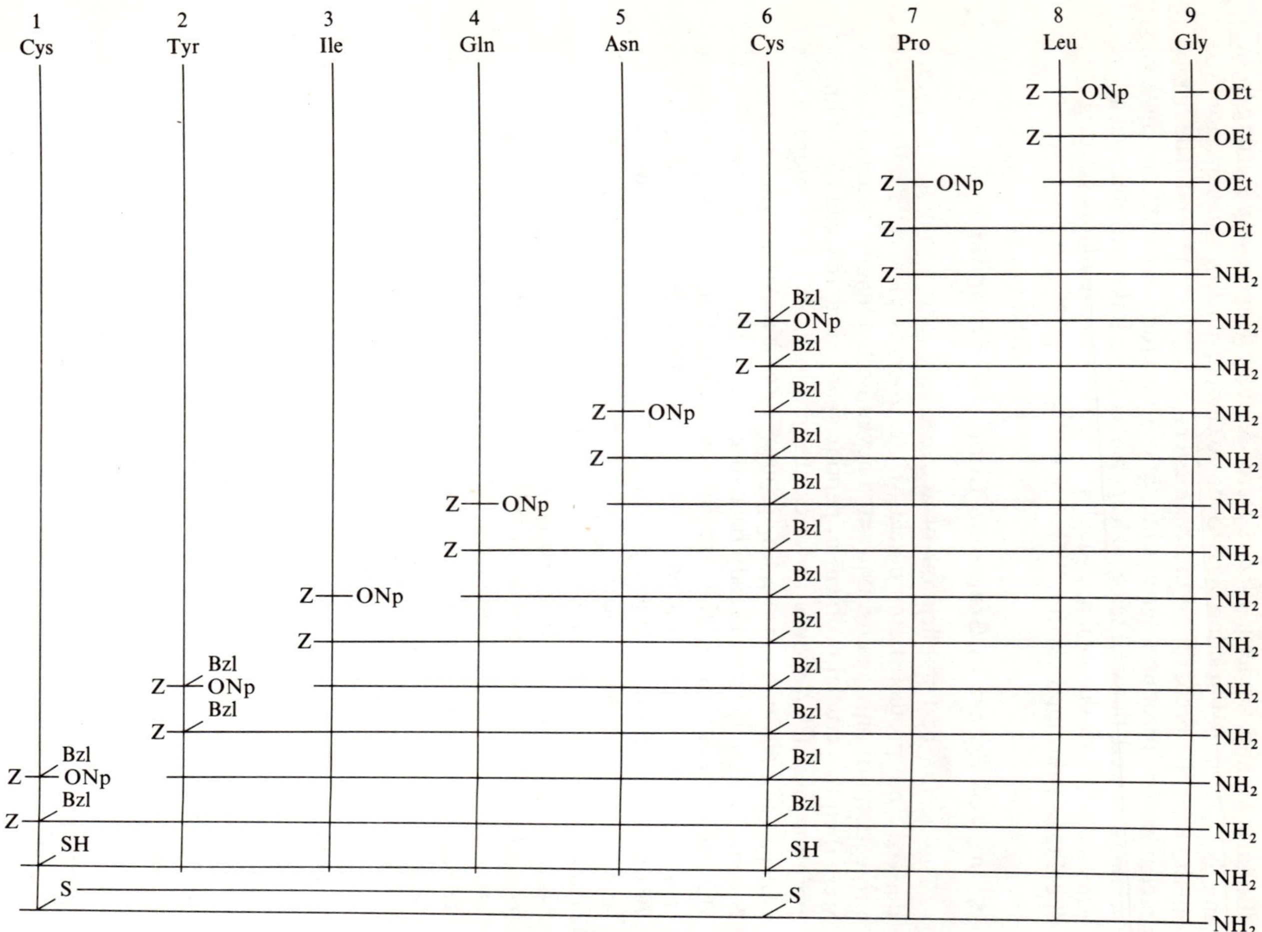
1 Cys
2 Tyr
3 Ile
4 Gln
5 Asn
6 Cys
7 Pro
8 Leu
9 Gly
Z
ONp
OEt
NH$_2$
Bzl
SH
S

Scheme 9

with active esters remains a valuable method. It allows the use of acylating agents in excess, and thereby eliminates, or at least reduces, the problem of diminishing reaction rates due to the increase in the molecular weight and concomitant decrease in the molar concentration of the amino component (Bodanszky, 1971). Furthermore, the excess active ester, the by-product formed from the leaving group (e.g., *p*-nitrophenol), and the salts of the tertiary bases can be removed with solvents in which the main product, a protected peptide, is insoluble. Also, the progress and completion of reactions are readily monitored on thin-layer chromatograms and can be quantitatively followed by measuring the amount of the eliminated hydroxyl component. These advantages render the application of active esters in the stepwise strategy a popular approach in peptide synthesis.

B. Active Esters in Fragment Condensation

It could be almost ideal to use active esters in the condensation of fragments, because this would allow an examination and, if necessary, purification of the activated carboxyl component prior to coupling. Also, the reactions could be carried out in exploratory experiments on a small scale before committing larger quantities to the preparation of the peptides on the desired scale. Other reactive intermediates, such as azides, are not sufficiently stable to permit a thorough examination. In most coupling procedures, activation and coupling occur in the same operation and therefore an analysis of the activated reactant is not possible. Notwithstanding the obvious advantage of active esters, they had very limited significance in fragment condensation, mainly because, as shown in Section V, in the preparation of active esters of protected peptides, racemization of their COOH-terminal residue has to be anticipated. This probably remains true also when active esters are prepared and used for coupling in a single step. The application of the reagent of Fujino and Hatanaka (1968) (compound **84**) is a good example, although not necessarily all the activation occurs in

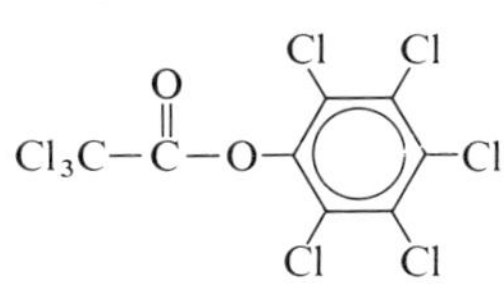

(**84**)

the form of an active ester. The acylation reaction might proceed, in part, through a mixed anhydride (Scheme 10) in which case, however, also some trichloroacetylation of the amino component has to be expected. Similarly, the coupling procedure of Mitin and Glinskaya (1969), based on the reaction of the carboxyl component with triphenylphosphite and imidazole, probably

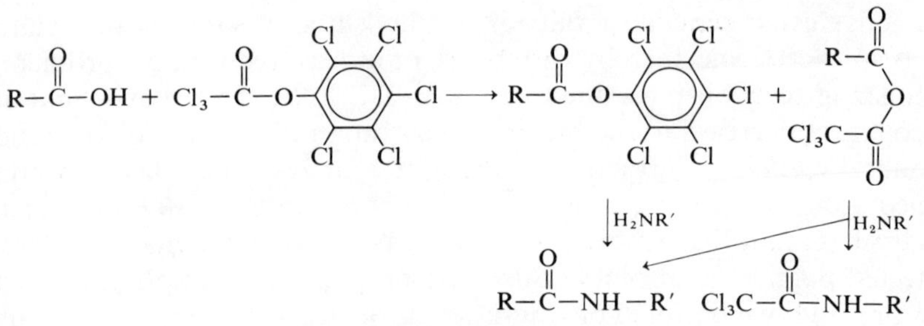

Scheme 10

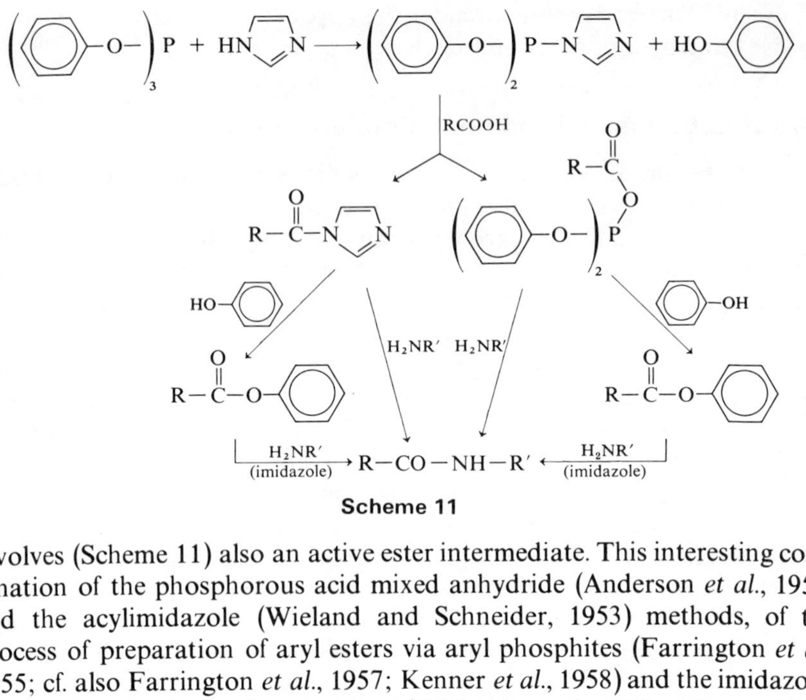

Scheme 11

involves (Scheme 11) also an active ester intermediate. This interesting combination of the phosphorous acid mixed anhydride (Anderson *et al.*, 1952) and the acylimidazole (Wieland and Schneider, 1953) methods, of the process of preparation of aryl esters via aryl phosphites (Farrington *et al.*, 1955; cf. also Farrington *et al.*, 1957; Kenner *et al.*, 1958) and the imidazole-catalyzed aminolysis of aryl esters (Wieland *et al.*, 1963; Mazur, 1963; Wieland and Vogeler, 1964) should be very efficient in the coupling of peptides, but it is unlikely that it would be immune to side reactions, particularly to racemization, known to occur in the contributing procedures.

Preparation of optically pure active esters of peptides was accomplished in an indirect way by acylation of amino acid active esters with a highly activated and therefore rapidly reacting protected amino acid or peptide (Goodman and Stueben, 1959). The amino acid active esters were obtained by acidolytic deprotection of benzyloxycarbonylamino acid active esters (reaction 57). A serious limitation of this approach is that it is based on the

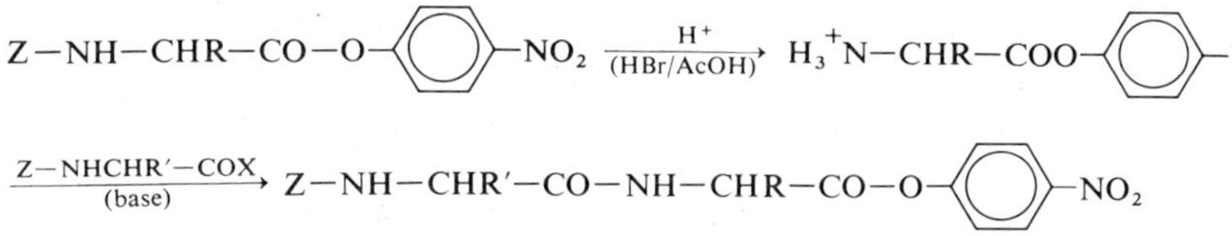

(57)

relative rates of the aminolysis reactions of two carboxyl-activated intermediates. This will be different from experiment to experiment and therefore the so-called "backing-off" procedure (Goodman and Stueben, 1959) will be satisfactory in some cases and not in others. The same principle, however, allowed the preparation of *N*-acetyl-*S*-benzyl-L-cysteine *p*-nitrophenyl ester in optically active, perhaps also optically pure, form (Cash, 1962).

It might be a worthwhile effort to prepare active esters of protected peptides and to purify them, prior to coupling to the amine component, until they are homogeneous, optically pure materials. If then the amino component is used as the free amine rather than a mixture of an amine salt and tertiary base, little or no formation of the undesired diastereomer is expected. The difficulties in the purification of protected peptide active esters are greatly reduced when the fragments can be so selected that their COOH-terminal residue is proline, which does not racemize readily, or glycine, which has no center of asymmetry. An example of such a protected peptide active ester is shown in compound (**85**) (Bodanszky *et al.*, 1972b).

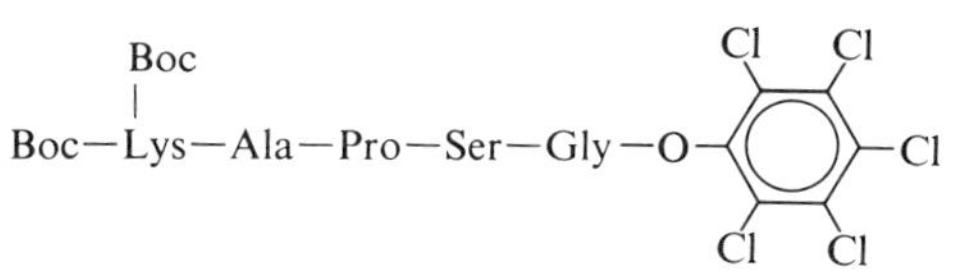

(**85**)

C. Active Esters in Cyclization

Linear chains of amino acids can form two kinds of cyclic peptides: homodetic peptides, in which all the bonds between the residues in the ring are peptide bonds, and heterodetic peptides, in which the ring is closed, e.g., by a lactone or a disulfide (Schwyzer *et al.*, 1956). Both types are common among microbial peptides, and several of these are biologically active as antibiotics, or toxic substances, e.g., the principles of poisonous mushrooms. Therefore, synthesis of cyclic peptides is an interesting and often challenging objective. A notable exception is the preparation of cyclic dipeptides (anhydro amino acids, diketopiperazines) which often form spontaneously, e.g., on distillation or even on storage of amino acid esters, and, particularly, of dipeptide esters. The driving force in the cyclization reaction, the thermodynamic stability of the six-membered ring is sufficient to overcome the barrier caused by the need for changing the more stable trans peptide bond to a cis bond. Diketopiperazines form with extreme ease from dipeptide esters in which the two residues have different configurations. In this case, the side chains lie on opposite sides of the general plane of the ring and hence do not interfere with the cyclization reaction (reaction 58). As seen from this exam-

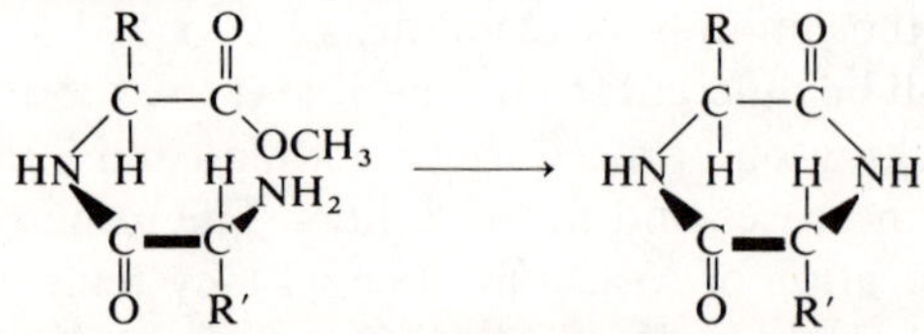

(58)

ple, the cyclization to diketopiperazines does not require more activation than that found in methyl or ethyl esters. This is not the case with larger cyclic peptides; in their preparation, active esters play an important role.

Homodetic Cyclic Peptides

The first successful synthesis of a biologically active cyclic peptide, gramicidin S (compound **86**), was accomplished by Schwyzer and Sieber (1957a,b). The crucial step, cyclization of the linear intermediate (Scheme 12), was

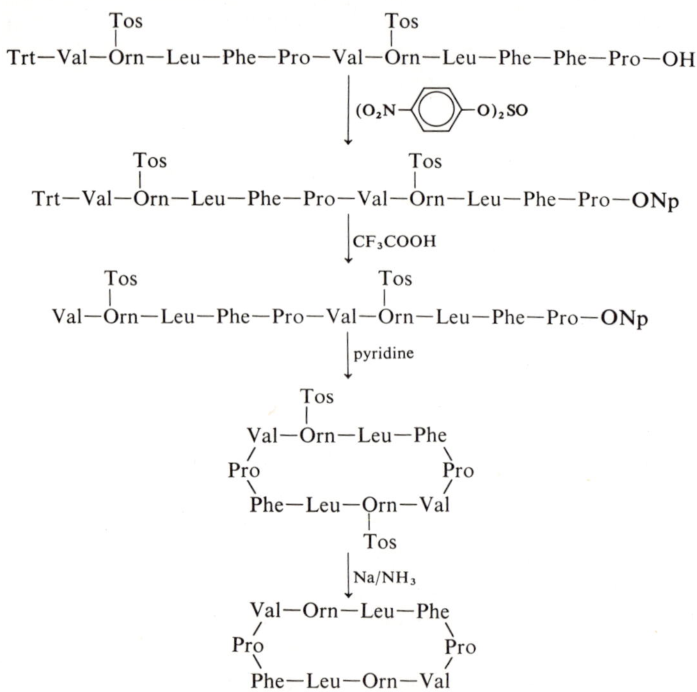

Scheme 12. Phe has D configuration.

achieved by the application of the *p*-nitrophenyl ester method (Bodanszky, 1955). An important feature of cyclization reactions is that, in order to avoid intermolecular reactions, they have to be carried out at high dilution. In the synthesis of gramicidin S, this principle was implemented by adding the decapeptide active ester, still protected by protonation, very slowly to warm pyridine. This technique was later adopted in numerous syntheses of cyclic peptides. On the other hand, a tendency for cyclodimerization (Schwyzer and Sieber, 1958) of the protected pentapeptide active ester intermediate H—Val—Orn(Tos)—Leu—DPhe—Pro—ONp was exploited for a new synthesis of gramicidin S. The dimer is stabilized by the intermolecular hydrogen bonds which, after cyclization, determine the conformation of the antibiotic. No cyclopentapeptide was found in these experiments, although such a peptide, semigramicidin S, was subsequently isolated from the products of similar cyclizations with active esters (Kato and Izumiya, 1974). Cyclodimerization is suppressed in some peptides which, due to intramolecular hydrogen bonds, have a quasi-cyclic conformation (Bodanszky and Henes, 1976). It can be further reduced when the final peptide bond of the ring is closed through the azide, rather than through an active ester.

An interesting and potentially important approach to cyclic peptides with active esters is the use of a COOH-terminal carboxyl protecting group that can be converted into an activating group just prior to cyclization. The principle of the active esters based on *p*-methanesulfonylphenol (Johnson and Jacobs, 1968; cf. also Johnson and Ruettinger, 1970; Schwyzer and Sieber, 1958) was applied (Marshall and Liener, 1970; Flanigan and Marshall, 1970) to a peptide anchored to polymeric support containing such a protecting-activating partial structure (Scheme 13). The nitrated phenolic resins of Fridkin *et al.* (1965) are similarly suitable for the preparation of cyclic peptides, except that carboxyl protection and activation in the intermediates of the chain lengthening procedure are not distinct, and hence the cyclization could occur prematurely.

D. Active Esters in Polymerization

Large molecular weight peptides are valuable models of proteins in many physicochemical investigations. Also, they could serve in medicine, e.g., as plasma substitutes. Therefore, the preparation of polymers of amino acids or of peptides with repeating sequences ("sequential peptides") is a challenging objective. Polymers of amino acids were most often obtained from *N*-carboxy anhydrides or Leuchs' anhydrides. Active esters were rarely used for this purpose. Not only are the *N*-carboxy anhydrides more convenient (their "protecting group" is spontaneously lost after coupling) but they also produce polymers of higher polymerization number. Active esters of amino

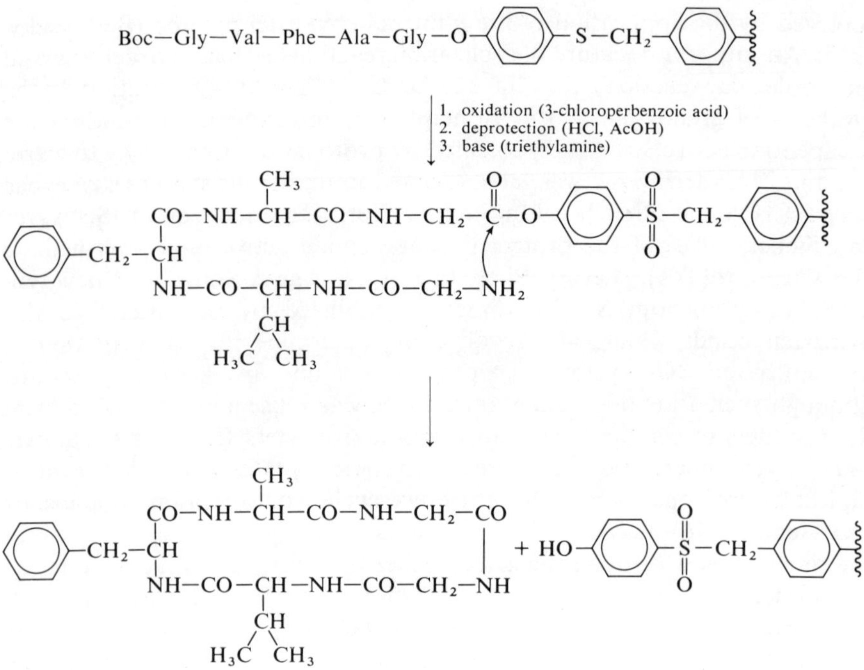

Scheme 13

acids show a tendency for the formation of diketopiperazines and other cyclic products. Only at high concentration of the esters is this tendency less dominant (Brack and Spach, 1971) and even then the average molecular weight of the polyamino acid formed is much lower than of those prepared from *N*-carboxy anhydrides.

Active esters play a major role in the synthesis of sequential polypeptides. An early example is the polymerization of the thiophenyl ester of Gly—DLVal—DLIle in acetone, with triethylamine added as the base to liberate the tripeptide active ester from its hydrochloride (Wieland and Bernhard, 1953). Yet, in this experiment only a low polymerization number, about 9, could be achieved, and the product was obviously a complex mixture of diastereoisomers. About ten years later, the polymerization of tripeptide *p*-nitrophenyl esters (DeTar *et al.*, 1963) produced optically homogeneous materials of higher molecular weight (reaction 59). The poly-

$$\text{Z—Asp(OCH}_3\text{)—Gly—Gly—ONp} \xrightarrow{\text{HBr/AcOH}} \text{HBr} \cdot \text{Asp(OCH}_3\text{)—Gly—Gly—ONp}$$
$$\xrightarrow{\text{triethylamine}} [\text{Asp(OCH}_3\text{)—Gly—Gly}]_n \qquad (59)$$

merization was monitored by nmr spectra: the CH_3O signal of the terminal residue is different from those in the side chain of mid-chain residues. The outcome of the polymerization is concentration dependent. Thus, at a concentration of 15% monomer only little polymer formed; the main reaction was probably cyclodimerization leading to a cyclic hexapeptide. In the same study, other polymers such as (His—Gly—Gly)$_n$ and (Phe—Gly)$_n$ were similarly prepared. An analogous approach to sequential polypeptides was developed by Stewart (1965, 1966). The collagen model poly-(Gly—LPro—Gly) ester was synthesized in the laboratory of Blout (Bloom *et al.*, 1966) from the tripeptide *p*-nitrophenyl ester hydrobromide. This, in turn, was prepared from benzyloxycarbonyl-glycyl-L-prolylglycine *p*-nitrophenyl ester with HBr in acetic acid. It should be noted here that selective removal of a benzyloxycarbonyl group from *p*-nitrophenyl esters of protected amino acids, without reduction of the nitro group, was accomplished by Kovacs and Rodin (1968); the presence of one mole of HCl prevents premature polymerizations. The application of the procedure to peptide derivatives was not always successful (Kovacs *et al.*, 1968a). The use of acid-labile protecting groups circumvents this problem. For example, active esters of *tert*-butyloxycarbonyl peptides can be applied as starting materials (DeTar and Vajda, 1967) or peptide esters protected by the *o*-nitrophenyl sulfenyl group (Johnson, 1969).

Different active esters might give different results with respect to the degree of polymerization. According to the studies of Kovacs and Kapoor (1965), pentachlorophenyl esters are particularly suitable as monomers because of their good crystallization properties. They also produce polymers of higher molecular weight (Kovacs *et al.*, 1968a) than other active esters. High molecular weight peptides were obtained from *N*-hydroxysuccinimide esters of tripeptides (Fridkin *et al.*, 1969) as well. Further increase in the polymerization number can be achieved by starting with very concentrated solutions of the peptide esters, or even from their suspensions (Brack and Spach, 1970).

VII. ACTIVE ESTERS IN VARIOUS TECHNIQUES OF PEPTIDE SYNTHESIS

A. Synthesis in Solution

1. *Introduction of Single Amino Acids into Naturally Occurring Peptides*

The advantages of acylation with active esters, such as the easy removal of excess reagent, the possibility of monitoring the chain-lengthening process and the often unequivocal course of the reaction, are more obvious in syntheses carried out in solution than in the building of chains which are anchored to a polymeric support. These attractive features of the active ester method prompted its use beyond the conventional area of peptide synthesis. A particularly interesting application of active esters is the lengthening of the chain of naturally occurring peptides. For example, insulin was treated with the *p*-nitrophenyl ester of *tert*-butyloxycarbonyl-L-methionine (Levy and Carpenter, 1966) to afford, after deprotection, trimethionyl insulin. One methionine residue was added to the NH_2-terminus of each chain and a third to the side chain of the single lysine residue in the molecule. In a more selective acylation, homogeneous *p*-iodophenylalanine insulin was prepared with the 2,4,5-trichlorophenyl ester of the protected iodoamino acid (Krail *et al.*, 1971) applied to a partially protected dephenylalanine insulin. In similar experiments, the A-chain was lengthened with bis-*tert*-butyloxycarbonyl-L-lysine 2,4,5-trichlorophenyl ester, and also with bis-*tert*-butyloxycarbonyl-L-histidine *p*-nitrophenyl ester, while the *p*-nitrophenyl ester of *tert*-butyloxycarbonyl-L-arginylglycine was used for the acylation of a deglycine A-chain (Weinert *et al.*, 1969).

2. *Introduction of Protecting Groups with Active Esters*

In addition to the activation of amino acids, the active ester principle was applied also for the preparation of protected amino acids. The reagent *tert*-butyl *p*-nitrophenyl carbonate was proposed by Anderson and McGregor (1957) for the introduction of the *tert*-butyloxycarbonyl group:

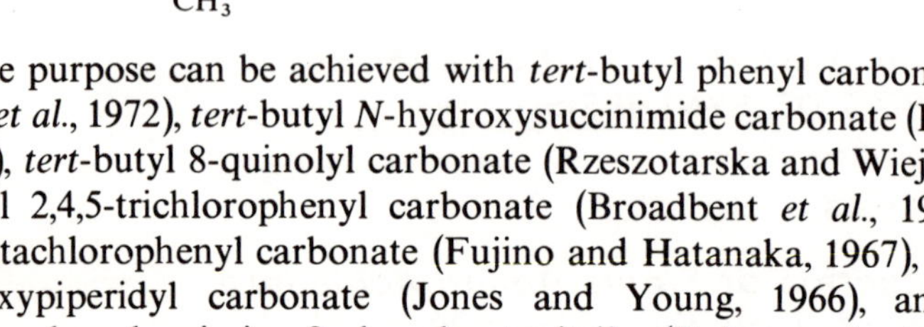

(60)

The same purpose can be achieved with *tert*-butyl phenyl carbonate (Ragnarsson *et al.*, 1972), *tert*-butyl *N*-hydroxysuccinimide carbonate (Frankel *et al.*, 1966), *tert*-butyl 8-quinolyl carbonate (Rzeszotarska and Wiejak, 1968), *tert*-butyl 2,4,5-trichlorophenyl carbonate (Broadbent *et al.*, 1967), *tert*-butylpentachlorophenyl carbonate (Fujino and Hatanaka, 1967), *tert*-butyl *N*-hydroxypiperidyl carbonate (Jones and Young, 1966), and 2-*tert*-butyloxycarbonyloxyimino-2-phenylacetonitrile (Itoh *et al.*, 1975). The active ester approach was used also for the introduction of the benzyloxycarbonyl group (Zahn and Falkenburg, 1960) and the acid labile *p*-methoxybenzyloxycarbonyl group (Weygand and Hunger, 1962). The application of reactive carbonates gradually became a standard procedure in the preparation of protected amino acids with a urethane-type protecting group. Thus, compounds **(87)**–**(90)** were prepared (Sieber and Iselin, 1968a,b;

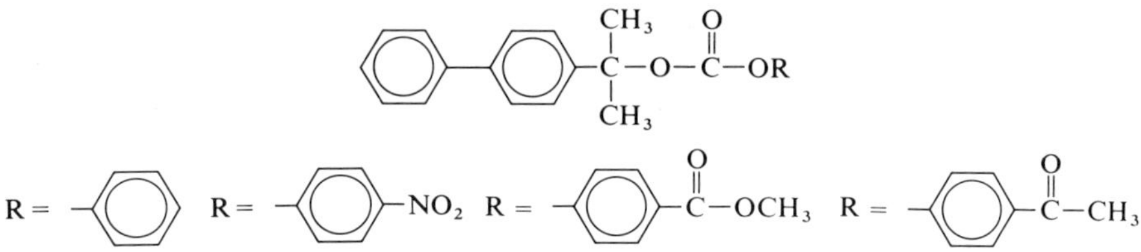

(87) **(88)** **(89)** **(90)**

Schnabel *et al.*, 1971a) in connection with the 2-(*p*-biphenylyl)propyl(2)oxycarbonyl (Bpoc) group. The active esters, at least, are less dangerous than the corresponding azides: explosions were reported in the use and storage of *tert*-butyl azidoformate, the reagent most commonly applied for the preparation of *tert*-butyloxycarbonylamino acids.

3. Introduction of Acetyl Groups, Radioactive and Fluorescent Groups and Biotin into Peptide Chains

The selectivity of active ester reactions allows the often unequivocal acetylation of amino groups with *p*-nitrophenyl acetate. For example, a pure acetyl-pentapeptide derivative was obtained from benzyloxycarbonyl-L-isoleucyl-D-alanyl-D-alanyl-L-valyl-D-leucine methyl ester by treatment with HBr in acetic acid, followed by a reaction of the resulting hydrobromide with *p*-nitrophenyl acetate in dimethylformamide, in the presence of a tertiary base (M. Bodanszky and B. J. Henes, unpublished, 1976). The same reagent, and also *p*-nitrophenyl esters of other acids, are popular substrates for studies on the mechanism of hydrolytic enzymes, probably because of the ease with which the progress or rate of the reactions can be determined with a spectrophotometer. The example just mentioned cautions against the indiscriminate use of *p*-nitrophenyl acetate and similar active esters: in addition to being hydrolyzed, they can be readily aminolyzed as well. Furthermore, a reactive ester is not necessarily the best substrate when models of enzymes, which catalyze the hydrolysis of unreactive esters and also of amides, e.g., chymotrypsin, are the objectives of the study.

To facilitate labeling with radioactive iodine, the deaminotyrosyl group can be incorporated into peptides or proteins (Bolton and Hunter, 1973) with the help of compound **(91)**, an *N*-hydroxysuccinimide ester. For the introduction of a fluorescent label, acylation with dansylglycine *N*-hydroxysuccinimide ester (compound **92**) was carried out by Mehlis and Apelt (1973). For the preparation of biocytin and biocytin-containing peptides, the *p*-nitrophenyl ester of biotin (compound **93**) can be used. Some

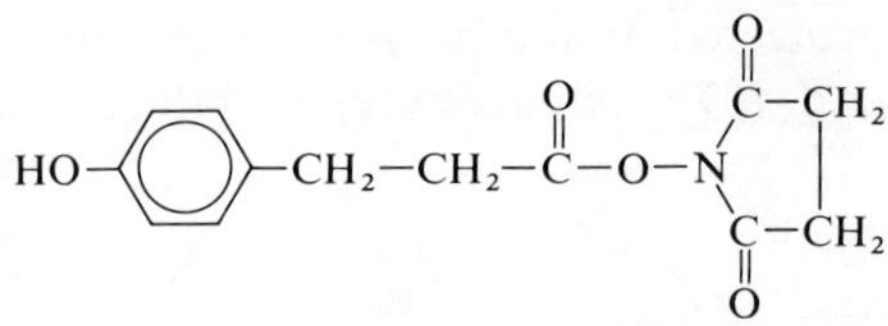

(91)

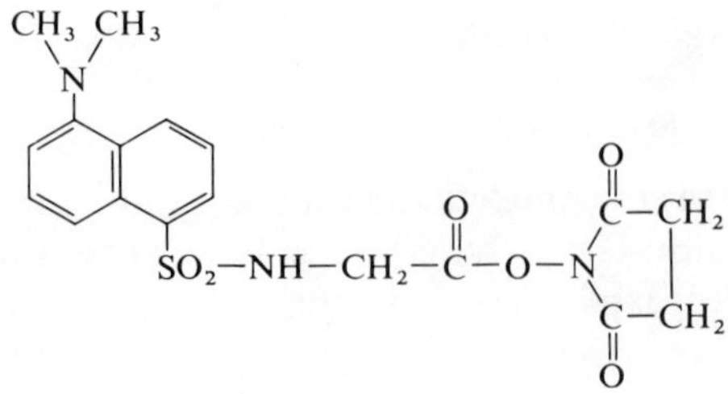

(92)

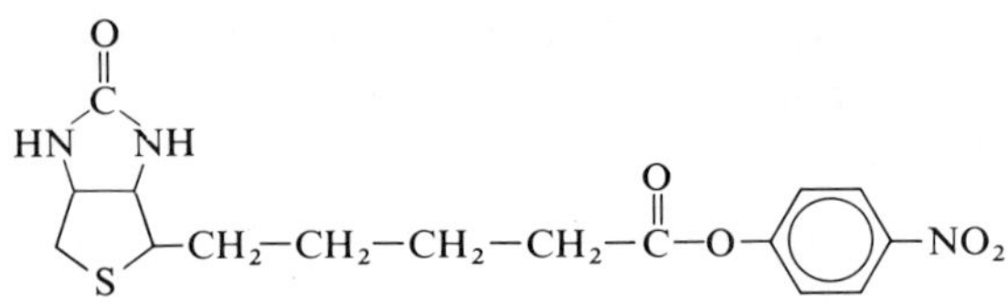

(93)

difficulties were encountered in the preparation of this active ester in satisfactory purity (McCormick, 1973) or in good yield (Becker *et al.*, 1971; Viswanatha *et al.*, 1975). The investigations of Bodanszky and Fagan (1977) revealed that the urea nitrogens in biotin are sufficiently reactive to interfere with various esterification procedures. In the reaction of biotin in pyridine with excess *p*-nitrophenyl trifluoroacetate (Sakakibara and Inukai, 1964), first an *N*-trifluoroacetyl derivative of compound (**93**) was formed. This could be converted, by alcoholysis, to the desired compound (**93** in reaction 61). On the other hand, once the biotin moiety is attached to the ε-amino

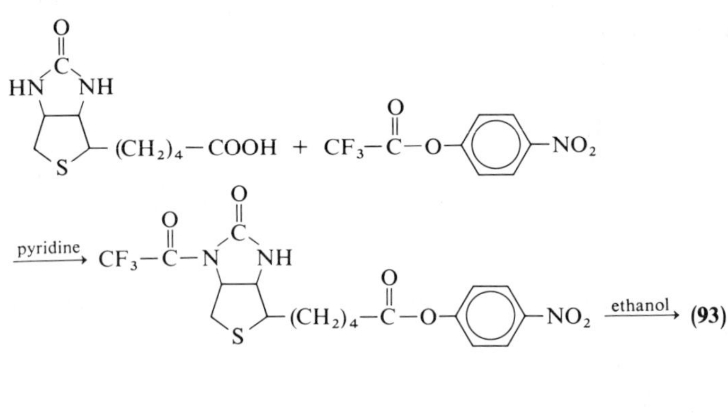

(61)

group of a lysine residue which is part of a peptide chain, the latter can be lengthened through acylation with moderately active esters that are less reactive than *p*-nitrophenyl trifluoroacetate in which the electron-withdrawing effect of the nitrophenyl group is superimposed upon the strong negative inductive effect in the acid component.

4. Crosslinking with Active Esters

An interesting utilization of active esters is the crosslinking of peptides and proteins with bifunctional acylating agents such as bisbenzyloxycarbonyl-L-cystine bis-*p*-nitrophenyl ester (compound **94**),

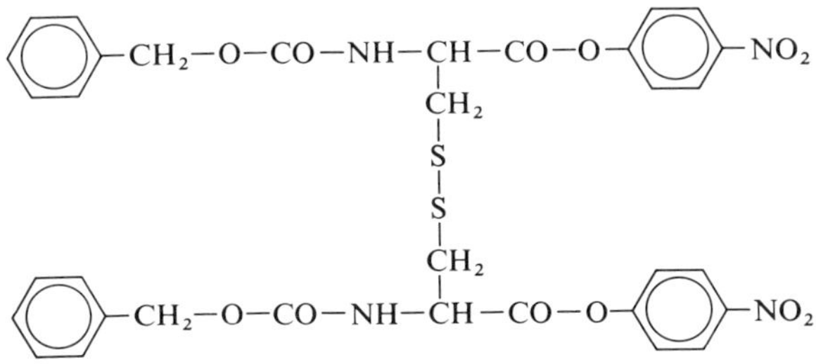

(94)

applied by Schade and Zahn (1962) for this purpose. Busse *et al.*, (1974) connected the two chains of insulin by acylation with carbonylbis-L-methionine *p*-nitrophenyl ester (compound **95**) (Busse and Carpenter, 1974),

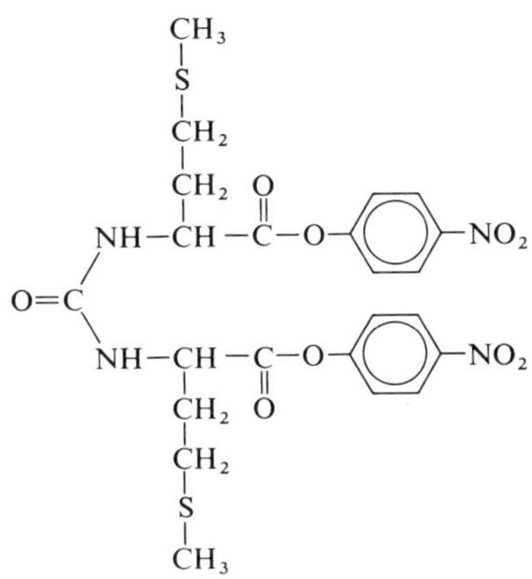

(95)

a method probably applicable also in other similar situations. A special advantage of the crosslink thus introduced is that it can be removed by cleavage with cyanogen bromide (Gross and Witkop, 1961).

Insulin is a favorite target of crosslinking experiments. The distance between the A and B chains can be "measured" with a series of crosslinking reagents such as compounds **(96)**–**(98)** (Brandenburg *et al.*, 1975).

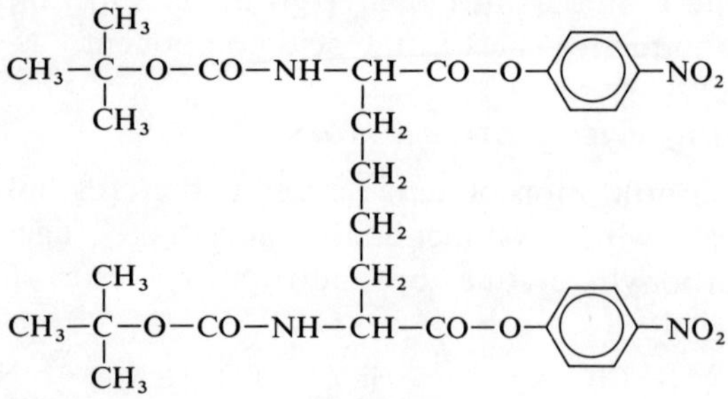

(96)

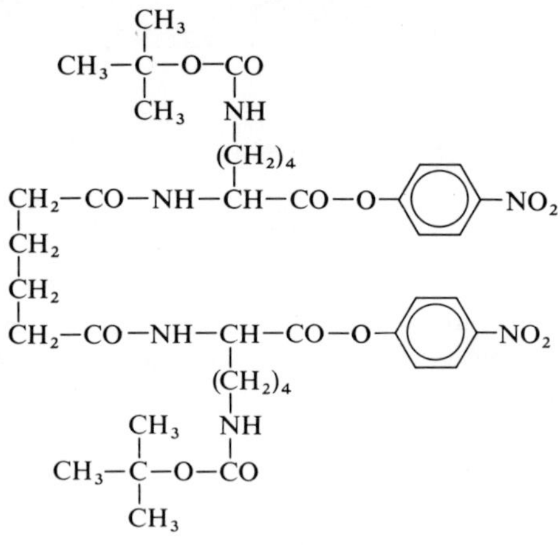

(97)

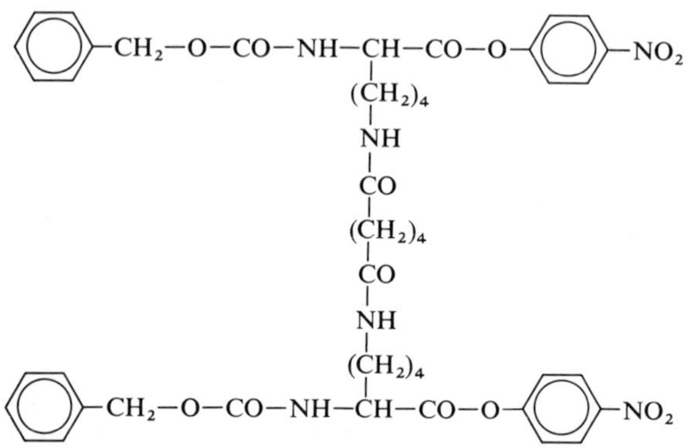

(98)

The laboratory of Zahn extended the application of active esters also to the crosslinking of wool and for the impregnation of wool by the polycondensation of bifunctional active esters and diamines on the surface of wool (reaction 62) (Rouette, 1967).

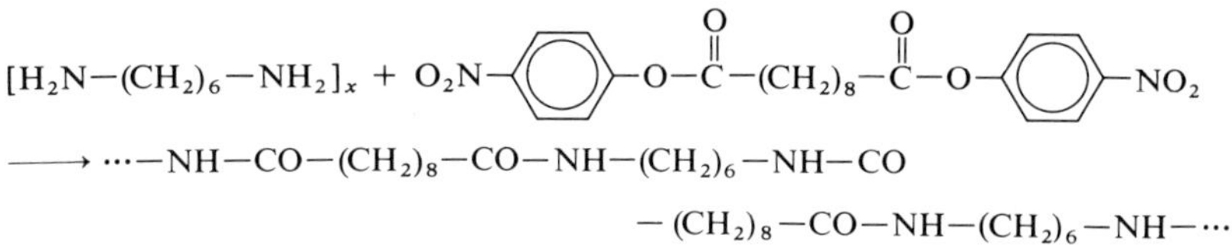

(62)

5. *In Situ* Peptide Synthesis

Facilitation of peptide synthesis is an obviously desirable objective. In fact, it might be considered as necessary when chains of considerable length have to be assembled. The processes which involve the anchoring of an ionizable protecting group (Wieland and Birr, 1967; Camble *et al.*, 1968; Wieland and Racky, 1968) can be executed with active esters as acylating agents, but also with other coupling methods. On the other hand, acylation with active esters is the method of choice in the *in situ* synthesis of peptides (Bodanszky *et al.*, 1973, 1974a, b). In this technique, a peptide (protected, e.g., by a *tert*-butyloxycarbonyl group) is placed in a centrifuge tube provided with a standard tapered joint. The Boc group is removed by brief treatment with trifluoroacetic acid; the tube is then attached to a rotary evaporator and the trifluoroacetic acid is removed *in vacuo*. The trifluoroacetate salt of the resulting amine is precipitated with ether, washed with ether by centrifugation and decantation, and dried. The material is dissolved in dimethylformamide and treated with a tertiary amine, such as diisopropylethylamine (Bodanszky *et al.*, 1973, 1974a, b). In this technique, a peptide (protected, e.g., by a *tert*-butyloxycarbonyl group) is placed in a centrifuge tube provided by spot tests with ninhydrin or fluorescamine. After completion of the reaction, the bulk of the solvent is removed by evaporation *in vacuo* and the protected peptide is precipitated by the addition of an appropriate solvent, usually ethyl acetate. The product is separated from the solution by centrifugation, washed with ethyl acetate by repeated suspension, centrifugation and decantation, and dried. The decanted solution contains the excess acylating agent, the displaced hydroxyl component, e.g., *o*-nitrophenol, and the salt of the tertiary base. (Diisopropylethylammonium trifluoroacetate is soluble in ethyl acetate, to a lesser extent even in ether.) The peptide, lengthened in this cycle by one residue, is ready for a new series of operations, that is, deblocking with trifluoroacetic acid, etc. The characteristic feature of the technique is that the intermediates need not be removed from the centrifuge tube throughout the chain building process. Sometimes it was possible to recrystallize or reprecipitate them, also *in situ*. Of course, it is quite possible to take samples of the trifluoroacetate salts or of the amino-protected intermediates for examinations on thin-layer chromatograms, for amino acid analysis, elemental analysis for the determination of melting point, specific rotation, U.V. absorption, etc. The *in situ* technique was also extended to peptides protected with the benzyloxycarbonyl group. In this case, HBr in acetic acid was used for deblocking; the hydrobromides were precipitated with ether and treated in dimethylformamide with a tertiary base and a benzyloxycarbonylamino acid active ester. Since the hydrobromides of the commonly used tertiary bases are not soluble in ethyl acetate or ether, 95% ethanol was used for precipitation and washing. The application of this technique is limited to peptides which, in their protected forms, are insoluble in a solvent that can keep the other components of the reaction mixture in solution. The practicality of the *in situ* procedure was demonstrated in the simultaneous synthesis of the COOH-terminal tricosapeptide of secretin and three of its analogues (Fink and Bodanszky, 1976).

B. Active Esters in Solid-Phase Peptide Synthesis

The application of insoluble active esters as acylating agents has already been discussed in this chapter. Therefore, we focus the attention here on the building of chains which are anchored to an insoluble support. In an early period of his investigations, Merrifield (1963) found active esters unsuitable in solid-phase synthesis. A reexamination of this question (Bodanszky and Sheehan, 1964) pointed to the role of the solvent in the reaction of active esters with polymer-bound amino components. In benzene, the solvent used by Merrifield, the poor swelling of the resin impeded the process. In dimethylformamide, however, both the reaction rates of the active esters and the swelling of the resin were satisfactory for practical results.* In a continuation of these studies (Bodanszky and Sheehan, 1966), a COOH-terminal part of the gastrointestinal hormone secretin was assembled on a modified solid support, a hydroxymethyl polymer. The latter allows the ester bond formation between the carboxyl of the COOH-terminal residue and the hydroxyl group in the polymer to be carried out under mild conditions. Also, the absence of alkylating properties, inherent in the original chloromethyl resin of Merrifield, eliminates several side reactions.

The way seemed to be open to solid-phase synthesis with active esters. For the incorporation of asparagine and glutamine, *p*-nitrophenyl esters of the protected forms of these amino acids are routinely applied. Here the advan-

* Toluene as a solvent and swelling agent (of polymers) is even more advantageous than dimethylformamide (Khan and Sivanandaiah, 1976).

tages of the active ester approach are obvious: with dicyclohexylcarbodiimide coupling, both the desired carboxamide derivative and an undesired residue with a cyano group in its side chain become part of the growing peptide chain. Since, in solid-phase synthesis, purification of intermediates is possible only after these are cleaved from the solid support, two species of chains will be built concurrently after the incorporation of an asparagine or glutamine residue, four species after the addition of a second amino acid with a carboxamide side chain, and so on. The active esters, on the other hand, can be purified by crystallization until they are free of the cyano derivatives. Their application is an easy way out of this dilemma. The *p*-nitrophenyl esters of *tert*-butyloxycarbonyl-L-asparagine and *tert*-butyloxycarbonyl-L-glutamine are commercially available.

In isolated cases, polymer-bound peptides were built entirely with active esters. Thus, the 21-residue long A-chains of ovine and bovine insulin were synthesized exclusively with *p*-nitrophenyl esters of *tert*-butyloxycarbonylamino acids (Hörnle, 1967; Weber *et al.*, 1967). On combination of the synthetic A-chains with natural B-chains, only a part of the biological activity of insulin could be produced. Yet, this activity was several times higher (Weitzel *et al.*, 1968) than that of preparations synthesized by dicyclohexylcarbodiimide couplings. Notwithstanding such advantages of active esters in acylation reactions, or the possibility of monitoring the progress of the reaction by the spectrophotometric determination of *p*-nitrophenol or other eliminated hydroxyl components, coupling with DCC remained the method of choice in solid-phase peptide synthesis. The obvious reason for the failure of active esters to play a more important role in solid phase synthesis is their moderate reactivity under hindered conditions (Karlsson *et al.*, 1970). Both the matrix of the resin and the peptide chain growing in the confined environment of the polymer cause serious steric hindrance. This renders the amino component less available for acylation. The same factors also further enhance the effect of any steric hindrance present in the acylating agent, because they reduce the chances of a favorable approach to the available area of the acylating group. Thus, for instance, acylation with active esters of protected isoleucine or valine often remains incomplete, even if the acylating agents are used in considerable excess and long times are allowed for the reactions (Lübke, 1971). Steric hindrance present in the activating group has a similar effect. Some active esters, e.g., pentachlorophenyl esters or *N*-hydroxysuccinimide esters, which react rapidly in solution, were found to be unsatisfactory in solid-phase peptide synthesis (Klostermeyer *et al.*, 1967; Klostermeyer, 1968). However, perhaps under different conditions or on a resin with a less dense matrix, pentachlorophenyl esters were used (Johnson, 1973) in the preparation of fairly long chains. The same kind of diverse observations can be found in the literature on the usefulness of 5-chloro-8-hydroxyquinoline esters; in the studies of

Rudinger and Gut (1967), slow and imperfect acylation was observed, while these esters were found superior to other active esters by Jakubke and Baumert (1974). A better agreement exists on the influence of the growing peptide chain. Two investigators (Bath, 1970; Hagenmaier, 1970) agree on the steric hindrance created by the peptide itself; it is first reduced when a few residues provide some distance between the free amino group and the polymer, but this is followed by an increase of reaction times when the length of the chain is sufficiently large to fill out a sizable portion of the space available for operations.

Under the encumbered conditions present in the inside of the spheres of the polymer, it is rather important to reduce any steric hindrance in the active ester. The observation (Bodanszky *et al.*, 1972a) that *o*-nitrophenyl esters are less sensitive to steric hindrance and to solvent effects than their *p*-isomers found explanation in the preferred conformation of most *o*-nitrophenyl esters. In the single-turn helix (see Fig. 1, Section IV, A) of these compounds, the intramolecular involvement of the nitro group allows less interaction with solvent molecules, as shown by the relatively low solubility of these compounds in solvents such as chloroform or tetrachloromethane: *o*-nitrophenyl esters are probably solvated to a lesser extent than the corresponding *p*-nitrophenyl esters. With their smaller solvate shell, these esters more readily penetrate the crowded environment in the polymeric support (Bodanszky *et al.*, 1974b). The applicability of *o*-nitrophenyl esters in solid-phase peptide synthesis was demonstrated (Bodanszky and Funk, 1973) in the preparation of the heptapeptide derivative L-leucyl-L-glutaminyl-L-asparaginyl-*S*-benzyl-L-cysteinyl-L-prolyl-L-leucylglycinamide that was obtained with better efficiency and in better quality than in the parallel experiment in which DCC was used for coupling. By no means should these results be interpreted as proof for the superiority of *o*-nitrophenyl esters over all kinds of active esters. For example, the observations of Kovács and Penke (1973) on the high rates of acylation of aminoacyl resins with pentafluorophenyl esters show that highly reactive and probably only slightly hindered acylating agents can be found for the purpose of solid-phase synthesis. The studies concerning *o*-nitrophenyl esters should rather be regarded as an attempt to introduce new considerations, such as steric hindrance, solvation or solvent dependence of aminolysis rates, in the search for the ideal acylating agent in the field of solid-phase peptide synthesis.

VIII. CONCLUSIONS

The advantages of active esters, such as less equivocal reactions, their relative stability (and hence their availability from commercial sources), the ease with which they allow monitoring the progress of the reaction, probably outweigh their disadvantages, e.g., their moderate aminolysis rates. This explains why, in the two decades that followed their introduction in peptide synthesis, the use of active esters has steadily increased. Initially, only esters of benzyloxycarbonylamino acids were prepared, then those of *tert*-butyloxycarbonylamino acid and *o*-nitrophenylsulfenylamino acids. Subsequently, however, the active ester principle was extended to tritylamino acids (Schnabel, 1964) and also to amino acids protected, as enamines (Dane *et al.*, 1962), or with β-dicarbonyl compounds (Balog *et al.*, 1970). The popularity of active esters could be best demonstrated by listing the active esters of protected amino acids and peptides reported in the literature. Such an exhaustive list seems to be both too difficult to prepare and too cumbersome to be used when information is needed quickly. Therefore, this chapter is concluded with tables (Tables I–XIII) of active esters selected on the basis of their general usefulness.

No attempt was made to evaluate the importance of one kind of active ester over others. The widely different conditions of peptide synthesis and particularly the *individuality of the amino acids* render such a choice difficult, if not impossible. It is conceivable that one kind of active ester is ideal for one protected amino acid, but a different activating group is more suitable for another. The only generalization valid in this area is that most generalizations cannot be substantiated. While no reliable predictions are possible in the field of peptide synthesis, the expectation that active esters will remain important tools in syntheses carried out in solution and eventually gain major significance in solid-phase synthesis as well, is probably justified.

Table I Benzyloxycarbonylamino Acid *p*-Nitrophenyl Esters

	mp (C°)	$[\alpha]_D^{20-25}$	Solvent $c = 1-2$	Reference
Ala	79–79.5	−38.1	EtOAc	Goodman and Stueben (1959)
Arg(Tos)	50–60 dec	−13.5	DMF	Guttmann *et al.* (1962)
Asn	158–159.5[a]	−31.5	DMF[b]	Bodanszky *et al.* (1963)
Asp(OBzl)	76	−16.6	DMF	Guttmann (1961)
Asp(OtBu)	86.5–87.5	+ 1.5	$CHCl_3$	Hylton *et al.* (1966)
Cys(Bzl)	93–94	−43	DMF[b]	Bodanszky and du Vigneaud (1959b)
Gln	155–156	−24	DMF[b]	Bodanszky *et al.* (1963)
Glu(OBzl)	114–115	−32.5	MeOH	Li *et al.* (1960)
Glu(OtBu)	60–61	−33.5	MeOH	Scoffone *et al.* (1964)
Gly	128	—	—	Bodanszky (1957a)
His(Z)	112–113	−13.8	$CHCl_3$	Meienhofer (1962)
Ile	60–62	−15.5	DMF[b]	Bodanszky and du Vigneaud (1959b)
Leu	95	−33.5	DMF	Bodanszky and du Vigneaud (1959b)
Lys(Z)	58–59	−18.5	DMF	Marchiori *et al.* (1962)
Lys(Boc)	88–91	−14.8	Acetone	Schwyzer and Rittel (1961)
Lys(Tos)	109–110	−16.5	DMF	Bodanszky *et al.* (1960)
Met	96–97	−33.8	DMF	Marchiori *et al.* (1962)
Phe	126–127	−24.7	DMF	Bodanszky and du Vigneaud (1959c)
Pro	94–96	−68	DMF	Bodanszky and du Vigneaud (1959b)
Ser(Bzl)	55–57	−12	DMF	Bodanszky *et al.* (1962)
Thr	93–95	−24	DMF	Bodanszky and Ondetti (1966a)
Trp	105	− 4.5	DMF[c]	Wilchek and Patchornik (1963)
Tyr(Bzl)	148–150	− 9.0	DMF	Bodanszky and du Vigneaud (1959b)
Tyr	158–159	−16.3	Acetone	Martin *et al.* (1959)
Val	66–67	−25	DMF	Marchiori *et al.* (1962)

[a] On rapid heating, 165°–166°C (Bodanszky and du Vigneaud, 1959b).
[b] Containing 1% AcOH.
[c] $c = 5$.

Table II *tert*-Butyloxycarbonylamino Acid *p*-Nitrophenyl Esters

	mp (C°)	$[\alpha]_D^{20-25}$	Solvent $c = 1-2$	References
Ala	83	−52.5	DMF	Sandrin and Boissonnas (1963)
Arg(NO_2)	126–128	−30.0	DMF[a]	Meienhofer *et al.* (1970)
Asn	178–180 dec	−36.0	DMF	Sandrin and Boissonnas (1963)
Asp(OBzl)	104–106[d]	−36	DMF[a]	Bodanszky and Williams (1967)
Asp(OtBu)	77–79	−42.5	DMF[a]	Ondetti *et al.* (1968)
Cys(Bzl)	95–96	−37.5	DMF	Beyerman *et al.* (1968)
Gln	154–155	−35	DMF	Beyerman *et al.* (1968)
Glu(OBzl)	123	−33.5	DMF	Sandrin and Boissonnas (1963)
Gly	70–71	—	—	Sandrin and Boissonnas (1963)
His(Z)	131–133	−19.2	Dioxane	Schaich *et al.* (1973)
His(Boc)	141–143	−22.8[b]	Dioxane	Schnabel *et al.* (1968)
Ile	65–66.5	−32.5	DMF	Wünsch (1974)
Leu	94–95	−48[c]	DMF	Beyerman *et al.* (1968)
Lys(Z)	83–85	−23.6	DMF	Lanzilotti *et al.* (1964)
Lys(Boc)	127	−26	DMF	Sandrin and Boissonnas (1963)
Lys(4CZ)	130–132	−13.5[b]	Dioxane	Schnabel *et al.* (1971b)
Lys(2CZ)	88–91	−23.8	DMF[a]	Bodanszky *et al.* (1976)
Met	99–100	−48	DMF	Scoffone *et al.* (1964)
Phe	132	−21	DMF	Sandrin and Boissonnas (1963)
Ser(Bzl)	Oil	—	—	Hörnle (1967)
Thr	98–100	−40	DMF[a]	Meienhofer *et al.* (1970)
Trp	114–116	+ 1.4	DMF[a]	Bodanszky and Natarajan (1975)
Tyr	164–166	− 7	DMF[a]	Meienhofer *et al.* (1970)
Tyr(Bzl)	145–146	− 3.0	DMF	Beyerman *et al.* (1968)
Val	64–66	−38.4[b]	Dioxane	Schnabel *et al.* (1971b)

[a] Containing 1% AcOH.
[b] $[\alpha]_{578}$.
[c] $c = 0.9$.
[d] Dimorphous; mp 80°–81°C (from EtOH).

Table III *o*-Nitrophenyl Esters of Protected Amino Acids

	Benzyloxycarbonyl derivatives[a]		*tert*-Butyloxycarbonyl derivatives[b]	
	mp (C°)	$[\alpha]_D^{25\,c}$	mp (C°)	$[\alpha]_D^{25\,c}$
Ala	94	−51	86–87	−79
Asp(OBzl)	74–75	−32	94–95	−42
Asn	162	−42	144.5–146.5	−55
Cys(Bzl)	101–102[f]	−105	103–105	−75
Glu(OBzl)	83–84	−83	123–125	−54
Gln	135–136	−39	149.5–151	−55
Gly	75	—	96.5–98	—
Ile	Oil	—	Oil[d]	−37
Leu	87–89	−47	56–57	−68
Lys(Boc)	76–78	−32	—	—
Lys(Dcz)[e]	—	—	135–137	−37
Lys(2CZ)[g]	—	—	108–110	−40
Met	65–67	−56	104–105	−73
Phe	109–110	−63	146–146.5	−65
Pro	Oil[d]	—	63–70	−84
Ser(Bzl)	55–57	−18	49–52	−30
Trp	116–118	−65	155–156	−62
Tyr(Bzl)	137	−55	139–140	−51
Val	Oil[d]	—	53–56	−44

[a] Bodanszky *et al.* (1974a).
[b] Bodanszky *et al.* (1973).
[c] $c = 2$, in DMF containing 1% AcOH.
[d] Crystals at low temperatures.
[e] Dcz = 2,6-dichlorobenzyloxycarbonyl.
[f] Bodanszky (1957a).
[g] M. Bodanszky and S. Natarajan, unpublished.

Table IV Benzyloxycarbonylamino Acid 2,4,5-Trichlorophenyl Esters

	mp (C°)	$[\alpha]_D^{20-25}$	Solvent $c = 1$–2	References
Ala	104	−24	DMF	Pless and Boissonnas (1963)
Asn	225	−30	DMF	Huguenin (1964)
Asp(OBzl)	92	−21	DMF	Pless and Boissonnas (1963)
Asp(OtBu)	73–74	−21.4	EtOAc	Handford *et al.* (1967)
Cys(Bzl)	100	−48	DMF	Pless and Boissonnas (1963)
Gln	180	−17	DMF	Pless and Boissonnas (1963)
Glu(OtBu)	107–108	−20	DMF	Pless and Boissonnas (1963)
Gly	107–108	—	—	Pless and Boissonnas (1963)
His(Bzl)	107–108	− 8	DMF	Pless and Boissonnas (1963)
Ile	79	− 9	DMF	Pless and Boissonnas (1963)
Leu	64–65	−29.1	MeOH	Handford *et al.* (1967)
Lys(Z)	113–114	−11	DMF	Pless and Boissonnas (1963)
Lys(Boc)	99	−11.3	DMF	Huguenin and Guttmann (1965)
Met	112.5–113.5	−31.4	DMF	Costopanagiotis *et al* (1966)
Phe	142	−51	DMF	Pless and Boissonnas (1963)
Ser	137	−25	DMF	Pless and Boissonnas (1963)
Ser(tBu)	70–72	− 2	DMF	Pless and Boissonnas (1963)
Thr(tBu)	Oil	—	—	Jones (1972)
Trp	115	−38	DMF	Pless and Boissonnas (1963)
Tyr	149–150	−30	DMF	Pless and Boissonnas (1963)
Tyr(Z)	143–144	−36	DMF	Pless and Boissonnas (1963)
Val	94	− 8	DMF	Pless and Boissonnas (1963)

Table V *tert*-Butyloxycarbonylamino Acid 2,4,5-Trichlorophenyl Esters

	mp (C°)	$[\alpha]_D^{20-25}$	Solvent $c = 1$–2	References
Ala	81–82	−39	DMF	Pless and Boissonnas (1963)
Asn	188	−35	DMF	Pless and Boissonnas (1963)
Asp(OtBu)	76–77	—	—	Broadbent *et al.* (1967)
Cys(Bzl)	77–78	—	—	Broadbent *et al.* (1967)
Gln	161–162	—	—	Broadbent *et al.* (1967)
Glu(OBzl)	106–108	−33.5	MeOH	Anderson *et al.* (1966)
Glu(OtBu)	114.5–115	−31	DMF	Anderson *et al.* (1966)
Gly	106–107	—	—	Pless and Boissonnas (1963)
His(Bzl)	123.5–124.5	− 9.9	$CHCl_3$	Handford *et al.* (1968)
Ile	69–70	—	—	Broadbent *et al.* (1967)
Leu	Oil	—	—	Broadbent *et al.* (1967)
Lys(Z)	98–99	—	—	Broadbent *et al.* (1967)
Met	89–90	−38.5	DMF	Bentley *et al.* (1964)
Phe	122	−27	DMF	Pless and Boissonnas (1963)
Pro	60–61.5	−40.8	MeOH	Fletcher and Jones (1972)
Ser	104–105	−41[a]	DMF	Bodanszky *et al.* (1977c)
Thr	85–88	−46.2	MeOH	Bodanszky *et al.* (1977b)
Thr(Bzl)	80–81	−20	DMF[b]	Bodanszky *et al.* (1977b)
Trp	132–133	−30.5	DMF	Bentley *et al.* (1964)
Tyr	156–158	−26.4	EtOAc[c]	Anderson *et al.* (1966)
Tyr(Bzl)	116–117	—	—	Broadbent *et al.* (1967)
Val	58–60	—	—	Broadbent *et al.* (1967)

[a] $[\alpha]_{578}$.

[b] Containing 1% AcOH.

[c] $c = 3.5$.

Table VI Benzyloxycarbonylamino Acid Pentachlorophenyl Esters

	mp (C°)	$[\alpha]_D^{20-25}$	Solvent c = 1–2	References
Ala	172–173	−25.1	$CHCl_3$	Kovacs *et al.* (1966)
Arg(NO_2)	109–111	−14.2	DMF	Kovacs and Ceprini (1965)
Asn	167–168.5	−26.7	DMF	Mukaiyama *et al.* (1970b)
Asp(OBzl)	151–152	—	—	Kovacs *et al.* (1966)
Cys(Bzl)	171–173	−34.3	$CHCl_3$	Kovacs *et al.* (1967a)
Gln	183–187	+11[a]	DMF	Kovacs *et al.* (1967a)
Glu(OtBu)	94–95	−12.0	$CHCl_3$[b]	Kovacs *et al.* (1966)
Gly	185	—	—	Kovacs *et al.* (1966)
Ile	122–124	− 5.8	$CHCl_3$	Kovacs *et al.* (1967a)
Leu	127–128	−33.3	MeOH	Kovacs *et al.* (1967a)
Lys(Z)	154–157	− 7.7	MeOH	Kovacs *et al.* (1967a)
Met	132–133	− 1.4	$CHCl_3$	Kovacs *et al.* (1967a)
Phe	158	−56	DMF	Pless and Boissonnas (1963)
Pro	93–95.5	−47.7	DMF	Kovacs *et al.* (1967a)
Ser	191.5–192.5	−23.6	DMF	Kovacs *et al.* (1967a)
Thr	162–164	−20.0	DMF	Visser *et al.* (1968)
Trp	167–169	−57.2	$CHCl_3$	Kovacs *et al.* (1967a)
Tyr(*t*Bu)	115–116	−43.2	DMF[c]	Johnson and Trask (1969)
Val	144–145	−21.0	$CHCl_3$	Kovacs *et al.* (1967a)

[a] $[\alpha]_D$ − 17.3 (in DMF) was reported by Mukaiyama *et al.* (1970b).
[b] $c = 3.4$.
[c] $c = 3$.

Table VII *tert*-Butyloxycarbonylamino Acid Pentachlorophenyl Esters

	mp (C°)	$[\alpha]_D^{20-25}$	Solvent $c = 1–2$	References
Ala	166	−22.2	$CHCl_3$[a]	Johnson and Trask (1968)
Asn	176	−41.7	DMF	Johnson (1973)
Asp(OBzl)	129–130	−28.9	DMF[b]	Johnson (1973)
Cys(Bzl)	154	−38.2	DMF	Johnson (1973)
Gln	151–152	−26.5	DMF	Visser *et al.* (1970)
Glu(OBzl)	138	−27.1	DMF[c]	Johnson (1973)
Glu(OtBu)	150–151	−19.2	$CHCl_3$	Visser *et al.* (1970)
Gly	142	—	—	Johnson and Trask (1968)
His(Dnp)	170	+ 3.9	DMF[d]	Johnson (1973)
Ile	119	−28.0	EtOAc[e]	Johnson and Trask (1968)
Leu	111	−35.3	EtOAc[f]	Johnson and Trask (1968)
Lys(Z)	140–141	−15.0	$CHCl_3$[g]	Johnson and Trask (1968)
Lys(Boc)	148	−20.6	DMF[h]	Johnson and Trask (1968)
Met	139	−37	DMF[i]	Johnson (1973)
Phe	148	−47.1	EtOAc[j]	Johnson and Trask (1968)
Pro	117	−32.5	DMF[k]	Johnson (1973)
Ser	157	−31.0	DMF	Johnson (1973)
Ser(Bzl)	76	−13.4	DMF[l]	Johnson (1973)
Thr	196	−29.0	DMF[m]	Johnson (1973)
Thr(Bzl)	126	−16.0	DMF[n]	Johnson (1973)
Trp	195	−49.2	DMF	Johnson (1973)
Tyr	170–172	−41	$CHCl_3$	Kamber (1973)
Tyr(Bzl)	142	−26.8	$CHCl_3$[o]	Johnson and Trask (1968)
Val	126	−38.1	EtOAc[p]	Johnson and Trask (1968)

[a] $c = 5.1$.
[b] $c = 2.6$.
[c] $c = 3$.
[d] $c = 2.7$.
[e] $c = 5.2$.
[f] $c = 5$.
[g] $c = 5$.
[h] $c = 4.6$.
[i] $c = 2.3$.
[j] $c = 4.8$.
[k] $c = 3.9$.
[l] $c = 2.2$.
[m] $c = 2.4$.
[n] $c = 2.4$.
[o] $c = 5$.
[p] $c = 5$.

Table VIII Pentafluorophenyl Esters of Protected Amino Acids

	Benzyloxycarbonyl derivatives[a]		*tert*-Butyloxycarbonyl derivatives[b]	
	mp (C°)	$[\alpha]_D^{25}$ ($c = 2$, EtOAc)	mp (C°)	$[\alpha]_D^{25}$ [$c = 1$, EtOAc (E)] [$c = 1$, Dioxane (D)]
Ala	84–85	−22.4	82–83	−31.2 D
Arg(NO_2)	98–99	−12.6	—	—
Asn	—	—	124–126	−16.7 E
Asp(OBzl)	95.5–96	− 9.9	—	—
Asp(OtBu)	76.5–77.5	−43.1	98	−10.7 E
Cys(Bzl)	85–86	−25.6	99–101	−42.0 E
Gln	116–118	−10.9	146–148	−25.8 DE
Glu(OtBu)	85–87	−16.5	90–91	−24.1 E
Gly	86–87	—	79–80	—
Ile	Oil	−13.5	48–49	−16.9 E
Leu	Oil	−12.7	48–50	−30.2 E
Lys(Z)	105–107	8.6	54–55	−17.1 D
Lys(Boc)	71–75	− 9.6	—	—
Met	104–106	−18.7	80–82	−22.5 D
Phe	98–100	−26.7	111–112	−26.9 D
Pro	Oil	−55.1	51–52	−53.0 D
Ser	142.5–143.5	−16.3	90–91	−29.8 D
Thr	—	—	105–106	−32.6 E
Trp	124–125	−26.4	109–111	−28.1 D
Val	50–51.5	−12.3	62–64	−18.4 D

[a] Kisfaludy *et al.* (1967, 1970; cf. also Kovacs *et al.*, 1967b, c).
[b] Kisfaludy *et al.* (1973).

Table IX Benzyloxycarbonylamino Acid *N*-hydroxyphthalimide Esters[a]

	mp (C°)	$[\alpha]_D^{22}$ [b]
Gly	124	—
Leu	112	−34.0
Pro	109	−81.9
Phe	117–119	−44.6
Cys(Bzl)	110–112	−46.8
Tyr(Z)	68–69[c]	—
Ile	122[c]	—
Asn	153–155[c]	—
Gln	132–134[c]	—

[a] Nefkens *et al.* (1962).
[b] In acetic acid, $c = 1$–2.
[c] Kaurov *et al.* (1970).

Table X Benzyloxycarbonylamino Acid *N*-Hydroxysuccinimide Esters

	mp (C°)	$[\alpha]_D^{20-25}$	Solvent $c = 1-2$	References
Ala	123–123.5	−37.2	Dioxane	Anderson *et al.* (1964)
Arg(Z_2)	85–86	− 9.7	Dioxane	Anderson *et al.* (1964)
Asn	129–130	−27.5	DMF	Sakakibara and Hase (1968)
Asp(OtBu)	151–152	−27.4	DMF	Hofmann *et al.* (1965)
Cys(Bzl)	90–91	−58.5	Dioxane	Kovacs *et al.* (1970)
Gln	134–136	−23.6	DMF[a]	Beacham *et al.* (1971)
Glu(OtBu)	101–103[b]	−29.9	EtOH[a]	Beacham *et al.* (1971)
Gly	113–114	—	—	Anderson *et al.* (1963)
His(Dnp)	90–93	+38.3	$CHCl_3$	Shaltiel and Fridkin (1970)
Ile	115.5–116	−15.5	Dioxane	Anderson *et al.* (1964)
Leu	116–117	−33.6	Dioxane	Anderson *et al.* (1964)
Lys(Z)	110–112	−16.4	Dioxane	Wünsch (1974)
Lys(Boc)	96–98[b]	−15.9	Dioxane[c]	Ruttenberg (1968)
Met	101–102	−15.9	Dioxane	Anderson *et al.* (1964)
Phe	140–140.5	−17.3	Dioxane	Anderson *et al.* (1964)
Pro	90	−54	Dioxane	Anderson *et al.* (1964)
Thr(tBu)	90	− 6.3	Dioxane	Mathur *et al.* (1966)
Tyr(tBu)	118[b]	−16.2	Dioxane	Mathur *et al.* (1966)
Tyr(Bzl)	144–145[b]	−50.5	DMF	Zahn *et al.* (1966)
Val	116–117	−25.1	Dioxane	Anderson *et al.* (1964)

[a] $c = 2.4$.
[b] Higher mp (no $[\alpha]_D$) reported by Löw and Kisfaludy (1965).
[c] $c = 3.5$.

Table XI *tert*-Butyloxycarbonylamino Acid *N*-Hydroxysuccinimide Esters

	mp (C°)	$[\alpha]_D^{20-25}$	Solvent $c = 1$–2	References
Ala	143–144[a]	−49	Dioxane	Anderson *et al.* (1964)
Arg(NO_2)	115–117	−24.7	DMF[b]	Ondetti *et al.* (1970)
Asp(OBzl)	103–104	−20	Dioxane	Laufer and Blout (1967)
Asp(OtBu)	104–106	−23.1	MeOH	Wünsch (1974)
Cys(Bzl)	117–117.5	−54.0	Dioxane	Anderson *et al.* (1964)
Gln	132–133	−35.6	AcOH	Lefrancier and Bricas (1967)
Glu(OBzl)	100–101	−22	DMF	Danho and Li (1971)
Gly	168–170	—	—	Anderson *et al.* (1964)
Ile	92–93	−26.5	Dioxane	Anderson *et al.* (1964)
Leu	116	−41.8	Dioxane	Anderson *et al.* (1964)
Met	128–129	−20.6	Dioxane	Anderson *et al.* (1964)
Lys(Boc)	92–96	−30.6	EtOH	Shimonishi (1970)
Lys(Z)	109–110	−21.1	Dioxane	Wünsch (1974)
Phe	152–153	−19.0	Dioxane	Anderson *et al.* (1964)
Pro	135–136	−55.3	Dioxane	Anderson *et al.* (1964)
Ser(Bzl)	112–113	+ 6.5	Dioxane[c]	Laufer and Blout (1967)
Thr	134–135	−33.5	EtOAc	Storey *et al.* (1972)
Thr(Bzl)	94	− 8.5	MeOH	Danho and Li (1971)
Trp	153–154	−22.4	Dioxane	Anderson *et al.* (1964)
Tyr	180–182	−12.5[d]	Dioxane	Schnabel *et al.* (1968)
Tyr(Boc)	95–97	+ 3.5[d]	Dioxane	Schnabel *et al.* (1968)
Tyr(Bzl)	149–150	− 6.3	Dioxane	Wünsch (1974)
Val	128–129	−37.0	Dioxane	Anderson *et al.* (1964)

[a] A second form has mp 167°C.
[b] $c = 2.3$.
[c] $c = 0.52$.
[d] $[\alpha]_{578}$.

Table XII Protected Amino Acid 1-Hydroxypiperidine Esters

	mp (C°)	$[\alpha]_D^{20-25}$	Solvent $c = 1$–2	References
Z-Ala	77–78	−23.1	DMF	Handford *et al.* (1965)
Z-Asn	—	−10.8	Ethyl Acetate	Jones *et al.* (1967)
Z-Cys(Bzl)	72–73	−43.1	Ethyl Acetate	Handford *et al.* (1965)
Z-Glu(OMe)	95–97	− 5.3	Dioxane	Jones *et al.* (1967)
Z-Gly	113–114	—	—	Handford *et al.* (1965)
Boc-Gly	60–63	—	—	Jones *et al.* (1967)
Z-Ile	Oil	−10.7	DMF[a]	Jones *et al.* (1967)
Z-Leu	67–68	−20.6	DMF	Handford *et al.* (1965)
Boc-Leu	98–100	−35.1	DMF	Jones *et al.* (1967)
Z-Lys(Z)	—	−13.6	DMF	Jones *et al.* (1967)
Z-Phe	69.5–71	+18.2	$CHCl_3$	Handford *et al.* (1965)
Z-Phe	69.5–71	+ 0.8	Ethyl Acetate	Handford *et al.* (1965)
Boc-Phe	100–102	−15.5	DMF	Jones *et al.* (1967)
Z-Ser	76–80	−22.6	DMF	Jones *et al.* (1967)
Z-Val	46–47	−13.9	DMF	Handford *et al.* (1965)

[a] $c = 3$.

Table XIII Benzyloxycarbonylamino Acid 5-Chloro-8-Hydroxyquinoline Esters[a]

	mp (C°)	$[\alpha]_D^{20-25}$	Solvent[b]
Ala	159.5–160.5	−62.2	DMF
Cys(Bzl)	128.5–129	−50.2	DMF
Gly	124–125	—	—
Leu	95–96	−43.3	DMF
Phe	146–147	−53.9	DMF

[a] Jakubke and Voigt (1966).
[b] $c = 2$.

REFERENCES

Anderson, G. W., and Callahan, F. M. (1958). *J. Am. Chem. Soc.* **80**, 2902–2903.

Anderson, G. W., and McGregor, A. C. (1957). *J. Am. Chem. Soc.* **79**, 6180–6183.

Anderson, G. W., Blodinger, J., Young, R. W., and Welcher, A. D. (1952). *J. Am. Chem. Soc.* **74**, 5304–5306.

Anderson, G. W., Zimmerman, J. E., and Callahan, F. M. (1963). *J. Am. Chem. Soc.* **85**, 3039.

Anderson, G. W., Zimmerman, J. E., and Callahan, F. M. (1964). *J. Am. Chem. Soc.* **86**, 1839–1842.

Anderson, G. W., Callahan, F. M., and Zimmerman, J. E. (1965). *Acta Chim. Acad. Sci. Hung.* **44**, 51–59.

Anderson, J. C., Kenner, G. W., MacLeod, J. K., and Sheppard, R. C. (1966). *Tetrahedron, Suppl.* **8**, 39–51.

Arens, J. F. (1955). *Recl. Trav. Chim. Pays-Bas* **74**, 769–770.

Balog, A., Vargha, E., Breazu, D., Bev, L., and Gönczy, F. (1970). *Rev. Roum. Chim.* **15**, 1391–1407.

Bankowski, K., and Drabarek, S. (1971). *Rocz. Chem.* **45**, 1205–1211.

Bankowski, K., and Drabarek, S. (1972). *Rocz. Chem.* **46**, 607–613.

Bankowski, K., and Drabarek, S. (1974). *Rocz. Chem.* **48**, 715–719.

Barth, A., and Losse, G. (1964). *Z. Naturforsch., Teil B* **19**, 264–265.

Bath, R. J. (1970). Ph.D. Dissertation, Case Western Reserve University, Cleveland, Ohio.

Beacham, J., Dupuis, G., Finn, F. M., Storey, H. T., Yanaihara, C., Yanaihara, N., and Hofmann, K. (1971). *J. Am. Chem. Soc.* **93**, 5526–5539.

Beaumont, S. M., Handford, B. O., and Young, G. T. (1965a). *Acta Chim. Acad. Sci. Hung.* **44**, 37–43.

Beaumont, S. M., Handford, B. O., Jones, J. H., and Young, G. T. (1965b). *Chem. Commun.* pp. 53–54.

Becker, J. M., Wilchek, M., and Katchalski, E. (1971). *Proc. Natl. Acad. Sci. U.S.A.* **68**, 2604–2607.

Bentley, P. H., Gregory, H., Laird, A. H., and Morley, J. S. (1964). *J. Chem. Soc.* pp. 6130–6138.

Beyerman, H. C., and Maassen van den Brink, W. (1963). *Proc. Chem. Soc., London* p. 266.

Beyerman, H. C., Maassen van den Brink, W., Weygand, F., Prox, A., König, W., Schmidhammer, L., and Nintz, E. (1965). *Recl. Trav. Chim. Pays-Bas* **84**, 213–231.

Beyerman, H. C., Boers-Boonekamp, C. A. M., and Maassen van den Brink-Zimmermannová, H. (1968). *Recl. Trav. Chim. Pays-Bas* **87**, 257–273.

Bittner, S., Knobler, Y., and Frankel, M. (1965). *Tetrahedron Lett.* pp. 95–99.

Bloom, S. M., Dasgupta, S. K., Patel, R. P., and Blout, E. R. (1966). *J. Am. Chem. Soc.* **88**, 2035–2041.

Bodanszky, M. (1955). *Nature (London)* **175**, 685–686.

Bodanszky, M. (1957a). *Acta Chim. Acad. Sci. Hung.* **10**, 335–346.

Bodanszky, M. (1957b). *Chem. Ind. (London)* p. 524.

Bodanszky, M. (1960). *Ann. N.Y. Acad. Sci.* **88**, 655–664.

Bodanszky, M. (1969). *In* "Peptides 1968, Proc. of the 9th European Peptide Symposium" (E. Bricas, ed.), p. 150. North Holland Publ., Amsterdam.

Bodanszky, M. (1971). *In* "Prebiotic and Biochemical Evolution" (A. P. Kimball and J. Oro, eds.), pp. 217–222. North Holland Publ., Amsterdam.

Bodanszky, M., and Bath, R. J. (1969). *Chem. Commun.* pp. 1259–1260.

Bodanszky, M., and Birkhimer, C. A. (1960). *Chimia* **14**, 368.

Bodanszky, M., and Birkhimer, C. A. (1962). *Chem. Ind. (London)* pp. 1620–1621.

Bodanszky, M., and Bodanszky, A. (1967). *Chem. Commun.* pp. 591–593.
Bodanszky, M., and du Vigneaud, V. (1959a). *Nature* (*London*) **183**, 1324–1325.
Bodanszky, M., and du Vigneaud, V. (1959b). *J. Am. Chem. Soc.* **81**, 5688–5691.
Bodanszky, M., and du Vigneaud, V. (1959c). *J. Am. Chem. Soc.* **81**, 6072–6075.
Bodanszky, M., and du Vigneaud, V. (1962). *Biochem. Prep.* **9**, 110–112.
Bodanszky, M., and Fagan, D. (1977). *J. Am. Chem. Soc.* **99**, 235–239.
Bodanszky, M., and Funk, K. W. (1973). *J. Org. Chem.* **38**, 1296–1300.
Bodanszky, M., and Henes, J. B. (1976). *Bioorg. Chem.* **4**, 212–218.
Bodanszky, M., and Klausner, Y. S. (1973). *In* "The Chemistry of Polypeptides" (P. G. Katsoyannis, ed.), pp. 21–41. Plenum, New York.
Bodanszky, M., and Natarajan, S. (1975). *J. Org. Chem.* **40**, 2495–2499.
Bodanszky, M., and Ondetti, M. A. (1966a). *Chem. Ind.* (*London*) pp. 26–27.
Bodanszky, M., and Ondetti, M. A. (1966b). *In* "Peptide Synthesis," pp. 162–172. Wiley (Interscience), New York.
Bodanszky, M., and Sheehan, J. T. (1960). *Chem. Ind.* (*London*) pp. 1268–1269.
Bodanszky, M., and Sheehan, J. T. (1964). *Chem. Ind.* (*London*) pp. 1423–1424.
Bodanszky, M., and Sheehan, J. T. (1966). *Chem. Ind.* (*London*) pp. 1597–1598.
Bodanszky, M., and Williams, N. J. (1967). *J. Am. Chem. Soc.* **89**, 685–689.
Bodanszky, M., Meienhofer, J., and du Vigneaud, V. (1960). *J. Am. Chem. Soc.* **82**, 3195–3198.
Bodanszky, M., Ondetti, M. A., Rubin, B., Piala, J. P., Fried, J., Sheehan, J. T., and Birkhimer, C. A. (1962). *Nature* (*London*) **194**, 485–486.
Bodanszky, M., Denning, G. S., Jr., and du Vigneaud, V. (1963). *Biochem. Prep.* **10**, 122–125.
Bodanszky, M., Ondetti, M. A., Levine, S. D., and Williams, N. J. (1967). *J. Am. Chem. Soc.* **89**, 6753–6757.
Bodanszky, M., Bath, R. J., Chang, A., Fink, M. L., Funk, K. W., Greenwald, S. M., and Klausner, Y. S. (1972a). *In* "Chemistry and Biology of Peptides" (J. Meienhofer, ed.), pp. 203–207. Ann Arbor Sci. Publ., Ann Arbor, Michigan.
Bodanszky, M., Chaturvedi, N., Hudson, D., and Itoh, M. (1972b). *J. Org. Chem.* **37**, 2303–2307.
Bodanszky, M., Funk, K. W., and Fink, M. L. (1973). *J. Org. Chem.* **38**, 3565–3570.
Bodanszky, M., Kondo, M., Lin, C. Y., and Sigler, G. F. (1974a). *J. Org. Chem.* **39**, 444–447.
Bodanszky, M., Fink, M. L., Funk, K. W., Kondo, M., Lin, C. Y., and Bodanszky, A. (1974b). *J. Am. Chem. Soc.* **96**, 2234–2240.
Bodanszky, M., Yang Lin, C., and Yiotakis, A. E. (1976). *Bioorg. Chem.* **5**, 339–350.
Bodanszky, M., Fink, M. L., Klausner, Y. S., Natarajan, S., Tatemoto, K., Yiotakis, A. E., and Bodanszky, A. (1977a). *J. Org. Chem.* **42**, 149–152.
Bodanszky, M., Henes, J. B., Yiotakis, A. E., and Said, S. (1977b). *J. Med. Chem.*, **20**, 1461–1464.
Bodanszky, M., Natarajan, S., Hahne, W., and Gardner, J. (1977c). *J. Med. Chem.*, **20**, 1047–1050.
Bolton, A. E., and Hunter, W. M. (1973). *Biochem. J.* **133**, 529–534.
Brack, A., and Spach, G. (1970). *C.R. Hebd. Seances Acad. Sci.* **271**, 916–919.
Brack, A., and Spach, G. (1971). *Bull. Soc. Chim. Fr.* pp. 4481–4489.
Brandenburg, D., Schermutzki, W., Wollmer, A., Vogt, H. P., and Gliemann, J. (1975). *In* "Peptides: Chemistry, Structure and Biology" (R. Walter and J. Meienhofer, eds.), pp. 497–502. Ann Arbor Sci. Publ., Ann Arbor, Michigan.
Brenner, M. (1967). *In* "Peptides, Proc. of the 8th European Peptide Symposium" (H. C. Beyerman, A. Van de Linde, and W. Maassen van den Brink, eds.), p. 1. North Holland Publ., Amsterdam.
Brenner, M., Zimmermann, J. P., Wehrmüller, J., Quitt, P., and Photaki, I. (1955). *Experientia* **11**, 397–399.
Brenner, M., Zimmermann, J. P., Wehrmüller, J., Quitt, P., Hartmann, A., Schneider, W., and Beglinger, U. (1957). *Helv. Chim. Acta* **40**, 1497–1517.

Broadbent, M., Morley, J. S., and Stone, B. E. (1967). *J. Chem. Soc. C* pp. 2632–2636.
Buijle, R., and Viehe, H. G. (1964). *Angew. Chem.* **76**, 572; *Angew. Chem., Int. Ed. Engl.* **3**, 582.
Busse, W. D., and Carpenter, F. H. (1974). *J. Am. Chem. Soc.* **96**, 5947–5949.
Busse, W. D., Hansen, S. R., and Carpenter, F. H. (1974). *J. Am. Chem. Soc.* **96**, 5949–5950.
Camble, R., Garner, R., and Young, G. T. (1968). *Nature (London)* **217**, 247–248.
Cash, W. D. (1962). *J. Org. Chem.* **27**, 3329–3331.
Corley, L., Sachs, D. H., and Anfinsen, C. (1972). *Biochem. Biophys. Res. Commun.* **47**, 1353–1355.
Costopanagiotis, A. A., Preston, J., and Weinstein, B. (1966). *J. Org. Chem.* **31**, 3398–3400.
Curtius, T. (1883). *Ber. Dtsch. Chem. Ges.* **16**, 753–757.
Curtius, T. (1904a). *J. Prakt. Chem.* **70**, 57–128.
Curtius, T. (1904b). *Ber. Dtsch. Chem. Ges.* **37**, 1284–1300.
Dane, E., Drees, F., Konrad, P., and Dockner, T. (1962). *Angew. Chem.* **74**, 873.
Danho, W., and Li, C. H. (1971). *Int. J. Pept. Protein Res.* **3**, 81–92.
DeTar, D. F., and Vajda, T. (1967). *J. Am. Chem. Soc.* **89**, 998–1004.
DeTar, D. F., Honsberg, W., Honsberg, U., Wieland, A., Gouge, M., Bach, H., Tahara, A., Brinigar, W. S., and Rogers, F. F., Jr., (1963). *J. Am. Chem. Soc.* **85**, 2873–2874.
Dutta, A. S., and Morley, J. S. (1971). *J. Chem. Soc. C* pp. 2896–2900.
du Vigneaud, V., Ressler, C., Swan, J. M., Roberts, C. W., Katsoyannis, P. G., and Gordon, S. (1953). *J. Am. Chem. Soc.* **75**, 4879–4880.
Dyckes, D. F., Creighton, T., and Sheppard, R. C. (1974). *Nature (London)* **247**, 202–204.
Eberle, A., Fauchere, J. L., Tesser, G. I., and Schwyzer, R. (1975). *Helv. Chim. Acta* **58**, 2106–2129.
Elliott, D. F., and Russell, D. W. (1957). *Biochem. J.* **66**, 49P.
Elmore, D. T., and Smyth, J. (1963). *Proc. Chem. Soc., London* p. 18.
Farrington, J. A., Kenner, G. W., and Turner, J. M. (1955). *Chem. Ind. (London)* pp. 601–602.
Farrington, J. A., Hextall, G. W., Kenner, G. W., and Turner, J. M. (1957). *J. Chem. Soc.* p. 1407.
Fink, M. L., and Bodanszky, M. (1976). *J. Am. Chem. Soc.* **98**, 974–977.
Fischer, E. (1902). *Ber. Dtsch. Chem. Ges.* **35**, 1095–1106.
Fischer, E. (1906). *Ber. Dtsch. Chem. Ges.* **39**, 453–474.
Flanigan, E., and Marshall, G. R. (1970). *Tetrahedron Lett.* pp. 2403–2406.
Fletcher, G. A., and Jones, J. H. (1972). *Int. J. Pept. Protein Res.* **4**, 347–371.
Fletcher, G. A., Löw, M., and Young, G. T. (1973). *J. Chem. Soc., Perkin Trans. I* pp. 1162–1164.
Frankel, M., Ladkany, D., Gilon, C., and Wolman, Y. (1966). *Tetrahedron Lett.* pp. 4765–4768.
Fridkin, M., Patchornik, A., and Katchalski, E. (1965). *J. Am. Chem. Soc.* **87**, 4646–4648.
Fridkin, M., Patchornik, A., and Katchalski, E. (1966). *J. Am. Chem. Soc.* **88**, 3164–3165.
Fridkin, M., Patchornik, A., and Katchalski, E. (1968). *J. Am. Chem. Soc.* **90**, 2953–2957.
Fridkin, M., Frenkel, A., and Ariely, S. (1969). *Biopolymers* **8**, 661–668.
Fridkin, M., Kalir, R., Warshawsky, A., and Patchornik, A. (1975). *In* "Peptides: Chemistry, Structure and Biology" (R. Walter and J. Meienhofer, eds.), pp. 395–401. Ann Arbor Sci. Publ., Ann Arbor, Michigan.
Fujino, M., and Hatanaka, C. (1967). *Chem. Pharm. Bull.* **15**, 2015–2016.
Fujino, M., and Hatanaka, C. (1968). *Chem. Pharm. Bull.* **16**, 929–932.
Fujino, M., and Nishimura, O. (1969). *Chem. Pharm. Bull.* **17**, 1937–1941.
Garner, R., Schafer, D. J., and Young, G. T. (1971). *In* "Peptides 1969, Proc. of the 10th European Peptide Symposium" (E. Scoffone, ed.), pp. 102–108. North Holland Publ., Amsterdam.
Geiger, R., König, W., Sandow, J., and Schally, A. V. (1974). *Hoppe-Seyler's Z. Physiol. Chem.* **355**, 1526–1534.
Gish, D. T., Katsoyannis, P. G., Hess, G. P., and Stedman, R. J. (1956). *J. Am. Chem. Soc.* **78**, 5954.

Glatthard, R., and Matter, M. (1963). *Helv. Chim. Acta* **46**, 795–804.
Gödicke, W., and Losse, G. (1967). *Z. Chem.* **7**, 232–233.
Goodman, M., and Stueben, K. C. (1959). *J. Am. Chem. Soc.* **81**, 3980–3983.
Gordon, M., Miller, J. G., and Day, A. R. (1948). *J. Am. Chem. Soc.* **70**, 1946–1953.
Gordon, M., Miller, J. G., and Day, A. R. (1949). *J. Am. Chem. Soc.* **71**, 1245–1250.
Govindachari, T. R., Nagarajan, K., and Rajapda, S. (1966). *Tetrahedron* **22**, 3367–3372.
Gross, E., and Witkop, B. (1961). *J. Am. Chem. Soc.* **83**, 1510–1511.
Guarneri, M., Ferroni, R., Giori, P., and Benassi, C. (1972). *In* "Chemistry and Biology of Peptides" (J. Meienhofer, ed.), pp. 213–216. Ann Arbor Sci. Publ., Ann Arbor, Michigan.
Guttmann, S. (1961). *Helv. Chim. Acta* **44**, 721–744.
Guttmann, S., Pless, J., and Boissonnas, R. A. (1962). *Helv. Chim. Acta* **45**, 170–177.
Hagenmaier, H. (1970). *Tetrahedron Lett.* pp. 283–286.
Handford, B. O., Jones, J. H., Young, G. T., and Johnson, T. F. N. (1965). *J. Chem. Soc.* pp. 6814–6827.
Handford, B. O., Hylton, T. A., Preston, J., and Weinstein, B. (1967). *J. Org. Chem.* **32**, 1243–1246.
Handford, B. O., Hylton, T. A., Wang, K. T., and Weinstein, B. (1968). *J. Org. Chem.* **33**, 4251–4255.
Heikens, D., Hermans, P. H., and van Velden, F. (1954). *Nature* (*London*) **174**, 1187–1188.
Hofmann, K., Haas, W., Smithers, M. J., and Zanetti, G. (1965). *J. Am. Chem. Soc.* **98**, 631–639.
Hörnle, S. (1967). *Hoppe-Seyler's Z. Physiol. Chem.* **348**, 1355–1358.
Huguenin, R. L. (1964). *Helv. Chim. Acta* **47**, 1934–1941.
Huguenin, R. L., and Guttmann, S. (1965). *Helv. Chim. Acta* **48**, 1885–1898.
Hylton, T. A., Preston, J., and Weinstein, B. (1966). *J. Org. Chem.* **31**, 3400–3402.
Iselin, B., and Schwyzer, R. (1960). *Helv. Chim. Acta* **43**, 1760–1766.
Iselin, B., Feurer, M., and Schwyzer, R. (1955). *Helv. Chim. Acta* **38**, 1508–1516.
Iselin, B., Rittel, W., Sieber, P., and Schwyzer, R. (1957). *Helv. Chim. Acta* **40**, 373–387.
Itoh, M. (1972). *In* "Chemistry and Biology of Peptides" (J. Meienhofer, ed.), pp. 365–367. Ann Arbor Sci. Publ., Ann Arbor, Michigan.
Itoh, M. (1973). *Bull. Chem. Soc. Jpn.* **46**, 2219–2221.
Itoh, M. (1974). *Bull. Chem. Soc. Jpn.* **47**, 471–475.
Itoh, M., Hagiwara, D., and Kamiya, T. (1975). *Tetrahedron Lett.* pp. 4393–4394.
Jäger, G., and Geiger, R. (1970). *Chem. Ber.* **103**, 1727–1747.
Jakubke, H. D. (1963). *Z. Chem.* **3**, 65–66.
Jakubke, H. D. (1964). *Chem. Ber.* **97**, 2816–2828.
Jakubke, H. D. (1965a). *Instus Justus Liebigs Ann. Chem.* **682**, 244–249.
Jakubke, H. D. (1965b). *Z. Chem.* **5**, 453–455.
Jakubke, H. D. (1966). *Z. Chem.* **6**, 52–67.
Jakubke, H. D., and Baumert, A. (1974). *J. Prakt. Chem.* **316**, 241–248.
Jakubke, H. D., and Voigt, A. (1966). *Chem. Ber.* **99**, 2944–2954.
Jeschkeit, H. (1968). *Z. Chem.* **8**, 20–22.
Jeschkeit, H. (1969). *Z. Chem.* **9**, 266–267.
Johnson, B. J. (1969). *J. Chem. Soc. C* pp. 1412–1416.
Johnson, B. J. (1973). *J. Pharm. Sci.* **62**, 1019–1021.
Johnson, B. J., and Jacobs, P. M. (1968). *Chem. Commun.* pp. 73–75.
Johnson, B. J., and Ruettinger, T. A. (1970). *J. Org. Chem.* **35**, 255–257.
Johnson, B. J., and Trask, E. G. (1968). *J. Org. Chem.* **33**, 4521–4524.
Johnson, B. J., and Trask, E. G. (1969). *J. Chem. Soc. C* pp. 2644–2647.
Jones, D. S. (1972). *J. Chem. Soc., Perkin Trans. I* pp. 1407–1415.
Jones, J. H., and Young, G. T. (1966). *Chem. Ind.* (*London*) p. 1722.
Jones, J. H., and Young, G. T. (1968a). *J. Chem. Soc. C* pp. 53–61.

Jones, J. H., and Young, G. T. (1968b). *J. Chem. Soc. C* pp. 436–441.
Jones, J. H., Liberek, B., and Young, G. T. (1967). *In* "Peptides, Proc. of 8th European Peptide Symposium" (H. C. Beyerman, A. van de Linde, and W. Maassen van den Brink, eds.), pp. 15–21. North Holland Publ., Amsterdam.
Kaczmarek, M., Kupryszewski, G., and Wajcht, J. (1967). *Zesz. Nauk. Wydz. Szk. Pedagog. Gdansku: Mat., Fiz., Chem.* **7**, 143; *Chem. Abstr.* **70**, 47840j (1969).
Kalir, R., Fridkin, M., and Patchornik, A. (1974). *Eur. J. Biochem.* **42**, 151–156.
Kamber, B. (1973). *Helv. Chim. Acta* **56**, 1370–1381.
Karlsson, S., Lindeberg, G., and Ragnarsson, U. (1970). *Acta Chem. Scand.* **24**, 337–338.
Karrer, P., and Heynemann, H. (1948). *Helv. Chim. Acta* **31**, 398–405.
Kato, T., and Izumiya, N. (1974). *In* "Chemistry and Biochemistry of Amino Acids, Peptides and Proteins" (B. Weinstein, ed.), Vol. 2, pp. 1–38. Dekker, New York.
Kaurov, O. A., Martynov, V. F., and Morozov, V. B. (1970). *Zh. Obshch. Khim.* **40**, 908–911; *Chem. Abstr.* **73**, 35748j (1970).
Kemp, D. S. (1967). *Tetrahedron* **23**, 2001–2015.
Kemp, D. S., and Chien, S. W. (1967). *J. Am. Chem. Soc.* **89**, 2743–2745.
Kemp, D. S., and Woodward, R. B. (1965). *Tetrahedron* **21**, 3019–3035.
Kenner, G. W. (1959). *Angew. Chem.* **71**, 741.
Kenner, G. W., and Seely, J. H. (1972). *J. Am. Chem. Soc.* **94**, 3259–3260.
Kenner, G. W., Thompson, P. J., and Turner, J. M. (1958). *J. Chem. Soc.* pp. 4148–4152.
Khan, S. A., and Sivanandaiah, K. M. (1976). *Synthesis* pp. 614–615.
Khorana, H. G. (1955). *Chem. Ind.* (*London*) pp. 1087–1088.
Kisfaludy, L., Ceprini, M. Q., Rakoczy, B., and Kovacs, J. (1967). *In* "Peptides, Proc. of 8th European Peptide Symposium" (H. C. Beyerman, A. van de Linde, and W. Maassen van den Brink, eds.), pp. 25–27. North Holland Publ., Amsterdam.
Kisfaludy, L., Roberts, J. E., Johnson, R. H., Mayers, G. L., and Kovacs, L. (1970). *J. Org. Chem.* **35**, 3563–3565.
Kisfaludy, L., Löw, M., Nyeki, O., Szirtes, R., and Schön, I. (1973). *Justus Liebigs Ann. Chem.* **1973**, 1421–1429.
Kisfaludy, L., Schön, I., Szirtes, T., Nyeki, O., and Löw, M. (1974). *Tetrahedron Lett.* pp. 1785–1786.
Klausner, Y. S., and Bodanszky, M. (1972). *Synthesis* pp. 453–463.
Klausner, Y. S., and Chorev, M. (1975). *J. Chem. Soc., Chem. Commun.* pp. 973–974.
Klieger, E., and Schröder, E. (1963). *Justus Liebigs Ann. Chem.* **661**, 193–201.
Klostermeyer, H. (1968). *Chem. Ber.* **101**, 2823–2831.
Klostermeyer, H., Halstrøm, J., Kusch, P., Föhles, J., and Lunkenheimer, W. (1967). *In* "Peptides, Proc. of 8th European Peptide Symposium" (H. C. Beyerman, A. van de Linde, and W. Maassen van den Brink, eds.), pp. 113–116. North Holland Publ., Amsterdam.
Koltun, W. L., and Gurd, F. R. N. (1959). *J. Am. Chem. Soc.* **81**, 301–305.
König, W., and Geiger, R. (1970a). *Chem. Ber.* **103**, 788–798.
König, W., and Geiger, R. (1970b). *Chem. Ber.* **103**, 2024–2034, and 2034–2040.
König, W., and Geiger, R. (1972). *In* "Chemistry and Biology of Peptides" (J. Meienhofer, ed.), pp. 343–350. Ann Arbor Sci. Publ., Ann Arbor, Michigan.
König, W., and Geiger, R. (1973). *Chem. Ber.* **106**, 3626–3635.
Kovacs, J., and Ceprini, M. Q. (1965). *Chem. Ind.* (*London*) p. 2100.
Kovacs, J., and Kapoor, A. (1965). *J. Am. Chem. Soc.* **87**, 118–119.
Kovacs, J., and Rodin, R. L. (1968). *J. Org. Chem.* **33**, 2418–2420.
Kovacs, J., Gianotti, R., and Kapoor, A. (1966). *J. Am. Chem. Soc.* **88**, 2282–2292.
Kovacs, J., Ceprini, M. Q., Dupraz, C. A., and Schmidt, G. N. (1967a). *J. Org. Chem.* **32**, 3696–3698.
Kovacs, J., Kisfaludy, L., and Ceprini, M. Q. (1967b). *In* "Peptides, Proc. of 8th European

Peptide Symposium" (H. C. Beyerman, A. van de Linde and W. Maassen van den Brink, eds.), pp. 23–24. North Holland Publ., Amsterdam.
Kovacs, J., Kisfaludy, L., and Ceprini, M. Q. (1967c). *J. Am. Chem. Soc.* **89**, 183–184.
Kovacs, J., Schmit, G. N., and Ghatak, U. R. (1968a). *Biopolymers* **6**, 817–836.
Kovacs, J., Mayers, G. L., Johnson, R. H., and Ghatak, U. R. (1968b). *Chem. Commun.* pp. 1066–1067.
Kovacs, J., Kisfaludy, L., Ceprini, M. Q., and Johnson, R. H. (1969). *Tetrahedron* **25**, 2555–2573.
Kovacs, J., Mayers, G. L., Johnson, R. H., Cover, R. E., and Ghatak, U. R. (1970). *J. Org. Chem.* **35**, 1810–1815.
Kovacs, J., Cortegiano, H., Cover, R. E., and Mayers, G. L. (1971). *J. Am. Chem. Soc.* **93**, 1541–1543.
Kovacs, J., Cover, R. E., Johnson, R. H., Kalas, T., Mayers, G. L., and Roberts, J. E. (1973). *J. Org. Chem.* **38**, 2518–2521.
Kovács, K., and Penke, B. (1973). *In* "Peptides 1972, Proc. of 12th European Peptide Symposium" (H. Hanson and H. D. Jakubke, eds.), pp. 187–188. North Holland Publ., Amsterdam.
Krail, G., Brandenburg, D., Zahn, H., and Geiger, R. (1971). *Hoppe-Seyler's Z. Physiol. Chem.* **352**, 1595–1598.
Kupryszewski, G. (1961a). *Rocz. Chem.* **35**, 595–600; *Chem. Abstr.* **55**, 27121i (1961).
Kupryszewski, G. (1961b). *Zesz. Nauk. Wydz. Szk. Pedagog. Gdansku: Mat., Fiz., Chem.* **1**, 95; *Chem. Abstr.* **60**, 654 (1964).
Kupryszewski, G., and Kaczmarek, M. (1961). *Rocz. Chem.* **35**, 931–936; *Chem. Abstr.* **56**, 6081c (1962).
Kupryszewski, G., and Muzalewski, F. (1967). *Zesz. Nauk. Wydz. Szk. Pedagog. Gdansku: Mat., Fiz., Chem.* **7**, 159; *Chem. Abstr.* **70**, 47839r (1969).
Kupryszewski, G., Muzalewski, F., and Przybylski, J. (1968). *Rocz. Chem.* **42**, 1009; *Chem. Abstr.* **70**, 4582j (1968).
Lanzilotti, A. E., Benz, E., and Goldman, L. (1964). *J. Am. Chem. Soc.* **86**, 1880–1882.
Laufer, D. A., and Blout, E. R. (1967). *J. Am. Chem. Soc.* **89**, 1246–1249.
Laufer, D. A., Chapman, T. M., Marlborough, D. I., Vaidya, V. M., and Blout, E. R. (1968). *J. Am. Chem. Soc.* **90**, 2696–2698.
Law, H. D. (1965). *J. Chem. Soc.* pp. 3897–3900.
Lefrancier, R., and Bricas, E. (1967). *Bull. Soc. Chim. Biol.* **49**, 1257–1271.
Levy, D., and Carpenter, F. H. (1966). *J. Am. Chem. Soc.* **88**, 3676–3677.
Li, C. H., Meienhofer, J., Schnabel, E., Chung, D., Lo, T. B., and Ramachandran, J. (1960). *J. Am. Chem. Soc.* **82**, 5761–5762.
Liberek, B. (1963). *Tetrahedron Lett.* pp. 1103–1108.
Liberek, B., and Grzonka, Z. (1964). *Tetrahedron Lett.* pp. 159–165.
Lloyd, K., and Young, G. T. (1968). *Chem. Commun.* pp. 1400–1401.
Lloyd, K., and Young, G. T. (1971). *J. Chem. Soc. C* pp. 2890–2896.
Losse, G., Barth, A., and Schatz, K. (1964). *Justus Liebigs Ann. Chem.* **677**, 185–190.
Losse, G., Hoffmann, K. H., and Hetzer, G. (1965). *Justus Liebigs Ann. Chem.* **684**, 236–242.
Löw, M., and Kisfaludy, L. (1965). *Acta Chim. Acad. Sci. Hung.* **44**, 61–66; *Chem. Abstr.* **63**, 13404h (1965).
Lubiewska-Nakonieczna, L., Rzeszotarska, B., and Taschner, E. (1970). *Justus Liebigs Ann. Chem.* **741**, 157–166.
Lübke, K. (1971). *In* "Peptides 1969, Proc. of 10th European Peptide Symposium" (E. Scoffone, ed.), pp. 154–156. North Holland Publ., Amsterdam.
Lynen, F., and Reichert, E. (1951). *Angew. Chem.* **63**, 47–48.
McCormick, D. B. (1973). *J. Heterocycl. Chem.* **10**, 235–237.

Marchiori, F., Rocchi, R., and Scoffone, E. (1962). *Ric. Sci., Parte 2: Sez A* **2**, 647–648; *Chem. Abstr.* **60**, 4245c (1964).
Marshall, G. R., and Liener, I. E. (1970). *J. Org. Chem.* **35**, 867–868.
Martin, G. J., Golubow, J., and Axelrod, A. E. (1959). *J. Biol. Chem.* **234**, 294–298.
Martinez, J., and Winternitz, F. (1972). *Bull. Soc. Chim. Fr.* 4707–4709.
Martinez, J., and Winternitz, F. (1975). *Tetrahedron Lett.* pp. 2631–2632.
Mathur, R. B., Klostermeyer, H., and Zahn, H. (1966). *Hoppe Seyler's Z. Physiol. Chem.* **346**, 60–68.
Mazur, R. H. (1963). *J. Org. Chem.* **28**, 2498.
Mehlis, B., and Apelt, H. (1973). *In* "Peptides 1972, Proc. of 12th European Peptide Symposium" (H. Hanson and H. D. Jakubke, eds.), p. 379. North Holland Publ., Amsterdam.
Meienhofer, J. (1962). *Chimia* **16**, 385–424.
Meienhofer, J., Sano, P., and Patel, R. P. (1970). *In* "Peptides, Chemistry and Biochemistry" (B. Weinstein and S. Lande, eds.), pp. 419–434. Dekker, New York.
Meienhofer, J., Cotton, R., and Atherton, E. (1973). *In* "Peptides 1971, Proc. of 11th European Peptide Symposium" (H. Nesvadba, ed.), pp. 60–64. North Holland Publ., Amsterdam.
Merrifield, R. B. (1963). *J. Am. Chem. Soc.* **85**, 2149–2154.
Mitin, Yu. V., and Glinskaya, O. V. (1969). *Tetrahedron Lett.* pp. 5267–5270.
Mitin, Yu. V., and Nadezhdina, L. B. (1968). *Zh. Org. Khim.* **4**, 1181; *Chem. Abstr.* **69**, 87227r (1968).
Mitin, Yu. V., and Vlasov, G. P. (1968). *Dokl. Akad. Nank SSSR* **179**, 353–355; *Chem. Abstr.* **69**, 77719 (1968).
Mukaiyama, T., Ueki, M., Maruyama, H., and Matsueda, R. (1968). *J. Am. Chem. Soc.* **90**, 4490–4491.
Mukaiyama, T., Matsueda, R., and Suzuki, M. (1970a). *Tetrahedron Lett.* pp. 1901–1904.
Mukaiyama, T., Goto, K., Matsueda, R., and Ueki, M. (1970b). *Tetrahedron Lett.* pp. 5293–5296.
Muzalewski, F. (1973). *In* "Peptides 1971, Proc. of 11th European Peptide Symposium" (H. Nesvadba, ed.), pp. 39–44. North Holland Publ., Amsterdam.
Nakamizo, N. (1969). *Bull. Chem. Soc. Jpn.* **42**, 1071–1077 and 1078–1082.
Natarajan, S., and Bodanszky, M. (1976). *J. Org. Chem.* **41**, 1269–1272.
Nefkens, G. H. L., and Tesser, G. I. (1961). *J. Am. Chem. Soc.* **83**, 1263.
Nefkens, G. H. L., Tesser, G. I., and Nivard, R. J. F. (1962). *Recl. Trav. Chim. Pays-Bas* **81**, 683–690.
Nicolaides, E. D., De Wald, H. A., Shorley, P. G., and Collier, H. O. J. (1960). *Nature (London)* **187**, 773–774.
Ohno, M., Tsukamoto, S., and Izumiya, N. (1972). *J. Chem. Soc., Chem. Commun.* pp. 663–664.
Ondetti, M. A., Narayanan, V. L., von Saltza, M., Sheehan, J. T., Sabo, E. F., and Bodanszky, M. (1968). *J. Am. Chem. Soc.* **90**, 4711–4716.
Ondetti, M. A., Pluscec, J., Sabo, E. F., Sheehan, J. T., and Williams, N. (1970). *J. Am. Chem. Soc.* **92**, 195–199.
Pacsu, E., and Wilson, E. J. (1942). *J. Org. Chem.* **7**, 117–125 and 126–135.
Panneman, H. J., Marx, A. F., and Arens, J. F. (1959). *Recl. Trav. Chim. Pays-Bas* **78**, 487–511.
Panse, G. T., and Laufer, D. A. (1970). *Tetrahedron Lett.* pp. 4181–4184.
Paquette, L. A. (1965). *J. Am. Chem. Soc.* **87**, 5186–5190.
Patchornik, A., Fridkin, M., and Katchalski, E. (1967). *Pept. Proc. Eur. Pept. 8th, 1966*, pp. 91–99.
Paul, R., Anderson, G. W., and Callahan, F. M. (1961). *J. Org. Chem.* **26**, 3347–3350.
Pless, J., and Boissonnas, R. A. (1963). *Helv. Chim. Acta* **46**, 1609–1625.
Ragnarsson, U., Karlsson, S. M., and Sandberg, B. E. (1972). *Acta Chem. Scand.* **26**, 2550–2552.

Rajappa, S., and Akerkar, A. S. (1966). *Chem. Commun.* pp. 826–827.

Rajappa, S., Nagarajan, K., and Iyer, V. S. (1967). *Tetrahedron* **23**, 4805–4809.

Ramachandran, J., and Li, C. H. (1963). *J. Org. Chem.* **28**, 173–177.

Ressler, C. (1956). *J. Am. Chem. Soc.* **78**, 5956–5957.

Ressler, C., and Kashelikar, D. V. (1966). *J. Am. Chem. Soc.* **88**, 2025–2035.

Ressler, C., and Ratzkin, H. (1961). *J. Org. Chem.* **26**, 3356–3361.

Rocchi, R., Marchiori, F., and Scoffone, E. (1963). *Gazz. Chim. Ital.* **93**, 823–833.

Rothe, M., and Kunitz, F. W. (1957). *Justus Liebigs Ann. Chem.* **609**, 88–102.

Rouette, H. K. (1967). *Z. Gesamte Textilind.* **69**, 325–328.

Rudinger, J., and Gut, V. (1967). *In* "Peptides, Proc. of 8th European Peptide Symposium" (H. C. Beyerman, A. van de Linde, and W. Maassen van den Brink, eds.), pp. 89–90. North Holland Publ., Amsterdam.

Ruttenberg, M. A. (1968). *J. Am. Chem. Soc.* **90**, 5598–5603.

Rzeszotarska, B., and Wiejak, S. (1968). *Justus Liebigs Ann. Chem.* **716**, 216–218.

Sakakibara, S., and Hase, S. (1968). *Bull. Chem. Soc. Jpn.* **41**, 2816.

Sakakibara, S., and Inukai, N. (1964). *Bull. Chem. Soc. Jpn.* **37**, 1231–1232.

Sakakibara, S., and Inukai, N. (1965). *Bull. Chem. Soc. Jpn.* **38**, 1979–1984.

Sandrin, E., and Boissonnas, R. A. (1963). *Helv. Chim. Acta* **46**, 1637–1669.

Sarantakis, D., Sutherland, J. K., Tortorella, C., and Tortorella, V. (1968). *J. Chem. Soc. C* pp. 72–73.

Schade, F., and Zahn, H. (1962). *Angew. Chem.* **74**, 904; *Angew. Chem., Int. Ed. Engl.* **1**, 661 (1962).

Schaich, E., Fretzdorff, M., and Schneider, F. (1973). *Hoppe-Seyler's Z. Physiol. Chem.* **354**, 897–902.

Schnabel, E. (1964). *Justus Liebigs Ann. Chem.* **673**, 171–175.

Schnabel, E., and Li, C. H. (1960). *J. Am. Chem. Soc.* **82**, 4576–4579.

Schnabel, E., Herzog, H., Hoffmann, P., Klauke, E., and Ugi, I. (1968). *Justus Liebigs Ann. Chem.* **716**, 175–185.

Schnabel, E., Schmidt, G., and Klauke, E. (1971a). *Justus Liebigs Ann. Chem.* **743**, 69.

Schnabel, E., Klostermeyer, H., and Berndt, H. (1971b). *Justus Liebigs Ann. Chem.* **749**, 90–108.

Schwyzer, R. (1953). *Helv. Chim. Acta* **36**, 414–424.

Schwyzer, R., and Iselin, B. (1955). *Ann. Acad. Sci. Fenn. Ser. A2* **60**, 181–189.

Schwyzer, R., and Rittel, W. (1961). *Helv. Chim. Acta* **44**, 159–169.

Schwyzer, R., and Sieber, P. (1957a). *Helv. Chim. Acta* **40**, 624–639.

Schwyzer, R., and Sieber, P. (1957b). *Chimia* **10**, 265.

Schwyzer, R., and Sieber, P. (1958). *Helv. Chim. Acta* **41**, 2186–2190.

Schwyzer, R., and Sieber, P. (1966). *Helv. Chim. Acta* **49**, 134–158.

Schwyzer, R., Iselin, B., and Feurer, M. (1955a). *Helv. Chim. Acta* **38**, 69–79.

Schwyzer, R., Feurer, M., Iselin, B., and Kägi, H. (1955b). *Helv. Chim. Acta* **38**, 80–91.

Schwyzer, R., Feurer, M., and Iselin, B. (1955c). *Helv. Chim. Acta* **38**, 83–91.

Schwyzer, R., Iselin, B., Rittel, W., and Sieber, P. (1956). *Helv. Chim. Acta* **39**, 872–883.

Scoffone, E., Rocchi, R., Vidali, G., Scatturin, V., and Marchiori, I. (1964). *Gazz. Chim. Ital.* **94**, 743–759; *Chem. Abstr.* **61**, 13409b (1964).

Shaltiel, S., and Fridkin, M. (1970). *Biochemistry* **9**, 5122–5127.

Sheehan, J. C., and Hess, G. P. (1955). *J. Am. Chem. Soc.* **77**, 1067–1068.

Sheehan, J. C., Goodman, M., and Hess, G. P. (1956). *J. Am. Chem. Soc.* **78**, 1367–1369.

Sheehan, J. C., Hasspacher, K., and Yeh, Y. L. (1959). *J. Am. Chem. Soc.* **81**, 6086.

Shimonishi, Y. (1970). *Bull. Chem. Soc. Jpn.* **43**, 3251–3255.

Sieber, P., and Iselin, B. (1968a). *Helv. Chim. Acta* **51**, 614–622.

Sieber, P., and Iselin, B. (1968b). *Helv. Chim. Acta* **51**, 622–632.

Smith, E. L., Spackmann, D. H., and Polglose, W. J. (1952). *J. Biol. Chem.* **199**, 801–817.

Staab, H. A. (1957). *Justus Liebigs Ann. Chem.* **609**, 75–83.
Stevens, C. L., and Munk, M. E. (1958). *J. Am. Chem. Soc.* **80**, 4069–4071.
Stewart, F. H. C. (1965). *Aust. J. Chem.* **18**, 887–901.
Stewart, F. H. C. (1966). *Aust. J. Chem.* **19**, 489–501, 1503–1509, and 2373–2380.
Stewart, F. H. C. (1968). *Aust. J. Chem.* **21**, 477–482 and 1639–1649.
Stich, K., and Leemann, H. G. (1963). *Helv. Chim. Acta* **46**, 1887–1895.
Storey, H. T., Beacham, J., Cernosek, S. F., Finn, F. M., Yanaihara, C., and Hofmann, K. (1972). *J. Am. Chem. Soc.* **94**, 6170–6178.
Taschner, E., Blotny, G., Bator, B., and Wasielewski, C. (1964). *Bull. Acad. Pol. Sci., Ser. Sci. Chim.* **12**, 755–759.
Taschner, E., Rzeszotarska, B., and Lubiewska, L. (1965a). *Angew. Chem., Int. Ed. Engl.* **4**, 594.
Taschner, E., Rzeszotarska, B., and Lubiewska, L. (1965b). *Justus Liebigs Ann. Chem.* **690**, 177–181.
Taschner, E., Rzeszotarska, M., and Kuziel, A. (1965c). *Acta Chim. Acad. Sci. Hung.* **44**, 67–70.
Taschner, E., Rzeszotarska, B., and Lubiewska, L. (1967). *Chem. Ind.* (*London*) pp. 402–404.
Taylor, E. C., Kinzle, F., and McKillop, A. (1970). *J. Org. Chem.* **35**, 1672–1674.
Treiber, R. L. (1973). *Experientia* **29**, 1335–1337.
Trudelle, Y. (1971). *Chem. Commun.* pp. 639–641.
Visser, S., Roeloffs, J., Kerling, K. E. T., and Havinga, E. (1968). *Recl. Trav. Chim. Pays-Bas* **87**, 559–571.
Visser, S., Raap, J., Kerling, K. E. T., and Havinga, E. (1970). *Recl. Trav. Chim. Pays-Bas* **89**, 865–875.
Viswanatha, R., Bayer, E., and Wilchek, M. (1975). *Biochim. Biophys. Acta* **401**, 152–156.
Weber, U., Hörnle, S., Grieser, G., Herzog, K. H., and Weitzel, G. (1967). *Hoppe-Seyler's Z. Physiol. Chem.* **348**, 1715–1717.
Weinert, M., Brandenburg, D., and Zahn, H. (1969). *Hoppe-Seyler's Z. Physiol. Chem.* **350**, 1556–1562.
Weitzel, G., Weber, U., Hörnle, S., and Schneider, F. (1968). *In* "Peptides 1968, Proc. of 9th European Peptide Symposium" (E. Bricas, ed.), pp. 171–172. North Holland Publ., Amsterdam.
Weygand, F., and Hunger, K. (1962). *Chem. Ber.* **95**, 1–6.
Weygand, F., and Ragnarsson, U. (1966). *Z. Naturforsch., Teil B* **21**, 1141–1144.
Weygand, F., and Steglich, W. (1961). *Angew. Chem.* **73**, 757.
Weygand, F., Hoffmann, D., and Wünsch, E. (1966). *Z. Naturforsch., Teil B* **21**, 426–428.
Wieland, T., and Bernhard, H. (1953). *Justus Liebigs Ann. Chem.* **582**, 218–224.
Wieland, T., and Birr, C. (1966). *Angew. Chem., Int. Ed. Engl.* **5**, 310; *Angew. Chem.* **78**, 303 (1966).
Wieland, T., and Birr, C. (1967). *Chimia* **21**, 581–582.
Wieland, T., and Bokelman, E. (1952). *Justus Liebigs Ann. Chem.* **576**, 20–34.
Wieland, T., and Heinke, B. (1958). *Justus Liebigs Ann. Chem.* **615**, 184–202.
Wieland, T., and Jaenicke, F. (1956). *Justus Liebigs Ann. Chem.* **599**, 125–130.
Wieland, T., and Kahle, W. (1966). *Justus Liebigs Ann. Chem.* **691**, 212–217.
Wieland, T., and Racky, W. (1968). *Chimia* **22**, 375–376.
Wieland, T., and Schäfer, W. (1951). *Angew. Chem.* **63**, 146–147.
Wieland, T., and Schäfer, W. (1952). *Justus Liebigs Ann. Chem.* **576**, 104–109.
Wieland, T., and Schneider, G. (1953). *Justus Liebigs Ann. Chem.* **580**, 159–168.
Wieland, T., and Vogeler, K. (1964). *Justus Liebigs Ann. Chem.* **680**, 125–132.
Wieland, T., Schäfer, W., and Bokelmann, E. (1951). *Justus Liebigs Ann. Chem.* **573**, 99–104.
Wieland, T., Heinke, B., Vogeler, K., and Morimoto, H. (1962). *Justus Liebigs Ann. Chem.* **655**, 189–194.
Wieland, T., Determann, H., and Kahle, W. (1963). *Angew. Chem., Int. Ed. Engl.* **2**, 154–155.

Wieland, T., Lewalter, J., and Birr, C. (1970). *Justus Liebigs Ann. Chem.* **740**, 31–47.
Wieland, T., Birr, C., and Fleckenstein, P. (1972). *Justus Liebigs Ann. Chem.* **756**, 14–19.
Wilchek, M., and Patchornik, A. (1963). *J. Org. Chem.* **28**, 1874–1875.
Woodward, R. B., Olofson, R. A., and Mayer, H. (1961). *J. Am. Chem. Soc.* **83**, 1010–1012.
Wu, Y., and Busch, D. H. (1972). *J. Am. Chem. Soc.* **94**, 4115–4122.
Wünsch, E., ed. (1974). "Houben-Weyl's Methoden der organischen Chemie," Vol. 15, Part II. Thieme, Stuttgart.
Wünsch, E. (1975). *Helv. Med. Acta* **37**, 307–320.
Wünsch, E., and Drees, F. (1966). *Chem. Ber.* **99**, 110–120.
Yamada, S., Terashima, S., and Wagatsuma, M. (1970). *Tetrahedron Lett.* pp. 1501–1504.
Yamazaki, N., Higashi, F., and Niwano, M. (1974). *Tetrahedron* **30**, 1319–1321.
Zahn, H., and Falkenburg, H. R. (1960). *Justus Liebigs Ann. Chem.* **636**, 117–131.
Zahn, H., Bremer, H., and Zabel, R. (1965). *Z. Naturforsch., Teil B* **20**, 653–660.
Zahn, H., Danho, W., and Gutte, B. (1966). *Z. Naturforsch., Teil B* **21**, 763–773.
Zimmerman, J. E., and Anderson, G. W. (1967). *J. Am. Chem. Soc.* **89**, 7151–7152.

Chapter 4

The Azide Method in Peptide Synthesis

JOHANNES MEIENHOFER

I.	Introduction	198
II.	Experimental Procedure	199
III.	The Formation of Hydrazides	200
	A. Strategy and Tactics	200
	B. Ester Hydrazinolysis	202
	1. Reaction Conditions	202
	2. Solid Phase Techniques	204
	3. Stability of Protective Groups and Amino Acids	205
	4. Reactivity of Esters	205
	5. Racemization	206
	6. Side Reactions	207
	C. The N'-Protected Hydrazide Approach	208
	1. Solution Synthesis	208
	2. Solid Phase Procedures	210
	D. Direct Coupling of α-Carboxyls with Hydrazine	210
	E. Alternative Procedures	213
IV.	The Generation of Azides	213
	A. Aqueous Acid–Sodium Nitrite Procedure	213
	B. Alkyl Nitrite in Organic Solvents: The Honzl–Rudinger Procedure	214
	C. The Diphenyl Phosphorazidate Method	216
V.	Azide Segment Condensation	217
	A. Reaction Conditions	217
	B. Racemization	219
	C. Side Reactions	225
	D. Recent Applications to Large Segments	226
	1. Bovine Ribonuclease A	227
	2. Porcine Proinsulin C-Peptide	228

THE PEPTIDES, VOL. 1

ISBN 0-12-304201-1

3. Ribonuclease T_1 228
4. Porcine Secretin 228
VI. Special Applications 228
VII. Opportunities and Constraints 232
A. Comparative Evaluation 232
B. Prognosis 233
References 233

I. INTRODUCTION*

The development of the azide method by Curtius (1902) was one of the starting points of peptide synthesis, see Fruton (1949) for a historical review. With the introduction of more efficient methods of peptide bond formation in the 1950's the use of N^α-protected amino acid azides declined sharply. At the same time, the azide method became the most valuable synthetic tool for peptide segment condensation, when it was recognized that its apparent "racemization-free" performance was not attained by other efficient coupling procedures. The need for "racemization-free" segment condensation increased when targets of synthesis such as β-MSH and ACTH reached a size range where the resolution power of existing purification procedures could no longer be relied upon to remove all stereomeric impurities. Extensive use of azide segment coupling was made by Hirschmann and collaborators (1969) in their remarkable synthesis of the 104-peptide, bovine ribonuclease A(21–124), the S-protein of RNase. The success of this synthesis as well as the difficulties encountered during its course exemplify the present scope and constraints of the azide method as compared to the carbodiimide, active ester, and anhydride procedures.

Attractive features of the azide method include (1) minimal danger of racemization in peptide segment condensation, (2) good performance in assembling peptide chains of up to sixty amino acid residues, (3) facile conversion of three important carboxyl protecting groups (methyl, ethyl, and benzyl esters) into activated azide via hydrazides, and (4) opportunity to employ minimal side chain protection.

* In this chapter the new peptide size nomenclature proposed by M. Bodanszky in the Pierce Award lecture at the Fifth American Peptide Symposium (Bodanszky, 1977) will be used. Greek prefixes are replaced by arabic numerals in peptides with more than 10 amino acid residues. Greek prefixes are retained for di- to decapeptides. Examples: 15-peptide, 31-peptide; also ACTH(1–24), β-LPH(61–91).

Peptide *segment* and *segment condensation* are used (Pettit, 1976) instead of the customary "fragment" and "fragment condensation."

No hyphen is used in D-amino acid symbols when these are part of a peptide structure, e.g., H-Gly-DPhe-Ala-OH.

The comparatively low tendency of racemization is the prime consideration of selecting the azide procedure for the activation of peptide segments having chiral COOH-termini. In fact, azide segment coupling under commonly applied experimental conditions was considered to be "racemization-free" (Heard and Young, 1963; Bodanszky and Ondetti, 1966) until Sieber *et al.*, (1970) observed 12% and 40% racemization in two azide segment condensations during a synthesis of calcitonin M. Subsequently, the influence of several factors and reaction conditions on racemization in these two azide condensations and in model reactions were carefully studied (Sieber *et al.*, 1970; Sieber and Riniker, 1973). Obviously, "freedom from racemization" cannot any longer be taken for granted but there is a consensus that the azide procedure is still the least racemization-prone among the known methods of peptide bond formation (Young, 1973; Wünsch, 1974a,b; Wünsch and Stelzel, 1974b, see esp., p. 313; Bodanszky *et al.*, 1976; Finn and Hofmann, 1976a; Kemp, this volume, Chapter 7).

This eminence of the azide procedure over other efficient methods is somewhat offset by several constraints: (1) it is experimentally more laborious and time-consuming by proceeding in three stages (hydrazide formation → azide formation → coupling), (2) many side reactions have been observed (Schnabel, 1962, 1963; Klausner and Bodanszky, 1974; Wünsch and Stelzel, 1974b, see esp., p. 313), and (3) its rate of acylation is considerably slower than those of the anhydride or carbodiimide reactions.

Previous reviews of the azide method deal predominantly with *azides of single amino acids* [Greenstein and Winitz, 1961; Wünsch and Stelzel, 1974b, pp. 296–322 (both provide illustrative procedures and tables of acylamino acid hydrazides); Schröder and Lübke, 1965 (tables); Klausner and Bodanszky, 1974 (chemistry, side reactions)]. This chapter is concerned with preparation and use of *azides of protected peptides* and applications to segment condensation in syntheses of large peptides.

II. EXPERIMENTAL PROCEDURE

The azide procedure is generally carried out in three consecutive stages, i.e., hydrazide formation, azide formation, and peptide bond formation (Scheme 1). Side reactions have been reported to occur in each of the 3 stages. Possibilities for minimizing some of these will be discussed. Side reactions occurring during hydrazide formation (stage I) are the least worri-

$$\underset{(\mathbf{1})}{\mathrm{R{-}CO{-}Y}} \xrightarrow[\mathrm{I}]{} \underset{(\mathbf{2})}{\mathrm{R{-}CO{-}NHNH_2}} \xrightarrow[\mathrm{II}]{\mathrm{H^+, NaNO_2}} \underset{(\mathbf{3})}{\mathrm{R{-}CO{-}N_3}} \xrightarrow[-\mathrm{HN_3}]{\mathrm{H_2NR'}} \underset{(\mathbf{4})}{\mathrm{R{-}CONH{-}R'}}$$

III

Scheme 1

some since protected peptide hydrazides are generally stable and may be purified before proceeding. This opportunity does not normally exist for the azides which are unstable and used immediately after their formation for the acylation reaction. Side reactions occurring during coupling (stage III) may be caused or influenced by the amine component to be acylated, e.g., its solubility, size, and steric hindrance. Extensive racemization (Sieber *et al.*, 1970) observed with the calcitonin M 16-phenylalanine and 20-histidine residues (see Section I) are examples of such unpredictable side reactions.

III. THE FORMATION OF HYDRAZIDES

A. Strategy and Tactics

Hydrazides of N^{α}- and side chain-protected peptides are commonly prepared by hydrazinolysis of their carboxyl-terminal alkyl esters, essentially following the procedure of Curtius (1902) for the preparation of simple *N*-protected amino acid hydrazides. However, with increasing chain length the rate of hydrazinolysis may slow to the extent that several days are required for complete reaction. Concomitant with longer reaction times and/or larger excess of hydrazine, the danger of undesired side reactions with certain sensitive side chain or protecting group functions will increase. Moreover, most peptides are likely to contain a greater variety of different functionalities with increasing size.

These complications are circumvented in the alternative approach of Hofmann *et al.*, (1950, 1952) by starting a peptide synthesis with the COOH-terminal amino acid hydrazide in the *N′*-protected form (compound **5**). The

$$X\text{—NHCHR}^1\text{CO—NHNH—Y}$$

X, N^{α}-protective group
Y, N'-hydrazide protective group

(**5**)

protected hydrazide function serves to mask the carboxyl group throughout the chain assembly and is converted to the "free" hydrazide by selective rapid cleavage of the *N′*-protective group (Y). This strategy has considerably enhanced the scope of the azide method. Not only is hydrazinolysis avoided in the later stages of a synthesis, but the larger number of selectively cleavable amine protecting groups is utilized for temporary carboxyl protection.

In both procedures the respective hydrazide precursors, i.e., the alkyl esters or *N′*-protected hydrazides are introduced at the beginning of a synthesis and carried through all subsequent synthetic steps until converted to hydrazide for anticipated azide segment condensation. Careful advance planning is required for synthesizing large peptides by azide segment condensation. Should an azide coupling fail or produce low yields the intermediate segment esters or hydrazides are generally not suitable for use with alternate methods (DCC, mixed anhydrides, active esters) that depend on free α-carboxyls. This is so, because alkaline ester saponification remains often incomplete, particularly with longer peptide chains, or leads to extensive side product formation if forced by extended reaction times, higher NaOH concentration or elevated temperature (MacLaren, 1958; Schröder and Lübke, 1965, pp. 55–57 and 61–63; Wünsch *et al.*, 1974). Moreover, selective hydrazide cleavage has not yet been explored (see below). Ideally, one should be able to choose freely between different coupling methods during any given stage of a synthesis. All it requires are mild and facile procedures for interconversion of intermediate peptide esters or hydrazides into acids and vice versa. Indeed, such procedures are presently becoming available.

An efficient procedure for esterification of protected peptide acids under very mild conditions has recently been described (Wang *et al.*, 1977). Titration of the α-carboxylate to pH 7 with cesium carbonate and treatment of the salt with alkyl halides affords a variety of different alkyl esters in high yields. The ability to introduce the methyl ester and, by subsequent hydrazinolysis, the hydrazide groups into peptide segments in the later stages of a synthesis makes protected peptide acids or their precursor forms (e.g., *tert*-butyl esters) accessible to azide coupling. More recently, an alternative procedure was developed (Wang *et al.*, 1978) for converting protected peptide acids directly into hydrazides by treatment with hydrazine in the presence of dicyclohexylcarbodiimide and 1-hydroxybenzotriazole (DCC—HOBt coupling) (König and Geiger, 1970). These procedures allow switching to azide coupling without resynthesis of segments in cases where unacceptably high racemization is encountered with the use of other methods.

The reverse switching, e.g., to DCC—HOBt coupling, when azide condensation is found to be very slow or failing would provide a highly desirable increase in synthetic strategic flexibility. The most promising procedure to prepare acids from hydrazides appears to be oxidation to diimides (Wolman *et al.*, 1962) initially designed for peptide bond formation (Scheme 2, reaction a). This reaction has indeed been used (Scheme 2, reaction b) for converting simple acid hydrazides (Tsuji *et al.*, 1975) or *N*-protected amino acid hydrazides (Schnyder and Rottenberg, 1975) into acids and for analytical peptide hydrazide determination (Medzihradszky-Schweiger, 1962; Kaufmann *et al.*, 1975), but it has not as yet been exploited for preparative-scale conversion of protected peptide hydrazides into acids.

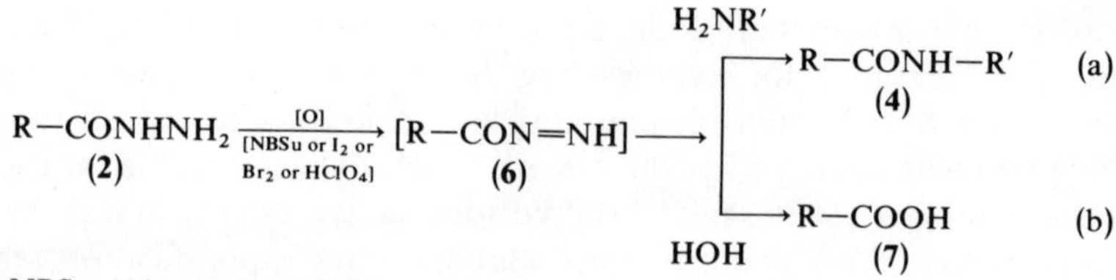

NBSu, *N*-bromosuccinimide

Scheme 2

As a consequence of the developments, described above, azide segment coupling may now be employed, when needed, at various stages of a larger synthesis, even if its use had not been planned in advance, thus providing increased strategic flexibility. Protected peptide hydrazides have become available from the following precursor forms: (1) methyl, ethyl, or benzyl esters by hydrazinolysis; (2) N′-protected hydrazides by selective cleavage of the protecting group; (3) peptide acids by cesium salt esterification followed by hydrazinolysis; and (4) peptide acids by direct hydrazide formation with DCC—HOBt. Experimental aspects of these procedures will be discussed in Sections III,B and C.

B. Ester Hydrazinolysis

1. Reaction Conditions

The most commonly used *solvents* for hydrazinolysis of protected peptide methyl, ethyl, or benzyl esters are dimethylformamide, methanol, or mixtures of both (Gillessen *et al.*, 1963; Strachan *et al.*, 1969). Considerable rate enhancement of peptide ester hydrazinolysis may be attained by using *n*-butanol (MacLaren *et al.*, 1958) or a 1:1 mixture of dimethylformamide-*n*-butanol versus pure dimethylformamide as a solvent (Camble *et al.*, 1973). The optimal promoting effect of *n*-butanol on the reaction rate of hydrazinolysis was first observed by Ferren *et al.*, (1957).

The esters are dissolved in the minimal amount of solvent, with warming if necessary. Excess hydrazine hydrate or anhydrous hydrazine is added. The *amounts* may vary over a wide range (4 to 100 equiv), as well as the *temperatures* (25 to 80°C) and the reaction times (several minutes, hours, days). It is therefore advisable to *monitor* the progress of hydrazide formation by thin layer chromatography in suitable solvent systems using hydrazine-specific spray reagents for identification (*vide infra*).

Work-up is often facilitated by the tendency of hydrazides to precipitate from the reaction mixture. Filtration and thorough washing with suitable solvents often yields directly homogeneous compounds. Alternatively, solvent evaporation and repeated precipitation or recrystallization is applied. Protected peptide hydrazides that cannot be obtained in homogeneous form in this manner may be purified by preparative high-performance liquid chromatography on silica gel 60 (Gabriel *et al.*, 1976, 1977) using, for example, a solvent system composed of $CHCl_3-CH_3OH-CH_3COOH$ (85:10:5).

Characterization by nmr spectroscopy (resonances at 8.8 to 9.0 ppm) and elemental nitrogen analysis distinguishes hydrazides from their ester precursors. However, with larger than 10- to 15-peptides the accuracy of these criteria becomes marginal and eventually inadequate. Specific oxidative hydrazide–nitrogen determination (Medzihradszky-Schweiger, 1962; Kaufmann *et al.*, 1975) should then be used.

Illustrative procedures have been listed by Wünsch and Stelzel (1974b, p. 298) and by Greenstein and Winitz (1961, p. 958). Reaction conditions vary widely. The type of solvent, concentrations, time, and temperature depend on solubility, size, protecting groups, and amino acid composition of each individual peptide ester.

Suitable solvents other than methanol, *n*-butanol, and dimethylformamide include ethanol, dioxane, pyridine, dimethylacetamide, hexamethylphosphoramide, dimethyl sulfoxide (Mitaki and Shimonishi, 1977), and phenol. Pure hydrazine (hydrate) without solvent may also be used (Schott *et al.*, 1947). To avoid formation of bis(*N*-acyl peptide) hydrazide (structure **8**) excess hydrazine is used. Excess also accelerates the reaction. As an

$$RCO-NHNH-COR$$

(8)

extreme, hundredfold excess of hydrazine (25% N_2H_4 in dimethylformamide) served to convert peptide methyl esters to hydrazides within 2–3 min at 25°C with peptides containing as many as 9, 20, 26, and 44 amino acid residues (Strachan *et al.*, 1969; Jenkins *et al.*, 1969). In contrast, successive treatment for 1 hr at 70°C, 15 hr at 50°C, and 3 days at 25°C with 35–70 equiv hydrazine hydrate in dimethylformamide was required to convert several different 12-peptide ester analogues of bovine pancreatic ribonuclease A(1–12) to the corresponding hydrazides (see, e.g., Scoffone *et al.*, 1967). Similarly, refluxing for 40 hr with a threefold excess of $N_2H_4 \cdot H_2O$ in EtOH was used in preparing Z-Val-Lys(Tos)-N_2H_3 from its methyl ester (Schwyzer and Sieber, 1958).

In view of the large and unpredictable differences in rates of peptide ester hydrazinolysis, it is advisable to monitor hydrazide formation by thin layer chromatography. Suitable solvent systems include, among many others $CHCl_3-MeOH-60\%$ aq. NH_4OH (60:32:10) (Strachan *et al.*, 1969), $CHCl_3-MeOH$ (5:1 or 9:1 or 19:1), $CHCl_3-MeOH-AcOH$

(85:10:5), *n*-BuOH—AcOH—pyridine—H_2O (15:3:10:12), and *n*-BuOH—AcOH—EtOAc—H_2O (1:1:1:1). Several spray reagents have been reported which facilitate identification of hydrazides on chromatograms, e.g., 1% picryl chloride in ethanol producing yellow spots (Scoffone *et al.*, 1967), 1% aq. $Na_2Cr_2O_7$ producing grayish-blue spots (Goren and Fridkin, 1970), or a 1:1 mixture of 0.3 *M* $FeCl_3$ and 0.2 *M* $K_3Fe(CN)_6$ in 0.1 *N* acetic acid producing blue spots (Hofmann *et al.*, 1965; see also Pataki, 1966; Brenner and Hofer, 1961).

With increasing size elemental nitrogen analysis becomes insufficient for assessing the homogeneity of peptide hydrazides. For example, in decapeptide hydrazides contamination with up to 25% of corresponding ester does not decrease elemental nitrogen values beyond the accepted 0.3% error margin. Moreover, tlc may not always separate these or larger peptide hydrazides for their parent esters. *Specific hydrazide determination* is essential for larger than decapeptide hydrazides. Manometric hydrazide—nitrogen analysis through oxidation (Wolman *et al.*, 1962) by iodine in dimethylformamide has been developed by Medzihradszky-Schweiger (1962). Facile oxidimetric titration (methyl red indicator) of the hydrazide function of peptides using $KBrO_3$—KBr in dimethylformamide has recently been described (Kaufmann *et al.*, 1975).

2. Solid Phase Techniques

Hydrazides of protected peptides are readily obtained by hydrazinolysis of acyl peptide—resin benzyl esters, prepared by solid phase procedures (Merrifield, 1969; Erickson and Merrifield, 1976). Reaction with 10–200 equiv of hydrazine in dimethylformamide at 20°C for 2–4 days provided high yields of hydrazides for peptides with up to 10 residues (review, Meienhofer, 1973).

This approach takes advantage of the speed and convenience of the solid phase method for rapid production of homogeneous (purified) protected peptide hydrazides which may be used in azide segment condensation for synthesis of larger peptides in solution (Wang and Kulesha, 1975; see discussion in Meienhofer, 1977). A fully automated procedure for solid phase synthesis of protected peptide hydrazides on recycled hydroxymethyl resin has been proposed (Chang *et al.*, 1976). Considerably faster hydrazinolysis than with the resin benzyl ester linkage was achieved (Wang, 1976) with modified solid supports which contain the α-methylphenacyl ester or the *o*-nitrobenzyl ester (Merrifield, 1963) anchoring bonds.

One of the most accomplished solid phase syntheses, yet, with respect to product yield (97%) and purity was the preparation of the crystalline protected decapeptide hydrazide intermediate (compound **9**) of gramicidin S via resin ester hydrazinolysis (Ohno *et al.*, 1971).

Moz—DPhe—Pro—Val—Orn(NO_2Z)—Leu—
DPhe—Pro—Val—Orn(NO_2Z)—Leu—N_2H_3

(9)

3. Stability of Protective Groups and Amino Acids

Most of the commonly used amine protective groups are stable during peptide ester hydrazinolysis even with prolonged reaction periods, elevated temperatures or very high excess of hydrazine. Stable groups include benzyloxycarbonyl (Z) and ring-substituted derivatives, *tert*-butyloxycarbonyl (Boc) and related *tert*-alkyl urethanes, 2-(*p*-biphenylyl)propyl(2)oxycarbonyl (Bpoc), toluenesulfonyl (Tos), trityl (Trt), and several recently developed blocking groups, e.g., isonicotinyloxycarbonyl (Inc; Hirschmann and Veber, 1973; see also Macrae and Young, 1975) α,α-dimethyl-3,5-dimethyloxybenzyloxycarbonyl (Ddz; Birr *et al.*, 1972), isobornyloxycarbonyl (Ibc; Fujino *et al.*, 1972; Jäger and Geiger, 1973), and *p*-dihydroxyboronatobenzyloxycarbonyl (Dobz; Kemp *et al.*, 1975). NH_2-Protective groups cleaved by hydrazinolysis include phthaloyl, trifluoroacetyl, *o*-nitrophenylsulfenyl (Nps; Zervas *et al.*, 1963; Zervas and Hamalidis, 1965), and, at elevated temperature, the formyl group (Geiger and Siedel, 1968; Yajima *et al.*, 1968) and the 9-fluorenylmethyloxycarbonyl group (Fmc; Carpino and Han, 1972). Alternative procedures of hydrazide preparation may be used in the presence of these protective groups, see Section III,C.

Of the widely used carboxyl blocking groups, only the *tert*-butyl esters are sufficiently stable to permit selective hydrazinolysis of methyl, ethyl and benzyl esters. The carboxamide groups of asparagine and glutamine residues are sufficiently stable in most cases. Numerous hydrazides have been prepared successfully from protected peptide esters that contained asparagine, glutamine, and/or ω-*tert*-butyl esters of aspartic and glutamic acid residues. However, in several cases, side reactions of these residues during hydrazinolysis have been observed, *vide infra.*

Hydroxyl and sulfhydryl protective groups of the ether type are stable toward hydrazinolysis, e.g., *tert*-butyl, benzyl, and halobenzyl ethers, while acyl type protective groups (e.g., *O*-benzyloxycarbonyltyrosine) are readily cleaved. The *S*-*tert*-butyl, *S*-acetamidomethyl, and *S*-benzyl groups are relatively stable.

4. Reactivity of Esters

Other esters than methyl, ethyl, or benzyl esters may be used for the preparation of hydrazides through hydrazinolysis, including benzhydryl esters (Hiskey and Adams, 1965), *p*-methyloxybenzyl esters, *o*-nitrobenzyl esters (Wang, 1976), phenacyl esters (see Wang, 1976), and 2-*p*-toluenesulfonylethyl esters (Miller and Stirling, 1968).

The use of activated esters for considerably accelerated hydrazinolysis has first been reported by Wieland and Heinke (1958) and emphasized by Klieger and Gibian (1962a,b) and Schröder and Lübke (1965, pp. 63 and 80). Hydrazides were formed rapidly under mild conditions (small excess of N_2H_4, 20°C), e.g., from phenyl, thioethyl, or thiophenyl esters. Hydrazide formation was selective in the presence of methyl esters. Various amino acid hydrazides of glutamic and aspartic acid ω-esters were prepared as well as a few small peptides, but the use of activated esters for facilitated hydrazinolysis of larger protected peptides has yet to be exploited.

5. Racemization

Racemization does not appear to be a problem in peptide ester hydrazinolysis (Liberek and Michalik, 1968), but its actual extent and frequency, if any, has not been subject to a thorough study, as yet. For one case, hydrazinolysis of the 2,3-dihydroxy-*N*-ethylbenzamide ester of benzyloxycarbonyl[1-^{14}C]glycyl-L-phenylalanine (compound **10**), Kemp *et al* (1970a) reported practically racemization-free hydrazide formation

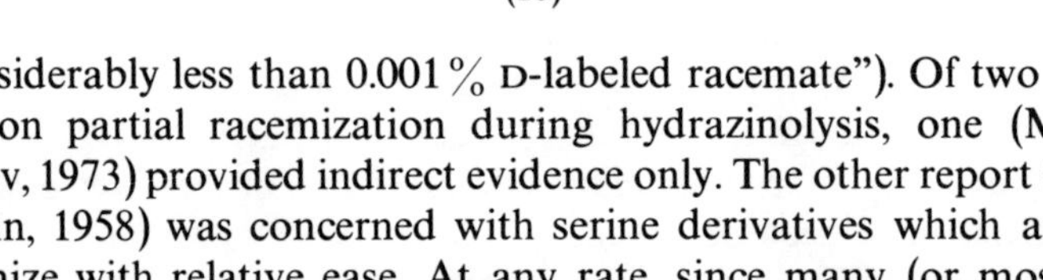

(10)

("... considerably less than 0.001% D-labeled racemate"). Of two literature reports on partial racemization during hydrazinolysis, one (Mitin and Maximov, 1973) provided indirect evidence only. The other report (Schnabel and Zahn, 1958) was concerned with serine derivatives which are known to racemize with relative ease. At any rate, since many (or most) hydrazides spontaneously precipitate and/or crystallize during hydrazinolysis, chances are good that small amounts of diastereoisomers, if formed, are going to be removed during work-up and/or purification and are thus unlikely to contaminate the product.

An observation which might explain the low levels or absence of racemization during peptide ester hydrazinolysis is the formation of chirally pure hydrazide from the chirally pure 5(4H)-oxazolone* of benzyloxycarbonyl-aminoisobutyryl-L-phenylalanine (compound **11**) [Goodman and McGahren, 1967; see review: Goodman and Glaser, 1970]. This indicates

* See Chapter 1, p. 39, footnote.

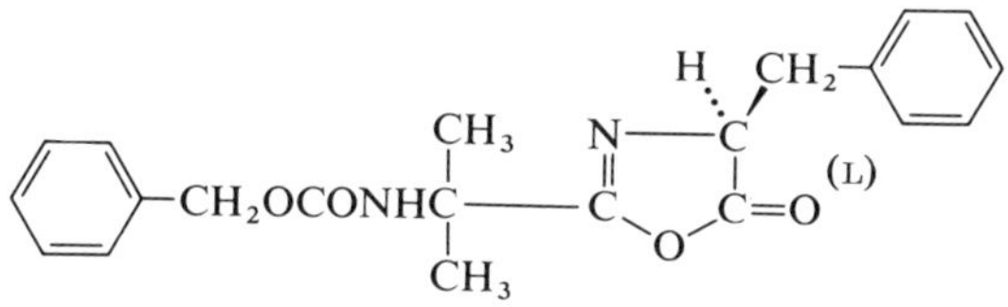

(11)

that hydrazide formation from 5(4H)-oxazolones may be faster than 5(4H)-oxazolone racemization.

6. Side Reactions

A considerable number of undesired side reactions have been observed during peptide ester hydrazinolysis (Klausner and Bodanszky, 1974; Wünsch and Stelzel, 1974b, pp. 300–302). Knowledge of the potential dangers (*vide infra*) frequently permits to minimize their extent by choosing appropriate reaction conditions (see, for example, Strachan *et al.*, 1969). On the other hand, contamination of protected peptide hydrazides by side products is no reason for alarm because the hydrazides are stable synthetic intermediates which can be purified prior to azide formation.

Side reactions reported in the literature are summarized below. Formation of bisacylated hydrazides (Woolley, 1948) can be avoided by using excess hydrazine. Other side reactions arise (*i*) from amino acid residues, or (*ii*) from protective groups.

(i) Formation of ornithine or aminoguanidine from *arginine*-containing peptides [Arg(NO_2), Arg(Tos)] (McKay, 1952; Paul *et al.*, 1961; Danho and Li, 1971) can be avoided by brief treatment with very large N_2H_4 excess (Strachan *et al.*, 1969). *Glutamine* (or *asparagine*) ω-carboxamides may occasionally react with hydrazine. MacLaren *et al.*, (1958) reported that the N^{α}-tosyldipeptide ester (**12**) formed a bishydrazide (**13** in reaction 1) but the corresponding N^{α}-benzyloxycarbonyldipeptide ester (**14**) did not (**15**) under identical conditions (reaction 2). Some aspartyl and asparaginyl residues,

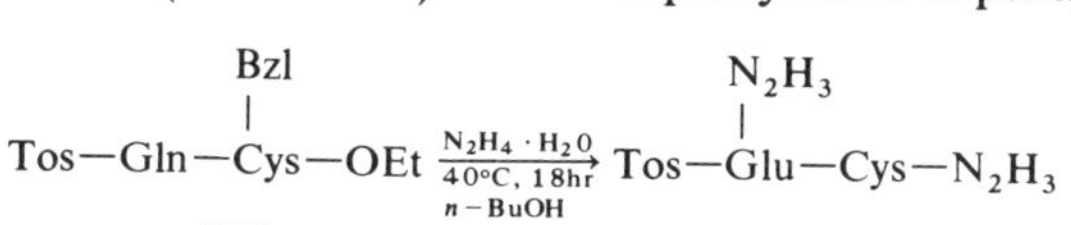

(12) **(13)** (1)

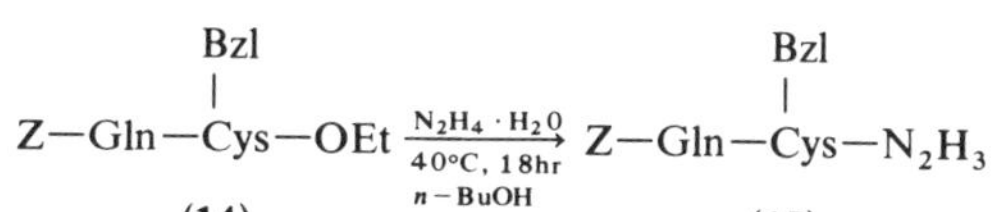

(14) **(15)** (2)

especially those followed by glycine residues, are prone to $\alpha \rightarrow \beta$ transpeptidation (Ondetti *et al.*, 1968) during ester hydrazinolysis (Mitsuyasu *et al.*, 1970; Naithani, 1973).

(ii) Elimination of benzylmercaptan from an *S*-benzyl*cysteine*-containing peptide has been reported (MacLaren *et al.*, 1958). The generally stable *tert*-butyl esters may suffer hydrazinolytic cleavage occasionally, e.g., Boc-Met-Asp(O*t*Bu)-OMe was converted to Boc-Met-Asp(N_2H_3)-N_2H_3 (Scoffone *et al.*, 1964). For other examples see Yajima and Kiso (1971); Wünsch and Stelzel (1974b, pp. 299–300); and Weinstein *et al.*, (1976).

C. The *N*′-Protected Hydrazide Approach

1. Solution Synthesis

Scheme 3 illustrates the sequence of reactions in which the hydrazide function is being kept in the *N*′-protected state throughout the synthesis of a peptide segment and the desired hydrazide obtained by selective cleavage of the blocking group (Hofmann *et al.*, 1950, 1952). Condensation of N^{α}-protected amino acids or peptides (**16**) with a monosubstituted hydrazine (compound **17**) by the carbodiimide or mixed anhydride procedure provides the dual function compound (**18**). The required peptide (**19**) is synthesized by chain elongation at its NH_2-terminus. Upon completion of the intermediate peptide segment the hydrazide protective group is selectively cleaved to provide the hydrazide (compound **20**) for azide segment

$$\underset{(\mathbf{16})}{\text{X—NHCHRCOOH}} + \underset{(\mathbf{17})}{\text{H}_2\text{NNH—Y}} \xrightarrow[\text{or carbodiimide}]{\text{mixed anhydride}} \underset{(\mathbf{18})}{\text{X—NHCHRCO—NHNH—Y}}$$

$$\xrightarrow{\text{Peptide synthesis}} \underset{(\mathbf{19})}{\text{X—[NHCHR}^{1,2,\ldots,n}\text{CO]}_n\text{—NHCHRCO—NHNH—Y}} \xrightarrow[\text{cleavage}]{\text{selective}}$$

$$\underset{(\mathbf{20})}{\text{X—[NHCHR}^{1,2,\ldots,n}\text{CO]}_n\text{—NHCHRCO—NHNH}_2} \longrightarrow \text{conversion to azide followed by segment condensation}$$

X, N^{α}-amine protective group
Y, *N*′-hydrazide protective group

Scheme 3

condensation through the COOH-terminus of the peptide. This approach permits synthesis of acyl peptide hydrazides which contain ω-benzyl esters, arginine and/or asparagine or other functions and residues that are sensitive to hydrazinolysis. With fewer restrictions in the choice of carboxyl protective groups this strategy is therefore better suited to the preparation of large size acyl peptide hydrazides.

It is essential to select a suitable combination of amine protective groups for the blocking of the α-amino, hydrazide and side chain functions. Table I

Table I Combinations of Hydrazide *N'*-Protective Groups and Peptide Chain Protective Groups

Hydrazide *N'*-protective Group		Formula RCO—NHNH—Y	Peptide protective groups in combination with Y: Amine protection	Peptide protective groups in combination with Y: ω-COOH protection	References
Benzyloxycarbonyl	Z	$-COOCH_2-$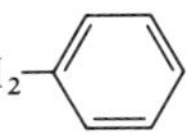	Boc, Bpoc, For, Pht	O*t*Bu	Hofmann *et al.* (1950, 1952, 1969); Storey *et al.* (1972); Wünsch and Wendlberger (1964)
tert-Butyloxycarbonyl	Boc	$-COOC(CH_3)_2-CH_3$	Z, Nps, For	OBzl, O*t*Bu	Schwyzer (1959); Boissonnas *et al.* (1960); Meienhofer (1966); Romovacek *et al.* (1974)
Triphenylmethyl	Trt	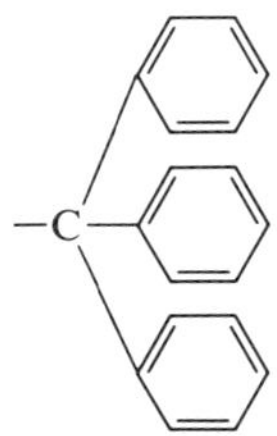	Bpoc, Z, Tfa	O*t*Bu	Weygand and Steglich (1959); Moroder *et al.* (1974)
Trifluoroacetyl	Tfa	$-COCF_3$	Nps, Z	—	Preston and Weinstein (1967)
Formyl	For	$-CHO$	Z	—	Biedrzycka and Biernat (1970)
Trichlorethyloxycarbonyl	Tcc	$-COOCH_2CCl_3$	Z, Boc, Moz	OBzl, O*t*Bu	Yajima and Kiso (1971); Tamura *et al.* (1977); Weinstein *et al.* (1976)
Picolyloxycarbonyl	Inc	$-COOCH_2-$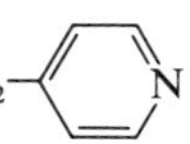	Boc	OBzl, OPic	Macrea and Young (1975)

shows examples of protected hydrazines used for α-carboxyl protection along with some appropriate α-amino and side chain carboxyl protective groups for each. Illustrative literature procedures for over 40 preparations in the use of various protected hydrazides are given by Wünsch and Stelzel (1974a).

2. *Solid Phase Procedures*

The *N*′-protected hydrazide approach has also been adapted to solid phase synthesis of peptide hydrazides as intermediates for azide segment condensation in solution (Wang and Merrifield, 1969; Wang, 1973, 1975). To this end, several modified resin supports with different protected hydrazide functional groups were prepared (Table II). Several crystalline tetra- and pentapeptide hydrazides have been synthesized with these supports in 40 to 75% yields. An excellent overall recovery of 36% was obtained in the preparation of the crystalline decapeptide hydrazide

Z—Ala—Val—Ser(Bzl)—Ile—Gln—Phe—Met(O)—Asn—Leu—$NHNH_2$

(21)

with the use of the *tert*-alkyloxycarbonylhydrazide resin (**22a**: Wang and Merrifield, 1972).

The 2-(*p*-biphenylyl)propyl(2)oxycarbonyl (Bpoc) group (Sieber and Iselin, 1968) has been used as temporary N^{α}-protective group in these solid phase syntheses. The conditions of Bpoc group cleavage (0.5% trifluoroacetic acid in methylene chloride for 12 min at 25°C) are mild enough to minimize rupture of the peptide to resin anchoring bonds (0.2% or less per step). The starting 2-(*p*-biphenylyl)propyl(2)oxycarbonyl amino acid is linked to the resin hydrazide by dicyclohexylcarbodiimide coupling in methylene chloride. The completed acyl peptide hydrazides (**23**) are cleaved from the solid supports (Table II) by 30-min treatment with 50% trifluoroacetic acid in methylene chloride at 25°C (Scheme 4).

D. Direct Coupling of α-Carboxyls with Hydrazine

Protected peptide hydrazides may conveniently be prepared from the corresponding acids by carbodiimide coupling with hydrazine in the presence of 1-hydroxybenzotriazole (Wang *et al.*, 1978). A series of N^{α}-and side chain-protected di- to nonapeptides have been converted in this manner to hydrazides in 65 to 75% yields without detectable racemization by the Manning and Moore (1968) test (sensitivity limit of ±0.1%). The advantages of this approach include: (i) α-carboxyl protection by either alkyl esters or *N*′-protected hydrazides is not required throughout the assembly of

Table II Resin Supports for Solid Phase Synthesis of Acyl Peptide Hydrazides[a]

Modified copolystyrene–1% divinylbenzene resin	Structure	References
tert-Alkyloxycarbonyl-hydrazide resin	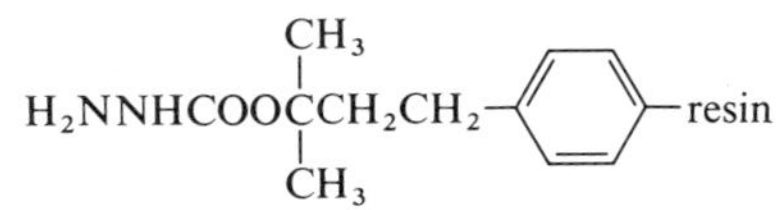 **(22a)**	Wang and Merrifield (1969)
p-Alkyloxybenzyloxycarbonyl-hydrazide resin	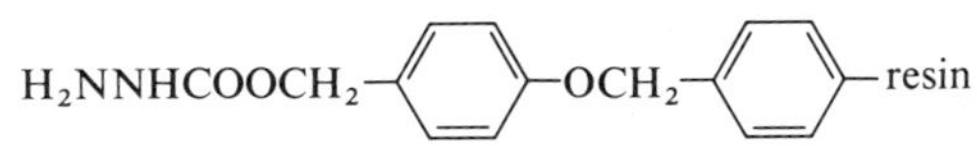 **(22b)**	Wang (1973)
3-(*p*-Benzyloxyphenyl)-1,1-dimethylpropyloxycarbonyl-hydrazide resin	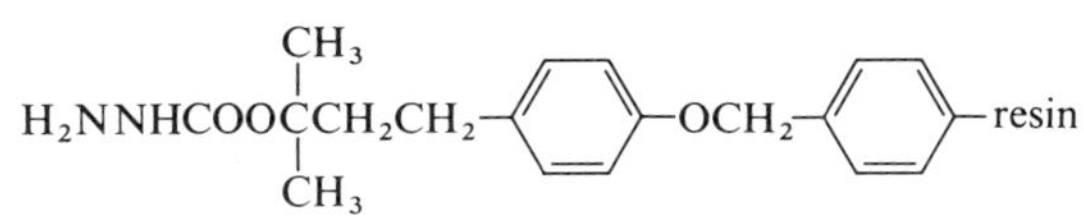 **(22c)**	Wang (1975)

[a] Attachment of starting acylamino acid by dicyclohexylcarbodiimide coupling; cleavage of the final acyl peptide hydrazide from solid support by treatment with 50% trifluoroacetic acid in CH_2Cl_2 for 30 min.

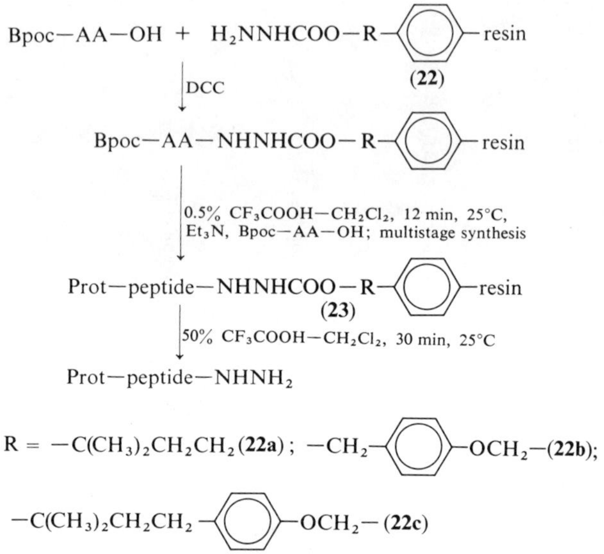

Scheme 4

a peptide—instead, *tert*-butyl ester or free α-carboxyl may be used, (ii) peptide segments with side chain benzyl or alkyl esters or any other base-sensitive protective groups may be converted to hydrazides, and (iii) protected peptide acids intended for segment condensation by other methods (e.g., dicyclohexylcarbodiimide, mixed anhydride, active ester) can be readily utilized for azide condensation in cases where unexpectedly high racemization or other side reactions should render the originally intended coupling method unacceptable.

Earlier hydrazide syntheses from peptide acids and hydrazine, e.g., by mixed anhydride or active ester coupling (Zahn *et al.*, 1965) faced racemization danger unless restricted, e.g., to peptides with COOH-terminal glycine. This restriction has been diminished by the racemization suppression afforded by the 1-hydroxybenzotriazole (König and Geiger, 1970) or *N*-hydroxysuccinimide (Weygand *et al.*, 1966a; Wünsch and Drees, 1966) additives to dicylohexylcarbodiimide couplings.

In a *typical preparation* a dimethylformamide solution of Boc-Val-Asp(OBzl)-Thr(Bzl)-Ser(Bzl)-Ser(Bzl)-Glu(OBzl)-OH was treated at 0°C with anhydrous hydrazine (1.1 equiv), 1-hydroxybenzotriazole (2 equiv), and dicyclohexylcarbodiimide (1.1 equiv) at pH 7.5 (adjusted by *N*-methylmorpholine, wet pH paper) for 2 hr at 0°C and for 18 hr at 25°C. Work-up by filtration (dicylohexylurea), evaporation, washing of ethyl acetate solution (or solid) with water, drying, evaporation, and recrystallization provided homogeneous protected hexapeptide hydrazide in 75% yield.

E. Alternative Procedures

The aminodiacylhydrazine rearrangement (Brenner and Hofer, 1961), reaction 3, albeit of theoretical interest, has not gained any practical significance for the preparation of protected peptide hydrazides in actual

$$\begin{array}{l} \text{Z—Gly—Gly—NH} \\ \qquad\qquad\quad | \\ \quad \text{H—Phe—NH} \end{array} \xrightarrow{\text{H}^+,\ 14\text{ hr at }40^\circ\text{C}} \begin{array}{c} \text{Z—Gly—Gly—Phe—NHNH}_2 \\ \\ (91\%) \end{array} \qquad (3)$$

synthesis. Similarly, selective acylation of the α-amino group of free amino acid hydrazides (Curtius and Levy, 1904) has been reported to be useful in one case (Cheung and Blout, 1965) but not in others (Zahn and Schnabel, 1957).

IV. THE GENERATION OF AZIDES

Prior to the major innovation by Honzl and Rudinger (1961) of using organic nitrites in the presence of acid in anhydrous organic solvents for the conversion of hydrazides into azides, there was hardly any alternative to the classical Curtius procedure, i.e., the use of sodium nitrite in aqueous acetic acid or HCl—acetic acid mixtures. The manipulation of this procedure was cumbersome and required experience and considerable technical skill. Originating from an investigation on the sometimes troublesome amide side reaction, the Honzl–Rudinger procedure proved to be so much easier and convenient to carry out that it totally replaced the aqueous sodium nitrite procedure within a few years. A recent novel variant of peptide azide formation directly from the acids with the use of diphenyl phosphorazidate, discovered by Yamada and collaborators (Shioiri *et al.*, 1972), has shown promising results in Yamada's laboratory, but has yet to be employed elsewhere.

A. Aqueous Acid–Sodium Nitrite Procedure

The use of sodium nitrite in aqueous mixtures of acetic and hydrochloric acid for the conversion of acyl peptide hydrazides into azides is by now mainly of historical interest. Nevertheless, the procedure has been thoroughly discussed by Wünsch and Stelzel (1974b, pp. 303–310) who list several illustrative procedures which may be consulted in addition to those given by Greenstein and Winitz (1961).

In general, the protected peptide hydrazide (10 Meq) is dissolved in a mixture of AcOH and HCl ($\geq$ 10 ml) and cooled to -5°C. A 1 *M* solution of $NaNO_2$ (10–11 ml) is added. When hydrazide tests (e.g., Hofmann *et al.*, 1965; see Section III,B,1) become negative (5–15 min) ice water (350 ml) is added. The azide precipitate is filtered off and washed repeatedly with ice cold 1 *N* HCl, ice water, ice cold 0.3 *M* $NaHCO_2$, and saturated NaCl or the azide is taken up in ethyl acetate and washed similarly. After rapid drying in vacuo over P_2O_5 at 0°C (or over Na_2SO_4), the azide is dissolved in a suitable solvent at 0°C or below for segment condensation.

Points to observe: Sparingly soluble hydrazides may be dissolved at elevated temperature in glacial AcOH, or suspensions may be used; low temperature (0°C or less) must be maintained throughout azide formation and coupling; 3 equiv or more of acid must be present; excess nitrous acid may be eliminated by washing with ammonium sulfamate (Schwyzer *et al.*, 1963); the entire azide formation procedure should be completed within 25–35 min.

Peptide hydrazides which are insoluble in aqueous AcOH—HCl can be converted into azides in dimethylformamide as a solvent to which aqueous HCl and aqueous $NaNO_3$ are added. Segment condensation with the amine component may be conducted directly in this mixture after neutralization with triethylamine. Although this was first described by Guttmann and Boissonnas (1958, 1960), it has recently (Klausner and Bodanszky, 1974; Wünsch and Stelzel, 1974b, p. 306, see also p. 303) been referred to as the "Medzihradszky method" (Medzihradsky *et al.*, 1962). Trifluoroacetic acid has also been used as a solvent for azide formation but it must be replaced (e.g., with dimethylformamide) for the subsequent segment condensation either by evaporation or by precipitation of the azide (e.g., with ether or water) (Schwyzer *et al.*, 1960; Gillessen *et al.*, 1963; Meienhofer, 1966).

Side reactions during azide formation (reviews, Schnabel, 1962; Klausner and Bodanszky, 1974; Wünsch and Stelzel, 1974b, pp. 307–310) include amide formation (Honzl and Rudinger, 1961); sulfoxide formation in *S*-benzylcysteine and methionine residues; and nitrosation of the indole ring of tryptophan (Zahn and Brandenburg, 1966; Agarwal *et al.*, 1969a,b). As with the observed side reactions during hydrazide formation, these reports are outnumbered by accounts on azide preparations without apparent complications. For the recently discovered promotion of racemization by the bicarbonate washing of azides (Kemp *et al.*, 1970b), see Section V,B and Table III.

B. Alkyl Nitrite in Organic Solvents: The Honzl–Rudinger Procedure

Realizing that amide formation (Prélog and Wieland, 1946; see also Klausner and Bodanszky, 1974) was not due to azide rearrangement but to a side reaction (Scheme 5, reaction a) during the conversion of hydrazide to azide (Scheme 5, reaction b), Honzl and Rudinger (1961) investigated reaction conditions for suppressing amide formation. They concluded that opti-

$$R\text{—CO—NHNH}_2 \xrightarrow[-H_2O]{HNO_2} R\text{—CO—NHNH—NO} \begin{cases} \xrightarrow{-NO_2} R\text{—CO—NH}_2 & \textbf{(a)} \\ \xrightarrow{-H_2O} R\text{—CO—N}_3 & \textbf{(b)} \end{cases}$$

Scheme 5

mal azide formation is obtained by working in anhydrous homogeneous solution (organic solvent) at high acidity with an organic nitrite at low temperature. The procedure was quickly recognized as being experimentally simpler, better to control and easier to carry out. Since it is also better suited and adaptable to larger peptides than the Curtius procedure, almost all peptide azide preparations are now conducted by the Honzl–Rudinger procedure.

In a *typical reaction* (Camble *et al.*, 1972) a solution of Z-Phe-Asp(O*t*Bu)-Phe-Ser-Val-Ser-Ser-Pro-$N_2H_3 \cdot CF_3COOH$ (642 mg) in dimethylformamide (3.5 ml) was cooled to −25°C and 6.9 *N* HCl in dioxane (0.4 ml) was added with stirring followed by *tert*-butyl nitrite (0.7 ml of a 10% solution in dimethylformamide). The solution was stirred for 15 min at −25°C and then cooled to −60°C for neutralization with triethylamine (0.53 ml) to pH 7.5–8.0. To this azide preparation a solution of the (decapeptide) amine component in dimethylformamide (3 ml) was added for the coupling reaction (3 days at 4°C). Workup and purification yielded 48% homogeneous 18-peptide.

Points to observe: Temperatures between −5° and −30°C have been used, depending on solubility. In the presence of acid-labile protective groups [*N*-trityl,*N*-2-(*p*-biphenylyl)propyl(2)oxycarbonyl (see Kamber *et al.*, 1976), *N*-*tert*-butyloxycarbonyl, *tert*-butyl ester, *tert*-butyl ether] temperatures below −20°C should be used. Activation times vary from a few minutes to 0.5 hr; disappearance of hydrazide reaction (Hofmann *et al.*, 1965; see Section III,B,1) should be monitored. Other nitrites such as *n*-butyl, *n*-amyl, and isoamyl nitrite and nitrosyl chloride (Honzl and Rudinger, 1961) have been used. For sparingly soluble peptide hydrazides more powerful solvents, e.g., dimethyl sulfoxide, hexamethylphosphoramide, or suspensions of the hydrazides may be used. If trifluoroacetic acid is used as a solvent it must be evaporated after azide formation and replaced by a neutral solvent for coupling. Phenol cannot be used, it reacts with azide, possibly forming phenyl ester.

Due to the high rate of azide formation, peptide hydrazides with unprotected (protonated) α-amino groups may be converted to azides without danger of deamination. Similarly, the carboxamide functions of asparagine and glutamine residues are not attacked. For possible side reactions during azide formations, see Section IV,A. Although the reaction conditions serve to suppress amide formation its total absence cannot always be taken for granted. Formation of Z-Ser-Ser-Ser-NH_2 from the corresponding hydrazide and isoamyl nitrite was observed (Waki *et al.*, 1968).

C. The Diphenyl Phosphorazidate Method

Reaction of diphenyl phosphorazidate (compound **25**) with a N^{α}-protected amino acid or peptide carboxylate (**24**) in dimethylformamide produces in the presence of base the respective azide (**27**), Scheme 6,

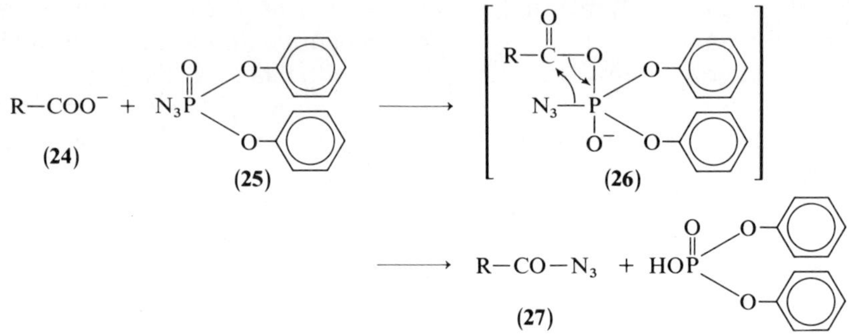

Scheme 6

probably via the intermediate structure **26** (Shioiri and Yamada, 1974). This method lends yet another dimension of experimental flexibility to the azide method. Hydrazides are altogether circumvented in this most direct route for the preparation of protected peptide azides. Moreover, as an additional convenience, diphenyl phosphorazidate may be added as a "coupling reagent" (review, Klausner and Bodanszky, 1972) to the mixture of amine and carboxyl components in dimethylformamide. Racemization was not apparent in the sensitive Young test (Williams and Young, 1963), but lower levels as observed by Kemp *et al.*, (1970a) for azide coupling are likely to occur.

Perhaps of too recent vintage to have been used by others, the procedure has been successfully applied by Yamada and collaborators to syntheses of motilin (Yamada *et al.*, 1975), secretin(1–6) (Ozawa *et al.*, 1977), vasoactive intestinal peptide(1–6) (Hamada *et al.*, 1977a), and gastric inhibitory polypeptide(1–10) (Hamada *et al.*, 1977b). While excellent coupling was obtained with acyl dipeptides, yields were low in segment condensation of a tri- to a hexapeptide (Hamada *et al.*, 1977b). Further data will have to be awaited for a more definitive assessment of the method's performance.

In a *typical experiment* (Hamada *et al.*, 1977a), diphenyl phosphorazidate (1.2 mmol) was added to an ice cold stirred mixture of Z-Ser-Asp(O*t*Bu)-OH (1.1 mmol), H-Ala-Val-Phe-OMe (1 mmol) and triethylamine (1.3 mmol) in dimethylformamide (27 ml). After stirring for 6 hr at 0°C and 20 hr at 25°C, workup and purification afforded crystalline protected pentapeptide in 77% yield.

Peptide synthesis with a related coupling reagent, azidotrisdimethylamino phosphonium hexafluorophosphate (**28**) (Castro and Dormoy, 1973a,b) has

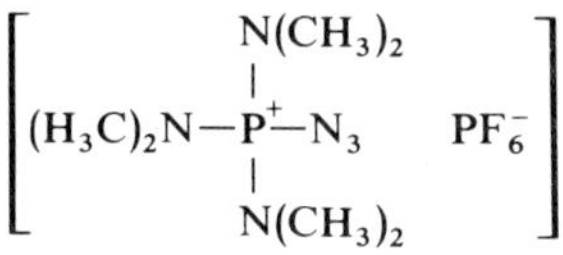

(28)

been proposed to proceed via an intermediate acyl azide. This was indicated by the formation of benzoyl azide in 65% yield when benzoic acid was treated with compound **28**.

V. AZIDE SEGMENT CONDENSATION

A. Reaction Conditions

The azide coupling is in general carried out immediately following the completion of azide formation. At low temperature (−10° to −30°C) the pH is adjusted by the addition of base (triethylamine, diisopropylethylyl-amine) to a range of 7 to 8 as judged by moist indicator paper. It is essential to maintain low reaction temperatures of 5°C or below. Consequently appropriate time must be allowed for completion of the reaction, which may require several days. It is important to work at the highest possible molar concentrations, i.e., utilizing the least amount of solvent necessary. Excess azide may be applied. Dimethylformamide is the solvent most commonly employed in peptide azide coupling. For larger peptide azides of decreasing solubility hexamethylphosphoramide, HMP (Meienhofer and Schnabel, 1965) or dimethyl sulfoxide may be used. It is essential that these solvents be highly purified and absolutely free of contaminating primary or secondary amines. The hydrazoic acid liberated during azide condensation is frequently neutralized by periodic additions of base during the reaction to maintain the pH between 7 and 8. Completion of azide coupling is usually monitored by the disappearance of the amine component.

The mild acylating property of azides (Determann and Wieland, 1963) permits to reduce the use of protective groups (Hofmann, 1971; Finn and Hofmann, 1976a). Protection is required only for the α-amino group of the peptide azide, for the ω-amino groups of lysine and other diamino acids, and for the sulfhydryl function of cysteine. For example in the synthesis of a 104-peptide, the *S*-protein of bovine pancreatic ribonuclease A, only three different protective groups were used for these functionalities (Hirschmann 1971a; Hirschmann and Veber, 1973). No protection is necessary for the side chain functions of arginine, aspartic and glutamic acid, histidine, serine, threonine and tyrosine, as well as for the α-carboxyl of the amine component. However, circumstances may occasionally demand additional protection of one or the other of these residues, in individual cases. For example, protection of the tyrosine phenolic function was needed in the presence of excess azide (Tilak and Hoffmann, 1977).

In a *typical experiment* (Borin *et al.*, 1977), a solution of Z-Lys(Tfa)-Thr-Glu-Arg-Glu-Asp-Leu-Ile-Ala-N_2H_3 (3.46g, 2.51 mmol) in dimethylformamide (40 ml) was cooled to $-15°C$ and 4.71 *N* HCl in dioxane (1.60 ml, 7.53 mmol) was added with stirring, followed by *tert*-butyl nitrite (0.33 ml, 2.76 mmol). The mixture was stirred at $-15°C$ for 10 min, then cooled to $-60°C$ and neutralized with triethylamine (2.19 ml, 15.55 mmol). To this azide-containing solution a chilled ($-10°C$) solution of the amine component, H-Tyr-Leu-Lys(Tfa)-Lys(Tfa)-Ala-Thr-Asn-Glu(O*t*Bu)-O*t*Bu · Ac OH (3.99 g, 3 mmol), in dimethylformamide (40 ml) was added and the pH adjusted to 7.5 by the addition of triethylamine. The reaction mixture was stirred for 10 days at 5°C, the pH being maintained at 7.5 by periodic additions of triethylamine. For workup the reaction mixture was concentrated, 0.1 *M* acetic acid was added, the solid precipitate collected and reprecipitated from dimethylformamide with H_2O to yield 4.38g (68.3%) of crude 17-peptide.

Points to observe: Peptide azides are not very stable and prone to undergo Curtius rearrangement (see Section V,C), even when isolated in crystalline form and stored at 0°C, as reported for Boc-Ser-Tyr-Ser-Met-N_3 (Schwyzer and Kappeler, 1961). This kind of limited stability and the long coupling times led to recommendations of adding the azide in several timed portions, however, such a manipulation must be weighed against the ensuing decrease in reaction rate due to reduced initial molar concentration. Acceleration of azide coupling by raising the temperature above 5°C cannot be recommended since it would promote the intramolecular Curtius rearrangement. Highest possible molar concentrations are most important: e.g., part of the solvent may be evaporated from the reaction mixture (at low temperature!) once the amine component has been added.

A wide range of solvents may be used. Alcohols and phenol should be avoided as they might give rise to urethane or ester formation (see Weygand *et al.*, 1966b). *N*-Protected peptide azides are more soluble in organic solvents than their precursor hydrazides. The amine components, frequently of larger size, present generally the more severe insolubility problems and require more powerful solvents such as hexamethylphosphoramide or dimethyl sulfoxide. Mixtures of these and dimethylformamide may be used. The ready removal of dimethyl sulfoxide by lyophilization facilitates workup in cases where the products cannot be precipitated.

Monitoring of azide couplings by the disappearance of the ninhydrin reaction of the amine component has not been satisfactory for larger peptides which do not exhibit adequate ninhydrin reactions in dimethylformamide, hexamethylphosphoramide, and dimethyl sulfoxide solutions and often do not migrate in discrete spots on thin layer chromatograms. Both false positives or false negatives may occur depending on concentration and solvent composition of the reaction mixture. Fluorescamine (Felix and Jimenez, 1974; DeBernardo *et al.*, 1974) or other fluorescent amine reagents (e.g., *o*-phthaldialdehyde or dansyl chloride) are more sensitive. When used in conjunction with analytical high performance liquid chromatography (HPLC), the monitoring of azide couplings might be conducted readily within 15–30 min per sample.

The azide method allows activation of histidine without N^{im}-protection (Holley and Sondheimer, 1954; Fischer and Whetstone, 1954). Observations of imidazolide formation from intramolecular reaction of the azide with imidazole nitrogen under basic conditions (Merrifield and Woolley, 1956) and the occurrence of unusually high racemization during syntheses of thyroliberin analogues, prompted Veber (1975) to study the behavior of pGlu-His-N_3 (reaction 4). An equilibrium was found to exist between the

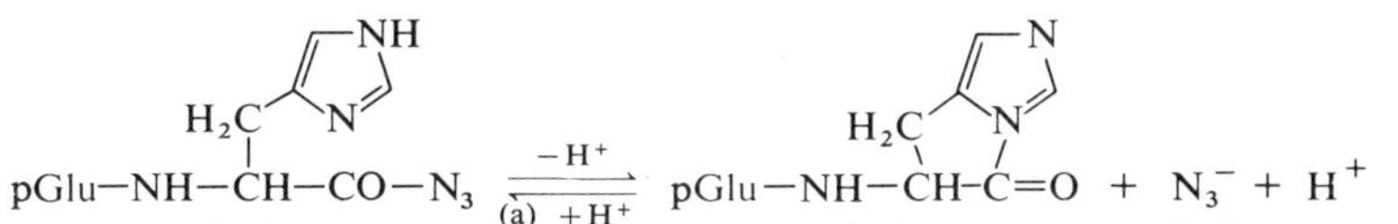

(29) (30) (31) (4)

acyl azide (**29**: ir, 2110 cm^{-1}) and the imidazolide (**30**: ir, 1770 cm^{-1}). Formation of compound **30** is promoted by base and reversed by acid and, to a limited extent, by azide ion (**31**: ir, 1980 cm^{-1}). The rate of coupling of the imidazolide (**30**) is considerably slower than that of the acyl azide (**29**). Thus, shift of equilibrium to compound **30** was observed to intensify racemization. For example, coupling with methyl pipecolate resulted in complete histidine racemization at pH 8–8.5, compared to 25% racemization at pH 7–7.5. It should be interesting to explore whether azide ion would also influence the rate of peptide bond formation from other intermediates, e.g., 5(4H)-oxazolones. In any event, it seems advisable to avoid segment condensations that involve azide activation of COOH-terminal histidine with unprotected imidazole function.

B. Racemization

The demise of the long-time notion of "racemization-free" coupling by the azide method (Anderson *et al.*, 1966; Weygand *et al.*, 1966b; Sieber *et al.*, 1970; Kemp *et al.*, 1970a) may have caused nostalgic resignation on one hand and on the other a tendency of overestimating now the risk of acyl azide racemization (Young, 1973). What we need instead, is a thorough understanding of all factors involved. Such knowledge may permit rational attempts of suppressing racemization (Young, 1967). This common aim has been emphasized in several recent studies (Kemp *et al.*, 1970b; Kemp and Rebek, 1970; Hirschmann, 1971b; Sieber *et al.*, 1971; Sieber and Riniker, 1973; Kisfaludy *et al.*, 1973) and literature reviews (Klausner and Bodanszky, 1974; Wünsch and Stelzel, 1974b, pp. 312–313; Finn and Hofmann, 1976b; Kemp, this volume, Chapter 7) but many questions still remain unanswered.

The initial observations of Anderson *et al.*, (1966) and Weygand *et al.*, (1966b) drew attention to the capability of strong base (trimethyl- or triethylamine), when present in excess, to cause racemization during azide coupling (Table III, preparations 1,2). These relevations were chilling but not alarming at the time because no racemization was detected in these studies when the same or other components were coupled under standard operating conditions, thus apparently confirming all earlier results. However, Weygand did already strongly warn against using any excess base in azide condensations (Weygand *et al.*, 1966b). Soon the relevance of this warning to "actual synthesis" became apparent when the CIBA-Geigy group, one of the most experienced teams in peptide synthesis, reported the occurrence of 12% and 40% racemization during two azide condensations (see Table III, preparations 3,4) under standard conditions in a synthesis of human calcitonin (Sieber *et al.*, 1970). It was pointed out (Sieber *et al.*, 1971) that the frequent necessity of adding base to the reaction mixture in synthesis of larger peptides by the azide method makes it very difficult to avoid the presence of excess base. In fact, this experimental problem persists even today since we still depend entirely on using moist pH indicator paper for measuring the pH in azide coupling—a "highly sophisticated" tool, indeed.*

Within a short period of time (1970–1973) evidence for racemization in azide coupling was reported by several laboratories. Kemp and collaborators (1970a) developed an isotopic dilution assay for racemate determination which improved the sensitivity of the Anderson and Young tests (Anderson and Callahan, 1958; Williams and Young, 1963) to such low levels as $\pm 0.001\%$ and $\pm 0.05\%$, respectively. Azide coupling under optimal conditions produced 0.2–0.7% racemization by the Young test (Table III, preparation 5). Bicarbonate washing, which is routinely used in the aqueous acid—$NaNO_2$ formation of azide, promoted racemization of Bz-Leu-N_3 from 0.1%, without washing, to 0.27–0.36% (Table III, preparation 6) (Kemp *et al.*, 1970b). Acyl azide preparations which had been freed of traces of acid (acetic acid) were particularly vulnerable to base-catalyzed racemiza-

* Perhaps a study of the titration characteristics of various bases in dimethylformamide might provide useful information, particularly in combination with some special electrode.

Table III Racemization during Acyl Azide Coupling

No.	Azide preparation[a]	Azide	Amine	Racemization[b] (%)	Cause	References
1	aq	Z-Gly-Phe-N_3	H-Gly-OEt	1.6	Excess base (Me_3N)	Anderson *et al.* (1966)
2	aq	Tfa-Val-N_3	H-Val-OMe · HCl	54.0	Excess base (Et_3N)	Weygand *et al.* (1966b)
3	HR	Z-Asn-Lys(Boc)-Phe-His-N_3	Octapeptide-OH	12.0	Et_3N	Sieber *et al.* (1970, 1971); Sieber and Riniker (1973)
				3.0	$(iPr)_2EtN$	
4	HR	Z-Thr(*t*Bu)-Tyr(*t*Bu)-Thr(*t*Bu)-Gln-Asp(O*t*Bu)-Phe-N_3	12-Peptide-OH	40.0	Standard reaction DMF–Et_3N	
5	aq	Bz-Leu-N_3	H-Gly-OEt · HCl	0.2–0.66	"Young Test" optimal conditions	Kemp *et al.* (1970a)
6	aq	Bz-Leu-N_3	H-Gly-OEt · HCl	0.27–0.36	Repeat $NaHCO_3$ wash	Kemp *et al.* (1970b); Kemp and Rebek (1970)
				0.1	No $NaHCO_3$ wash	
7	aq	Z-Gly-Phe-N_3	None for 15 min then H-Gly-OEt	50.3	Excess base (Et_3N)	
8	HR	Boc-Ser-Thr-Met-Ser-Ile-N_3	H-Thr-Asp-Cys(Acm)-Arg-OEt	2.8	Standard reaction DMF–Et_3N	Hirschmann (1971b)
9	HR	Z-Gly-Ala-N_3	H-Leu-OBzl	9.5	Excess base (Et_3N)	Kisfaludy and Nyeki (1972)
				0	1-Diethylamino-2-propanol	Kisfaludy *et al.* (1973)

[a] aq, aqueous $NaNO_2$—HCl; HR, Honzl–Rudinger procedure.
[b] DL (50 % D) = 100 % Racemization.

tion: after 15 min at 3°C in dimethylformamide containing 0.2% triethylamine the Anderson azide (Z-Gly-Phe-N_3) was racemized to the extent of 50.3% (Table III, preparation 7). Kemp strongly emphasized that excess of tertiary amines should be avoided in azide couplings (Kemp *et al.*, 1970b). It was left open whether azide racemization by triethylamine proceeds via 5(4H)-oxazolone or direct enolization (Kemp and Rebek, 1970) but ir evidence in favor of intermediate 5(4H)-oxazolone has been mentioned by Taschner and collaborators (Dzieduszycka *et al.*, 1973). Hirschmann (1971b) observed 2.8% racemization in a coupling of Boc-Ser-Thr-Met-Ser-Ile-N_3 with a tetrapeptide ester under standard conditions of the Honzl-Rudinger procedure (Table III, preparation 8). Kisfaludy *et al.* (1973) studied the influence of amount and type of base in the reaction of Z-Gly-Ala-N_3 with H-Leu-OBzl. Confirming the strong racemizing effect of excess triethylamine, they found that the use of 1-diethylamino-2-propanol, even in excess, did not give rise to racemization, Table III, preparation 9, as judged by the Izumiya test (0.1% sensitivity limit; Izumiya and Muraoka, 1969; Izumiya *et al.*, 1971). Solvent also appeared to have a significant influence on azide racemization, the danger being largest in chloroform.

The most detailed study yet is that by Sieber *et al.*, (1970, 1971; Sieber and Riniker, 1973). The influence of type (strength) and concentration of base was investigated as well as the nature of both the amine and carboxyl components. Using as a model the calcitonin(11–28) sequence which had suffered 40% racemization during the synthesis (Scheme 7, reaction a) of the hormone, substantial influence of the nature of both the amine and the azide component on the extent of racemization were observed. With valine methyl ester 30% racemization occurred (Scheme 7, reaction b) under standard

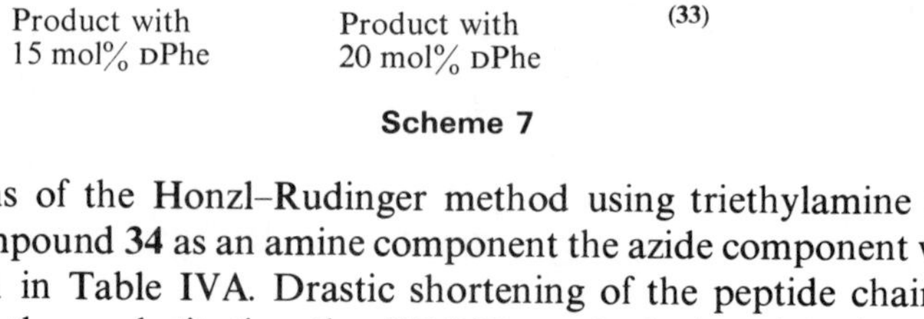

Scheme 7

conditions of the Honzl–Rudinger method using triethylamine as a base. With compound **34** as an amine component the azide component was varied as shown in Table IVA. Drastic shortening of the peptide chain had less influence than substituting the COOH-terminal phenylalanine by leucine which reduced racemization to almost one quarter. This demonstrated that

Table IV Influence of Component Structure on Racemization during Azide Coupling[a]

No.	Azide component	Amine component	Excess Et_3N (mmol)	Racemization (%)
A. Effects of varying the azide component				
1	Z-Thr(*t*Bu)-Tyr(*t*Bu)-Thr(*t*Bu)-Gln-Asp(O*t*Bu)-Phe-N_3	H-Val-OMe	3	31.0
2	Z-Asp(O*t*Bu)-Phe-N_3	H-Val-OMe	3	26.4
3	Boc-Gly-Phe-N_3	H-Val-OMe	3	16.8
4	Z-Thr(*t*Bu)-Tyr(*t*Bu)-Thr(*t*Bu)-Gln-Asp(O*t*Bu)-Leu-N_3	H-Val-OMe	3	8.0
B. Effects of varying the amine component				
5	Z-Asp(O*t*Bu)-Phe-N_3	H-Val-OtBu	3	7.2
6	Z-Asp(O*t*Bu)-Phe-N_3	H-Val-OMe	2	19.0
7	Z-Asp(O*t*Bu)-Phe-N_3	H-Val-Gly-Ala-Pro-NH_2	3	11.8
8	Boc-Gly-Phe-N_3	H-Val-OMe	2	2.4
9	Boc-Gly-Phe-N_3	H-Val-OMe	3	8.4
10	Boc-Gly-Phe-N_3	H-Gly-OMe	2	3.2
11	Boc-Gly-Phe-N_3	H-Gly-OMe	3	22.4

[a] Selected data from Sieber and Riniker (1973).

the COOH-terminal amino acid exerts a very large influence on racemization and was in line with the order of increasing racemization rates for certain amino acids [Val < Leu < Ala < Phe] determined in a different system (Sato *et al.*, 1970). With Z-Asp(O*t*Bu)-Phe-N_3 whose racemization rate was similar to that of compound **32** (see Table IV,A, reactions 2,3) the amine component was further varied as shown in Table IVB. Considerably less racemization occurred with valine *tert*-butyl ester than with the methyl ester, or with a tetrapeptide amide. Examples 8 to 11 showed, unexpectedly, a greater degree of racemization with glycine methyl ester than with valine methyl ester.

Influences of the type and amount of different bases on racemization are depicted in Table V. Using excess base in the model reaction of Z-Asp(O*t*Bu)-Phe-N_3 with H-Val-OMe, diisopropylethylamine caused considerably less racemization than triethylamine. The weaker bases, *N*-methyl- and *N*-ethylmorpholine, performed still better but reaction rates were three times slower than those obtained with the strong bases. This finding should discourage the use of *N*-methylmorpholine in condensations of large segments which are already slower, per se. From the available data it appears that the extent of racemization during azide coupling depends largely on the same qualitative factors as racemization in active ester (see Williams and Young, 1972), carbodiimide, and mixed anhydride couplings (see Kemp *et al.*, 1970a).

Recommendations: Based on present information, the following precautions are recommended for azide segment condensation: (1) avoiding even the slightest excess of base, even for very short periods of time, (2) using diisopropylethylamine or diethylamino-2-propanol, (3) neutralizing with stoichiometric amounts of strong base, then maintaining the pH during the reaction by the addition of weak base (e.g., *N*-methylmorpholine), (4) adding acetic acid if reaction rate and yield do not severely deteriorate, and (5) avoiding of coupling at histidine positions.

Table V Effects of Type and Amount of Base Added on the Extent of Racemization in the Coupling of Z-Asp(O*t*Bu)-Phe-N_3 with H-Val-OMe[a]

Base	Racemization (%)	
	No excess base	Excess base (2 mmol)
Triethylamine	0.8	19.0
N,N-Diisopropylethylamine	1.4	4.4
N-Methylmorpholine	1.8	2.8
N-Ethylmorpholine	1.2	1.6

[a] Selected data from Sieber and Riniker (1973).

Certainly, more model studies will be required for gaining a better understanding of the main influences on racemization during azide coupling and means for its suppression. If one wonders why racemization was discovered only 60 years after the introduction of the method, there may be several reasons. First of all, the levels are indeed lower than with other methods and the early racemization tests were not very sensitive. Moreover, in actual synthesis it is the purified products that are naturally checked for chiral integrity. To detect racemization one must, however, examine the crude reaction product, since retention of the stereoisomeric contaminant in the product throughout work-up is exceptional (see, for example, Sieber *et al.*, 1977).

Although, in conclusion, the azide method is not free from racemization, the levels occurring in actual synthesis are comparatively lower than in the active ester, carbodiimide, and anhydride methods. There are several reports where rigorous examination of crude azide condensation products by sensitive racemization tests failed to detect stereoisomeric contamination (e.g., Weygand *et al.*, 1966b; Izumiya and Muraoka, 1969; Izumiya *et al.*, 1971; Kisfaludy *et al.*, 1973).

C. Side Reactions

The major side reaction of concern in azide coupling (review, Schnabel, 1962) is the Curtius rearrangement to isocyanates (Curtius, 1890) (Scheme 8) which may lead to several side products. The isocyanate formation (**3** → **35**)

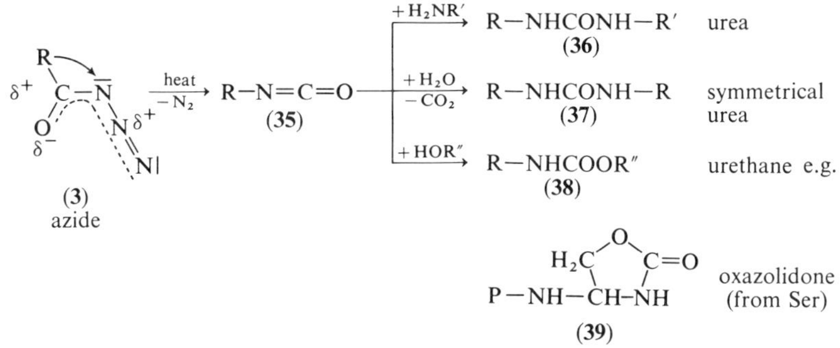

Scheme 8

is an unimolecular reaction depending on temperature, structure, and solvent but not on the azide concentration, and can be conveniently monitored by infrared spectroscopy (Schwyzer and Kappeler, 1961). Azides have a characteristic band at 2100 cm^{-1} while isocyanates have a sharp band at 2220 cm^{-1}. The rate of the Curtius rearrangement varies widely with the structure of acyl peptide azides: a large amount of isocyanate was formed within 24 to 48 hr at 0°C from isolated crystallized Boc-Ser-Tyr-Ser-Met-N_3 (Schwyzer and Kappeler, 1961). In contrast Z-Gly-Ile-Val-Glu-Gln-Cys(Bzl)-Cys(Bzl)-Ala-Gly-N_3 was very stable, showing only a trace of isocyanate after 72 hr at 0°C, and had to be kept at 25°C for 5 days to undergo major rearrangement (Katsoyannis *et al.*, 1966). It is therefore, advisable to monitor azide couplings by IR spectroscopy at 2220 cm^{-1}. If isocyanate is detected, lowering of the reaction temperature to subzero levels may suppress further Curtius rearrangement. On the other hand, reaction times may be extended for increased yields in cases (Katsoyannis *et al.*, 1966) where unreacted azide and no isocyanate is present.

Isocyanate reacts rapidly with amines or alcohols, present in azide coupling mixtures, to form secondary contaminants. Reaction with the amine component leads to urea derivatives (**36**) which differ from the target peptide by an inserted NH. This difference cannot be detected by elemental analysis in larger molecules. However, amino acid analysis serves to identify the urea derivative by the absence of the azide-activated amino acid due to decomposition of the urea residue during acid hydrolysis (Hofmann *et al.*, 1960). Two isocyanate molecules can form symmetrical urea derivatives, **37** (Schnabel, 1962) whose amino acid analysis resembles that of the azide component except for the missing COOH-terminal residue. The urethane formation (**38**, Scheme 8) has been most noticeable with the oxazolidone (**39**) [ir, 1770 cm^{-1}] obtained from acylserine azide (Fruton, 1942; Mitsuyasu *et al.*, 1969), which can be detected by its IR absorption (1770 cm^{-1}).

Another, occasionally observed, side reaction (Schnabel, 1962) is the formation of symmetrical bisacyl hydrazides (compound **40** in reaction 5) from

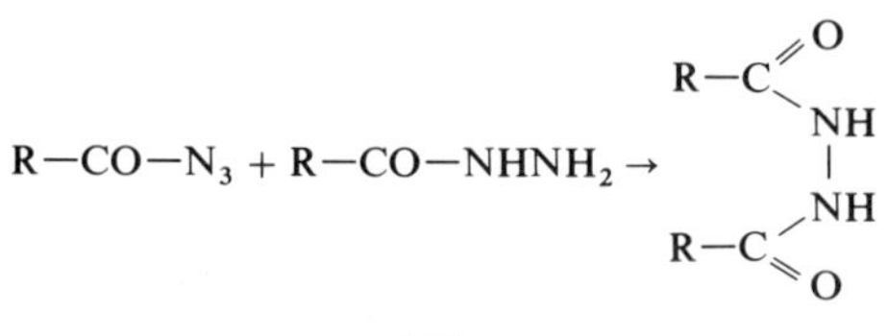

(5)

(**40**)

azide and unreacted hydrazide (Harris and Fruton, 1951; Schneider, 1960), probably due to insufficient acidity during azide formation.

D. Recent Applications to Large Segments

Segment condensation by the azide method appears to have good chances of success at reasonable product yields (10–50%) with components containing up to 40 amino acid residues. Beyond that size, the yield was 3% in the single example recorded in the literature, i.e., the condensation of a 44-

peptide azide with a 60-peptide amine component by Hirschmann *et al.*, (1969). A few selected examples of successful azide condensation of segments in the size range of 10-peptides to 40-peptides are summarized below for reference.

1. *Bovine Ribonuclease A*

In the ribonuclease synthesis (see Hirschmann, 1971a,b), reaction of N^{α}-Z-(21–40)-peptide azide with (41–64)-peptide α-methyl ester for several days at +5°C in a DMF—HMP mixture yielded 25% protected 44-peptide by forming an Arg → Lys(Z) bond (Strachan *et al.*, 1969), see struc-

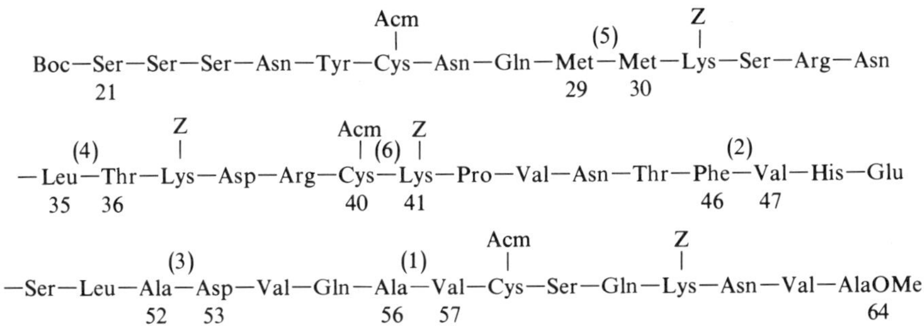

Synthetic bovine ribonuclease A(21–64) segment. Numerals indicate the consecutive order of azide couplings.

(41)

ture **41**. Under similar conditions, N^{α}-Boc-(65–85)-peptide azide was coupled with α-COOH unprotected (86–124)-peptide in DMF—HMP to provide a 40% recovery of the (65–124)-segment with formation of an Arg → Glu

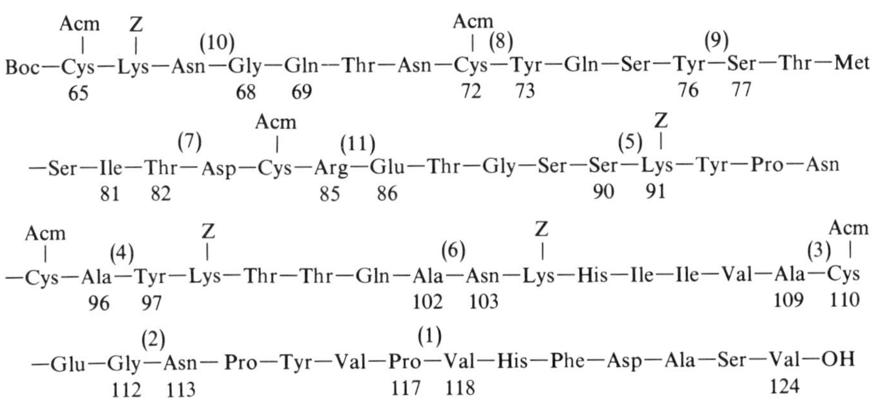

Synthetic bovine ribonuclease A(65–124) segment

(42)

bond (Jenkins *et al.*, 1969), see structure **42**. In the final coupling of that synthesis, the N^{α}-Z-(21–64)-peptide azide was incubated for 96 hr at 5°C in DMF—HMP solution with the 60-peptide amine component forming an Ala → Cys(Acm) bond and providing the target 104-peptide sequence in ca. 3% yield (Hirschmann *et al.*, 1969).

2. Porcine Proinsulin C-Peptide

(1–10)-Peptide azide was coupled with the (11–33)-peptide amine component in DMF at −5°C for 48 hr forming a Gly → Ala bond and providing a 33-peptide in 53% yield (Yanaihara *et al.*, 1970).

3. Ribonuclease T_1

Hofmann *et al.* have been pursuing the synthesis of this 104-peptide for several years. Reaction of N^{α}-For-(1–11)-peptide azide with (12–47)-peptide Boc-hydrazide for 48 hr at 4°C in DMF—DMSO (1:1) provided a 17% yield of protected 47-peptide by forming a Tyr → Ser bond (Storey *et al.*, 1972). N^{α}-Z-(48–65)-peptide azide was coupled with (66–80)-peptide Boc—hydrazide for 72 hr at 4°C in DMF—DMSO (6:1) to afford a 39% yield of N^{α}-Z-33-peptide Boc-hydrazide by forming a Gly → Asp(O*t*Bu) bond (Camble *et al.*, 1972). Modifying the strategy, N^{α}-Z-(66–88)-peptide azide was treated with α-COOH-unprotected (89–104)-peptide in DMF—DMSO (1:1) for 96 hr at 4°C forming a Gly → Val bond and producing a N^{α}-Z-39-peptide in ca. 10% yield (Romovacek *et al.*, 1974).

4. Porcine Secretin

The use of azide segment condensation in a stepwise fashion, starting with the COOH-terminal amine component and elongating the sequence successively toward the NH_2-terminus is illustrated in Scheme 9 (Yanaihara *et al.*, 1977). Protected peptide azides were prepared both by ester hydrazinolysis and from Boc-hydrazides. For another recent example of azide segment chain elongation in a synthesis of camel β_1-MSH, see Tamura *et al.* (1977).

VI. SPECIAL APPLICATIONS

Special applications of the azide method in peptide synthesis include (i) peptide cyclization, (ii) preparation of polyamino acids, (iii) use for incremental (stepwise) chain elongation, and (iv) protective group introduction.

1. The azide procedure has been used frequently for cyclization of linear peptides for the same reason it is being used in segment condensation, i.e., its relatively lower danger of racemization compared to other efficient methods. Azide formation and cyclization are carried out in separate successive stages,

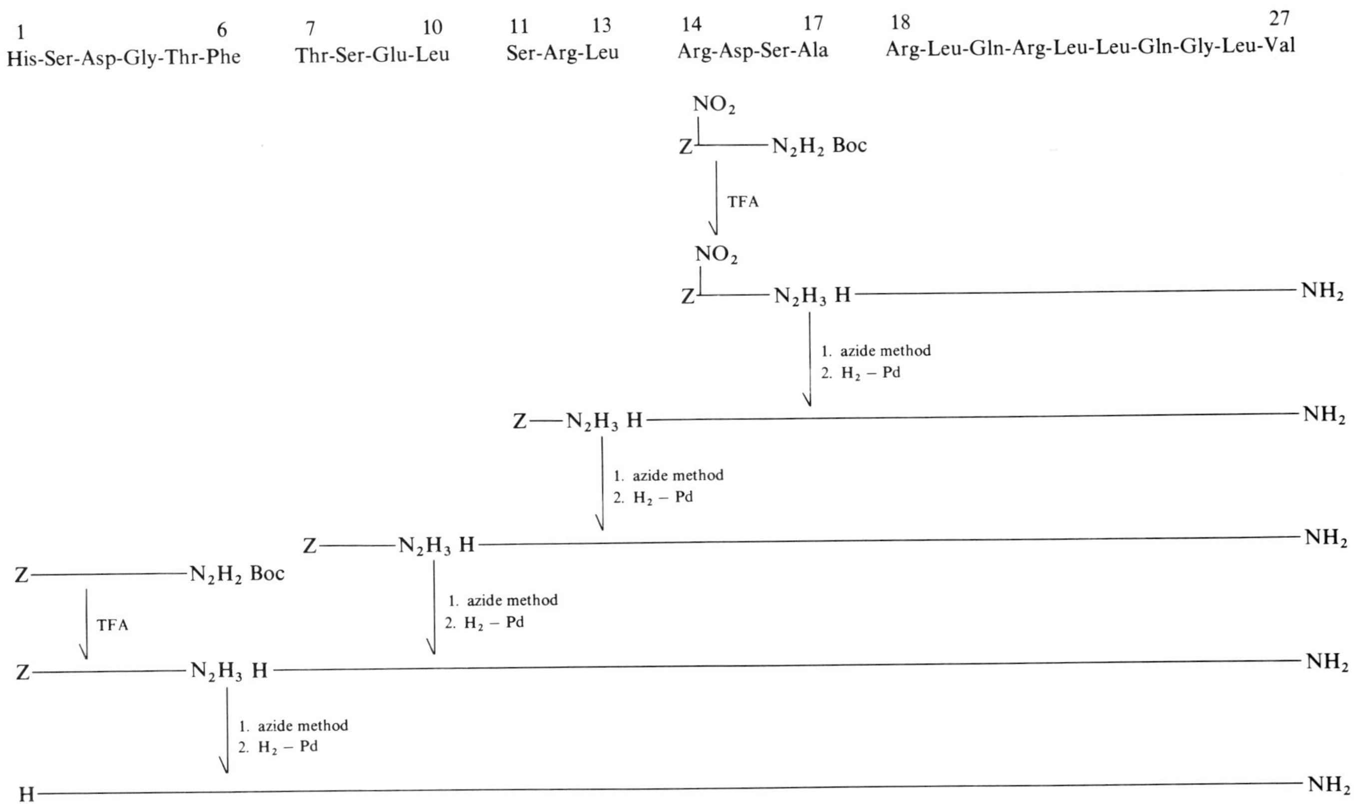

1 6 7 10 11 13 14 17 18 27
His-Ser-Asp-Gly-Thr-Phe Thr-Ser-Glu-Leu Ser-Arg-Leu Arg-Asp-Ser-Ala Arg-Leu-Gln-Arg-Leu-Leu-Gln-Gly-Leu-Val
NO2
Z N2H2 Boc
TFA
NO2
Z N2H3 H NH2
1. azide method
2. H2 – Pd
Z N2H3 H NH2
1. azide method
2. H2 – Pd
Z N2H3 H NH2
Z N2H2 Boc
TFA
1. azide method
2. H2 – Pd
Z N2H3 H NH2
1. azide method
2. H2 – Pd
H NH2

Scheme 9

following the removal of the N^{α}-protective group of a peptide hydrazide (**43**) (Scheme 10, reaction a). Azide formation (**44** → **45**) proceeds so fast that the α-amine function is not affected with deamination. Cyclization (**45** → **46**) is conducted at high dilution. For applications and references several recent reviews may be consulted (Kopple, 1972; Klausner and Bodanszky, 1974; Ovchinnikov and Ivanov, 1975; Bodanszky *et al.*, 1976; Deber *et al.*, 1976).

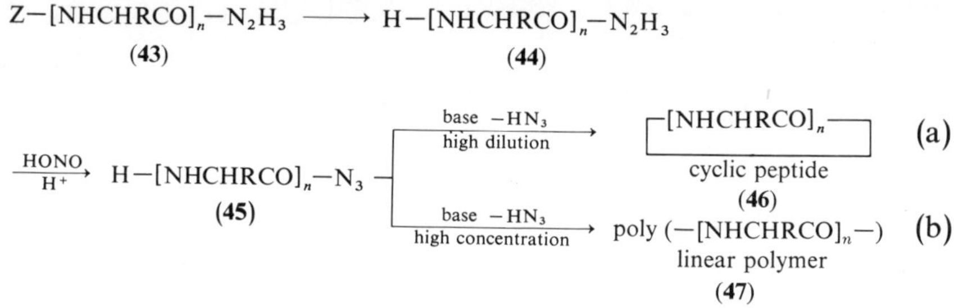

Scheme 10

2. Linear sequential polypeptides (structure **47**, Scheme 10, reaction b) can be prepared from peptide azides (**45**) when they are neutralized in concentrated solution (Magee and Hofmann, 1949; Rydon and Smith, 1955; see Blout *et al.*, 1974, for a recent treatise and earlier literature). In an innovative approach, Bailey and collaborators have prepared crystalline aminoacyl azide and peptide azide hydrobromides (**50**) from the respective N^{α}-*tert*-butyl- or *tert*-amyloxycarbonyl protected hydrazides (Scheme 11: **48** → **49** → **50**). Polymerization of these purified monomers to form polyamino acids or sequential polymers appears to proceed without diketopiperazine formation, Scheme 11, reaction a (Bailey and Reinert, 1965;

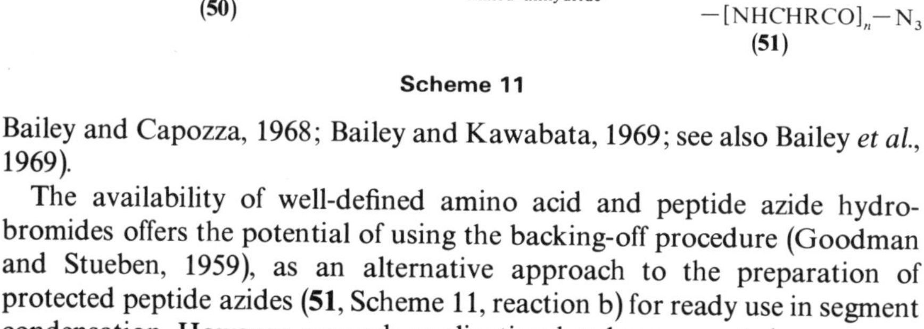

Scheme 11

Bailey and Capozza, 1968; Bailey and Kawabata, 1969; see also Bailey *et al.*, 1969).

The availability of well-defined amino acid and peptide azide hydrobromides offers the potential of using the backing-off procedure (Goodman and Stueben, 1959), as an alternative approach to the preparation of protected peptide azides (**51**, Scheme 11, reaction b) for ready use in segment condensation. However, no such application has been reported as yet.

3. Azide coupling was used for solid-phase reverse incremental (stepwise) chain elongation starting at the NH_2-terminus (Felix and Merrifield, 1970) where minimization of racemization was essential. In this strategy, Scheme 12, an amino acid *N'*-*tert*-butyloxycarbonylhydrazide was attached to the

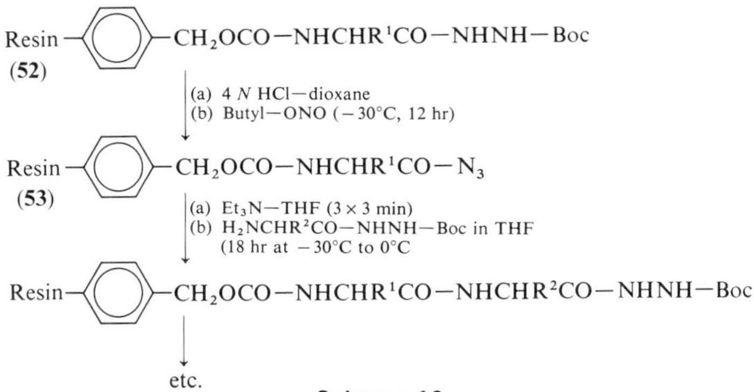

Scheme 12

resin by a benzyloxycarbonyl bond (**52**). A cycle for the incorporation of one amino acid residue consisted of *tert*-butyloxycarbonyl group cleavage, conversion of the ensuing hydrazide to azide (**53** by the Honzl–Rudinger procedure and coupling with the next amino acid *tert*-butyloxycarbonyl hydrazide. The tetrapeptide Leu-Ala-Gly-Val was successfully prepared but this variant of the solid phase procedure did not become attractive for routine synthesis due to the long reaction times and less than quantitative yields of the azide couplings.

Very recently, the use of excess *N*-alkyloxycarbonyl amino acid azide for solution synthesis of peptides by incremental chain elongation has been studied (Tilak and Hoffmann, 1977), inspired by the success of the repetitive excess mixed anhydride procedure (REMA; Tilak, 1970; van Zon and Beyerman, 1976). High yields (85–90%) of pure products were obtained with several model di- and tripeptides.

4. Some of the most useful urethane-type amine protective groups, such as the *tert*-butyloxycarbonyl (**54**) or the 4-methyloxybenzyloxycarbonyl groups, have been introduced into amino acids by azide condensation, Scheme 13 (review, Carpino, 1973) in cases where the corresponding chlor-

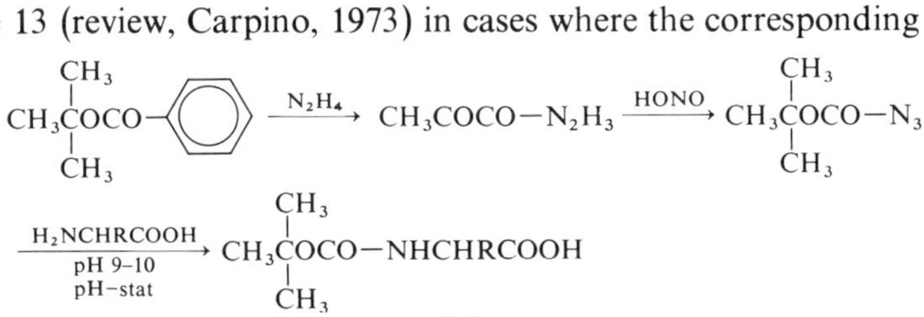

(**54**)
Boc—amino acid

Scheme 13

ides were either unstable or unaccessible. Table VI lists azides used for the blocking of amine functions. The danger of explosion has led to the development of alternative reagents, e.g., di-*tert*-butyl dicarbonate, compound **55** (Moroder *et al.*, 1976).

Table VI Azide Derivatives Used in Amine Protective Group Introduction

Derivative	References
tert-Butyloxycarbonyl azide	Carpino (1957); see also Wünsch (1974c)
4-Methyloxybenzyloxycarbonyl azide	Weygand and Hunger (1962); see also Wünsch (1974d)
4-Decyloxybenzyloxycarbonyl azide	Brechbühler *et al.* (1965)
Diphenylmethyloxycarbonyl azide	Hiskey *et al.* (1967)
tert-Amyloxycarbonyl azide	Honda *et al.* (1967)
2-(*p*-Biphenylyl)propyl(2)oxycarbonyl azide	Sieber and Iselin (1968)
9-Fluorenylmethyloxycarbonyl azide	Carpino and Han (1972)
Diisopropylmethyloxycarbonyl azide	Sakakibara *et al.* (1970)
α,α-Dimethyl-3,5-dimethyloxybenzyloxycarbonyl azide	Birr *et al.* (1972)

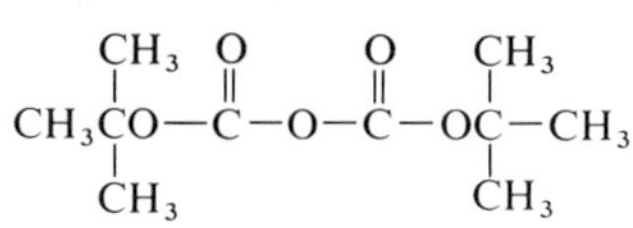

(55)

VII. OPPORTUNITIES AND CONSTRAINTS

A. Comparative Evaluation

The azide procedure is a versatile method. Scheme 14 presents a summary of the main procedural pathways and applications.

It may be of interest at this point to evaluate the overall effectiveness of the azide procedure in comparison with those of the active ester (AE), carbodiimide (DCC), and mixed anhydride (MA) methods by considering such main experimental criteria as chiral integrity, frequency of side reactions, ability of segment condensation, average yield, and time requirement. In this context, the azide method is (a) superior to all in maintaining chiral integrity, (b) on par with DCC and MA, but inferior to AE, by giving rise to side reactions, (c) superior to AE and MA and on par with DCC—HOBt in large-segment condensation, and (d) inferior to all in product yield and time requirement.

Points in favor of the azide method are: the convenient routes of hydrazide preparation, the opportunity of facile hydrazide purification, the convenient azide formation by the Honzl–Rudinger (1961) procedure, the possibility of using minimal side-chain protection. Certain potential opportunities have not as yet been utilized, such as increased flexibility in the strategic approach to acid-hydrazide interconversion, azide preparation by the backing-off procedure, or refined analytic monitoring of reactions, e.g., by high performance liquid chromatography.

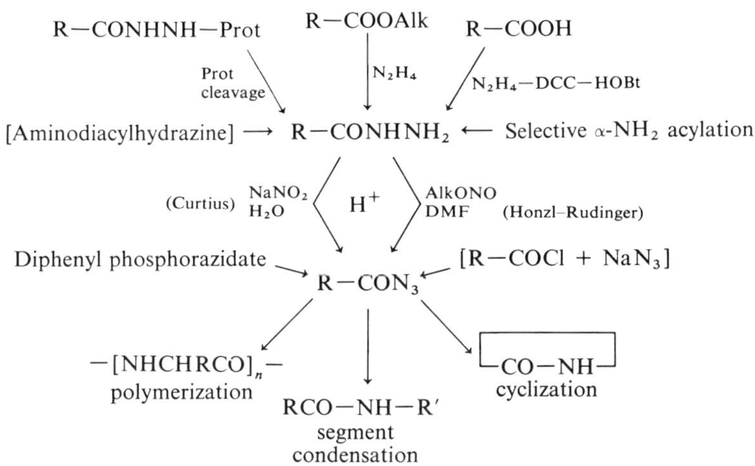

Scheme 14 Large arrows, routine procedures; small arrows, occasional application; and brackets, not used in peptide synthesis.

Among the main constraints one has to count the poor control of pH and amounts of (excess) base present during coupling, the as yet unclear role of the azide anion concentration on reaction rate, and the rapidly declining yields in preparing peptides with over 50 amino acid residues.

B. Prognosis

Interestingly, the above relative rating of the overall performance of the azide method is almost identical with a rating done 17 years ago (Meienhofer, 1962). This reflects the fact that the considerable improvements made with the azide procedure since that time have been paralleled by similarly effective improvements or expansions of the other efficient methods. If this trend continues, drastic changes in the above relative merit rating are unlikely to occur in the foreseeable future. It appears to be predictable that azide segment condensation will continue to play an important role in peptide synthesis. One hopes that further improvements of the method may make it suitable for syntheses of small proteins.

ACKNOWLEDGMENTS

I wish to thank Dr. S. S. Wang for suggestions and criticism, Mr. R. C. Makofske for experimental data, and Mrs. Debra Dabaghian for excellent help in the preparation of the manuscript.

REFERENCES

Agarwal, K. L., Kenner, G. W., and Sheppard, R. C. (1969a). *J. Chem. Soc. C* pp. 954–958.
Agarwal, K. L., Kenner, G. W., and Sheppard, R. C. (1969b). *Experientia* **25**, 346–348.
Anderson, G. W., and Callahan, F. M. (1958). *J. Am. Chem. Soc.* **80**, 2902–2903.
Anderson, G. W., Zimmerman, J. E., and Callahan, F. M. (1966). *J. Am. Chem. Soc.* **88**, 1338–1339.
Bailey, W. J., and Capozza, R. C. (1968). *Polym. Prepr.* **9**, 1261–1265.
Bailey, W. J., and Kawabata, N. (1969). *Polym. Prepr.* **10**, 181–185.
Bailey, W. J., and Reinert, G. E. (1965). *Polym. Prepr.* **6**, 740–746.
Bailey, W. J., Kawabata, N., and Capozza, R. C. (1969). *In* "Kinetics and Mechanism of Polyreactions," pp. 195–197. Akademiai Kiado, Budapest.
Biedrzycka, D., and Biernat, J. F. (1970). *Rocz. Chem.* **44**, 2477–2481; *Chem. Abstr.* **75**, 36625 (1971).
Birr, C., Lochinger, W., Stahnke, G., and Lang, P. (1972). *Justus Liebigs Ann. Chem.* **763**, 162–172.
Blout, E. R., Bovey, F. A., Goodman, M., and Lotan, N., eds. (1974). "Peptides, Polypeptides and Proteins." Wiley, New York.
Bodanszky, M. (1977). *In* "Peptides, Proceedings of 5th American Peptide Symposium" (M. Goodman and J. Meienhofer, eds.), pp. 1–13. Wiley, New York.
Bodanszky, M., and Ondetti, M. A. (1966). "Peptide Synthesis." Wiley, New York.
Bodanszky, M., Klausner, Y. S., and Ondetti, M. A. (1976). "Peptide Synthesis," 2nd ed. Wiley, New York.
Boissonnas, R. A., Guttmann, S., and Jaquenoud, P.-A. (1960). *Helv. Chim. Acta* **43**, 1349–1358.
Borin, G., Filippi, B., Cavaggion, F., and Marchiori, F. (1977). *Int. J. Pept. Protein Res.* **10**, 95–101.
Brechbühler, H., Büchi, H., Hatz, E., Schreiber, J., and Eschenmoser, A. (1965). *Helv. Chim. Acta* **48**, 1746–1771.
Brenner, M., and Hofer, W. (1961). *Helv. Chim. Acta* **44**, 1794–1798.
Camble, R., Dupuis, G., Kawasaki, K., Romovacek, H., Yanaihara, N., and Hofmann, K. (1972). *J. Am. Chem. Soc.* **94**, 2091–2100.
Camble, R., Cotton, R., Dutta, A. S., Gormley, J. J., Hayward, C. F., Morley, J. S., and Smithers, M. J. (1973). *In* "Peptides 1972" (H. Hanson and H. D. Jakubke, eds.), pp. 200–209. North-Holland Publ., Amsterdam.
Carpino, L. A. (1957). *J. Am. Chem. Soc.* **79**, 4427–4431.
Carpino, L. A. (1973). *Acc. Chem. Res.* **6**, 191–198.
Carpino, L. A., and Han, G. Y. (1972). *J. Org. Chem.* **37**, 3404–3409.
Castro, B., and Dormoy, J. R. (1973a). *Tetrahedron Lett.* pp. 3243–3246.
Castro, B., and Dormoy, J. R. (1973b). *Bull. Soc. Chim. Fr.* pp. 3359–3361.
Chang, J. K., Shimizu, M., and Wang, S. S. (1976). *J. Org. Chem.* **41**, 3255–3258.
Cheung, H. T., and Blout, E. R. (1965). *J. Org. Chem.* **30**, 315–316.
Curtius, T. (1890). *Ber. Dtsch. Chem. Ger.* **23**, 3023–3033.

Curtius, T. (1902). *Ber. Dtsch. Chem. Ger.* **35**, 3226–3228.
Curtius, T., and Levy, L. (1904). *J. Prakt. Chem.* **70**, 89–108.
Danho, W., and Li, C. H. (1971). *Int. J. Protein Res.* **3**, 81–92.
Deber, C. M., Madison, V., and Blout, E. R. (1976). *Acc. Chem. Res.* **9**, 106–113.
DeBernardo, S., Weigele, M., Toome, V., Manhart, K., Leimgruber, W., Böhlen, P., Stein, S., and Udenfriend, S. (1974). *Arch. Biochem. Biophys.* **163**, 390–399.
Determann, H., and Wieland, T. (1963). *Justus Liebigs Ann. Chem.* **670**, 136–140.
Dzieduszycka, M., Smulkowski, M., and Taschner, E. (1973). *In* "Peptides 1972" (H. Hanson and H. D. Jakubke, eds.), pp. 102–107. North-Holland Publ., Amsterdam.
Erickson, B. W., and Merrifield, R. B. (1976). *In* "The Proteins" (H. Neurath and R. L. Hill, eds.), 3rd ed., Vol. 2, pp. 255–527. Academic Press, New York.
Felix, A. M., and Jimenez, M. H. (1974). *J. Chromatogr.* **89**, 361–364.
Felix, A. M., and Merrifield, R. B. (1970). *J. Am. Chem. Soc.* **92**, 1385–1391.
Ferren, R. A., Miller, J. G., and Day, A. R. (1957). *J. Am. Chem. Soc.* **79**, 70–73.
Finn, F. M., and Hofmann, K. (1976a). *In* "The Proteins" (H. Neurath and R. L. Hill, eds.), 3rd ed., Vol. 2, pp. 105–253. Academic Press, New York.
Finn, F. M., and Hofmann, K. (1976b). *In* "The Proteins" (H. Neurath and R. L. Hill, eds.), 3rd ed., Vol. 2, pp. 128–130 and 190. Academic Press, New York.
Fischer, R. F., and Whetstone, R. R. (1954). *J. Am. Chem. Soc.* **76**, 5076–5080.
Fruton, J. S. (1942). *J. Biol. Chem.* **146**, 463–470.
Fruton, J. S. (1949). *Adv. Protein Chem.* **5**, 1–82.
Fujino, M., Shinagawa, S., Nishimura, O., and Fukuda, T. (1972). *Chem. Pharm. Bull.* **20**, 1017–1020.
Gabriel, T. F., Michalewsky, J., and Meienhofer, J. (1976). *J. Chromatogr.* **129**, 287–293.
Gabriel, T. F., Jimenez, M. H., Felix, A. M., Michalewsky, J., and Meienhofer, J. (1977). *Int. J. Pept. Protein Res.* **9**, 129–136.
Geiger, R., and Siedel, W. (1968). *Chem. Ber.* **101**, 3386–3391.
Gillessen, D., Schnabel, E., and Meienhofer, J. (1963). *Justus Liebigs Ann. Chem.* **667**, 164–171.
Goodman, M., and Glaser, C. (1970). *In* "Peptides: Chemistry and Biochemistry" (B. Weinstein and S. Lande, eds.), pp. 267–335.
Goodman, M., and McGahren, W. J. (1967). *Tetrahedron* **23**, 2031–2050.
Goodman, M., and Stueben, K. (1959). *J. Am. Chem. Soc.* **81**, 3980–3983.
Goren, H. J., and Fridkin, M. (1970). *J. Chromatogr.* **47**, 519–523.
Greenstein, J. P., and Winitz, M. (1961). "Chemistry of the Amino Acids," Vol. 2, pp. 949–965. Wiley, New York.
Guttmann, S., and Boissonnas, R. A. (1958). *Helv. Chim. Acta* **41**, 1852–1867.
Guttmann, S., and Boissonnas, R. A. (1960). *Helv. Chim. Acta* **43**, 200–216.
Hamada, Y., Shioiri, T., and Yamada, S. (1977a). *Chem. Pharm. Bull.* **25**, 221–223.
Hamada, Y., Rishi, S., Shioiri, T., and Yamada, S. (1977b). *Chem. Pharm. Bull.* **25**, 224–230.
Harris, J. I., and Fruton, J. S. (1951). *J. Biol. Chem.* **191**, 143–151.
Heard, A. L., and Young, G. T. (1963). *J. Chem. Soc.* pp. 5807–5810.
Hirschmann, R. (1971a). *In* "Peptides 1969" (E. Scoffone, ed.), pp. 138–144. North-Holland Publ., Amsterdam.
Hirschmann, R. (1971b). *Intra-Sci. Chem. Rep.* **5**, 203–228.
Hirschmann, R., and Veber, D. F. (1973). *In* "The Chemistry of Polypeptides" (P. G. Katsoyannis, ed.), pp. 125–142. Plenum, New York.
Hirschmann, R., Nutt, R. F., Veber, D. F., Vitali, R. A., Varga, S. L., Jacob, T. A., Holly, F. W., and Denkewalter, R. G. (1969). *J. Am. Chem. Soc.* **91**, 507–508.
Hiskey, R. G., and Adams, J. B. (1965). *J. Am. Chem. Soc.* **87**, 3969–3673.

Hiskey, R. G., Mizoguchi, T., and Smithwick, E. L. (1967). *J. Org. Chem.* **32**, 97–102.
Hofmann, K. (1971). *In* "Peptides 1969" (E. Scoffone, ed.), pp. 130–137. North-Holland Publ., Amsterdam.
Hofmann, K., Magee, M. Z., and Lindenmann, A. (1950). *J. Am. Chem. Soc.* **72**, 2814–2815.
Hofmann, K., Lindenmann, A., Magee, M. Z., and Khan, N. H. (1952). *J. Am. Chem. Soc.* **74**, 470–476.
Hofmann, K., Thompson, T. A., Yajima, H., Schwartz, E. T., and Inouye, H. (1960). *J. Am. Chem. Soc.* **82**, 3715–3721.
Hofmann, K., Schmiechen, R., Wells, R. D., Wolman, Y., and Yanaihara, N. (1965). *J. Am. Chem. Soc.* **87**, 611–619.
Hofmann, K., Visser, J. P., and Finn, F. M. (1969). *J. Am. Chem. Soc.* **91**, 4883–4887.
Holley, R. W., and Sondheimer, E. (1954). *J. Am. Chem. Soc.* **76**, 1326–1328.
Honda, I., Shimonishi, Y., and Sakakibara, S. (1967). *Bull. Chem. Soc. Jpn.* **40**, 2415–2418.
Honzl, J., and Rudinger, J. (1961). *Collect. Czech. Chem. Commun.* **26**, 2333–2344.
Izumiya, N., and Muraoka, M. (1969). *J. Am. Chem. Soc.* **91**, 2391–2392.
Izumiya, N., Muraoka, M., and Aoyagi, H. (1971). *Bull. Chem. Soc. Jpn.* **44**, 3391–3395.
Jäger, G., and Geiger, R. (1973). *Justus Liebigs Ann. Chem.* **1973**, 1535–1544 and 1928–1933.
Jenkins, S. R., Nutt, R. F., Dewey, R. S., Veber, D. F., Holly, F. W., Paleveda, W. J., Jr., Lanza, T., Jr., Strachan, R. G., Schoenewaldt, E. F., Barkemeyer, H., Dickinson, M. J., Sondey, J., Hirschmann, R., and Walton, E. (1969). *J. Am. Chem. Soc.* **91**, 505–506.
Kamber, B., Riniker, B., Sieber, P., and Rittel, W. (1976). *Helv. Chim. Acta* **59**, 2830–2840.
Katsoyannis, P. G., Tometsko, A. M., Zalut, C., and Fukuda, K. (1966). *J. Am. Chem. Soc.* **88**, 5625–5635.
Kaufmann, K.-D., Raddatz, F., and Bauschke, S. (1975). *Z. Chem.* **15**, 449–450.
Kemp, D. S., and Rebek, J., Jr. (1970). *J. Am. Chem. Soc.* **92**, 5792–5793.
Kemp, D. S., Wang, S. W., Busby, G., III, and Hugel, G. (1970a). *J. Am. Chem. Soc.* **92**, 1043–1055.
Kemp, D. S., Bernstein, Z., and Rebek, J., Jr. (1970b). *J. Am. Chem. Soc.* **92**, 4756–4757.
Kemp, D. S., Roberts, D., Hoyng, C., Grattan, J., Vellaccio, F., and Reczek, J. (1975). *In* "Peptides: Chemistry, Structure and Biology" (R. Walter and J. Meienhofer, eds.), pp. 295–305. Ann Arbor Sci. Publ., Ann Arbor, Michigan.
Kisfaludy, L., and Nyéki, O. (1972). *Acta Chim. Acad. Sci. Hung.* **72**, 75–76.
Kisfaludy, L., Nyéki, O., and Szirtes, T. (1973). *In* "Peptides 1971" (H. Nesvadba, ed.), pp. 54–59. North-Holland Publ., Amsterdam.
Klausner, Y. S., and Bodanszky, M. (1972). *Synthesis* pp. 453–463.
Klausner, Y. S., and Bodanszky, M. (1974). *Synthesis* pp. 549–559.
Klieger, E., and Gibian, H. (1962a). *Justus Liebigs Ann. Chem.* **651**, 194–205.
Klieger, E., and Gibian, H. (1962b). *Justus Liebigs Ann. Chem.* **655**, 195–210.
König, W., and Geiger, R. (1970). *Chem. Ber.* **103**, 788–798.
Kopple, K. D. (1972). *J. Pharm. Sci.* **61**, 1345–1356.
Liberek, B., and Michalik, A. (1968). *Colloq. Int. C.N.R.S.* **175**, 159–161; see also: *In* "Peptides 1968" (E. Bricas, ed.), pp. 159–161. North-Holland Publ., Amsterdam.
McKay, A. F. (1952). *Chem. Rev.* **51**, 301–346.
MacLaren, J. A. (1958). *Aust. J. Chem.* **11**, 360–365.
MacLaren, J. A., Savige, W. E., and Swan, J. M. (1958). *Aust. J. Chem.* **11**, 345–359.
Macrea, R., and Young, G. T. (1975). *J. Chem. Soc., Perkin Trans. 1* pp. 1185–1191.
Magee, M. Z., and Hofmann, K. (1949). *J. Am. Chem. Soc.* **71**, 1515–1516.
Manning, J. M., and Moore, S. (1968). *J. Biol. Chem.* **243**, 5591–5597.
Medzihradszky, K., Bruckner, V., Kajtar, M., Löw, M., Bajusz, S., and Kisfaludy, L. (1962). *Acta Chim. Acad. Sci. Hung.* **30**, 105–108.

Medzihradszky-Schweiger, H. (1962). *Acta Chim. Acad. Sci. Hung.* **34**, 213–216.
Meienhofer, J. (1962). *Chimia* **16**, 385–414.
Meienhofer, J. (1966). *Acta Chim. Acad. Sci. Hung.* **48**, 171–180.
Meienhofer, J. (1973). *In* "Hormonal Proteins and Peptides" (C. H. Li, ed.), Vol. 2, pp. 45–267. Academic Press, New York (see pp. 140–142, Table XV).
Meienhofer, J. (1977). *In* "Antibodies in Human Diagnosis and Therapy" (E. Haber and R. M. Krause, eds.), pp. 215–218. Raven, New York.
Meienhofer, J., and Schnabel, E. (1965). *Z. Naturforsch. Teil B* **20**, 661–665.
Merrifield, R. B. (1963). *J. Am. Chem. Soc.* **85**, 2149–2154.
Merrifield, R. B. (1969). *Adv. Enzymol.* **32**, 221–296.
Merrifield, R. B., and Woolley, D. W. (1956). *J. Am. Chem. Soc.* **78**, 4646–4649.
Miller, A. W., and Stirling, C. J. M. (1968). *J. Chem. Soc., C.* pp. 2612–2617.
Mitaki, T., and Shimonishi, Y. (1977). *Bull. Chem. Soc. Jpn.* **50**, 991–997.
Mitin, Yu. V., and Maximov, E. E. (1973). *Zh. Obshch. Khim.* **43**, 203–207; *J. Gen. Chem. USSR (Engl. Transl.)* **43**, 199–203.
Mitsuyasu, N., Waki, M., Makisumi, S., and Izumiya, N. (1969). *Mem. Fac. Sci., Kyushu Univ., Ser. C* **6**, 145–152; *Chem. Abstr.* **73**, 77605 (1970).
Mitsuyasu, N., Waki, M., Kato, T., and Izumiya, N. (1970). *Mem. Fac. Sci., Kyushu Univ., Ser. C* **7**, 97–101; *Chem. Abstr.* **73**, 88154 (1970).
Moroder, L., Borin, G., Marchiori, F., and Scoffone, E. (1974). *Justus Liebigs Ann. Chem.* **1974**, 213–224.
Moroder, L., Hallett, A., Wünsch, E., Keller, O., and Wersin, G. (1976). *Hoppe-Seyler's Z. Physiol. Chem.* **357**, 1651–1653.
Naithani, V. K. (1973). *Hoppe-Seyler's Z. Physiol. Chem.* **354**, 659–672.
Ohno, M., Kuromizu, K., Ogawa, H., and Izumiya, N. (1971). *J. Am. Chem. Soc.* **93**, 5251–5254.
Ondetti, M. A., Deer, A., Sheehan, J. T., Pluscec, J., and Kocy, O. (1968). *Biochemistry* **7**, 4069–4075.
Ovchinnikov, Yu. A., and Ivanov, V. T. (1975). *Tetrahedron* **31**, 2177–2209.
Ozawa, K., Shioiri, T., and Yamada, S. (1977). *Chem. Pharm. Bull.* **25**, 122–125.
Pataki, G. (1969). "Techniques of Thin-Layer Chromatography in Amino Acid and Peptide Chemistry," 2nd ed., p. 109. Ann Arbor Sci. Publ., Ann Arbor, Michigan.
Paul, R., Anderson, G. W., and Callahan, F. M. (1961). *J. Org. Chem.* **26**, 3347–3350.
Pettit, G. R. (1976). "Synthetic Peptides," Vol. 4, pp. 22–32. Elsevier, Amsterdam.
Prélog, V., and Wieland, P. (1946). *Helv. Chim. Acta* **29**, 1128–1132.
Preston, J., and Weinstein, B. (1967). *Experientia* **23**, 265–266.
Romovacek, H., Drabarek, S., Kawasaki, K., Dowd, S. R., Obermeier, R., and Hofmann, K. (1974). *Int. J. Pept. Protein Res.* **6**, 435–445.
Rydon, H. N., and Smith, P. W. G. (1955). *J. Chem. Soc.* pp. 2542–2549.
Sakakibara, S., Fukuda, T., Kishida, Y., and Honda, I. (1970). *Bull. Soc. Chem. Jpn.* **43**, 3322.
Sato, M., Tatsuno, T., and Matsuo, H. (1970). *Chem. Pharm. Bull.* **18**, 1794–1798.
Schnabel, E. (1962). *Justus Liebigs Ann. Chem.* **659**, 168–184.
Schnabel, E. (1963). *In* "Peptides, Proceedings of 5th European Peptide Symposium" (G. T. Young, ed.), pp. 77–82. Pergamon Press, London.
Schnabel, E., and Zahn, H. (1958). *Justus Liebigs Ann. Chem.* **614**, 141–148.
Schneider, F. (1960). *Hoppe-Seyler's Z. Physiol. Chem.* **320**, 82–91.
Schnyder, J., and Rottenberg, M. (1975). *Helv. Chim. Acta* **58**, 521–523.
Schott, H. F., Larkin, J. B., Rockland, L. B., and Dunn, M. S. (1947). *J. Org. Chem.* **12**, 490–495.
Schröder, E., and Lübke, K. (1965). "The Peptides," Vol. 1, pp. 79–85. Academic Press, New York.
Schwyzer, R. (1959). *Angew. Chem.* **71**, 742.

Schwyzer, R., and Kappeler, H. (1961). *Helv. Chim. Acta* **44**, 1991–2002.
Schwyzer, R., and Sieber, P. (1958). *Helv. Chim. Acta* **41**, 1582–1587.
Schwyzer, R., Surbeck-Wegmann, E., and Dietrich, H. (1960). *Chimia* **14**, 366.
Schwyzer, R., Iselin, B., Kappeler, H., Riniker, B., Rittel, W., and Zuber, H. (1963). *Helv. Chim. Acta* **46**, 1975–1996.
Scoffone, E., Marchiori, F., Tamburro, A. M., and Rocchi, R. (1964). *Gazz. Chim. Ital.* **94**, 695–709.
Scoffone, E., Rocchi, R., Marchiori, F., Moroder, L., Marzotto, A., and Tamburro, A. M. (1967). *J. Am. Chem. Soc.* **89**, 5450–5455.
Shioiri, T., and Yamada, S. (1974). *Chem. Pharm. Bull.* **22**, 849–854, 855–858, and 859–863.
Shioiri, T., Ninomiya, K., and Yamada, S. (1972). *J. Am. Chem. Soc.* **94**, 6203–6205.
Sieber, P., and Iselin, B. (1968). *Helv. Chim. Acta* **51**, 614–622 and 622–632.
Sieber, P., and Riniker, B. (1973). *In* "Peptides 1971" (H. Nesvadba, ed.), pp. 49–53. North-Holland Publ., Amsterdam.
Sieber, P., Riniker, B., Brugger, M., Kamber, B., and Rittel, W. (1970). *Helv. Chim. Acta* **53**, 2135–2150.
Sieber, P., Brugger, M., and Rittel, W. (1971). *In* "Peptides 1969" (E. Scoffone, ed.), pp. 60–61. North-Holland Publ., Amsterdam.
Sieber, P., Kamber, B., Hartmann, A., Jöhl, A., Riniker, B., and Rittel, W. (1977). *Helv. Chim. Acta* **60**, 27–37.
Storey, H. T., Beacham, J., Cernosek, S. F., Finn, F. M., Yanaihara, C., and Hofmann, K. (1972). *J. Am. Chem. Soc.* **94**, 6170–6178.
Strachan, R. G., Paleveda, W. J., Jr., Nutt, R. F., Vitali, R. A., Veber, D. F., Dickinson, M. J., Garsky, V., Deak, J. E., Walton, E., Jenkins, S. R., Holly, F. W., and Hirschmann, R. (1969). *J. Am. Chem. Soc.* **91**, 503–505.
Tamura, F., Ogawa, H., Fujii, N., Yajima, H., Miyata, K., Nakamura, M., and Tanaka, A. (1977). *Chem. Pharm. Bull.* **25**, 767–774.
Tilak, M. A. (1970). *Tetrahedron Lett.* pp. 849–854.
Tilak, M. A., and Hoffmann, J. A. (1977). *J. Org. Chem.* **42**, 2098–2100.
Tsuji, J., Hayakawa, S., and Takayanagi, H. (1975). *Chem. Lett. Jpn.* pp. 437–438.
van Zon, A., and Beyerman, H. C. (1976). *Helv. Chim. Acta* **59**, 1112–1126.
Veber, D. F. (1975). *In* "Peptides: Chemistry, Structure and Biology" (R. Walter and J. Meienhofer, eds.), pp. 307–316. Ann Arbor Sci. Publ., Ann Arbor, Michigan.
Waki, M., Mitsuyasu, N., Kato, T., Makisumi, S., and Izumiya, N. (1968). *Bull. Chem. Soc. Jpn.* **41**, 669–672.
Wang, S. S. (1973). *J. Am. Chem. Soc.* **95**, 1328–1333.
Wang, S. S. (1975). *J. Org. Chem.* **40**, 1235–1239.
Wang, S. S. (1976). *J. Org. Chem.* **41**, 3258–3261.
Wang, S. S., and Kulesha, I. D. (1975). *J. Org. Chem.* **40**, 1227–1234.
Wang, S. S., and Merrifield, R. B. (1969). *J. Am. Chem. Soc.* **91**, 6488–6491.
Wang, S. S., and Merrifield, R. B. (1972). *Int. J. Pept. Protein Res.* **4**, 309–318.
Wang, S. S., Gisin, B. F., Winter, D. P., Makofske, R., Kulesha, I. D., Tzougraki, C., and Meienhofer, J. (1977). *J. Org. Chem.* **42**, 1286–1290.
Wang, S. S., Kulesha, I. D., Winter, D. P., Makofske, R., Kutny, R., and Meienhofer, J. (1978). *Int. J. Pept. Protein Res.* **11**, 297–300.
Weinstein, B., Cook, R. M., and Heffernan, J. T. (1976). *Int. J. Pept. Protein Res.* **8**, 21–25.
Weygand, F., and Hunger, K. (1962). *Chem. Ber.* **95**, 1–6.
Weygand, F., and Steglich, W. (1959). *Chem. Ber.* **92**, 313–319.
Weygand, F., Hoffmann, D., and Wünsch, E. (1966a). *Z. Naturforsch.* **21**, 426–428.
Weygand, F., Prox, A., and König, W. (1966b). *Chem. Ber.* **99**, 1451–1460.

Wieland, T., and Heinke, B. (1958). *Justus Liebigs Ann. Chem.* **615**, 184–202.

Williams, A. W., and Young, G. T. (1972). *J. Chem. Soc., Perkin Trans. I*, pp. 1194–1200.

Williams, M. W., and Young, G. T. (1963). *J. Chem. Soc.*, pp. 881–889.

Wolman, Y., Gallop, P. M., Patchornik, A., and Berger, A. (1962). *J. Am. Chem. Soc.* **84**, 1889–1892.

Woolley, D. W. (1948). *J. Biol. Chem.* **172**, 71–81.

Wünsch, E., ed. (1974a). "Houben-Weyl's Methoden der organischen Chemie," Vol. 15, Part I. Thieme, Stuttgart.

Wünsch, E., ed. (1974b). "Houben-Weyl's Methoden der organischen Chemie," Vol. 15, Part II. Thieme, Stuttgart.

Wünsch, E., ed. (1974c). "Houben-Weyl's Methoden der organischen Chemie," Vol. 15, Part I, p. 119, Thieme, Stuttgart.

Wünsch, E., ed. (1974d). "Houben-Weyl's Methoden der organischen Chemie," Vol. 15, Part I, p. 71. Thieme, Stuttgart.

Wünsch, E., and Drees, F. (1966). *Chem. Ber.* **99**, 110–120.

Wünsch, E., and Stelzel, P. (1974a). *In* "Houben-Weyl's Methoden der organischen Chemie" (E. Wünsch, ed.), Vol. 15, Part I, pp. 430–450. Thieme, Stuttgart.

Wünsch, E., and Stelzel, P. (1974b). *In* "Houben-Weyl's Methoden der organischen Chemie" (E. Wünsch, ed.), Vol. 15, Part II, pp. 296–322. Thieme, Stuttgart.

Wünsch, E., and Wendlberger, G. (1964). *Chem. Ber.* **97**, 2504–2514.

Wünsch, E., Deimer, K. H., Thamm, P., and Stelzel, P. (1974). *In* "Houben-Weyl's Methoden der organischen Chemie" (E. Wünsch, ed.), Vol. 15, Part I, pp. 334–338. Thieme, Stuttgart.

Yajima, H., and Kiso, Y. (1971). *Chem. Pharm. Bull.* **19**, 420–423.

Yajima, H., Kawasaki, K., Okada, Y., Minami, H., Kubo, K., and Yamashita, I. (1968). *Chem. Pharm. Bull.* **16**, 919–928.

Yamada, S., Ikota, N., Shioiri, T., and Tachibana, S. (1975). *J. Am. Chem. Soc.* **97**, 7174–7175.

Yanaihara, N., Hashimoto, T., Yanaihara, C., and Sakura, N. (1970). *Chem. Pharm. Bull.* **18**, 417–420 (1970).

Yanaihara, N., Kubota, M., Sakagami, M., Sato, H., Mochizuki, T., Sakura, N., Hashimoto, T., Yanaihara, C., Yamaguchi, K., Zeze, F., and Abe, K. (1977). *J. Med. Chem.* **20**, 648–655.

Young, G. T. (1967). *In* "Peptides" (H. C. Beyerman, A. van de Linde, and W. Maassen van den Brink, eds.), pp. 55–66. North-Holland Publ., Amsterdam.

Young, G. T. (1973). *In* "Peptides 1972" (H. Hanson and H. D. Jakubke, eds.), p. 132. North-Holland Publ., Amsterdam.

Zahn, H., and Brandenburg, D. (1966). *Justus Liebigs Ann. Chem.* **692**, 220–230.

Zahn, H., and Schnabel, E. (1957). *Justus Liebigs Ann. Chem.* **605**, 212–232.

Zahn, H., Meienhofer, J., and Schnabel, E. (1965). *Acta Chim. Acad. Sci. Hung.* **44**, 109–128.

Zervas, L., and Hamalidis, C. (1965). *J. Am. Chem. Soc.* **87**, 99–104.

Zervas, L., Borovas, D., and Gazis, E. (1963). *J. Am. Chem. Soc.* **85**, 3660–3666.

Chapter 5

The Carbodiimide Method

DANIEL H. RICH AND JASBIR SINGH

I. Introduction 242
II. Mechanism of Carboxyl Activation 243
A. *O*-Acylisourea as Intermediate 243
B. Reaction in the Presence of Amines 246
C. Formation of Symmetrical Anhydrides 246
D. *p*-Nitrophenol as a Trapping Agent 247
E. Pentachlorophenol as a Trapping Agent 248
F. *N*-Hydroxysuccinimide as a Trapping Agent 248
G. 1-Hydroxybenzotriazole as a Trapping Agent 250
H. Advantages of the Combination of Dicyclohexylcarbodiimide with Hydroxybenzotriazole 251
III. Reaction Conditions for Dicyclohexylcarbodiimide-Mediated Peptide Bond Formation 252
A. Preferred Reaction Conditions 252
B. Conditions Favoring Symmetrical Anhydrides 254
C. Application of Dicyclohexylcarbodiimide in Combination with 1-Hydroxybenzotriazole 256
D. Reactions with α-Substituted and Secondary Amino Acids 256
IV. Reactions of Carbodiimides with Amines 258
A. Reaction with Amino Acid Esters 258
B. Reaction with Imidazole; Hydroxybenzotriazole Catalysis 258
References 260

THE PEPTIDES, VOL. 1

ISBN 0-12-304201-1

I. INTRODUCTION

The carbodiimide method (Khorana, 1953) was introduced to peptide synthesis by Sheehan and Hess (1955) with the use of dicyclohexylcarbodiimide (DCC; **1**) to affect dehydration and peptide bond formation (Eq. 1).

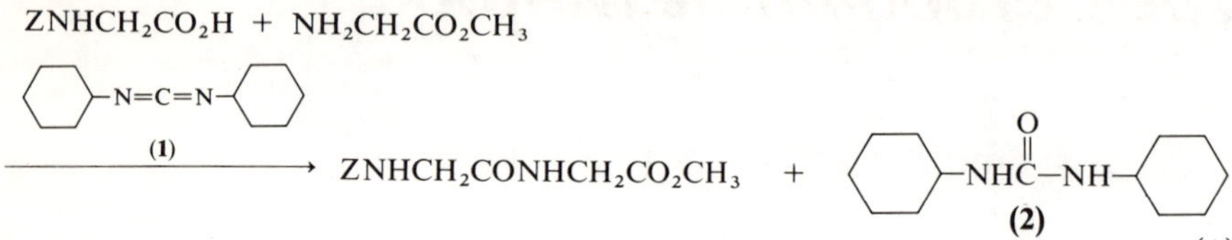

(1)

N,N′-Dicyclohexylurea (DCU, **2**) is formed as a by-product. The *procedure* consisting of adding DCC to a solution of the N-protected amino acid and amino acid or peptide ester won wide acceptance because of its simplicity, speed, and compatibility with water. It was soon used to synthesize oxytocin (du Vigneaud *et al.*, 1953, 1954, vasopressin, and numerous other biologically active peptides. Esters were also prepared using DCC and the success of using active esters in peptide synthesis, e.g., *p*-nitrophenyl esters (Bodanszky, 1956), is due to the ease with which they are obtained via DCC-activation. "Water-soluble" carbodiimides (Sheehan and Hlavka, 1956) were used to further expand the method by facilitating the work-up of some peptides. Dicyclohexylcarbodiimide has become the most widely used activating agent in peptide synthesis. Virtually every solid phase synthesis and many solution syntheses employ DCC, either directly or through the use of active esters.

The DCC method in its original form is accompanied by *side reactions.* Coupling of acyl peptides to amino acid or peptide esters can be accompanied by considerable racemization (Hofmann *et al.*, 1958; Anderson and Callahan, 1958) which occurs when 5(4H)-oxazolones are formed (Goodman and Glaser, 1970). Activation of glutamine or asparagine carboxyl groups can cause partial dehydration of the ω-carboxamide groups to nitriles (Ressler, 1956; Gish *et al.*, 1956). Activation of amino acids or peptide acids with DCC, especially in the presence of base, yields *N*-acylureas which decrease yields and frequently complicate workup of products. These problems stimulated studies to determine mechanisms of carboxyl activation and acyl peptide racemization and also led to the practice of adding nucleophiles (e.g., *N*-hydroxysuccinimide, HOSu; or 1-hydroxybenzotriazole, HOBt) to DCC reaction mixtures to suppress side reactions.

The purpose of this chapter is to describe and evaluate the current use of DCC in peptide synthesis with special emphasis on relating mechanistic studies with the successful use of additives. DCC has been reviewed previously by Schröder and Lübke (1965), Wünsch and Wendlberger (1974), Finn and Hofmann (1976), and Erickson and Merrifield (1976). The chemistry of carbodiimides has been reviewed (Kurzer and Douraghi-Zadeh, 1967).

II. MECHANISM OF CARBOXYL ACTIVATION

A. *O*-Acylisourea as Intermediate

DCC reacts with N-protected acylamino acids to give a variety of isolable products (Scheme 1). Depending on the reaction conditions and the type of N-protection, anhydrides (**4**), *N*-acylureas (**5**), 5(4H)-oxazolones (**7**), and

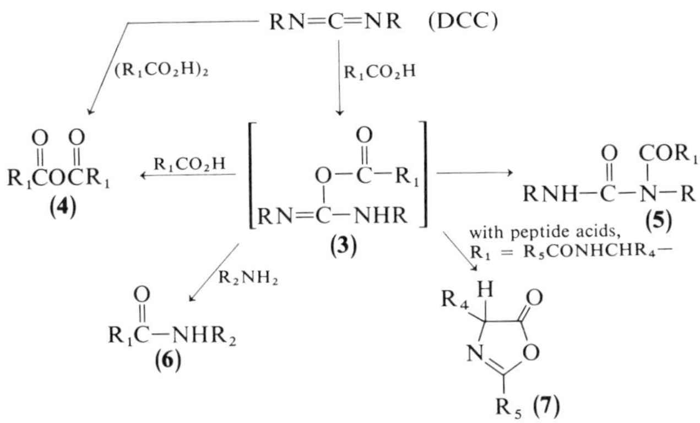

Scheme 1. Major products from reaction of DCC with peptide and amino acids.

peptides (**6**) can be formed. Carbodiimides are reactive molecules known to add nucleophiles rapidly under acidic conditions (Kurzer and Douraghi-Zadeh, 1967). In the case of peptides, DCC reacts first with a carboxyl group to form the *O*-acylisourea (**3**), a highly activated acylating agent. Intermediate **3** has never been isolated and its existence is based on analogy to stable products, e.g., (**8**), derived from the reaction of other oxygen nucleophiles with DCC (Kovacs *et al.*, 1967). More direct evidence for the *O*-acylisourea (**3**) structure was provided by Doleschall and Lempert (1963) who isolated a cyclic *O*-acylisourea (**11**) formed by desulfurization of **9**. Cyclic *O*-acylisourea (**11**) reacts slowly with water or methanol but is hydrated to the urea rapidly in the presence of mineral acid. The chemical reactivity of **11** is related to that of DCC. Hegerty and Bruice (1970) studied the hydrolysis and aminolysis of **11** in water and showed that the protonated species **11b** aminolyzed about 10^5 faster than the unprotonated form **11a** in

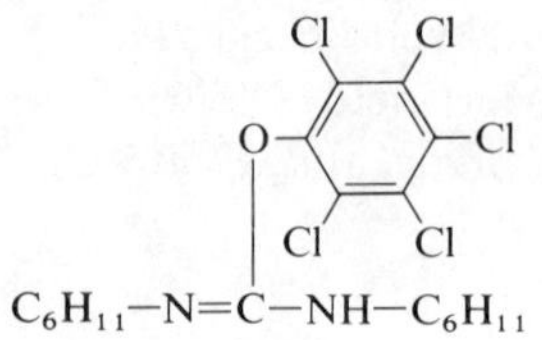

(8)

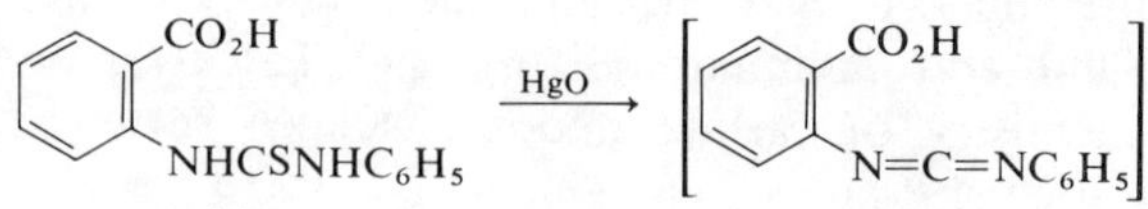

(9) **(10)**

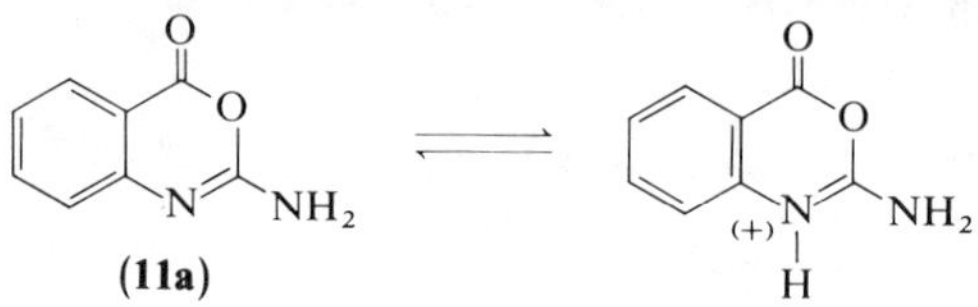

(11a) **(11b)**

the pH range of ca. 4–7. Reaction of carbodiimides with carboxylic acids is also catalyzed by acid in organic solvents. DCC reacts with the dimer of acetic acid about 30 times faster than with the monomer (DeTar and Silverstein, 1966a). Another similarity between **11** and DCC was found in that hydrolysis of **11** is fastest at pH ca. 9.9 but very slow between pH 4–8 where aqueous DCC coupling reactions are normally carried out.

O-Acylisoureas (**3**) derived from aliphatic acids or protected amino acids are less stable than the model **11** and their enhanced reactivity causes them to react more rapidly with available nucleophiles. Their fate in solution depends upon the reaction conditions to which they are subjected (Scheme 1). A feature of highly activated esters such as **3** is their lack of selectivity with respect to various basic amine and oxygen nucleophiles (Jencks and Gilchrist, 1968). Aminolysis is accompanied by competing reactions from other nucleophiles proximate to the *O*-acylisourea **3**, e.g., carboxylates (anhydride formation), intramolecular amide groups (5(4H)-oxazolone formation and racemization or nitrile formation with Asn), and the isourea nitrogen (*N*-acylurea formation). Reactions leading to 5(4H)-oxazolones or *N*-acylureas are intramolecular and will be formed, as shown in Scheme 2, at a constant rate unless some other nucleophile intervenes. This is a key point because many of the problems encountered in peptide synthesis, e.g., 5(4H)-oxazolone and *N*-acylurea formation, are caused by these side reactions. *N*-Acylureas (**18**) are formed by the 4-center reaction shown in Scheme 2 and not by acylation of DCC or DCU by an intermediate anhydride (DeTar and Silverstein, 1966a). One exception is the reaction of dimethyl malonic acid (**19**) with DCC; the anhydride **20**, formed initially, reacts further to give the barbiturate **21** (Resofszki *et al.*, 1974). This pathway has not been observed in peptide synthesis.

Reaction conditions that favor intermolecular nucleophilic attack on the

(a) Dehydration of asparagine

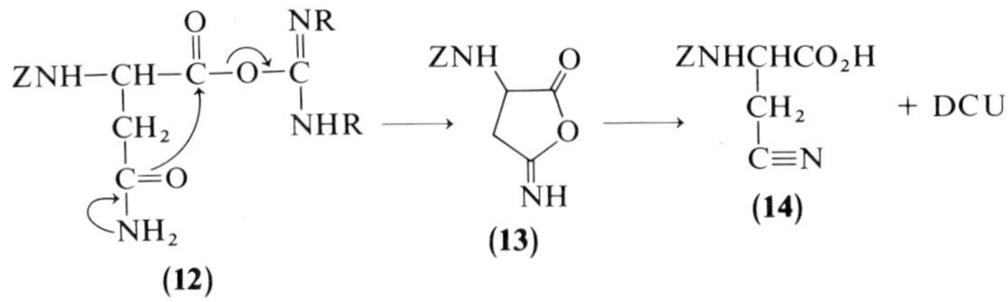

(b) 5(4H)-Oxazolone formation from peptide acids

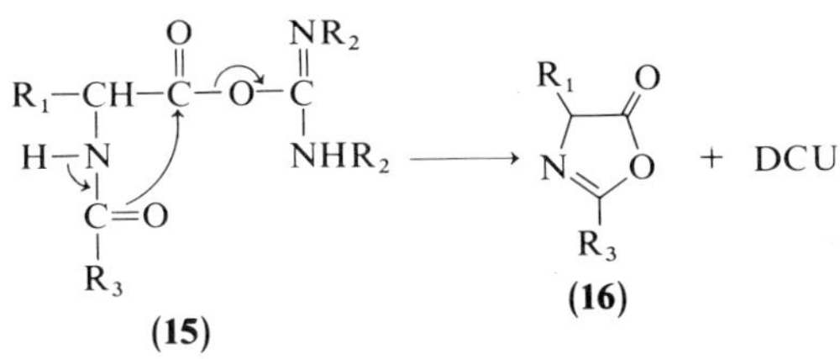

(c) *N*-Acylurea formation

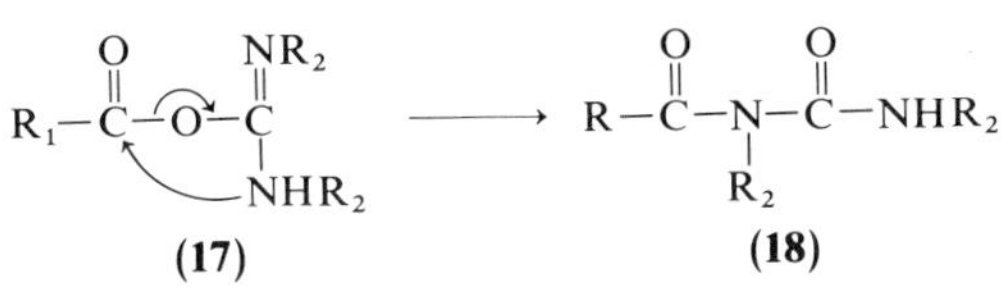

Scheme 2. Alternate reaction pathways for decomposition of *O*-acylisoureas.

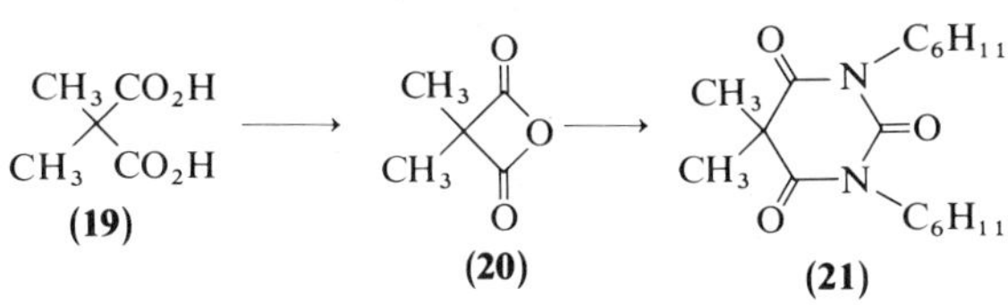

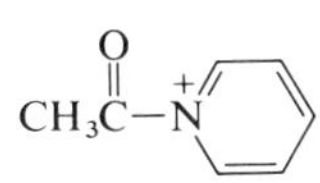

(22)

O-acylisourea (**3**) lead to fewer side reactions and to cleaner products because 5(4H)-oxazolone and *N*-acylurea formation are suppressed. For example, *N*-acylurea (**5**) formation is less likely to occur when reactions are run in solvents with low dielectric constants (CCl_4, CH_2Cl_2, C_6H_6) than in high dielectric constant solvents (DMF, acetonitrile, DMSO, H_2O). The former solvents favor dimerization of carboxylic acids, e.g., the association constant of acetic acid is about 2000 in carbon tetrachloride but only 0.5 in acetonitrile. As mentioned previously the dimers react much faster with DCC to form high yields of anhydrides and the net effect is that much less

N-acylurea is formed (DeTar and Silverstein, 1966a,b). Addition of triethylammonium acetate to the solution did not increase anhydride formation indicating that the second acetic acid molecule is bound tightly to and traps the *O*-acylisourea (see Scheme 1, **3** → **4**) (DeTar and Silverstein, 1966b). Other nucleophiles, e.g., *p*-nitrophenol and pyridine, also appear to prevent *N*-acylurea formation by intercepting the acylisourea to form *p*-nitrophenyl acetate and the highly reactive pyridinium ion (**22**). In general (in the absence of amines or phenols) N-protected amino acids will react rapidly ($t_{1/2} \simeq$ 1–5 min) with DCC in nonpolar solvents to form symmetrical anhydrides in good yield (DeTar and Silverstein, 1966b). Anhydride formation occurs whether the amino acid to DCC ratio is 2:1 or 1:1 because the reaction of a carboxylic acid with **3** is faster than with DCC. The latter ratio is less desirable because DCC is basic and catalyzes decomposition of symmetrical anhydrides.

B. Reaction in the Presence of Amines

Addition of amines to a DCC coupling mixture alters the reaction mechanism. DCC reacts 30-fold slower with acids in the presence of triethylamine, and consequently more *N*-acylurea is formed (DeTar and Silverstein, 1966b). Because both triethylamine and benzylamine produce the same amount of *N*-acylurea from the reaction with acetic acid and DCC, DeTar and Silverstein (1966b) concluded that benzylamine reacts directly with the *O*-acylisourea (**3**) rather than with an intermediate anhydride (**4**). Rebek and Feitler (1973) developed a method for measuring acylation proceeding by way of either **3** or **4** in a single reaction and determined that no more than 55% of the amide bonds were formed via the anhydride in solution. Thus when the traditional DCC method is used for solution coupling at least half of the peptide bonds (and possibly more) are formed via the *O*-acylisourea (**3**).

C. Formation of Symmetrical Anhydrides

It is important to evaluate the results of DeTar and Silverstein in terms of the use of DCC with Boc-amino acids in solid phase synthesis (Erickson and Merrifield, 1976). Dichloromethane or chloroform, which are low dielectric constant solvents, are usually employed as solvents for coupling with DCC, Boc-amino acids and the amino peptide resin and it would be predicted that much of the Boc-amino acid will be converted to the symmetrical anhydride. Rebek and Feitler (1974) developed a method to test whether the acylating agent in solid phase synthesis was the *O*-acylisourea (**3**) or the symmetrical anhydride, and their results showed that under reaction conditions similar but not identical to those employed in solid phase synthesis, symmetrical anhydrides are the acylating agents formed by the reaction of DCC with protected amino acids. However, the length of time that the Boc-amino acids and the resin are equilibrated prior to addition of DCC in solid phase synthesis may alter the mechanism of acylation. Esko and Karlsson (1970) have shown that a 4-hr equilibration of an excess of Boc-amino acid with the aminoacyl polystyrene resin causes one equivalent of the Boc-amino acid to become tightly bound to the resin in a form not removable by extensive washing with dichloromethane. In a related study Elliott *et al.* (1973) found that only a 30 min equilibration was needed for comparable results. If these conditions are used, then the reaction of DCC with the ionic complex between the Boc-amino acid and the amino resin may produce the *O*-acylisourea as the predominant acylating species. The experiments of Rebek and Feitler (1974) do not appear to have tested these conditions, but their results are probably valid for solid phase synthesis in which the Boc-amino acid and resin are equilibrated less than 5 min before addition of DCC and the excess Boc-amino acid is not removed by washing prior to addition of DCC (see also Chapter 6, Section VI,A for symmetrical anhydrides).

D. *p*-Nitrophenol as a Trapping Agent

Addition of reagents that react more rapidly than intramolecular nucleophiles with the *O*-acylisourea would be expected to reduce the amount of side products formed. Synthesizing *p*-nitrophenyl esters of protected amino acids (Bodanszky, 1956) is one well-established method for accomplishing this. *p*-Nitrophenol reacts directly with and traps the *O*-acylisourea before competing side reactions occur. Addition of pyridine to the solution or its use as solvent increases the rate of ester formation presumably by forming the acyl pyridinium ion (**22**) (DeTar and Silverstein, 1966b). This method works well for protected amino acids but cannot be used for coupling peptides with racemizable carboxyl-terminal residues because synthesis of *p*-nitrophenyl esters of peptide acids is still accompanied by extensive, and sometimes complete, racemization. This indicates that *p*-nitrophenol is insufficiently reactive to prevent 5(4H)-oxazolone formation. One way optically pure peptide *p*-nitrophenyl esters can be prepared by the carbodiimide method is through a strategy designed by Goodman and Stueben (1959) which was later named the "backing-off procedure" (Kovacs *et al.*, 1967). Amino acid active ester hydrochlorides or hydrobromides can be condensed with acylamino acids according to Eq. 2 because the amine reacts faster with the *O*-acylisourea intermediate than with the active ester. The optical purity

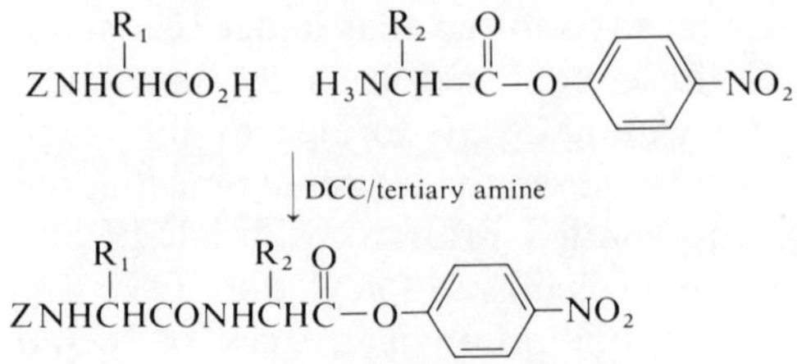

(2)

of peptide *p*-nitrophenyl esters obtained in this way establishes that racemization of the carboxyl-terminal residue in peptides that are activated with the use of DCC occurs before the peptide *p*-nitrophenyl ester is formed.

E. Pentachlorophenol as a Trapping Agent

Pentachlorophenol (PCP) reacts faster than *p*-nitrophenol with protected amino acids and DCC but the reaction is not fast enough to suppress 5(4H)-oxazolone formation with DCC activated peptide acids (DeTar and Silverstein, 1966c). The DCC–PCP complex (**23**), which is prepared by addition of PCP to an ethyl acetate solution of DCC, reacted with **24** to give ester **25**

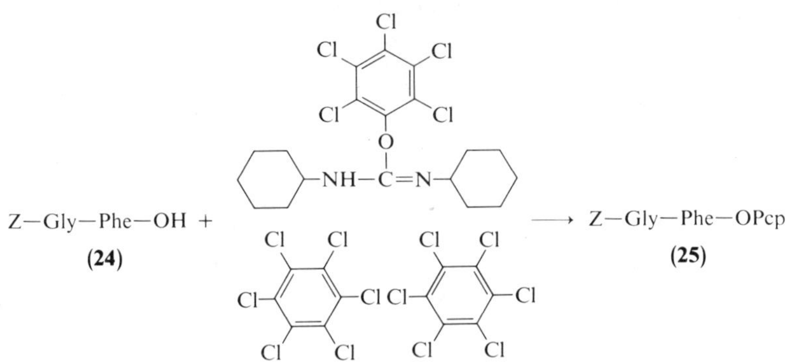

(24) (25)

(23)

(88% yield) which was more optically pure (90–100%) than a sample obtained using DCC and PCP separately (Kovacs *et al.*, 1967). However ester **25** was formed by reaction with an intermediate 5(4H)-oxazolone. Kovacs *et al.* suggested that the greater acidity of PCP (p*K* 5.3) versus *p*-nitrophenol (p*K* 7.2) suppressed base-catalyzed racemization of the 5(4H)-oxazolone.

F. *N*-Hydroxysuccinimide as a Trapping Agent

The first significant improvement of the carbodiimide method originated from the discovery (Wünsch and Drees, 1966) that addition of *N*-hydroxysuccinimide (**26**) to the reaction mixture gave a much higher yield

(75%) of the protected octapeptide Pht-Phe-Val-Gln-Leu-Met-Asn-Thr(*t*Bu)-O*t*Bu (glucagon sequence 22–29) than obtained from a standard DCC synthesis (31% yield). Further investigations of this effect revealed that the two most serious shortcomings of the carbodiimide method, racemization during fragment condensation and *N*-acylurea formation, can be almost completely suppressed (Weygand *et al.*, 1966) by addition of 2 equivalents of *N*-hydroxysuccinimide to a peptide coupling mixture in tetrahydrofuran or dimethylformamide at −20°C. The absence of measurable racemization (<1%) and the complete absence of *N*-acylurea was confirmed by gas chromatography. According to Weygand *et al.* (1968), it is essential to work at temperatures below 0°C; at 0°C low levels of racemization (about 1% to 1.6%) still occurred. However, Zimmerman and Anderson (1967) found complete suppression of racemization in dicyclohexylcarbodiimide couplings at room temperature by addition of 1.1 equivalent of *N*-hydroxysuccinimide by both the Anderson test and the very sensitive Young test. They concluded that the favorable effect of added *N*-hydroxysuccinimide **(26)** must

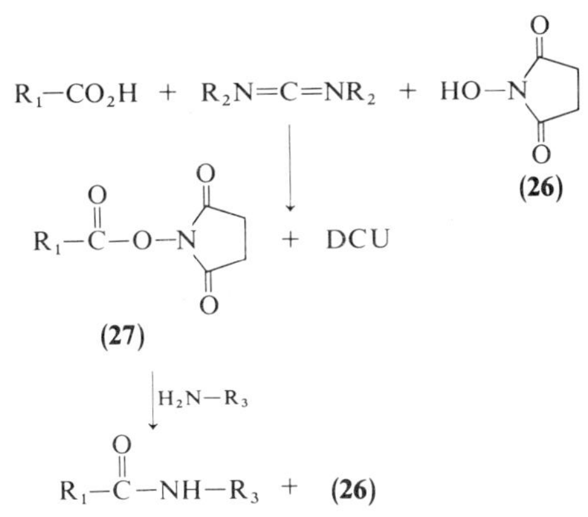

be primarily due to the intermediate formation of *N*-hydroxysuccinimide esters of the carboxyl component and that neutralization of dicyclohexylcarbodiimide is only a minor factor since addition of pivalic acid which has a similar pK_a did not always suppress racemization. A labeling study by Morimoto *et al.* (1975) showed that oxygen from a labeled carboxyl group (^{18}O) is transferred to DCU and not to the succinimide ester **(27)** and is consistent with the esterification reaction expected from DCC. The Wünsch–Weygand procedure was used successfully in the first synthesis of glucagon (Wünsch, 1967) and in syntheses of porcine calcitonin (Riniker *et al.*, 1969).

However, a side reaction can occur with *N*-hydroxysuccinimide (Eq. 3) (Löw and Kisfaludy, 1965; Gross and Bilk, 1968) which is especially serious when active ester formation is slowed by steric hindrance (Weygand *et al.*,

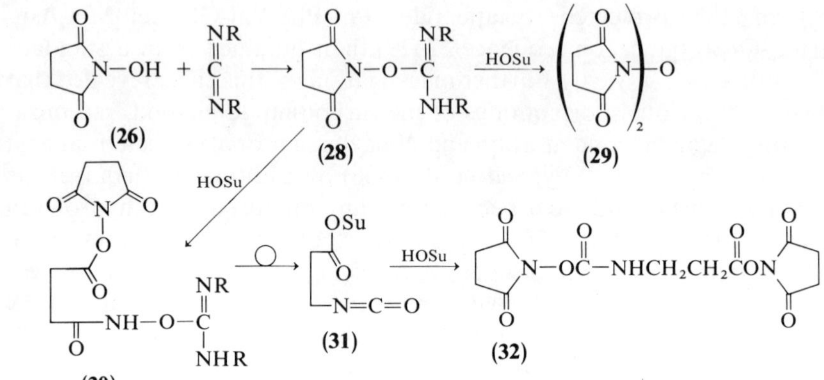

(26) (28) (29) (30) (31) (32) (3)

1968). This rearrangement, which is related to the Lossen rearrangement (Lossen, 1872; Smith, 1963) could occur with any hydroxamic acid derivative capable of forming a species such as **30**. To circumvent the reaction of **28** to **30**, Fujino *et al.* (1974) prepared the bicyclic derivative, *N*-hydroxy-5-norbornene-2,3-dicarboximide (HONB) (**33**) for use with DCC in place of HOSu. When HONB (**33**) is stirred with DCC in THF, no dicyclohexylurea or β-alanine by-products were formed within at least 10 hr. Synthesis of Z—Gly—Phe—Gly—OEt and Z—Phe—Ile—Gly—OBzl by a one step reaction gave quantitative yields of product with less than 1% racemization.

G. 1-Hydroxybenzotriazole as a Trapping Agent

Probably the best method for coupling acyl peptides using DCC was developed by König and Geiger (1970a) who found that 1-hydroxybenzotriazole (HOBt; **34**) and a number of related substituted 1-hydroxybenzotriazoles were suitable additives for the synthesis of peptides by the carbodiimide method, and were highly effective in suppressing racemization and *N*-acylurea formation under certain conditions. HOBt reacts rapidly with *O*-acylisoureas to form HOBt peptide esters which rapidly acylate amino acid esters.

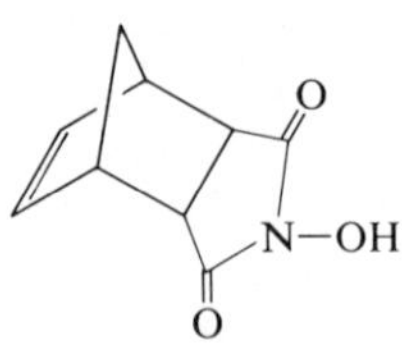

(33)

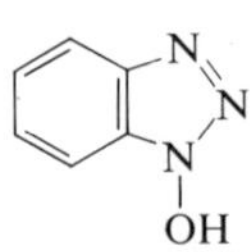

(34)

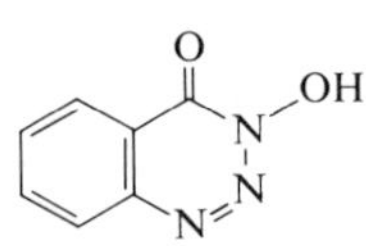

(35)

The basicity of the reaction medium is important. Racemization is increased when free amino acid esters, rather than hydrochlorides, are used or when more basic residues such as Pro, are being used as the amino component. The use of DCC with 3-hydroxy-4-oxo-3,4-dihydro-1,2,3-benzotriazine (**35**) was the most effective reagent, among the series that König and Geiger studied, for reducing the amount of racemization observed during couplings especially those requiring more basic conditions. For example, coupling of Z-Pro-Val-OH with HCl · Pro-O*t*Bu in DMF gave 19% of the L-D-L tripeptide using HOBt but only 1.3% using triazine **35**. However triazine **35** is a hydroxamic acid derivative that rearranges via the Lossen type reaction to the acylating agent **36**. By-products formed from reaction of **36** with

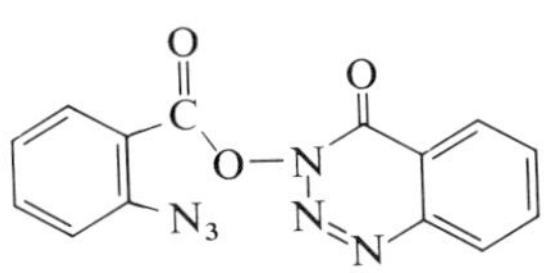

(36)

amino acid esters limit the utility of triazine **35** (König and Geiger, 1970c). The low levels of racemization observed for HOBt under mildly basic conditions have been confirmed by Kemp *et al.* (1974) using the isotope dilution variant (Kemp *et al.*, 1970) of the Anderson test.

H. Advantages of the Combination of Dicyclohexylcarbodiimide with Hydroxybenzotriazole

The DCC–HOBt combination is an extremely efficient method for coupling peptides in solution. HOBt apparently reacts so rapidly with *O*-acylisourea (**3** in Scheme 1) that the competing intramolecular reactions, *N*-acylurea formation or dehydration of asparagine are not observed (König and Geiger, 1970a). The DCC–HOBt combination was used with Boc-Asn-OH and Boc-Gln-OH, in place of the corresponding *p*-nitrophenyl esters, for synthesis of the A (14–21) sequence of ovine insulin by the solid phase method (Wolters *et al.*, 1974). Similarly, Z-Leu-Leu-Gln-OH was condensed with H-Gly-Leu-Val-NH_2 using DCC–HOBt to give the secretin 22–27 fragment, Z-Leu-Leu-Gln-Gly-Leu-Val-NH_2 in 83% yield without dehydration of the glutamine carboxamide group (Jäger *et al.*, 1974).

The DCC–HOBt combination suppresses other side reactions. Formation of peptide bonds with DCC in the presence of trifluoroacetate ion can cause substantial trifluoroacetylation of the amine component (Fletcher *et al.*, 1973). Coupling of Boc-Pro-OH with TFA · Pro-Gly-OPic · TFA with DCC gave a 21% yield of TFA-Pro-Gly-O-Pic. When DCC–HOBt was used, no trifluoroacetylated product was detected.

Extensive racemization of Boc-*Im*-benzyl-L-histidine can occur with

DCC-activated couplings during solid phase synthesis (Jorgensen *et al.*, 1970) and to a lesser extent with amino acid esters in solution. Addition of HOBt to these reaction mixtures reduced racemization to acceptably low levels without impairing coupling efficiencies (Windridge and Jorgensen, 1971). The imidazole of histidine may catalyze racemization in several ways: (1) intramolecular base catalyzed formation of 5(4H)-oxazolone; (2) base catalysis of direct abstraction of the α-proton; and, (3) formation of a cyclic acylimidazole of the type **37** which has a higher rate of racemization (Veber, 1976). Treatment of Boc-L-His-OH with DCC produces racemic **37** in good yield (Sheehan *et al.*, 1959). The high reactivity of *N*-methylimidazole as an acyl transfer agent suggests that a species such as **38** could be formed from reaction of Boc-His(Bzl)-OH and DCC and that HOBt suppresses formation of **38** by forming the hydroxybenzotriazole ester quickly.

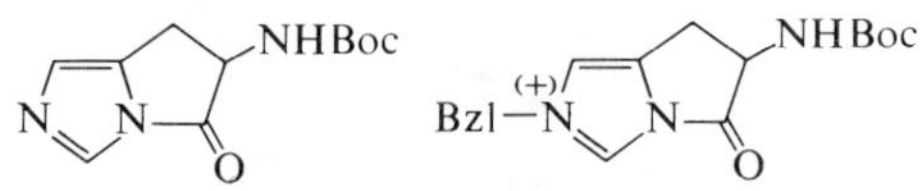

(37) **(38)**

In addition to *N*-hydroxysuccinimide and 1-hydroxybenzotriazole, many *N*-hydroxy compounds (**39–48**) have been proposed as additives for DCC mediated couplings (Table I). Most of these compounds are hydroxamic acid derivatives and are potentially susceptible to Lossen rearrangements leading to undesirable side products of the type obtained from *N*-hydroxysuccinimide (**26** → **32**) or from the benzotriazine (**35** → **36**). None of these additives appears superior to 1-hydroxybenzotriazole for the *in situ* reaction of coupling acyl peptides without racemization, although several are comparable to HOSu.

III. REACTION CONDITIONS FOR DICYCLOHEXYL-CARBODIIMIDE-MEDIATED PEPTIDE BOND FORMATION

A. Preferred Reaction Conditions

The successful use of symmetrical anhydrides and *in situ*-formed 1-hydroxybenzotriazole esters in peptide synthesis in recent years is an indication that for most coupling reactions a species as highly reactive as the *O*-acylisourea (**3**) is not necessary. Indeed, it is undesirable because its high chemical reactivity leads to so many side reactions. The studies described in Section II indicate that for most purposes DCC should be used as a reagent to prepare HOBt esters *in situ* or symmetrical anhydrides *in situ* depending on the objective of synthesis. Either procedure is more likely to give higher

Table 1 *N*-Hydroxy Compounds Used as Additives for DCC-Activated Peptide Bond Synthesis

Compound	Structure No.	Name	Reference
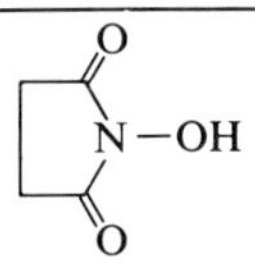	**26**	*N*-Hydroxysuccinimide	Wünsch and Drees, 1966; Weygand *et al.*, 1966
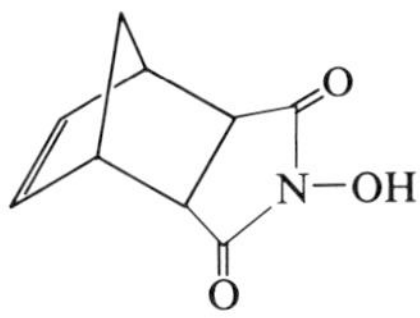	**33**	*N*-Hydroxy-5-norbornene -2,3-dicarboximide	Fujino *et al.*, 1974
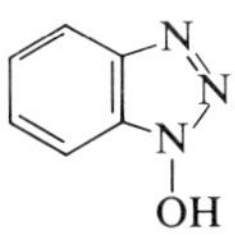	**34**	1-Hydroxybenzotriazole	König and Geiger, 1970a,b,c
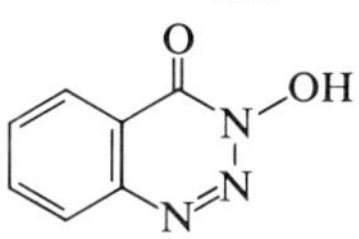	**35**	3-Hydroxy-4-oxo-3,4-dihydro-1,2,3-benzotriazine	König and Geiger, 1970a,b,c
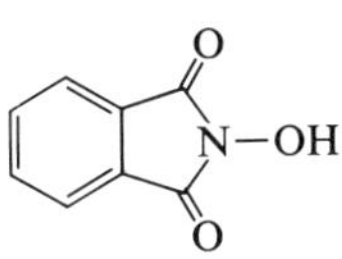	**39**	*N*-Hydroxyphthalimide	Nefkens and Tesser, 1961
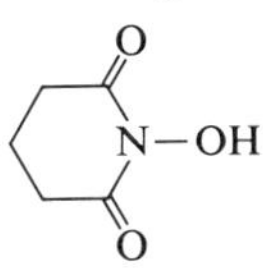	**40**	*N*-Hydroxyglutarimide	Jeschkeit, 1968, 1969
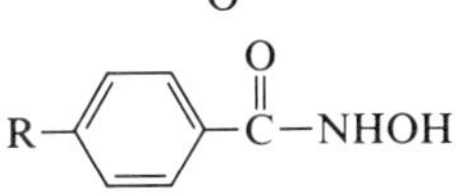 R = H	**41**	Benzhydroxamic acid	Lubiewska *et al.*, 1970
R = Cl	**42**	*p*-Chlorobenzhydroxamic acid	Govindachari *et al.*, 1966
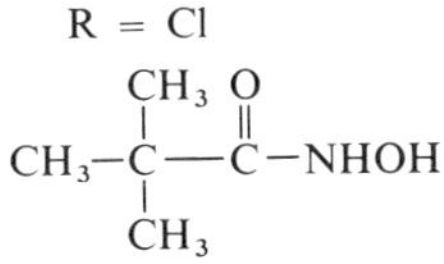	**43**	Pivalylhydroxamic acid	Rajappa *et al.*, 1967
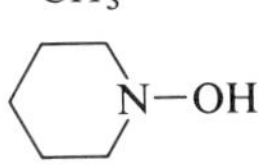	**44**	*N*-Hydroxypiperidine	Beaumont *et al.*, 1965
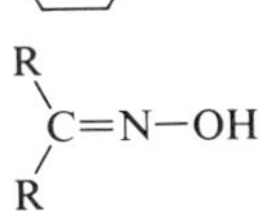	**45**	Acetoxime	Bittner *et al.*, 1965

(*continued*)

Table I (*Continued*)

Compound	Structure No.	Name	Reference
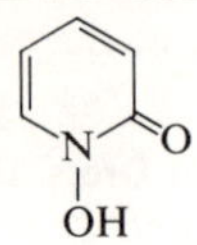	**46**	1-Hydroxy-2(1H)-pyridone	Paquette, 1965
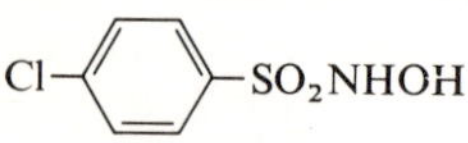	**47**	*p*-chlorobenzene-sulfolhydroxamic acid	Yajima *et al.*, 1973
$EtO_2C{-}C(C{\equiv}N){=}N{-}OH$	**48**	Ethyl-2-hydroximino-2-cyanoacetate	Itoh, 1973

yields of a peptide with less contamination by *N*-acylurea, nitrile, and racemate when possible, etc., compared to standard DCC couplings. Exceptions to this generalization will undoubtedly be found because each peptide has unique physical and chemical properties that determine the optimal reaction conditions with respect to solvent, temperature, coupling method, etc. A large number of successful experimental conditions for preparing peptides using DCC without additives have been summarized (Schröder and Lübke, 1965; Wünsch and Wendlberger, 1974) and these attest to the fact that each peptide may require individualized treatment. Conditions common to most procedures include using solvents with as low a dielectric constant as possible consistent with the solubility of each component, using weakly basic media to suppress *N*-acylurea formation and using temperatures at or below 0°C. The reaction of DCC with acids is highly exothermic (Anderson and Callahan, 1958) and even lower initial temperatures (−70° to −10°C) can be useful or necessary.

B. Conditions Favoring Symmetrical Anhydrides

The decision to use DCC to prepare symmetrical anhydrides *in situ* or 1-hydroxybenzotriazole esters *in situ* depends on the synthesis to be executed. For solid phase synthesis the available data suggest that preformed symmetrical anhydrides may give higher yields of purer products than the standard use of DCC and Boc-amino acid in 1:1 ratios. There are two ways to prepare symmetrical anhydrides. Wieland *et al.* (1971) prepared symmetrical anhydrides by reacting the sodium salts of Boc-amino acids with phosgene. Hagenmaier and Frank (1972) developed a "pre-mix" procedure in which two equivalents of the Boc-amino acid are mixed with one equivalent of DCC in chloroform or dichloromethane at 0°C for 30 min. DCU precipitates from solution quickly while the symmetrical anhydride remains in solution to be added to the peptide resin. (This procedure differs from using two equivalents of Boc-amino acid to one of DCC when the amino acid is equilibrated with the resin prior to addition of DCC.) Yamashiro and Li (1976) reported that for the synthesis of a 19-peptide segment of ovine prolactin preformed symmetrical anhydrides prepared by the pre-mix procedure (Hagenmaier and Frank, 1972) gave better results than standard couplings using DCC and Boc-amino acids. This was especially true when the Boc—Leu-resin was prepared using triethylamine instead of tetramethylammonium salts.

Rebek and Feitler (1974) have shown that under experimental conditions where aminoacyl resin and Boc-amino acid are not extensively equilibrated prior to addition of DCC (see Section II), symmetrical anhydrides are the predominant and, perhaps, exclusive acylating agent in solid phase peptide synthesis. This result is not surprising in view of (a) the substantial excess of Boc-amino acid over resin-amine normally used, (b) the strong tendency of acids to dimerize in dichloromethane, and (c) the 30-fold faster reaction of acid dimer with DCC (DeTar and Silverstein, 1966a,b). Thus, the symmetrical anhydride is formed in solution whether the amino acid to DCC ratio is 2:1 or 1:1. In the latter case varying amounts of DCC would persist in solution throughout the reaction period.

Under normal solid phase or solution coupling conditions the ratio of Boc-amino acid to DCC is not critical when coupling is completed quickly. However, in cases requiring long reaction times to achieve complete coupling, excess DCC can cause problems. Although a solution of a symmetrical anhydride is stable in the absence of base, the addition of DCC (which has about one-fifth the basicity of pyridine) causes rapid decomposition via intramolecular acylation (Eq. 4) (DeTar and Silverstein, 1966c). Boc and

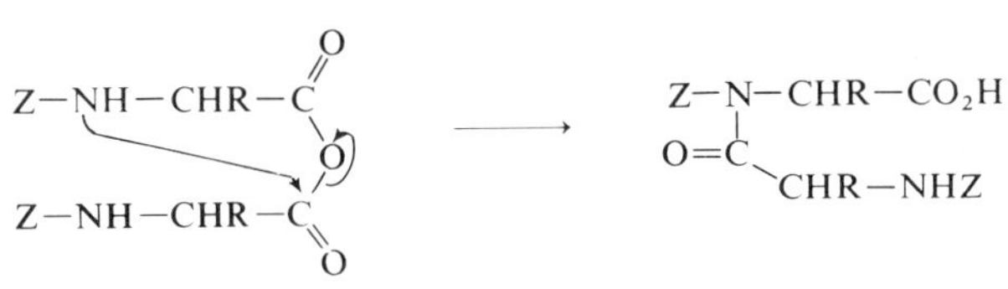

(4)

Bpoc derivatives also rearrange in this way and the rate is increased in the presence of triethylamine hydrochloride (Merrifield *et al*., 1974). A careful study by Merrifield *et al*. (1974) showed that the insertion products were formed in less than 0.1% using normal DCC ratios (1:1) in solid phase synthesis. However, not all coupling reactions are completed quickly and for those requiring extended reaction times the intramolecular rearrangement of symmetrical anhydrides could become significant in the last 1–5% of coupling and lead to insertion by-products. Under these circumstances an amino acid/DCC ratio of 2:1 would be preferred. Studies have shown that the amount of activated amino acid remaining in a solid phase mixture decreases with time (Tometsko, 1973; Dorman, 1974). These results are consistent with a slow decomposition or rearrangement of the symmetrical anhydride, and are not caused by loss of the *O*-acylisourea. Another example of a problem that might be encountered using equimolar amounts of DCC and Boc-amino acid was found in a recent synthesis of Leu-enkephalin (Christensen *et al.*, 1977). The authors found that irreversible blocking of the aminoacyl resin occurred when it was in excess and when the Boc-amino acid/DCC ratio was 1:1. This irreversible blocking did not occur when the corresponding ratio was 2:1. The nature of the possible termination reaction has not been identified.

The available data suggest that for solid phase synthesis, the use of DCC to prepare symmetrical anhydrides from Boc-amino acids may be superior to the usual method in which the Boc-amino acid is equilibrated with the amine resin and then activated with 0.5 to 1.0 equivalents of DCC. However, the DCC–HOBt method, and not the symmetrical anhydride method, should be used with Boc—His(Bzl)—OH (Windridge and Jorgensen, 1971).

C. Application of Dicyclohexylcarbodiimide in Combination with 1-Hydroxybenzotriazole

For most solution syntheses with normal amino acids the DCC—HOBt combination appears to provide higher yields of product with fewer side reactions than DCC alone. No major side reaction has been reported yet for the DCC—HOBt combination and for this reason HOBt is preferred over HOSu for DCC mediated couplings of acyl peptides. As noted earlier, couplings between hindered acyl peptides and amino-terminal prolyl-peptides may require the use of 3-hydroxy-4-oxo-3,4-dihydro-1,2,3-benzotriazine (**35**) to suppress racemization (König and Geiger, 1970c). HOBt cannot be used in the presence of the *o*-nitrophenylsulfenyl protection because it cleaves this group from the amine function (Geiger *et al.*, 1973). HOBt may catalyze the addition of DCC to the imidazole group of unprotected histidine (see Section IV for a detailed discussion of this reaction).

D. Reactions with α-Substituted and Secondary Amino Acids

Acylation of α-substituted amino acids (e.g., α-methylalanine, i.e., α-aminoisobutyric acid, abbreviated Aib) is often difficult to accomplish in good yield and this is particularly true when DCC is used in the traditional way. For example, Leplawy *et al.* (1960) reported that dipeptide **51** was not formed when **49** and **50** were coupled using DCC (Eq. 5). Jones *et al.* (1965)

$$\text{For—Aib—OH} + \text{H—Aib—OCH}_3 \xrightarrow{\text{DCC}} \text{For—Aib—Aib—OCH}_3 \quad (5)$$

(49) **(50)** **(51)**

reported that this condensation could be achieved using oxazolone **53** prepared by reaction of **52** with DCC. Reaction of **53** with H—Aib—OMe (**50**) gives the highly hindered dipeptide **54** in 90% yield. This procedure

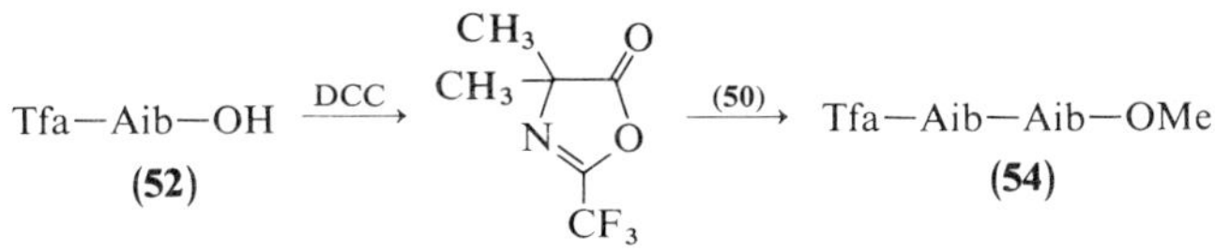

(52) **(53)** **(54)**

illustrates another reactive intermediate that can be generated by capture of the *O*-acylisourea (**3**).

Acylation of peptides containing an amino-terminal α-benzylphenylalanine residue has been reported to give erratic results with DCC (Barrett *et al.*, 1972). Although Z—Pro—OH (**55**) and H—Phe(2—Bzl)—OEt gave the protected dipeptide in 75% yield, the corresponding reaction of **56** with the dipeptide ester **57** gave only trace amounts of protected tripeptide **58**, and *N*-acylurea was the major product. Some of these problems may be

$$\text{Nps—Pro—OH} + \text{H—Phe(2—Bzl)—Arg(NO}_2\text{)—ONb} \xrightarrow{\text{DCC}}$$

(56) **(57)**

$$\text{Nps—Pro—Phe(2—Bzl)—Arg(NO}_2\text{)—ONb}$$

(58)

overcome by using the DCC-"pre-mix" procedure (Hagenmaier and Frank, 1972) to prepare symmetrical anhydrides *in situ*. R. Jasensky and D. H. Rich (unpublished experiments, 1977) obtained excellent yields (95%) of dipeptide **59** using the "pre-mix" procedure.

$$\text{Boc—Ala—OH} + \text{H—Aib—OMe} \xrightarrow[\text{premix}]{\text{DCC}} \text{Boc—Ala—Aib—OMe}$$

(50) **(59)**

Acylation of several secondary amino acids (Pro, MePhe, MeAla, thiazolidine-2-carboxylic acid, Thz) with symmetrical anhydrides using the DCC-"pre-mix" procedure has been studied (P. Bhatnagar, J. Singh, and D. H. Rich, unpublished experiments, 1977). Coupling of Boc—Leu—OH with MePhe(3—SBzl)—Gly—OEt gave the tripeptide **60** in 84% yield, and coupling of Boc—Thz—OH with Pro—OMe (**61**) gave the dipeptide **62** in 93%

Boc—Leu—MePhe(3—S—Bzl)—Gly—OEt (**60**)

Pro—OMe (**61**)

Boc—Thz—Pro—OMe (**62**)

yield. When the DCC—HOBt rather than the "pre-mix" procedure was used, the yields of **60** and of **62** were decreased to 32% and 68%, respectively.

IV. REACTIONS OF CARBODIIMIDES WITH AMINES

A. Reaction with Amino Acid Esters

Carbodiimides react with aliphatic amines to form guanidines. The addition requires high reaction temperatures or acid catalysis to occur. Amino acid ester hydrochlorides are sufficiently acidic to catalyze this reaction (Kurzer and Douraghi-Zadeh, 1967). Glycine ethyl ester hydrochloride or its *p*-nitrophenyl ester react rapidly with DCC at room temperature ($t_{1/2}$ = 0.5 hr) to form guanides **63** and/or **64** (Muramatsu, 1963; DeTar *et al.*, 1966).

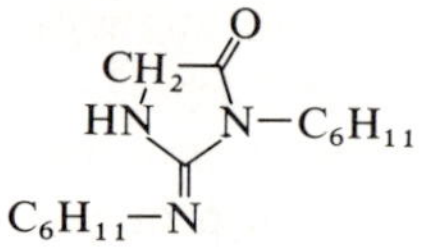

(64)

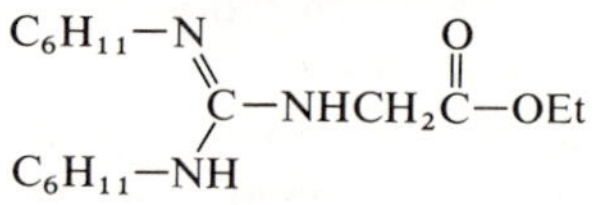

(63)

Reaction with peptide hydrochlorides, e.g., HCl·H—Ser—Gly—ONp, also occurs. These reactions are fast enough to be a source of byproducts in peptide synthesis when the system is subjected to specific reaction conditions. Merrifield, *et al.* (1977) studied this reaction in detail using radioactive dicyclohexylcarbodiimide and found no measurable reaction (< 0.2%) between DCC and unprotonated valine which was esterified to a polystyrene–divinyl benzene resin. Similarly, no reaction occurred during normal DCC couplings and therefore the addition of the neutralized amine group to DCC is not a significant side reaction in solid phase synthesis. However, the labeled DCC did react with the protonated form of Val–resin in the same way that DCC reacts with the hydrochloride of glycine ethyl ester. Thus caution should be taken when DCC is added to an amine resin or peptide in the presence of an acid.

B. Reaction with Imidazole; 1-Hydroxybenzotriazole Catalysis

DCC has been shown to react with the imidazole ring of peptides which contain unprotected histidine (Rink and Riniker, 1974) (Eq. 6). Addition is strongly catalyzed by HOBt. During the synthesis of fragment (1–34) of parathyrine, the N^{im}-amido-His^{32} (18–34) derivative **(64)** was formed in 13% yield during the DCC–HOBt condensation of fragments (18–24) and

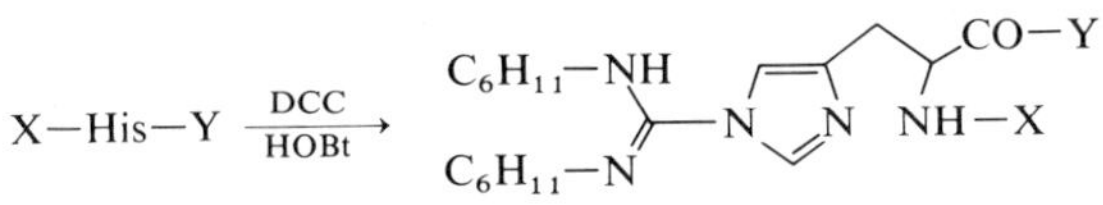

(64) (6)

(25–34). The DCC group can be removed by warming **64** in methanol which reacts to form *O*-methyl-*N*,*N*′-dicyclohexylisourea (**65**). However the isourea **65** is not an inert molecule but can be a good methylating agent. Musich and Rapoport (1977) studied the reaction of the diisopropyl analog **66** which converted proline or *N*-methylproline into the corresponding betaines in good yield at room temperature. Similar methylations of the other

$$R{-}NH{-}\overset{\displaystyle OCH_3}{\overset{|}{C}}{=}N{-}R \qquad \begin{array}{l} \textbf{(65)}\ R = C_6H_{11}{-} \\ \textbf{(66)}\ R = {-}CH(CH_3)_2 \end{array}$$

nucleophilic groups which are found in His, Met, Cys, or Tyr are possible with **65** but have not been reported to date. *N*-Hydroxysuccinimide may also catalyze the addition of DCC to an imidazole ring. The use of DCC/HOSu was reported to decrease the amount of unprotected histidine during a synthesis of a protected (21–47) analogue of staphylococcal nuclease (Zeiger and Anfinsen, 1973). Deprotection with HF did not regenerate the missing His but addition of imidazole to the coupling mixture prevented loss of His in the product.

N-Acylureas derived from DCC usually are not reactive toward nucleophiles. An exception to this was reported for the reaction of aziridine **67** with *N*-acylurea **68** which formed the unsymmetrical urea **69** and the amide of cyclohexylamine **70** (Fenwick, 1973).

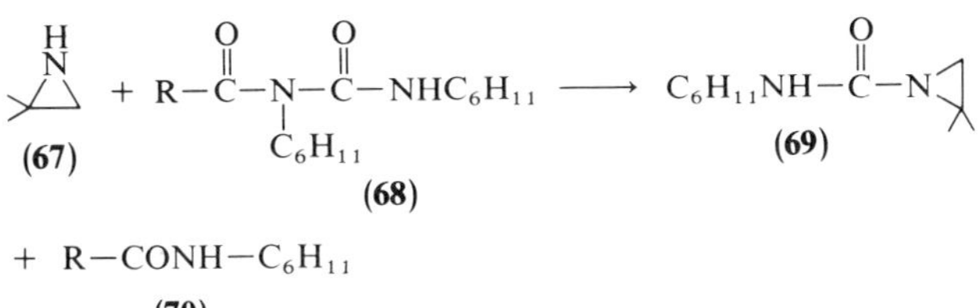

(67) (68) (69) (70)

DCC has been reported to react with more acid labile *N*-protected amino acids (Boc, Bpoc) in dichloromethane to form *N*-carboxyanhydrides (Eq. 7) (Bodanszky *et al.*, 1975). The extent of this undesired side reaction appears to be very small.

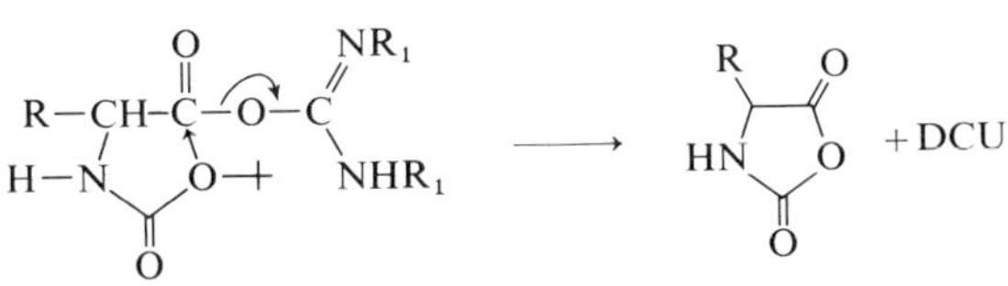

(7)

REFERENCES

Anderson, G. W., and Callahan, F. M. (1958). *J. Am. Chem. Soc.* **80**, 2902–2903.

Barrett, G. C., Hardy, P. M., Harrow, T. A., and Rydon, H. N. (1972). *J. Chem. Soc.* pp. 2634–2638.

Beaumont, S. M., Handford, B. O., Jones, J. H., and Young, G. T. (1965). *Chem. Commun.* pp. 53–54.

Bittner, S., Knobler, Y., and Frankel, M. (1965). *Tetrahedron Lett.* pp. 95–99.

Bodanszky, M. (1956). *Acta Chim. Acad. Sci. Hung.* **10**, 335–346.

Bodanszky, M., Klausner, A., and Bodanszky, A. (1975). *J. Org. Chem.* **40**, 1507–1508.

Christensen, T., Villemoes, P., and Brunfeldt, K. (1977). *In* "Peptides: Chemistry, Structure and Biology" (M. Goodman and J. Meienhofer, eds.), pp. 569–571. Wiley, New York.

DeTar, D. F., and Silverstein, R. (1966a). *J. Am. Chem. Soc.* **88**, 1013–1019.

DeTar, D. F., and Silverstein, R. (1966b). *J. Am. Chem. Soc.* **88**, 1020–1023.

DeTar, D. F., Silverstein, R., and Rogers, F. F., Jr. (1966). *J. Am. Chem. Soc.* **88**, 1024–1030.

Doleschall, G., and Lempert, K. (1963). *Tetrahedron Lett.* pp. 1195–1199.

Dorman, L. C. (1974). *Biochem. Biophys. Res. Commun.* pp. 318–325.

du Vigneaud, V., Ressler, C., Swan, J. M., Roberts, C. W., Katsoyannis, P. G., and Gordon, S. (1953). *J. Am. Chem. Soc.* **75**, 4879–4880.

du Vigneaud, V., Ressler, C., Swan, J. M., Roberts, C. W., and Katsoyannis, P. G. (1954). *J. Am. Chem. Soc.* **76**, 3115–3121.

Elliott, D. F., Moritz, P., and Wade, R. (1973). *In* "Peptides 1971" (H. Nesvada, ed.), pp. 168–171. North-Holland Publ., Amsterdam.

Erickson, B., and Merrifield, R. B. (1976). *In* "The Proteins" (H. Neurath and R. L. Hill, eds.), Vol. 2, pp. 349–351. Academic Press, New York.

Esko, K., and Karlsson, S. (1970). *Acta Chem. Scand.* **24**, 1415–1422.

Fenwick, G. R. (1973). *Chem. Ind. (London)* pp. 637–638.

Finn, F. M., and Hofmann, K. (1976). *In* "The Proteins" (H. Neurath and R. L. Hill, eds.), pp. 131–137. Academic Press, New York.

Fletcher, G. A., Low, M., and Young, G. (1973). *J. Chem. Soc.* pp. 1162–1164.

Fujino, M., Kobayashi, S., Obayashi, M., Tsunehiko, F., Shinagawa, S., and Nishimura, O. (1974). *Chem. Pharm. Bull. (Tokyo)* **22**, 1857–1863.

Geiger, R., Jäger, G., and König, W. (1973). *Chem. Ber.* **106**, 2347–2352.

Gish, D. T., Katsoyannis, P. G., Hess, G. P., and Stedman, R. J. (1956). *J. Am. Chem. Soc.* **75**, 5954.

Goodman, M., and Glaser, C. (1970). *In* "Peptides: Chemistry and Biochemistry" (B. Weinstein and S. Lande, eds.), pp. 267–336. Dekker, New York.

Goodman, M., and Stueben, K. C. (1959). *J. Am. Chem. Soc.* **81**, 3980–3983.

Govindachari, T. R., Nagarajan, K., and Rajappa, S. (1966). *Tetrahedron* **22**, 3367–3372.

Gross, H., and Bilk, L. (1968). *Tetrahedron* **24**, 6935–6939.

Hagenmaier, H., and Frank, H. (1972). *Hoppe-Seyler's Z. Physiol. Chem.* **353**, 1973–1976.

Hegerty, A. F., and Bruice, T. C. (1970). *J. Am. Chem. Soc.* **92**, 6568–6574.

Hofmann, K., Woolner, M. E., Spühler, G., and Schwartz, E. T. (1958). *J. Am. Chem. Soc.* **80**, 1486–1489.

Itoh, M. (1973). *Bull. Chem. Soc. Jpn.* **46**, 2219–2221.

Jäger, G., König, W., Wissmann, H., and Geiger, R. (1974). *Chem. Ber.* **107**, 215–231.

Jencks, W. P., and Gilchrist, M. (1968). *J. Am. Chem. Soc.* **90**, 2622–2637.

Jeschkeit, H. (1968). *Z. Chem.* **8**, 20–21.

Jeschkeit, H. (1969). *Z. Chem.* **9**, 266–267.

Jones, D. S., Kenner, G. W., Preston, J., and Sheppard, R. C. (1965). *J. Chem. Soc.* pp. 6227–6232.

Jorgensen, E. C., Windridge, G. C., and Lee, T. C. (1970). *J. Med. Chem.* **13**, 352–356.

Kemp, D., Wang, S. W., Busby, G., and Hugel, G. (1970). *J. Am. Chem. Soc.* **92**, 1043–1055.
Kemp, D. S., Trangle, M., and Trangle, K. (1974). *Tetrahedron Lett.* pp. 2695–2696.
Khorana, H. G. (1953). *Chem. Rev.* **53**, 145–166.
König, W., and Geiger, R. (1970a). *Chem. Ber.* **103**, 788–798.
König, W., and Geiger, R. (1970b). *Chem. Ber.* **103**, 2024–2033.
König, W., and Geiger, R. (1970c). *Chem. Ber.* **103**, 2034–2040.
Kovacs, J., Kisfaludy, L., and Ceprini, M. Q. (1967). *J. Am. Chem. Soc.* **89**, 183–184.
Kurzer, F., and Douraghi-Zadeh, K. (1967). *Chem. Rev.* **67**, 107–152.
Leplawy, M. T., Jones, D. S., Kenner, G. W., and Sheppard, R. C. (1960). *Tetrahedron* **11**, 39–51.
Lossen, W. (1872). *Justus Liebigs Ann. Chem.* **161**, 347–362.
Löw, M., and Kisfaludy, L. (1965). *Acta Chim. Acad. Sci. Hung.* **44**, 61–66.
Lubiewska-Nakonieczna, L., Rzeszotarska, B., and Taschner, E. (1970). *Justus Liebigs Ann. Chem.* **741**, 157–166.
Merrifield, R. B., Mitchell, A. R., and Clarke, J. E. (1974). *J. Org. Chem.* **39**, 660–668.
Merrifield, R. B., Gisin, B. F., and Bach, A. N. (1977). *J. Org. Chem.* **42**, 1291–1295.
Morimoto, H., Kawakami, T., and Hayashi, N. (1975). *Chem. Pharm. Bull.* (*Tokyo*) **23**, 2492–2495.
Muramatsu, I. (1963). *Nippon Kagaku Zasshi* **00**, 861; *Chem. Abstr.* **60**, 653 (1964).
Musich, J. A., and Rapoport, H. (1977). *J. Org. Chem.* **42**, 139–141.
Nefkens, G. H. L., and Tesser, G. I. (1961). *J. Am. Chem. Soc.* **83**, 1263–1264.
Paquette, L. A. (1965). *J. Am. Chem. Soc.* **87**, 5186–5190.
Rajappa, S., Nagarajan, K., and Iyer, V. S. (1967). *Tetrahedron* **23**, 4805–4809.
Rebek, J., and Feitler, D. (1973). *J. Am. Chem. Soc.* **95**, 4052–4053.
Rebek, J., and Feitler, D. (1974). *J. Am. Chem. Soc.* **96**, 1606–1607.
Resofszki, G., Huhn, M., Hegedüs, B., Dvortsak, P., and Kaloy, K. (1974). *Tetrahedron Lett.* pp. 3091–3094.
Ressler, C. (1956). *J. Am. Chem. Soc.* **78**, 5956–5957.
Riniker, B., Brugger, M., Kamber, B., Sieber, P., and Rittel, W. (1969). *Helv. Chim. Acta* **52**, 1058–1073.
Rink, H., and Riniker, B. (1974). *Helv. Chim. Acta* **57**, 831–835.
Schröder, E., and Lübke, K. (1965). *In* "The Peptides," Vol. 1, Academic Press, New York.
Sheehan, J. C., and Hess, G. P. (1955). *J. Am. Chem. Soc.* **77**, 1067–1068.
Sheehan, J. C., and Hlavka, J. J. (1956). *J. Org. Chem.* **21**, 439–441.
Sheehan, J. C., Hasspacher, K., and Yeh, Y. L. (1959). *J. Am. Chem. Soc.* **81**, 6086.
Smith, P. A. S. (1963). *In* "Molecular Rearrangements" (P. De Mayo, ed.), Vol. 1, p. 528, Academic Press, New York.
Tometsko, A. (1973). *Biochem. Biophys. Res. Commun.* **50**, 886–891.
Veber, D. F. (1976). *In* "Peptides: Chemistry, Structure and Biology" (R. Walter and J. Meienhofer, eds.), pp. 307–316. Ann Arbor Sci. Publ., Ann Arbor, Michigan.
Weygand, F., Hoffmann, D., and Wünsch, E. (1966). *Z. Naturforsch., Teil B* **21**, 426–428.
Weygand, F., Hoffmann, D., and Prox, A. (1968). *Z. Naturforsch., Teil B* **23**, 279–281.
Wieland, T., Birr, C., and Flor, F. (1971). *Angew. Chem.* **83**, 333–334.
Windridge, G. C., and Jorgensen, E. C. (1971). *J. Am. Chem. Soc.* **93**, 6318–6319.
Wolters, E. T. M., Tesser, G. I., and Nivard, R. J. F. (1974). *J. Org. Chem.* **39**, 3388–3397.
Wünsch, E. (1967). *Z. Naturforsch., Teil B* **22**, 1269–1276.
Wünsch, E., and Drees, F. (1966). *Chem. Ber.* **99**, 110–120.
Wünsch, E., and Wendlberger, G. (1974). *In* "Houben-Weyl's Methoden der organischen Chemie" (E. Wünsch, ed.), Vol. 15, Part II, pp. 103–117. Thieme, Stuttgart.
Yajima, H., Kurobe, M., and Koyama, K. (1973). *Chem. Pharm. Bull.* **21**, 2566–2567.
Yamashiro, D., and Li, C. H. (1976). *J. Am. Chem. Soc.* **98**, 2324–2328.
Zeiger, A. R., and Anfinsen, C. B. (1973). *J. Am. Chem. Soc.* **95**, 880–886.
Zimmerman, J. E., and Anderson, G. W. (1967). *J. Am. Chem. Soc.* **89**, 7151–7152.

Chapter 6

The Mixed Carbonic Anhydride Method of Peptide Synthesis

JOHANNES MEIENHOFER

I. Introduction 264
II. The Experimental Procedure 266
III. Reaction Conditions 267
A. Objectives and Requirements 267
B. Anhydride Formation (Activation) 269
1. The Reagent 270
2. The Tertiary Amine 270
3. The Acyl Substituent and Side Chain of the Carboxyl Component 271
4. The Solvent 272
5. Temperature and Activation Time 273
C. Coupling with the Amine Component 274
1. Coupling Conditions 274
2. Isolation of Products 276
D. Racemization 276
1. Nature of the N^{α}-Acyl Substituent 277
2. Nature and Amount of Tertiary Base 280
3. Time of Activation 281
4. Influence of Solvents 281
5. Length of Coupling Time 282
6. Resume of the Anderson Studies and Confirmatory Data 282
E. Side Reactions 283
1. Urethane Formation 285
2. Disproportionation 288
3. Diacylimide Formation 288

THE PEPTIDES, VOL. 1

ISBN 0-12-304201-1

IV. Coupling Reagents: EEDQ and IIDQ 289
V. Applications 293
A. Incremental (Stepwise) Chain Elongation 293
B. The Repetitive Excess Mixed Anhydride Procedure (REMA) 295
1. The REMA Concept 295
2. Experimental Aspects 296
3. Scope and Limitations 299
C. Peptide Segment Condensation 299
D. Special Applications 301
VI. Related Methods 304
A. Symmetrical Anhydrides of Alkyloxycarbonylamino Acids 305
B. Other Related Methods 306
VII. Opportunities and Constraints 308
A. Comparative Evaluation 308
B. Prognosis 308
References 309

I. INTRODUCTION*

The formation of peptide bonds by mixed anhydrides of N^{α}-protected amino acids or peptides with carbonic acid monoesters (compound **1**) was

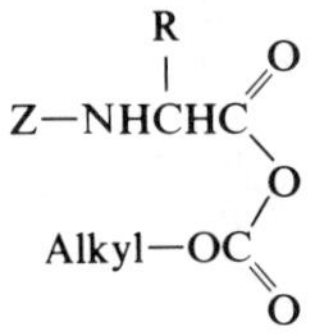

(1)

simultaneously developed in 1951 by Wieland and Bernhard (1951), Boissonnas (1951), and Vaughan (1951). The method became at once very popular and has been widely used since then.

In this chapter the term "mixed anhydride method" (MA) refers to this particular type of anhydride (**1**) unless otherwise specified. Many mixed anhydrides with different second component acids have been investigated for their use in peptide synthesis, as reviewed by Jones in Chapter 2, this

* The new peptide size nomenclature (Bodanszky, 1977) is used in which Greek prefixes are replaced by arabic numerals in peptides with more than 10 amino acid residues (see Chapter 4, p. 198, footnote). The terms *segment* and *segment condensation* (Pettit, 1976) replace the earlier *fragment* and *fragment condensation*.

volume. Historical perspectives have been discussed by Greenstein and Winitz (1961), Bodanszky and Ondetti (1966), and Bodanszky *et al.* (1976). Specifics on the chemistry of mixed carbonic anhydrides and their application to peptide synthesis are compiled in the reviews by Albertson (1962), Schröder and Lübke (1965), Tarbell (1969), Wünsch and Stelzel (1974), and Finn and Hofmann (1976).

Several features of the mixed anhydride method make it very attractive for peptide synthesis, in particular the high reaction rates at low temperatures, the ease of manipulation, the comparatively high purity of the products obtained, the generally satisfactory yields and the ready availability of reagents (Vaughan and Osato, 1952). Compared to other efficient methods of peptide synthesis, i.e., the azide, carbodiimide, or active ester methods, the mixed anhydride procedure is also by far the least expensive process of peptide synthesis, which becomes an important consideration for economic peptide production (see Feurer, 1977).

The major shortcoming has been the high tendency of mixed anhydrides to undergo racemization, due to the strong activation of the carboxyl carbonyl. As a consequence, mixed anhydride activation became largely confined to the safe use of single amino acids with N^{α}-urethane protective groups (i.e., benzyloxycarbonyl or *tert*-butyloxycarbonyl groups) which were known to suppress racemization sufficiently for practical purposes. For the same reason, the application of the method to peptide segment condensation became restricted to acyl peptide carboxyl components with COOH-terminal glycine or proline residues (see, for example, Zahn *et al.*, 1965a; Meienhofer and Schnabel, 1965).

It was only when Anderson and collaborators recognized the manner in which *steric factors and basicity of the tertiary base* influence the balance between 5(4H)-oxazolone* formation of the mixed anhydride and attack by the nucleophile, that precise conditions were defined for the coupling of acyl peptides with minimum risk of racemization (Anderson *et al.*, 1967b). At the same time, Wieland and collaborators (1969b, 1971a) elaborated similar well-defined conditions for mixed anhydride coupling, segment condensation, and peptide cyclization with optimum preservation of chiral integrity. Although the scope of the mixed anhydride method was thus enormously increased, it has not as yet regained its original popularity.† However, the recent success in rapid syntheses of gonadoliberin, substance P and secretin by the repetitive excess mixed anhydride or "REMA" procedure (see Beyer-

* See Chapter 1, p. 39, footnote.

† This is largely due to the success of dicyclohexylcarbodiimide coupling in the presence of racemization suppressing additives such as 1-hydroxybenzotriazole (König and Geiger, 1970), which has become the preferred procedure for peptide segment condensation in contemporary peptide chemistry.

man *et al.*, 1976) bodes well for the continued importance of the mixed anhydride method in peptide synthesis.

II. THE EXPERIMENTAL PROCEDURE

Peptide bond formation by the mixed anhydride procedure involves two separate stages: (1) "activation" of the carboxyl component, i.e., generation of a mixed anhydride of a N^{α}-protected amino acid or peptide, and (2) "coupling" with the amine component.

The activation step (Scheme 1) is carried out by the addition of carbonic acid monoalkyl ester chloride (alkyl chlorocarbonate, **3**) to a cold solution of the carboxyl component anion (**2**) in anhydrous organic solvent in the

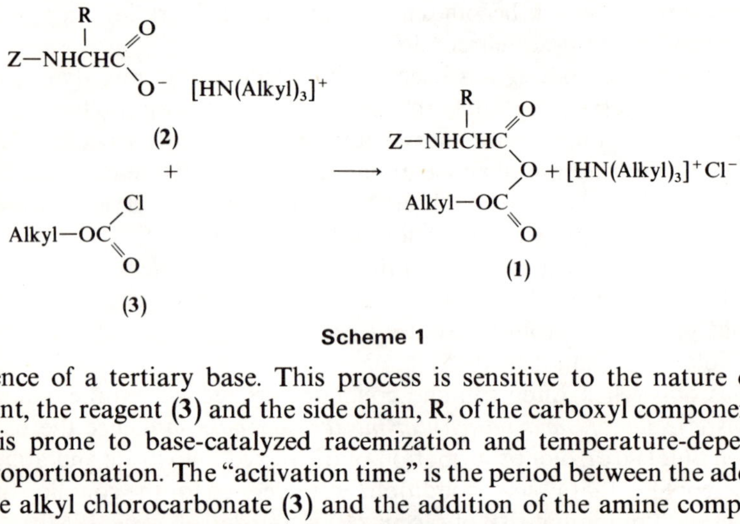

Scheme 1

presence of a tertiary base. This process is sensitive to the nature of the solvent, the reagent (**3**) and the side chain, R, of the carboxyl component (**2**), and is prone to base-catalyzed racemization and temperature-dependent disproportionation. The "activation time" is the period between the addition of the alkyl chlorocarbonate (**3**) and the addition of the amine component (**4**), once the mixed anhydride formation is complete.

Peptide bond formation (coupling, Scheme 2) usually proceeds rapidly with liberation of carbon dioxide gas and a volatile alcohol (**6**) as by-

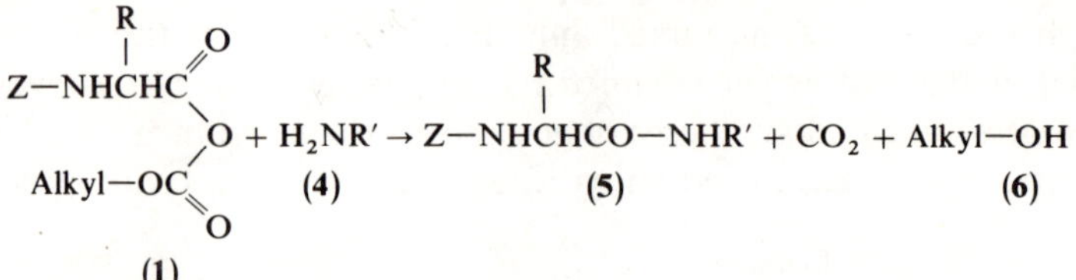

Scheme 2

products. This process is not very sensitive to the nature of the solvent, but it is prone to a few side reactions, of which only the formation of a second acylation product, by attack of the amine component at the carbonic acid carbonyl, is of practical concern.

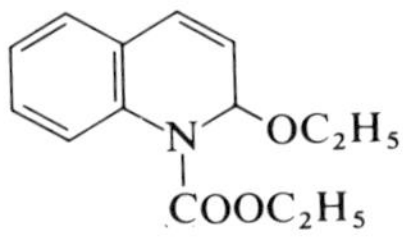

(7)

The coupling reagent, *N*-ethyloxycarbonyl-2-ethyloxy-1,2-dihydroquinoline (EEDQ, **7**) introduced by Belleau and Malek (1968), has been proposed to produce ethyl carbonic acid mixed anhydrides. Peptide bond formation is carried out simply by the addition of EEDQ (1 equiv) to an equimolar mixture of the carboxyl and amine components.

III. REACTION CONDITIONS

A. Objectives and Requirements

Principal objectives in peptide synthesis by the mixed anhydride method (Scheme 3) are (1) to generate the mixed anhydride (**8**) with the carboxyl

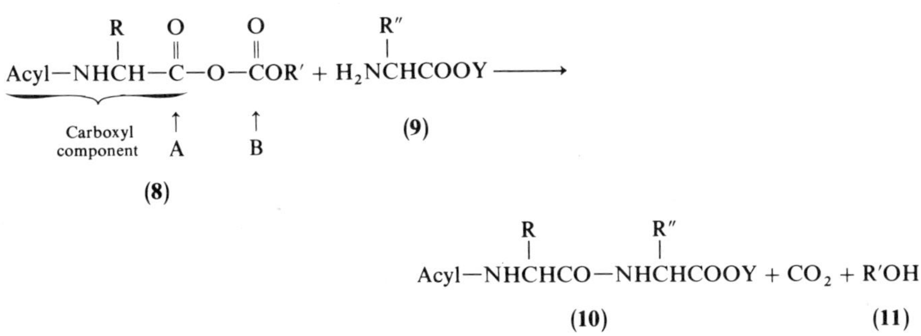

Scheme 3

component both *rapidly* and *quantitatively*, and (2) to achieve a complete or *close to complete reaction* of the anhydride with the amine component (**9**), thus to provide the desired peptide (**10**) in homogeneous form. The procedure is superbly suited to accomplish both aims, because mixed carbonic anhydrides can be prepared at low temperature under practically neutral conditions, they are fairly stable compounds which may be obtained analytically pure (Tarbell, 1969), and peptide bond formation usually proceeds rapidly with liberation of a gas (CO_2) and volatile alcohol (**11**) as the only by-products.

Yet, the mixed carbonic anhydrides (**8**) belong to the most highly activated acyl derivatives in common use in peptide synthesis and, as a consequence, they are prone to undergo a variety of rearrangements and decomposition reactions. To prevent or minimize these undesired complications, it is essential to conduct mixed anhydride synthesis under the carefully defined and controlled conditions described in the literature (Anderson *et al.*, 1967b; Wieland *et al.*, 1968a,b; 1969b; 1971a,c; 1974; van Zon and Beyerman, 1973).

The most important factors influencing *formation of mixed* anhydrides (activation stage) are (1) nature of reagent, (2) nature of tertiary amine, (3) acyl substituent and side chain (R in compound **8**) of the carboxyl component, (4) solvent, and (5) temperature and activation time. Steric bulk plays a pivotal role in factors 1–3 and affects yield (extent of anhydride formation), as does the solvent. Low temperature and short activation time suppress side reactions and racemization. The essentials of mixed anhydride generation will be discussed in Section III,B.

Important parameters that affect the stability of the mixed anhydride intermediate and its uniform *reaction with the amine component* include (1) temperature, (2) concentration of mixed anhydride (excess), (3) amount and nature of tertiary amine, if present, (4) steric bulk of the NH_2-terminus of amine component, (5) size and concentration of amine component, and (6) absence of unreacted alkyl chlorocarbonate reagent.

The requirements for optimum coupling of anhydride with amine component will be further discussed in Section III,C.

The high tendency of mixed anhydrides to racemize has been the most serious drawback for the mixed anhydride method. Factors influencing racemization include (1) nature and amount of tertiary base, (2) activation time, (3) solvent, (4) acyl substituent, and indirectly (5) length of coupling time.

The most important contributions to the understanding of the factors which influence *racemization* have been the studies by George W. Anderson and collaborators (Anderson *et al.*, 1965, 1966, 1967b; Anderson, 1970),* and those by Wieland and collaborators (1968a,b, 1969b, 1971a,c, 1974). The precise experimental conditions, determined by Anderson, for maximum suppression of racemization in mixed anhydride coupling of N^{α}-acyl peptides will be discussed in detail in Section III,D.

Potential side reactions of mixed anhydrides include formation of 5(4H)-oxazolone, urethane, diacyl imide, or ester, and disproportionation. Reaction conditions which are liable to raise the danger of potential *side reactions*

* The paper by Anderson *et al.* (1967b) must be considered required reading for anyone planning to use the mixed anhydride method.

to occur, are (1) steric bulk of the NH_2-terminus of amine component, (2) prolonged activation time, (3) temperature rise (inadequate cooling), and (4) incomplete mixed anhydride formation.

Fortunately, most side reactions are sufficiently reduced at low temperature and short reaction time and do not interfere with actual peptide synthesis. Only oxazolone formation, leading to racemization, and urethane formation, resulting in irreversible blocking of amine component, occur with a frequency as to be of concern. The requirements for suppressing side reactions will be examined in Section III,E.

B. Anhydride Formation (Activation)

Anhydride preparation (Scheme 1) must be carried out in a manner that assures (a) *complete reaction*, because unreacted alkyl chlorocarbonate (**3**) would attack the amine component with irreversible urethane formation, and (b) *fast reaction*, to suppress decomposition, disproportionation, oxazolone formation and racemization, etc., before its coupling to the amine component.

Activation is influenced by the nature of (1) the reagent, (2) the tertiary amine, (3) the *N*-acyl and side chain group of the carboxyl component, (4) the solvent, and (5) the temperature and reaction time (for reviews, see Albertson, 1962; Tarbell, 1969). A careful systematic reinvestigation of all factors, with emphasis on full suppression of racemization in the model reaction of Z-Gly-Phe-OH with H-Gly-OEt, led Anderson *et al.* (1967b) to recommend strict observation of the reaction conditions ("Anderson conditions") specified in the following illustrative procedure.

Illustrative procedure: Preparation of N^{α}-benzyloxycarbonylglycyl-L-phenylalanyl-glycine ethyl ester, Z-Gly-Phe-Gly-OEt (modified from Anderson *et al.*, 1967b). A solution of 1.78 g (5 mmol) of benzyloxycarbonylglycyl-L-phenylalanine in 25 ml of tetrahydrofuran (distilled from calcium hydride and stored over calcium hydride) was stirred and chilled to −15°C* using a dry ice–acetone bath. *N*-Methylmorpholine (0.56 ml, 5 mmol) was added, followed by isobutyl chlorocarbonate (0.67 ml, 5 mmol) producing a precipitate. After 1 to 2 min a solution of ethyl glycinate hydrochloride (0.7 g, 5 mmol) and *N*-methylmorpholine (0.56 ml, 5 mmol) in 10 ml of dry dimethylformamide (made by dissolving the hydrochloride by warming, cooling to room temperature in a water bath, adding the *N*-methylmorpholine and swirling) was added, using wash dimethylformamide. Stirring was continued

* A temperature of −15°C can be conveniently maintained by a cold bath mixture of ice and ethanol or ice and salt.

for 15 to 30 min at $-15°C$, then the cold bath was removed, and the mixture was allowed to warm to room temperature. After 60 to 90 min the solvent was removed under vacuum, and the residue dissolved in ethyl acetate (75 ml). Washing with 5% $NaHCO_3$ (25 ml), water (25 ml), 1 *N* hydrochloric acid (25 ml), and finally with water (25 ml), drying over anhydrous Na_2SO_4, and evaporation yielded 2.12 g of crude Z-Gly-Phe-Gly-OEt (96% yield), mp 118°–120°C.

1. The Reagent

Isobutyl chlorocarbonate (compound **12**) is the best reagent over-all (Anderson *et al.*, 1967b; Vaughan and Osato, 1952). It provides for the highest yields. The branched alkyl group increases the electron density at the adjacent carbonyl in the mixed anhydride (B in compound **8**) and reduces attack by the nucleophile (amine compound). Consequently, the carbonyl of the amino acid (A in compound **8**) becomes the predominant point of attack. Isobutyl chlorocarbonate is stable on storage in contrast to chlorocarbonates derived from secondary alcohols. The use of ethyl chlorocarbonate (compound **13**, Wieland and Bernhard, 1951; Boissonnas, 1951), frequently

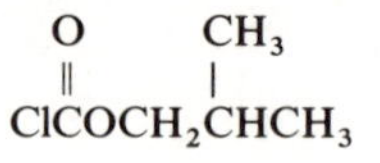

(12)

(13)

employed in early work (see Albertson, 1962, Table 2), has declined in recent years.

2. The Tertiary Amine

N-Methylmorpholine (compound **14**) is the preferred tertiary amine (Anderson *et al.*, 1967b). As a weak base its effect on racemization is minimal (see Section III,D); yet it is not sterically inhibited and sufficiently

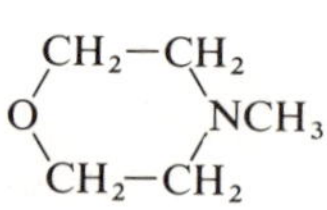

(14)

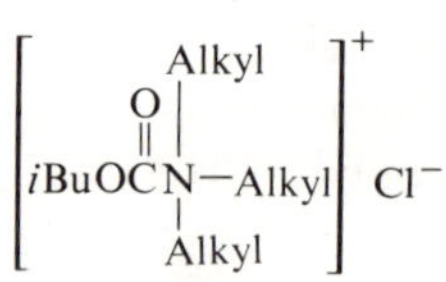

(15)

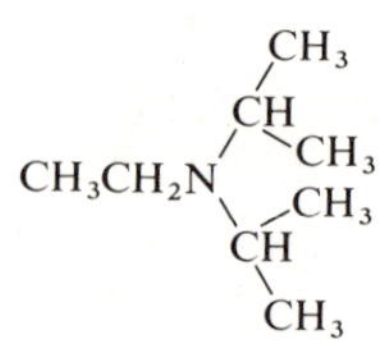

(16)

strong for rapid anhydride formation. Moreover, *N*-methylmorpholine and its hydrochloride are very water-soluble.

Evaluation of the effects of many bases on anhydride formation led

Anderson *et al.* to the conclusion that the tertiary amine is not merely a hydrogen chloride acceptor, but that it reacts with the chlorocarbonate to form an alkyloxycarbonylammonium ion complex (compound **15**) which subsequently reacts with the carboxylate. Support for this mechanism was provided by the low yields obtained with the sterically hindered ethyl diisopropylamine (compound **16**) and the high yields obtained with the sterically least hindered triethylamine (see Table II, Section III,D). It became obvious from the result of these studies that the commonly used triethylamine is not the most suitable base, and that the tertiary amine used for mixed anhydride formation should contain *at least one methyl group* (Anderson *et al.*, 1967b).

3. The Acyl Substituent and Side Chain of the Carboxyl Component

Compatibility of the carboxyl component's acyl substituent with mixed anhydride formation is discussed here, while aspects of racemization will be dealt with in Section III,D. Compatible N^{α}-acyl substituents include the entire family of urethane protective groups, e.g., benzyloxycarbonyl, *tert*-butyloxycarbonyl, 9-fluorenylmethyloxycarbonyl, furthermore the phthaloyl group (Boissonnas, 1951; King *et al.*, 1957), the formyl group (Heard and Young, 1963), and the *o*-nitrophenylsulfenyl and related groups (Zervas *et al.*, 1963; Stewart, 1966). The N^{α}-trityl group in peptides is compatible, but mixed anhydrides of N^{α}-tritylamino acids fail to react due to steric hindrance, except for Trt-Gly-OH and Trt-Ala-OH (Hillmann-Elies *et al.*, 1953; Amiard *et al.*, 1955; Zervas and Theodoropoulos, 1956).

Incompatible are the N^{α}-tosyl and trifluoroacetyl groups. N^{α}-Tosylamino acids fail to yield the expected products (Boissonnas, 1951; Vaughan, 1951; Berse *et al.*, 1961; see also Beecham, 1957). Zaoral and Rudinger (1961) isolated and characterized many of the unexpected side products obtained from tosylglycine. Only mixed anhydrides of N^{α}-tosylamino acids with pivaloyl chloride in pyridine gave satisfactory results (see also Zaoral, 1959, 1962) (see Section VI,B). Trifluoroacetylamino acids are converted (Scheme 4) to 2-trifluoromethylpseudooxazolone-(5) derivatives (structure **17**, Weygand *et al.*, 1963a,b, 1964).

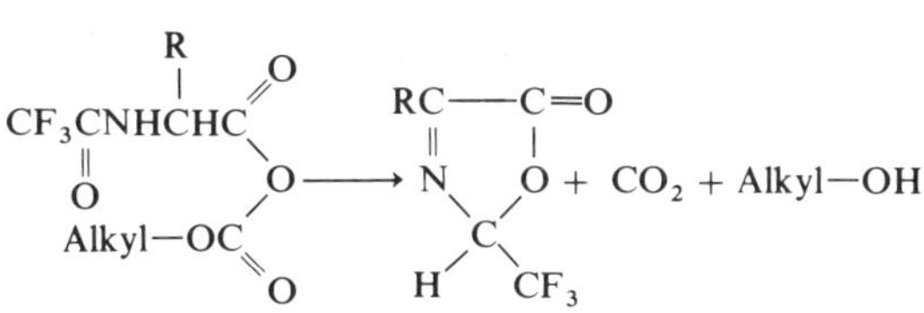

(17)

Scheme 4

The influence of the amino acid side chain of the carboxyl component (R in compound **8**) has not been systematically studied. However, steric bulk, as in isoleucine, leucine, or valine may slow down anhydride formation (Anderson, 1970). Functional groups in the side chain should preferably be protected to avoid undesired side reactions. Most carboxyl, hydroxyl, or amine protective groups, that are commonly used in peptide synthesis, may be employed. For larger acyl peptide components with several different side chain functionalities maximum protection should be applied.

4. The Solvent

Solvents have been reported to affect reactions of some simple unsymmetrical anhydrides (Albertson, 1962) but their influence in peptide synthesis has not been studied very thoroughly or systematically.

Ethyl acetate, tetrahydrofuran (Anderson *et al.*, 1967b), toluene (Vaughan, 1952), *tert*-butanol (Wieland *et al.*, 1962) and acetonitrile (Stewart, 1965) are recommended as most suitable solvents for mixed anhydride formation in providing optimum overall yields. (Solvent effects on racemization will be discussed in Section III,D.) Also considered as good (better than 90% yields) are dimethyloxymethane and -ethane (Anderson *et al.*, 1967b). There seems to be agreement that the use of chloroform and methylene chloride results in considerably lower yields and impure products (Vaughan, 1952; Stewart, 1965; Anderson *et al.*, 1967b).

When limited solubility of the carboxyl component requires the use of more powerful polar solvents, the preferred amide solvent is dimethylacetamide (DMAc). Yields were reported to be lower with dimethylformamide (DMF) or hexamethylphosphoramide (HMP) (Anderson *et al.*, 1967b). Albertson (1962) pointed out that dimethylformamide reacts with ethyl or isobutyl chlorocarbonate in a (slow) exothermic reaction, eventually consuming most of the reagent. However, excellent results have been obtained with dimethylformamide as a solvent for mixed anhydride formation in the synthesis of several peptides by the REMA method (van Zon and Beyerman, 1973; Beyerman *et al.*, 1976; Izeboud and Beyerman, 1978). Since it is now known to be difficult to obtain pure dimethylformamide or hexamethylphosphoramide, the unfavorable earlier results were perhaps due to solvent impurities. The solvent has no major influence on coupling, thus permitting a wide choice for dissolving the amine component (see Section III,C).

Strictly anhydrous solvents of high purity are essential for successful mixed anhydride formation. For efficient purification and drying procedures consult Perrin *et al.* (1966) or Vaughn (1967). Commercial grades of the amide solvents (DMAc, DMF, HMP) contain small amounts of dimethylamine which will react with the mixed anhydride. For example, the use of

100 ml of dimethylformamide which contains 0.1% dimethylamine would result in the formation of as much as 2 mmol of undesired dimethylamide of the carboxyl component. Distillation of such solvents from ninhydrin efficiently removes contaminations of primary and secondary amines.

Concentration effects have not been investigated systematically. With small carboxyl components (*N*-protected amino acids, di-, and tripeptides) 0.1 *M* to 0.4 *M* solutions are commonly used. Employing less solvent than 25 ml per 100 mmol of carboxyl component is not advisable because precipitating hydrochloride of the tertiary base may interfere with efficient stirring which, in turn, is essential for effective heat dissipation.

For rapid reaction with the amine component the concentration should obviously be kept as high as possible. Wieland *et al.* (1962) reported a 50% yield of Z-Gly-Gly-SPh obtained from the mixed anhydride of Z-Gly-Gly-OH and thiophenol in 0.05 *M* solution and a yield of 80% in ca. 0.25 *M* solution. The use of excess anhydride (Weygand *et al.*, 1967; Weygand and Di Bello, 1969; Tilak, 1970; Beyerman, 1972) assures rapid and high-yield coupling (see Section V,B for details).

5. Temperature and Activation Time

An activation time of 1–2 min at −15°C is recommended (Anderson, 1970). A few minutes longer time of anhydride formation may be required for carboxyl components which have bulky side chains such as isoleucine, leucine, or valine (Anderson, 1970).

Temperatures far below −20°C are not recommended (Greenstein and Winitz, 1961) because of marked decrease in overall yields. On the other hand, the temperature must not normally exceed 0°C since decomposition of mixed anhydrides via disproportionation becomes a serious danger (Wieland and Bernhard, 1951; see also Longosz and Tarbell, 1961).

Mixed anhydride synthesis is highly exothermic, both during activation and coupling. Adequate heat dissipation is therefore one of the chief experimental problems in larger scale preparations of > 100 mmol. The temperature of the reaction mixture may rise considerably above 0°C if the isobutyl chlorocarbonate reagent is added too rapidly. Temperature control inside the reaction vessel is essential. Reagent addition may require up to 20 min in order to prevent the temperature from rising above −5° to 0°C. Anhydrous conditions must be strictly maintained. Therefore the use of a jacketed reaction vessel fitted with overhead stirrer and inside thermometer, as shown in Fig. 1, is recommended for preparations at a 0.1 mole or larger scale. Two jacketed dropping funnels (one shown) are used to add precooled reagent and, subsequently, precooled amine component. The cooling mantles are connected to a circulating cryostat set at −20° to −30°C, as needed.

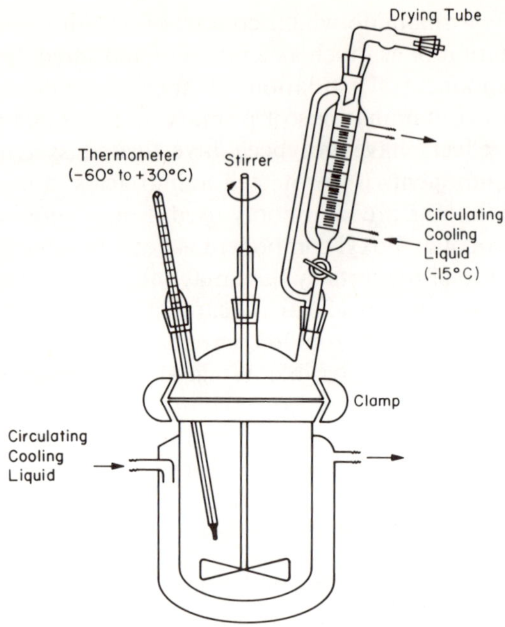

Figure 1. Reaction vessel for large-scale laboratory mixed anhydride synthesis (> 30 mmol scale) fitted with cooling jacket, thermometer, and jacketed dropping funnels (one shown) for the addition of precooled reagent and amine component.

C. Coupling with the Amine Component

1. Coupling Conditions

Mixed carbonic anhydrides are usually not isolated but, once formed, are immediately reacted *in situ* with the amine component. The coupling reaction may be conducted in aqueous medium or in organic solvent systems.

Amino acid or peptide derivatives with free carboxyls may be dissolved in aqueous or mixed aqueous–organic solvent systems and coupled as alkali salts under Schotten–Baumann conditions. However the yields are frequently inferior to those obtained with esters in organic solvent probably due to competing hydrolysis of the anhydride (Vaughan and Osato, 1952). The use of organic bases or quarternary tetraalkylammonium salts (Triton B) in organic or aqueous–organic solvent systems seems to be advisable. The preferred tertiary base is *N*-methylmorpholine (Anderson *et al.*, 1967b).

Amine components with protected carboxyls (ester or amide) are dissolved in suitable organic solvents comprising the entire range commonly used in peptide chemistry, except primary alcohols or phenol. For poorly soluble amine components, e.g., larger peptide derivatives, dimethylformamide, hexamethylphosphoramide or dimethyl sulfoxide may be safely used, since there is no major solvent influence on the coupling reaction (Anderson *et al.*, 1967b). Small amounts of water may be added if it should facilitate solubilization, however, better yields of purer products are obtained with solutions of the amine component in dry solvents without any competing hydrolysis.

If base is required to neutralize hydrogen halides of amine components, 1 equiv of *N*-methylmorpholine should be used and care be taken to avoid any excess. Since tertiary base is known to promote anhydride racemization (see Fig. 2 and Section III,D for details) it must not be added to the mixed anhydride solution prior to the addition of the amine component. Instead, it should be combined with the cooled solution of the amine component very shortly before the mixtures is to be added to the anhydride. If necessary, amino acid or peptide ester hydrohalides may be added to the mixed anhydride followed by slow addition of the tertiary amine (Stewart, 1966); this should not be done with amine components that contain free α-carboxyls. It is important to assure complete neutralization of ester hydrochlorides (see Anderson *et al.*, 1967b) especially when working with sparingly soluble compounds, such as glycine methyl or ethyl ester hydrochloride. Full neutralization is best accomplished by complete solubilization in powerful solvents such as *tert*-butanol (Wieland *et al.*, 1962), dimethylformamide or hexamethylphosphoramide.

The temperature is usually maintained at the same level at which the anhydride was prepared (−10° to −15°C). A thermometer inside of the reaction flask and slow addition of the amine component over a period of several minutes are required with amounts exceeding 0.02 to 0.03 moles since the temperature may rise above 0°C if the amine component is added at once. Isobutanol and carbon dioxide are by-products of the reaction and provision has to be made to let the gas escape (see drying tube in Fig. 1).

Low temperatures are generally maintained for 30–60 min after which the reaction mixture is allowed to attain room temperature for a few hours before workup. Coupling of amino acids or small peptides is frequently complete within less than 15 min. However, if the molar concentration of the components should decrease significantly when larger peptide segments are to be condensed, the low-temperature coupling period may have to be extended to many hours.

The completion of the reaction should always be monitored, e.g., by ninhydrin, fluorescamine (Udenfriend *et al.*, 1972; Weigele *et al.*, 1973), or *o*-phthalaldehyde (Benson and Hare, 1975), preferably in combination with thin layer chromatography or high-performance liquid chromatography (Gabriel *et al.*, 1976, 1977).

2. *Isolation of Products*

Workup of products prepared by mixed carbonic anhydride synthesis is very convenient, since the by-products, carbon dioxide and isobutanol, are volatile. If the solvent is tetrahydrofuran, dimethylformamide or another water-miscible solvent, the product is simply precipitated by dilution with large volumes of water or concentrated sodium chloride solution, acidification being necessary when amine components with unprotected carboxyl groups were used. The ensuing crude precipitates are collected, washed with dilute acid and water, and, in addition, with 1 *N* sodium bicarbonate when the carboxyl groups are protected. Crude oils are dissolved in ethyl acetate and washed, dried over $MgSO_4$ or Na_2SO_4 and concentrated *in vacuo*. This procedure is also employed if the reaction solvent is water-immiscible.

Crystallization or other purification steps follow workup. The purity of the crude products is often so high that crystallization occurs during workup, e.g., when adding water to a reaction mixture in tetrahydrofuran.

Yields are generally good ranging from 40% to 100% for syntheses of smaller peptides by stepwise chain elongation using benzyloxycarbonyl- or *tert*-butyloxycarbonylamino acids. Yields in condensations of larger segments range between 30% and 70%.

D. Racemization

The mixed carbonic anhydrides belong to the most highly activated acylating species in common use in peptide synthesis. Strong electron withdrawal from the carbonyl carbon creates a high risk of racemization at the chirality center of the carboxyl component's COOH-terminal amino acid.* Racemization of mixed anhydrides may proceed either via 5(4H)-oxazolone (**18**) formation† (Scheme 5) or may involve α-hydrogen abstraction. Possibly both processes may operate simultaneously (see Kemp, this volume, Chapter 7).

In practice, the loss of chiral integrity can be reduced to safe levels ($< 1\%$)

* For results of early studies on mixed anhydride racemization by tests which were designed to exaggerate loss of chirality, see Vaughan (1952), Vaughan and Eichler (1953), Erlanger *et al.* (1954), Sheehan and Yang (1958), Schwarz and Bumpus (1959), Smart *et al.* (1960), Williams and Young (1963), Heard and Young (1963), Determann and Wieland (1963), Applewhite and Nelson (1964), Antonovics *et al.* (1966), and Rzeszotarska *et al.* (1966).

† For studies on the 5(4H)-oxazolone mechanism of racemization in peptide synthesis, see Goodman and Stueben (1962), Goodman and Levine (1964), Williams and Young (1964), Antonovics and Young (1965), 1967), Goodman and McGahren (1965), Antonovics *et al.* (1966), Anderson *et al.* (1967b), Kemp and Chien (1967), Steglich *et al.* (1967), Young (1967), Goodman and Glaser (1970), Kemp and Rebek (1970), and Kemp (1973, and this volume, Chapter 7).

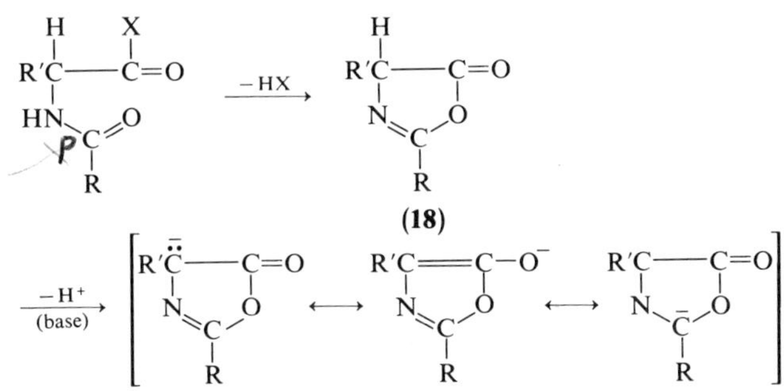

Scheme 5

by the selection of suitable N^{α}-acyl substituents. Thus the danger of racemization is smallest when N^{α}-urethane-protected amino acids are activated (Vaughan, 1952; Vaughan and Osato, 1952; Weygand *et al.*, 1963a,b; Determann, 1967) and used in sequential synthesis of peptides by incorporation of one amino acid at a time. However, considerable racemization may occur when the COOH-terminal amino acid residue of an acyl peptide is activated by mixed anhydride formation. Vaughan's (1952) investigations showed influence of solvent, nature of reagent, and activation time and temperature on the extent of racemization. Typical results of some of the early racemization studies are compiled in Table I.

It was only after Anderson *et al.* (1967b; Anderson, 1970) and Determann and Wieland (1963) determined reaction conditions for the suppression of racemization in acyl peptide activation that these mixed anhydrides could be used with more advantage. Anderson and collaborators elucidated the effects on racemization of (1) the nature of the N^{α}-acyl substituent, (2) the nature and amount of tertiary amine, (3) the activation time, (4) the solvent, and (5) the coupling time. In this work they used the "Anderson test" (Anderson and Callahan, 1958), Scheme 6, i.e., coupling of Z-Gly-Phe-OH (compound **19**) with H-Gly-OEt (compound **20**) to give Z-Gly-Phe-Gly-OEt (compound **21**), and a "Young test" [Bz-Leu-OH + H-Gly-OEt

$$\underset{(\mathbf{19})}{\text{Z—Gly—Phe—OH}} \xrightarrow[i\text{BuOCOCl}]{\text{activation}} \underset{\text{Mixed Anhydride}}{\text{Z—Gly—Phe—OCO}i\text{Bu}} \xrightarrow[\text{H—Gly—OEt }(\mathbf{20})]{\text{coupling}} \underset{(\mathbf{21})}{\text{Z—Gly—Phe—Gly—OEt}}$$

Scheme 6

(Williams and Young, 1963)] that was modified (Anderson *et al.*, 1966) to detect as little as 0.1% racemate.

1. Nature of the N^{α}-Acyl Substituent

As mentioned, N^{α}-benzyloxycarbonyl- and *tert*-butyloxycarbonyl- and related amino acids can be activated without detectable racemization

Table I Racemization Studies on Acyl Peptide Coupling by the Carbonic Mixed Anhydride Method, prior to the Study of Anderson *et al.* (1967b)[a]

No.	Test peptide[b]	Activation conditions	Variable	Yield (%)	Racemization (%)	References
1	Z-Gly-Leu-+-Gly-OMe	*i*BuOCOCl, toluene, Et_3N, 0.1 *M*, 5–10 min, −5°C		60	10.5	Vaughan (1952)
2	Z-Gly-Phe-+-Gly-OEt	EtOCOCl, Et_3N, 5 min, −5°C	Solvent			Vaughan (1952)
		(a) THF		60	<1	
		(b) Toluene-$CHCl_3$ (8:1)		65	30	
		(c) Toluene-$CHCl_3$ (2:1)		82	56	
		(d) $CHCl_3$		31	>90	
3	Z-Lys(Z)-Val-+-Phe-Gly-OEt	*i*BuOCOCl, THF, Et_3N, 5 min, −5°C		54	"Very little"	Vaughan and Eichler (1953)
4	Z-Gly-Phe-+-Gly-OEt[c]	Time course of activation (1–60 min)	Workers			
		(a) EtOCOCl, THF, Et_3N, 0.2 *M*, −15°C		—	<5–67	Determann and Wieland (1963)
		(b) Same as (a)		—	<1–56	Anderson *et al.* (1965)
5	Z-Gly-Phe-+-Gly-OEt	"Inverse Addition,"[d] 0.4 *M*, 10 min, −5°C	Molar ratios			Applewhite and Nelson (1964)
		(a) equiv of Acid + Et_3N, EtOCOCl, H-Gly-OEt		84	11	
		(b) 1 equiv Acid + Et_3N, 2 equiv EtOCOCl, 3 equiv H-Gly-OEt		90	<1	
6	Z-Leu-Phe-+-Val-O*t*Bu	EtOCOCl, THF, Et_3N, 10 min, −10°C			9.8	Weygand *et al.* (1966a); see also Prox *et al.* (1966)

[a] Abbreviations: EtOCOCl, ethyl chlorocarbonate; Et_3N, triethylamine; *i*BuOCOCl, isobutyl chlorocarbonate; THF, tetrahydrofuran.
[b] Site of peptide bond formation is indicated by dash with vertical bar.
[c] Z-Gly-Phe-OH + Et_3N + EtOCOCl incubated for periods of 1 min to 60 min prior to addition of H-Gly-OEt.
[d] A solution of Z-Gly-Phe-OH (Acid) and Et_3N in THF is slowly added to a vigorously stirred, cooled solution (−5°C) of EtOCOCl.

(Vaughan, 1952; North and Young, 1955; Weygand *et al.*, 1963a,b). Interestingly, the protective effect of the N^{α}-*tert*-butyloxycarbonyl group against racemization is even ca. fourfold stronger than that of the benzyloxycarbonyl group (Kovacs *et al.*, 1975; Kovacs, 1979). With mixed anhydrides of *o*-nitrophenylsulfenyl amino acids (Zervas *et al.*, 1963; Stewart, 1966) racemization remained below the 1% limit of detection. Phthaloylamino acids may be used similarly safely (Boissonnas, 1951). However, trifluoroacetyl- and formylamino acids (Heard and Young, 1963; Antonovics *et al.*, 1966; Rzeszotarska *et al.*, 1966) are prone to extensive racemization.

When *N*-acyl peptides, such as Z-Gly-Phe-OH, are activated in the presence of triethylamine, racemization begins immediately (Anderson *et al.*, 1965) and follows first-order kinetics for the initial 30 min or so of activation time (Fig. 2). Since the overall yield of L + DL (of Z-Gly-Phe-Gly-OEt) at 2 min was as good (78.5%) as with longer activation times, it became obvious that the extent of racemization in peptide coupling can be minimized (<1%) by keeping the activation time as short as possible. This confirmed an earlier report by Vaughan (1952), see Section III,D,3.

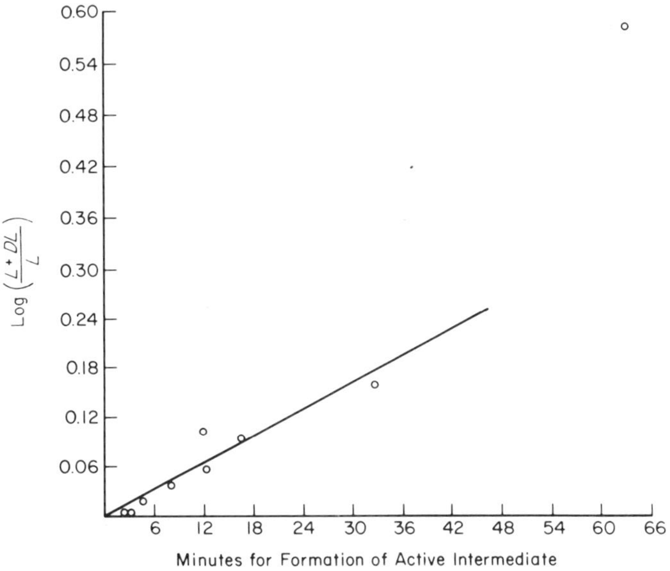

Figure 2. First order plot of the progress of racemization of the mixed anhydride Z-Gly-Phe-OCOOEt (0.2 *M* in anhydrous tetrahydrofuran at −15°C), during a synthesis of Z-Gly-Phe-Gly-OEt with the use of triethylamine. (From Anderson *et al.*, 1965, reprinted with permission of the Hungarian Academy of Sciences.)

2. Nature and Amount of Tertiary Base

The most critical factor in the mixed anhydride procedure is the nature and amount of the tertiary amine because it affects two separate processes, i.e., mixed anhydride generation and racemization (Anderson *et al.*, 1966, 1967b; Anderson 1970, 1972). The initial quarternary ion complex formation between the tertiary amine and isobutyl chlorocarbonate is so sensitive to steric factors that the strongly hindered ethyl diisopropylamine yielded only 3.2% of tripeptide **21** (containing 6% racemate). Unreacted amine, in turn, will promote racemization of the generated mixed anhydride (see Fig. 2), at a rate which depends on the basicity of the tertiary amine. Results obtained with different amines in mixed anhydride coupling of Z-Gly-Phe-OH **(19)** with H-Gly-OEt **(20)** are listed in Table II. Lack of detectable racemization with the use of one equivalent of trimethylamine indicated that the base was fully utilized in forming the mixed anhydride. On the other hand, complete racemization with the use of 2 equiv showed that trimethylamine is also a good racemizer of the formed mixed anhydride. Even with only a few percent excess over 1 equiv complete racemization occurred.

Anderson *et al.* (1966, 1967b) then discovered that racemization could be avoided with less basic tertiary amines. To achieve both, complete and rapid anhydride formation and suppression of racemization, amines with at least one methyl group turned out to be best suited. Optimum results were obtained with *N*-methylmorpholine which provided high yields of product

Table II Influence of Type and Amount of Tertiary Amine on Racemization during Z-Gly-Phe-Isobutylcarbonic Acid Mixed Anhydride Coupling with H-Gly-OEt[a,b]

Amine	Amine equiv.	Yield (%) Crude	DL	L
Triethyl	1	97	8	82
Triethyl	2	84	16	59
Methyldiethyl	1	98	—	94
Methyldiethyl	2	90	18	68
Trimethyl	1	100	—	90
Trimethyl	2	76	68	Trace
N-Methylmorpholine	1	96	—	92
N-Methylmorpholine	2	98	—	93

[a] 0.2 *M* in THF, using a 12-minute activation time at −15°C. An activation period (time for mixed anhydride formation) of 12 min was used to exaggerate racemization when it occurred. Recommended activation time for practical peptide coupling is 1–2 min ("Anderson conditions").

[b] From Anderson *et al.* (1967b) and Anderson (1970).

with no detectable racemate even when present in 100% excess (see Table II).

Anderson (1970) recommends the use of exactly one equivalent of *N*-methylmorpholine and a 1 to 2 min activation time at −15°C. This should avoid racemization in peptide segment condensation by the mixed anhydride method, even under stringent conditions.

3. Time of Activation

Vaughan (1952) already observed that a shorter activation period resulted in reduced racemization. Anderson *et al.* (1967a) found that with *N*-methylmorpholine as a base anhydride formation by isobutyl chlorocarbonate in ethyl acetate or tetrahydrofuran at −15°C is complete in less that 30 sec and that racemization of the anhydride, Z-Gly-Phe-$OCOCH_2CH(CH_3)_2$ (compound **22**) via 5(4H)-oxazolone formation (see also Kemp and Rebek, 1970) begins immediately, increasing steadily with time (see Fig. 2) at rates which depend on the basicity and amount of the tertiary amine present in the reaction mixture.* Since coupling of the anhydride **22** with H-Gly-OEt seems to proceed essentially racemization-free, it is obvious that racemization can be minimized by curtailing the activation time and thereby the 5(4H)-oxazolone formation as tightly as possible.†

4. Influence of Solvents

Solvent effects on racemization (and yield, see Section III,B,4) are large in the activation period (a), but not in the coupling stage (b).

(a) Anhydride formation (12 min activation time at −15°C) in the presence of *1 equiv of N-methylmorpholine* proceeded without detectable racemization (and in excellent yield) in ethyl acetate, dioxane, tetrahydrofuran (see also Vaughan, 1952), and acetonitrile, and in good yield in *N,N*-dimethylacetamide (DMAc) and triphenyl phosphate. In dichloromethane, *N*-methylpyrrolidone, *N,N*-dimethylformamide (DMF), and hexamethylphosphoramide (HMP) 2% to 10% racemization and lower yields occurred. In the presence of *2 equiv of N-methylmorpholine*, anhydride formation proceeded without detectable racemization only in ethyl acetate, tetrahydrofuran and dioxane.

* Ideally, no residual tertiary amine should be present in unprotonated form if anhydride formation was 100% complete.

† However, premature addition of the amine component to an incomplete anhydride preparation must also be carefully avoided since unreacted alkyl chlorocarbonate would rapidly react with the amine component to form a urethane derivative (see Section III,B). In the absence of a reliable test for determining the end point of anhydride formation the decision about the optimal length of a given activation in practice is left to the judgment of the investigator.

Thus, solvent-effected danger of racemization increases in the approximate order THF, EtOAc < triphenyl phosphate, DMAc, < CH_2Cl_2 < *N*-methylpyrrolidone, DMF, HMP, when *N*-methylmorpholine is employed as a base. Similar observations were made by Vaughan (1952) and by Goodman *et al.* (1977). However, the order may change with the base used (Kemp, this volume, Chapter 7, p. 358) and it is unknown at which stage of the activation process the solvent effect occurs.

(b) Once the anhydride is formed, solvents such as dimethylformamide can be added without detrimental effects, and may, therefore, be used as solvents for the amine component. Thus, Anderson *et al.* (1967b) recommend as optimal solvent combinations the use of ethyl acetate or tetrahydrofuran for the mixed carbonic anhydride formation with the carboxyl component and a variety of powerful solvents for dissolving the amine component.

5. Length of Coupling Time

Although the actual coupling reaction may not be accompanied by racemization, its speed is important in reducing the time span for unreacted anhydride to form oxazolone (see Fig. 2). The coupling speed varies with the chain length and steric bulk of individual components but it can generally be accelerated considerably by increasing molar concentrations. In practice this may be achieved by partial solvent evaporation after the addition of the amine component to the mixed anhydride.

6. Resume of the Anderson Studies and Confirmatory Data

Steric properties and basicity of the tertiary amine are of profound influence on racemization. To effect rapid and complete mixed anhydride formation, the base must not be sterically hindered, i.e., it must contain at least one methyl group. Racemization of the mixed anhydride occurs via base-promoted rearrangement into 5(4H)-oxazolone. Thus, if the base is quantitatively consumed during mixed anhydride formation, the danger of racemization will be minimal. But racemization can also be reduced by using a weaker base. The weaker the base, the less it will promote 5(4H)-oxazolone formation; however, too weak a base would be inadequate for efficient anhydride generation. *N*-Methylmorpholine appears to possess the right combination of properties. It promotes rapid generation of anhydride which contains only minute traces of racemate (see below), even when used in 100% excess. Moreover, if care is taken to avoid the presence of any excess of *N*-methylmorpholine additional safety from racemization will be gained.

Anderson *et al.* (1967b) concluded that the mixed anhydride method should be useful in coupling acyl peptides to amino acid or peptide derivatives with minimum racemization when properly applied.

Subsequently, Kemp and collaborators confirmed the findings of Anderson *et al.* "in every detail" (Kemp *et al.*, 1970b) (see Table III, Nos. 3, 4). This was particularly important because Kemp *et al.* (1970a) had developed highly sensitive racemization tests to measure levels of racemate in the range of 0.001% to 1%, i.e., far below the levels of Anderson's tests. This was achieved by applying isotopic dilution to the Anderson and Young tests. In Kemp's test, a low level of 0.2% racemate was detected to be present in Z-Gly-Phe-Gly-OEt **(21)** when prepared in 98% yield under "Anderson conditions," i.e., with an equivalent amount of *N*-methylmorpholine. Interestingly, a further 20-fold reduction down to a very low 0.01% level of racemate was obtained "with very carefully weighed equivalents of acid (Z-Gly-Phe-OH, **19**) and *N*-methylmorpholine" (Kemp *et al.*, 1970b). This illustrates again the enormous racemizing effects of excess tertiary base, even when present only in trace amounts. The use of a small excess of acid (carboxyl component) over isobutyl chlorocarbonate and *N*-methylmorpholine might alleviate the problem.

The ability to suppress racemization in peptide coupling effectively by applying conditions of mixed anhydride formation very similar to those of Anderson was further confirmed by the results of Sakakibara and Itoh (1967), see Table III. The loss of chirality of Z-Gly-Phe-Gly-OEt was suppressed below the detection limit (~1%) of the Anderson test when the weak base *N*-ethylmorpholine was employed. Izumiya *et al.* (1971) achieved a fourfold reduction in the level of racemization when Z-Gly-Ala-OH was coupled with H-Leu-OBzl in the presence of *N*-methylmorpholine instead of triethylamine (Table III, No. 4), as determined in a sensitive chromatographic test (0.1% detection limit, Izumiya and Muraoka, 1969). A further decrease to a low 0.2% of epimer was observed in the presence of *N*-hydroxysuccinimide (Table III, No. 5b).*

E. Side Reactions

Mixed anhydrides are prone to a number of side reactions, including (1) urethane formation, (2) anhydride disproportionation, (3) diacylimide

* The reports on racemization suppressing effects of *N*-hydroxysuccinimide in mixed anhydride synthesis are conflicting. Anderson *et al.* (1967a) obtained intermediate Z—Gly—Phe—OSu which on reaction with H—Gly—OEt provided the tripeptide (**18**) without detectable racemate content by the Anderson test and with 2% racemate content by the Young test. Kemp *et al.* (1970b) found no effect on racemate levels when HOSu was added to the mixed anhydride reaction.

Table III Confirmation of Racemization Suppression under "Anderson Conditions"[a,b] of Acyl Peptide Coupling by the Mixed Carbonic Anhydride Method[c]

No.	Test peptide[d]	Activation conditions[e]	Variable	Yield (%)	Racemization (%)	References
1	Z-Gly-Phe—┼—Gly-OEt	MeMorph, 4 min		80	<0.2	Taschner *et al.* (1968)
2	Z-Gly-Phe—┼—Gly-OEt	Et_3N	Base	30	1.5	Kemp *et al.* (1970b); see also Kemp and Rebek (1970)
		MeMorph		98	0.2	
		MeMorph (exact equiv "carefully weighed")	Weighing accuracy	98	0.01	
3	Bz-Leu—┼—Gly-OEt	1 equiv Et_3N	Type and amount of base	—	1.4	
		2 equiv Et_3N		—	39	
		1 equiv MeMorph		—	0.38	
		1.1 equiv MeMorph		—	2.4	
		2 equiv MeMorph		—	15.7	
4	Z-Gly-Ala—┼—Leu-OBzl	Et_3N	Base	87	9.5	Izumiya *et al.* (1971)
		MeMorph		91	2.4	
5	Z-Gly-Phe—┼—Gly-OEt	Et_3N, $CHCl_3$, 25 min, −5°C	Base	38	69.4	Sakakibara and Itoh (1967)
		EtMorph, $CHCl_3$, 25 min, −5°C	Solvent	52	2.1	
		EtMorph, THF, 10 min, −10°C		61	<0.1	
6	Z-Gly-Ala—┼—Leu-OBzl[f]	Et_3N + HOSu, THF, 1 min, −15°C	Base	55	1.1	Izumiya *et al.* (1971)
		MeMorph + HOSu, THF, 1 min, −15°C		63	0.2	

[a] See Anderson *et al.* (1976b): isobutyl chlorocarbonate THF, activation at −15°C.

[b] Experiments 5 and 6 were conducted under very similar conditions.

[c] Abbreviations: EtMorph, *N*-ethylmorpholine; Et_3N, triethylamine; HOSu, *N*-hydroxysuccinimide; MeMorph, *N*-methylmorpholine; THF, tetrahydrofuran.

[d] Site of peptide bond formation is indicated by dash with vertical bar.

[e] One equivalent of tertiary base is used unless indicated otherwise.

[f] As experiment No. 4, but with the addition of 1 equiv of HOSu.

formation, (4) 5(4H)-oxazolone formation, and (5) ester formation with CO_2 liberation (for reviews, see Albertson, 1962; Tarbell, 1969). Of these only urethane and 5(4H)-oxazolone formation have been of concern in actual peptide synthesis under the commonly used low-temperature reaction conditions. 5(4H)-Oxazolone formation has been discussed in the preceding section on racemization. Disproportionation, diacylimide, and ester formation have seldom been observed.

1. Urethane Formation

Attack of the amine component (**4**) at the carbonyl of the alkyl carbonate part of the mixed anhydride (**1**), an intrinsic side reaction (Scheme 7), leads

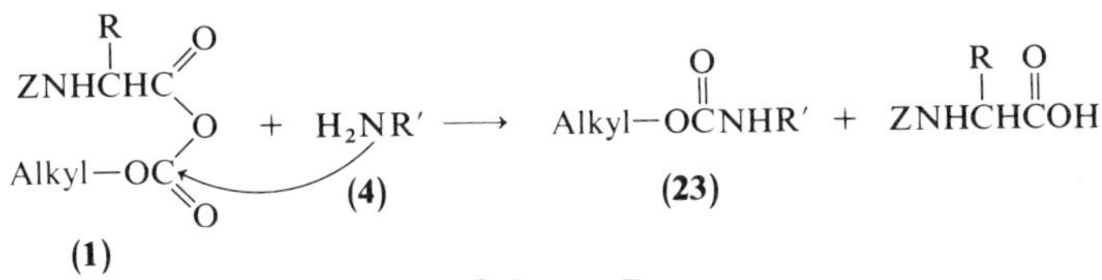

Scheme 7

to the formation of a urethane side product (**23**). For many years, experimental proof for the occurrence of this side reaction (Scheme 7) in actual peptide synthesis had been scant. Furthermore, residual chlorocarbonate (**3**) from incomplete anhydride formation will also react with amine component **4** (Scheme 8) to give identical urethane (**23**) side product (Battersby and

$$\underset{(\mathbf{3})}{\text{Alkyl—O}\overset{\text{O}}{\overset{\|}{\text{C}}}\text{Cl}} + \underset{(\mathbf{4})}{H_2NR'} \rightarrow \underset{(\mathbf{23})}{\text{Alkyl—O}\overset{\text{O}}{\overset{\|}{\text{C}}}\text{NHR}'}$$

Scheme 8

Robinson, 1955; Albertson, 1962). This reaction (Scheme 8) might account for indirectly observed side reactions (Vaughan and Eichler, 1953; Gillessen *et al.*, 1963) and for the isolated side product (**24**) (Ondetti and Thomas,

$$(CH_3)_2CHCH_2OCO\text{—MePhe—Pro—O}t\text{Bu}$$

(24)

1965). However, Yajima *et al.* (1969b) showed that the isolated urethane side product (compound **25**; 61% yield) was derived from the mixed anhydride

$$CH_3CH_2OCO\text{—Pro—Lys(For)—Asp—OH}$$

(25)

(Scheme 7), since quantitative formation of the anhydride was ascertained in a separate control reaction.

The urethane derivatives, **24** and **25** arose from attempts to acylate amine components at NH_2-terminal proline and *N*-methylphenylalanine residues. Difficulties of this kind and a tendency of secondary amines to produce these side products were previously reported (Rydon and Smith, 1956; Albertson, 1962). Subsequently, Beyerman (1972) described similar difficulties. The urethane side product **(26)** was obtained from reaction of Boc-Ile-isobutyl-

$(CH_3)_2CHCH_2OCO$—Pro—Leu—Ser(Bzl)—Arg(NO_2)—Leu—Phe—OMe

(26)

carbonic acid mixed anhydride with the hexapeptide methyl ester during a REMA synthesis. The problem was solved by coupling Bic-Ile-Pro-OH to the pentapeptide H-Leu-Ser(Bzl)-Arg(NO_2)-Leu-Phe-OMe.

Yajima and collaborators also warned that there might be an increased tendency of urethane side product formation when larger peptides are involved as amine component. Thus, mixed anhydride coupling of Boc-Arg(NO_2)-OH with the 12-peptide H-Met-Glu-His-Phe-Arg-Trp-Gly-Ser-Pro-Pro-Lys(For)-Asp-OH (Yajima *et al.*, 1969a,b), and of Z-Glu(OBzl)-OH with the octapeptide amide, H-His-Phe-Arg-Trp-Gly-Lys(For)-Pro-Val-NH_2, or the pentapeptide, H-His-Phe-Arg-Trp-Gly-OH (Yajima *et al.*, 1968, 1969b) gave rise to considerable urethane formation.

Furthermore, in a recent synthesis of the 27-peptide hormone secretin by the repetitive excess mixed anhydride (REMA) procedure (van Zon and Beyerman, 1976) the crude product was contaminated to the extent of about 40% with truncated sequences, presumably urethane side products.

Only recently has urethane side product formation (Scheme 7) been systematically studied (Bodanszky and Tolle, 1977). Mixed anhydrides of *tert*-butyloxycarbonylamino acids were coupled with readily detectable amino acid β-naphthylamides. Although mixed anhydrides, prepared from isobutyl chlorocarbonate are less prone to give the second acylation product (**23** in Scheme 7), than those prepared from ethyl or other alkyl chlorocarbonates,* the formation of some undesired urethane side product (i.e., isobutyloxycarbonylamino acid β-naphthylamide, **23a**) was detected in each experiment as shown in Table IV (see also Stewart, 1965). The amounts varied considerably depending on the steric bulk in the carboxyl component, whereas the nature of the amine components examined (two) had no major influence on the amount of **23a** formed during acylation. Since the results remained

* The branched alkyl group of isobutyl chlorocarbonate increases the electron density at the adjacent carbonyl of the mixed anhydride which diminishes attack by the nucleophile and thus the tendency toward urethane formation.

Table IV Urethane Side Product Formation in Mixed Anhydride Coupling of *tert*-Butyloxycarbonylamino Acids and Amino Acid β-naphthylamides Using Isobutyl Chlorocarbonate[a]

Expt.	Boc-amino acid	Amino acid β-naphthylamide	Urethane side product (%)[b]
1	Glycine	Glycine	0.5
2	L-Leucine	Glycine	0.6
3	L-Phenylalanine	Glycine	0.5
4	*O*-Bzl-L-serine	Glycine	0.7
5	L-Valine	Glycine	5.8, 6.6
6	L-Isoleucine	Glycine	8.2
7	Glycine	L-Valine	0.5
8	L-Valine	L-Valine	3.8
9	L-Valine	L-Valine	3.4[e]
10	Glycine	L-Leucine	0.5, 0.6
11	Glycine	Glycine	0.08[c,e]
12	L-Valine	Glycine	5.1[d,e]
13	L-Valine	Glycine	4.5[c,e]
14	L-Leucine	L-Leucine	0.4[c,e]

[a] Reaction conditions were those used in the repetitive excess mixed anhydride procedure (REMA: Tilak, 1970; Beyerman, 1972). The urethane side products (i.e., isobutyloxycarbonylamino acid β-naphthylamides) were determined by their absorption at 248 nm after trifluoroacetic acid treatment (for the Boc group cleavage from the dipeptide β-naphthylamides) and separation by countercurrent distribution. Modified from Bodanszky and Tolle (1977).

[b] Where two values are reported, parallel experiments were performed.

[c] The amounts of Boc-amino acid and *N*-methylmorpholine were twice that used in the standard experiments.

[d] The amount of Boc-amino acid was twice that used in the standard experiments.

[e] The time required for formation of the mixed anhydride was increased to 30 min.

unchanged with longer activation times, any direct reaction between isobutyl chlorocarbonate and amine component was considered unlikely.

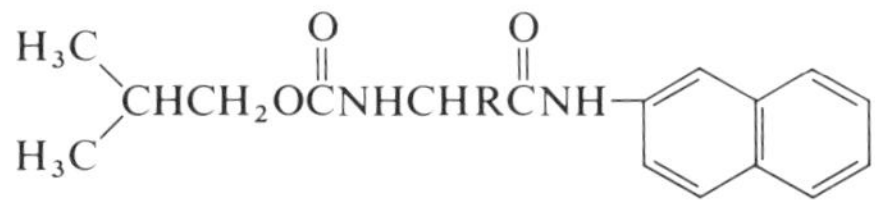

(23a)

Attempts to reduce the amount of urethane side product (**23a**) by using twofold excess of protected amino acid were of little avail. While the formation of small amounts of contaminating urethane side product might be acceptable in actual synthesis, Bodanszky and Tolle (1977) caution against the use of the mixed anhydride method for the coupling of sterically hindered carboxyl components.

2. *Disproportionation*

Another intrinsic side reaction is disproportionation of the mixed anhydride (**1**) to give two symmetrical anhydrides, the acylamino acid or peptide anhydride (**27**) and bisalkyl pyrocarbonate (**28**) (Scheme 9) (Wieland *et al.*,

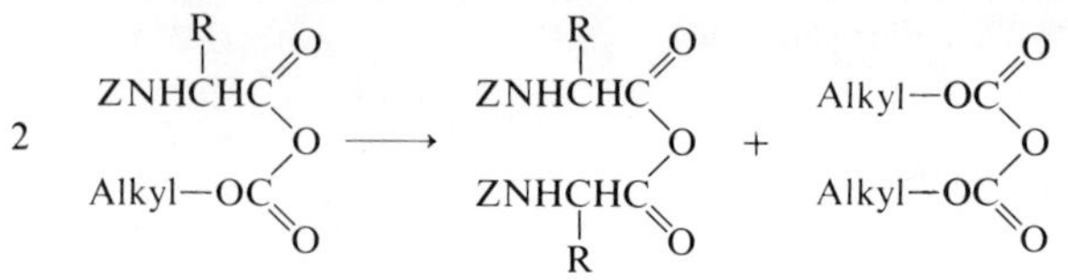

(**1**) (**27**) (**28**)

Scheme 9

1950; Albertson, 1962). Concerning the theory of disproportionation, see Longosz and Tarbell (1961) and Tarbell (1969). Since the symmetrical anhydride (**27**) can acylate only half as much amine as the mixed anhydride (**1**), disproportionation results in reduced yield. Because the desired product is still obtained, this side reaction may mostly go unnoticed in actual synthesis. It has also been determined that short activation time (von Brunn-Leube and Schramm, 1956) and low activation temperature, e.g., $-15°C$ (Wieland *et al.*, 1962), effectively suppress disproportionation.

3. *Diacylimide Formation*

This side reaction, which gives rise to products of the general structure (**29**) appears to be limited, fortunately, to acylglycine activation under forcing conditions, such as the use of excess acylating component (Kopple and

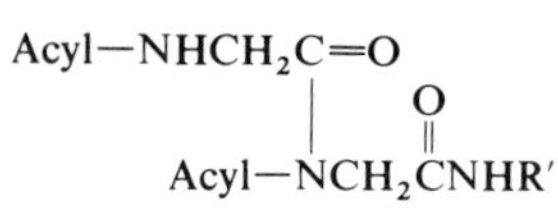

(**29**)

Renick, 1958; Schellenberg and Ullrich, 1959; Zaoral and Rudinger, 1961; Goodman and Stueben, 1962; Merrifield *et al.*, 1974), except for one report of diacylimide formation from benzyloxycarbonylphenylalanine (Determann, 1967; Determann and Kahle, 1969).

In summary, the mixed anhydride method, when carried out under well-defined reaction conditions, is considerably less afflicted with undesired side reactions than, for example, the azide or carbodiimide procedures.

IV. COUPLING REAGENTS: EEDQ AND IIDQ

A novel reagent for peptide bond formation, *N*-ethyloxycarbonyl-2-ethyloxy-1,2-dihydroquinoline (EEDQ, compound **7**) was introduced by Belleau and Malek (1968; see also Belleau *et al.*, 1968) who proposed that it reacts with the carboxyl component (**30**) to form an intermediate mixed anhydride (**32**) and quinoline (**33**) as shown in Scheme 10. The anhydride

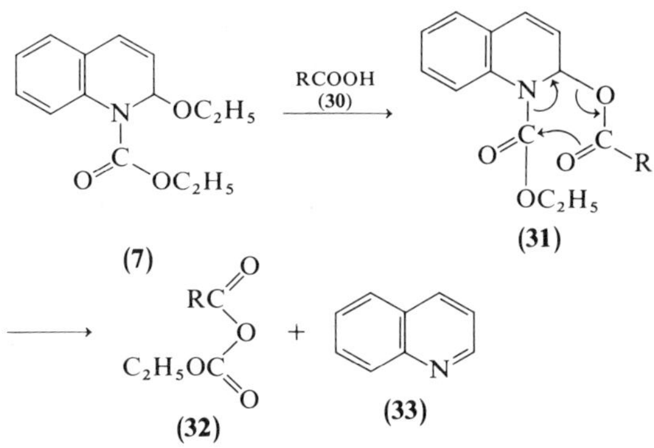

Scheme 10

reacts with the amine component to generate the peptide bond with liberation of ethanol and carbon dioxide, as discussed before (Scheme 2). Belleau and Malek proposed that "the use of EEDQ as a substitute in the classical mixed carbonic anhydride method is indicated because the slow formation and rapid consumption of intermediate mixed anhydride (**31**) precludes its accumulation and thus minimizes the possibility of side reactions (such as racemization)." Appealing as this concept is, an evaluation of the, as yet relatively infrequent, applications of EEDQ actually indicates that both racemization and urethane side product formation do occur. Hence, the advantage of the reagent may primarily consist in the convenience of its use as a coupling reagent (Klausner and Bodanszky, 1972), i.e., in the simple addition to a mixture of the carboxyl and amine component at ambient temperature. However, a systematic side by side comparison of EEDQ with the classical mixed anhydride procedure has not as yet been made.

The known superiority of isobutyl chlorocarbonate in providing mixed anhydrides that are less prone to wrong-way reaction giving urethane side product (see Section III,E,1) led Kiso and Yajima (1972) to prepare *N*-isobutyloxycarbonyl-2-isobutyloxy-1,2-dihydroquinoline (**34**). The use of this reagent provided excellent results in a solid phase segment condensation study (Kiso *et al.*, 1973).

The reagents have been applied (see Table V) to the synthesis of a variety

Table V Selected Examples of Peptide Synthesis with *N*-Ethyloxycarbonyl-2-Ethyloxy-1,2-Dihydroquinoline (EEDQ) and *N*-Isobutyloxycarbonyl-2-Isobutyloxy-1,2-Dihydroquinoline (IIDQ)[a]

Peptide[b]		Solvent	Reaction time (hr)	Yield (%)	Racemization (%)	References
A. Solution Synthesis						
Bz-Leu-┼-Gly-OEt	EEDQ	C_6H_6–EtOH (4:1)	12	95	<0.2	Belleau and Malek (1968)
Z-Ala-┼-Gly-OEt	EEDQ	C_6H_6	3	90		Belleau and Malek (1968)
Z-Gly-Ala-┼-Leu-OBzl	EEDQ	THF	7	97	0.2	Izumiya and Muraoka (1969)
Boc-Cys(Bzl)-┼-Tyr(Bzl)-Ile-Gln-Asn-Cys(Bzl)-OBzl	EEDQ	DMF–THF (3:1)	15	67	—	Mühlemann *et al.* (1972)
Z-Ala-┼-Tyr-OMe	IIDQ	DMF	24	88	—	Kiso and Yajima (1972); Kiso *et al.* (1973)
Z-Gln-┼-Thr-OMe	IIDQ	DMF	24	84	—	Kiso and Yajima (1972); Kiso *et al.* (1973)
Z-Ser-┼-Ala-OMe	IIDQ	DMF	24	80	—	Kiso and Yajima (1972); Kiso *et al.* (1973)
Moz-Asn-┼-Ala-OMe	IIDQ	DMF	24	95	—	Kiso and Yajima (1972); Kiso *et al.* (1973)
Z-Ala-Leu-┼-Gly-OBzl	EEDQ	THF	18	78	0.5	McDermott and Benoiton (1973)
Z-Ala-MeLeu-┼-Gly-OBzl[a]	EEDQ	THF	18	74	7.7	McDermott and Benoiton (1973)
Z-Gly-┼-Ala-MeLeu-O*t*Bu[a]	EEDQ	THF	18	95	—	McDermott and Benoiton (1973)
Boc-Thr-┼-Arg(NO_2)-Leu-OMe	EEDQ	DMF–CH_2Cl_2 (1:5)	15	45.6[c]	—	Klausner *et al.* (1973)

B. Solid Phase Synthesis						
Boc-Arg(NO_2)-Pro-Pro-Gly-Phe-Ser-Pro-Phe-Arg(NO_2)-OCH_3	EEDQ (sequential)	CH_2Cl_2	2–12	46 (7 steps)	—	Sipos and Gaston (1971)
Moz-Val-Lys(Z)-Val-Tyr-Pro-—|—Asp(OBzl)-Gly-Resin	EEDQ	DMF	48	90	—	Yajima and Kawatani (1971)
Moz-Val-Lys(Z)-Val-Tyr-Pro-—|—Ala-Gly-Resin	EEDQ	DMF	48	99.8	—	
Z-Gly-Ala—|—Leu-OBzl	Poly-Q[a]	DME[a]	22	72	6.0	Lauren and Williams (1972); see also Brown *et al.* (1972)
Boc-Gly-Ala—|—Leu-Resin	(a) EEDQ	DMF	24	83	8.5	Barton *et al.* (1973)
	(b) EEDQ	Dioxane	24	91	1.3	
Moz-Val-Lys(Z)-Val-Tyr-Pro-—|—Gly-Ala-Resin	IIDQ	DMF	48	99	—	Kiso *et al.* (1973)
pGlu-Gln-Ala-NH_2	EEDQ (sequential)	CH_2Cl_2-EtOH (10:1)	24	75	—	Wang *et al.* (1974)
pGlu-Asn-Ala-NH_2		CH_2Cl_2-DMF (1:1)	24	39	—	

[a] Abbreviations: DME, 1,2-dimethoxyethane; DMF, dimethylformamide: EEDQ, *N*-ethyloxycarbonyl-2-ethyloxy-1,2-dihydroquinoline; IIDQ, *N*-isobutyloxycarbonyl-2-isobutyloxy-1,2-dihydroquinoline; MeLeu, *N*-methylleucine; Moz, *p*-methyloxybenzyloxycarbonyl; Poly-Q, polymeric EEDQ reagent; THF, tetrahydrofuran.

[b] Site of peptide bond formation is indicated by dash with vertical bar.

[c] Urethane side product was also isolated.

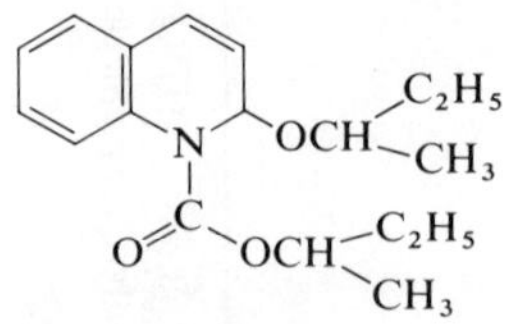

(34)

of small model peptides both by solution and solid phase procedures. Coupling may be carried out in a wide range of solvents, including tetrahydrofuran, dichloromethane, dioxane, dimethylformamide, or even ethanol (Belleau and Malek, 1968).

In a *typical procedure* (modified from Kiso *et al.*, 1973) equimolar amounts of the carboxyl and the amine component are dissolved in dimethylformamide at 0.1 to 0.4 *M* concentrations. EEDQ or IIDQ is added in 5–10% excess. The mixture is stirred for 15 to 24 hr at room temperature. After evaporation of the solvent, the residue is dissolved in ethyl acetate, which is washed with 1 *N* $NaHCO_3$, 10% citric acid, and H_2O, dried (Na_2SO_4), and evaporated. Products are recrystallized or chromatographically purified as indicated. Hydroxyamino acids have been activated without side chain protection. High yields of pure products have been obtained with small model peptides. It remains to be seen whether this will also hold for syntheses of larger peptides. For example, in the condensation of the pentapeptide Moz-Val-Lys(Z)-Val-Tyr-Pro-OH with H-Ala-Gly-Resin 3 equivalents of the peptide and 6 equivalents of EEDQ were required to obtain a high yield (Yajima and Kawatani, 1971) and maximum protection of side chain functionalities is probably necessary under such conditions.

The incidence of racemization (see Table V) appears to be similar to that observed for the classical mixed anhydride procedure under "Anderson conditions" (compare Table III). The fact that tertiary amine is not required for activation is a distinct advantage in the use of EEDQ or IIDQ. On the other hand, the reagents lack the rapidity of the mixed anhydride reaction. The tendency of racemization is increased in acylation of *N*-methylamino acids (McDermott and Benoiton, 1973) and in the synthesis of acylamino acid anilides and *p*-nitroanilides (Chen and Benoiton, 1978).

Formation of the urethane side product, ethyloxycarbonyl-nitroarginyl-leucine methyl ester, has been observed by Klausner *et al.* (1973) when EEDQ was used to couple Boc-Thr-OH with H-Arg(NO_2)-Leu-OMe.

At present it appears that the advantages of EEDQ and IIDQ may be seen in the ease and convenience of their use rather than in superior performance compared to the classical mixed anhydride procedure, but a more balanced judgment has to await a detailed direct comparison.

V. APPLICATIONS

The most frequent use of the mixed anhydride procedure is made in coupling single N^{α}-benzyloxycarbonyl- or N^{α}-*tert*-butyloxycarbonyl-protected amino acids to amino acids or peptides because the danger of racemization is negligible. Selected examples of incremental (stepwise) peptide chain elongation by mixed anhydrides are described in Section V,A.

The use of excess mixed anhydride became possible with the development of mild procedures for its rapid elimination at the end of the acylation reaction. Rapid peptide synthesis without intermediate purification by the repetitive excess mixed anhydride (REMA) procedure is discussed in Section V,B.

Application of the mixed anhydride method to segment condensation under the Anderson conditions of racemization suppression has not yet been utilized except in the laboratory of T. Wieland, as outlined in Section V,C.

Description of special applications in Section V,D includes the use of mixed anhydrides for (1) ester bond formation, (2) preparation of active esters of acylamino acids, (3) preparation of active esters of acyl peptides by "backing off," and (4) peptide cyclization.

A. Incremental (Stepwise) Chain Elongation

Numerous examples of mixed anhydride coupling of benzyloxycarbonyl- or *tert*-butyloxycarbonylamino acids with various amine components (amino acids or peptides) have been described. Earlier work, up to 1958, has been tabulated by Albertson (1962). For a recent application, the paper by Rzeszotarska and Kmiecik-Chmura (1977) may be consulted, which contains a list of 20 protected dipeptides, prepared by mixed anhydride coupling. Small peptides are readily prepared. The possibility of racemization is very low with the urethane-type N^{α}-protective groups, and may be further reduced by observing the "Anderson conditions" (Anderson *et al.*, 1967b). Many of the smaller peptides may be readily prepared without having to protect certain side chain functionalities, e.g., the hydroxyl groups, as discussed in detail by Albertson (1962; see also Stewart, 1966). One of the few restrictions concerns acylation of NH_2-terminal proline or *N*-methylamino acid residues in which a high tendency of urethane side-product formation exists (see Section III,E,1).

A recent application of the mixed anhydride procedure ("Anderson conditions") for incremental peptide chain elongation has been described by Anantharamaiah and Sivanandaiah (1974) in a synthesis of fibrinopeptide-B. Starting from a COOH-terminal tetrapeptide derivative, H-Phe-Gly-Gly-

Arg(NO_2)-OMe, five successive mixed anhydride couplings provided the crystalline protected nonapeptide (compound **35**). Similar applications in syntheses of gramicidin A (Sarges and Witkop, 1965) and gramicidin S analogues (Abe *et al.*, 1976) have been described.

OBzl OBzl
Z—Asn—Glu—Glu—Gly—Leu—Phe—Gly—Gly—Arg(NO_2)—OMe

(35)

However, the most extensive application of the mixed anhydride procedure in every conceivable strategic approach has been made by Wieland and collaborators (1968a,b, 1969b, 1971a,c, 1974; Heber *et al.*, 1974; Munekata *et al.*, 1977). A synthesis of antamanide, cyclo(-Val-Pro-Pro-Ala-Phe-Phe-Pro-Pro-Phe-Phe) (Wieland *et al.*, 1969a; see also Prox *et al.*, 1969) is shown in Fig. 3 in the linear form prior to cyclization.

A further development of the mixed carbonic anhydride method, distinguished by the use of excess acylating agent, is discussed in detail in Section V,B.

Application of the mixed anhydride procedure to solid phase synthesis has been infrequent. Tilak and Hollinden (1968a,b) reported a satisfactory synthesis of Phe-Phe-Tyr-Thr-Pro-Lys-Thr (human insulin B24–30) using excess benzyloxycarbonylamino acid isobutylcarbonic acid anhydrides (2–3 hr at −15°C), but low yields in attempts to prepare other insulin peptides in this manner. Solid phase synthesis of a series of pteroyloligo-γ-glutamates

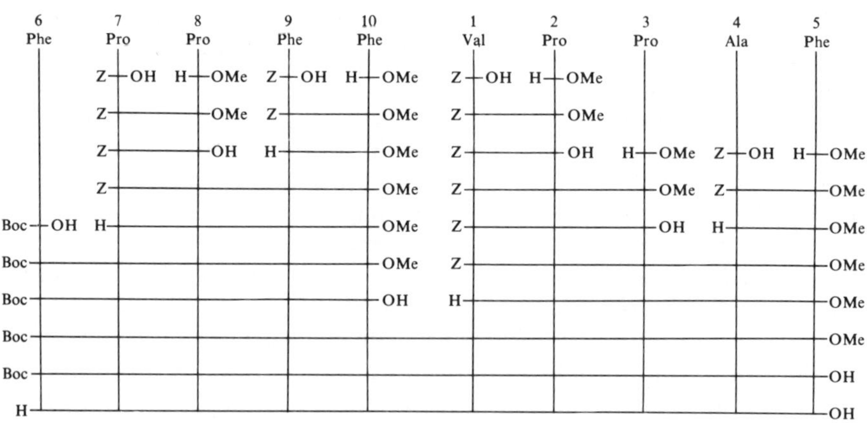

Figure 3. Schema of a synthesis of antamanide (Wieland *et al.*, 1969a; see also Prox *et al.*, 1969) by exclusive use of the mixed carbonic anhydride method. The final decapeptide depicts the linear form prior to cyclization. Boc, *tert*-butyloxycarbonyl; Me, methyl; Z, benzyloxycarbonyl. (From Wieland *et al.*, 1969b, reprinted with permission of Verlag Chemie).

(Krumdieck and Baugh, 1969) gave superior results with mixed anhydride coupling of Boc-Glu-OBzl than the standard dicyclohexylcarbodiimide procedure (see also Meienhofer *et al.*, 1970). In a comparative study (Jakubke and Baumert, 1974), solid phase synthesis of the model peptide Val-Gly-Phe-Ala-Leu-Ala produced better results with dicyclohexylcarbodiimide coupling than with mixed anhydrides of *tert*-butyloxycarbonylamino acids in threefold excess. Acylations were less complete and there was indication for urethane side-product formation. Diacylimide formation, i.e., acylation of the protected α-amino group of the mixed anhydride (compound **36**; see

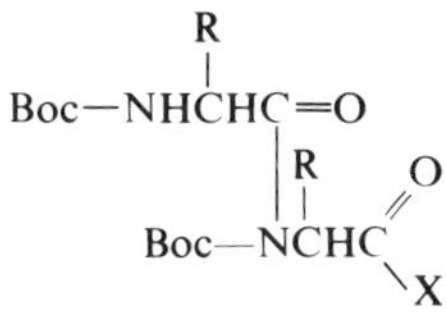

(**36**)

also **18**, Section III,E,3), was recently observed in solid phase synthesis of the model peptide Leu-Ala-Gly-Val by Merrifield *et al.* (1974). These and similar findings by others (see, e.g., Wieland *et al.*, 1971b) appear counter-indicative for mixed anhydrides in solid phase synthesis, in contrast to EEDQ (Section IV, Table V) and especially symmetrical anhydrides which are used increasingly with excellent results (see Section VI,A).

B. The Repetitive Excess Mixed Anhydride Procedure (REMA)

In 1970 Tilak described a rapid synthesis of a protected glucagon (7–15) nonapeptide by incremental or stepwise synthesis from the carboxyl terminus without intermittant purification. Excess mixed anhydride of N^{α}-alkyloxycarbonylamino acids was used to ensure complete acylation of the amine component in each step, followed by hydrolysis of the remaining anhydride by potassium bicarbonate. The synthetic cycle, deprotection followed by acylation with a mixed anhydride, can be repeated without purification of intermediates. The procedure has since been further expanded by Beyerman and collaborators who gave it the name REMA for repetitive excess mixed anhydride method. The most impressive achievement obtained with the REMA procedure has been a recent synthesis of biologically fully active secretin, the 27-residue gastrointestinal hormone (van Zon and Beyerman, 1976).

1. The REMA Concept

The success of Merrifield's solid phase synthesis (Merrifield, 1963, 1969) and in particular its dramatic time-saving and ease of operation have provided great incentive to develop the proven stepwise synthesis in solution (Bodanszky and du Vigneaud, 1959; Bodanszky, 1960) into similarly effective repetitive procedures. To obtain the essential quantitative reaction in each step, excess acylating agent must be used (Bodanszky, 1971) which in turn mandates maximum side chain protection (Bodanszky *et al.*, 1977; Finn and Hofmann, 1976). Weygand *et al.* (1967; Weygand and Di Bello, 1969) were the first to use anhydrides for this purpose. Rapid and complete peptide acylation was obtained with excess symmetrical *tert*-butyloxycarbonylamino acid anhydrides.

Tilak (1970) obtained similarly efficient peptide acylation with benzyloxycarbonyl- or *tert*-butyloxycarbonylamino acid mixed carbonic anhydrides and developed the process into a rapid repetitive procedure. He found that excess anhydride is speedily hydrolyzed by aqueous $KHCO_3$ at pH 8 and 0°C, once the acylation reaction is complete. The purity of the products, obtained in high yields, permitted repetitive peptide chain elongation without intermittant purification. The validity of the concept was demonstrated by a synthesis of the difficult glucagon sequence (7–15) which required protection of 8 side chain functionalities, see Table VI.

2. Experimental Aspects

In a *typical procedure* (van Zon and Beyerman, 1973), a 14-fold excess of mixed anhydride over amine component is prepared by adding isobutyl chlorocarbonate (1.4 equiv) to a cold solution (−15°C) of *N*-protected amino acid (1.5 equiv) and *N*-methylmorpholine (1.5 equiv) in tetrahydrofuran (or dimethylformamide). After a 2-min period of anhydride formation a solution of the amine component (1 equiv) in dimethylformamide is added at once and the mixture is stirred at a constant temperature of −15°C, which is maintained by a cryostat (see Fig. 1, p. 274).

For monitoring the progress of the reaction, samples (1 μl) are spotted on tlc plates and treated with fluorescamine (Udenfriend *et al.*, 1972; Weigele *et al.*, 1973). The intensity at 390 nm is compared with those obtained from known quantities of the amine component. Once fluorescamine indicates the presence of less than 0.1% of the free amine component, coupling is continued for 1 hr, but not beyond a total for 3 hr reaction time.

For workup, the temperature is raised to 0°C and a 2 *M* solution of $KHCO_3$ is added to obtain pH 8 which is maintained with stirring for 30 min. The peptide is then precipitated with a 4-fold volume of 75% saturated aq. NaCl at 0°C, and the precipitate washed thoroughly with water and dried to constant weight *in vacuo* at room temperature.

N^{α}-Benzyloxycarbonyl groups are cleaved in each step by hydrogenolysis or HBr acidolysis. Catalytic hydrogenation is carried out in methanol,

Table VI Peptides Synthesized by the Repetitive Excess Mixed Anhydride [REMA] Procedure (Protected State)[a]

Name	Synthetic protected peptide	Overall yield (%)[b]	References
Glucagon (7–15)	*t*Bu *t*Bu O*t*Bu *t*Bu *t*Bu Boc *t*Bu O*t*Bu Z-Thr-Ser-Asp- Tyr-Ser-Lys-Tyr-Leu-Asp-OEt 5	79	Tilak (1970); Tilak *et al.* (1972)
Human somatotropin (1–10)	Bzl Bzl NO_2 Boc-Phe-Pro-Thr-Ile-Pro-Leu-Ser-Arg-Leu-Phe-OMe 5 10	59	Beyerman (1972); Beyerman *et al.* (1973)
Gonadoliberin	Tos Bzl Bzl NO_2 pGlu-His-Trp-Ser-Tyr-Gly-Leu-Arg-Pro-Gly-NH_2 5 10	13	Beyerman *et al.* (1976)
Secretin	Z Bzl OBzl Bzl Bzl Bzl OBzl Z-His-Ser-Asp-Gly-Thr-Phe-Thr-Ser-Glu-Leu- 5 10 Bzl NO_2 NO_2 OBzl Bzl NO_2 Ser-Arg-Leu-Arg- Asp- Ser-Ala-Arg-Leu-Gln- 15 20 NO_2 Arg-Leu-Leu-Gln-Gly-Leu-Val-NH_2 25	22	van Zon and Beyerman (1973, 1976)
Substance P	NO_2 Z O Z-Arg-Pro-Lys-Pro-Gln-Gln-Phe-Phe-Gly-Leu-Met-NH_2 5 10	20	Izeboud and Beyerman (1978)

[a] See also Hoffmann and Tilak (1975) and Beddell *et al.* (1975) for applications of REMA not listed here.
[b] Based on the COOH-terminal amino acid.

dimethylformamide, acetic acid, or mixtures of these solvents using 100–500 mg of freshly prepared Pd-black (Willstätter and Waldschmidt-Leitz, 1921; Wieland, 1912) per mmol of peptide. CO_2 Evolution is tested with $Ba(OH)_2$ solution. With dipeptides and NH_2-terminal glutamine 1 equiv of HCl is added to prevent side reactions. After removal of the catalyst the filtrate is evaporated exhaustively *in vacuo* (10^{-4} cm Hg) at $< 30°C$ adding benzene to accomplish azeotropic removal of methanol. The peptide is then ready for the next coupling (after neutralizing HCl, if present).

For acidolysis of the N^{α}-benzyloxycarbonyl group, the peptide is dissolved in glacial acetic acid (with warming if necessary). An equal volume of 4 *N* hydrogen bromide in acetic acid is added and the solution is kept at room temperature for 1.5 hr. The peptide is precipitated by the addition of a large volume of anhydrous peroxide-free ether. Alternatively, HBr—HOAc may be evaporated *in vacuo*. N^{α}-*tert*-Butyloxycarbonyl groups are removed by treatment of the peptide with trifluoroacetic acid at room temperature for 30 min followed by workup, as above.

Care has to be taken to observe the guidelines of Anderson *et al.* (1967b) for optimum reaction conditions and suppression of potential racemization. Maximum protection of all side chain functionalities is necessary to prevent undesired side reactions (Beyerman *et al.*, 1974).

An unresolved problem for the REMA procedure remains the urethane side product formation resulting from coupling at the carbonate side of the mixed anhydride (Section III,E,1) which cannot be suppressed by lowering the reaction temperature. The side reaction becomes more pronounced with increased steric hindrance at the NH_2-terminus of the peptide to be acylated. With NH_2-terminal proline urethane side product formation may become the predominant reaction. For example, in the coupling of Boc-Ile-OH to Pro-Leu-Ser(Bzl)-Arg(NO_2)-Leu-Phe-OMe (Beyerman, 1972; Beyerman *et al.*, 1973) the urethane, isobutyloxycarbonyl-Pro-Ser(Bzl)-Arg(NO_2)-Leu-Phe-OMe (**26**) was exclusively formed (in 95% yield). In this, and all subsequent cases, the problem was solved by coupling dipeptides with COOH-terminal proline to the amine component, i.e., in this case Boc-Ile-Pro-OH to the pentapeptide H-Leu-Ser(Bzl)-Arg(NO_2)-Leu-Phe-OMe. In view of the results of the recent study of Bodanszky and Tolle (1977) (see Table IV) small amounts of urethane side product might be formed in every step. Indeed, in the REMA synthesis of secretin (van Zon and Beyerman, 1973, 1976), see Table VI, substantial amounts of urethane-type blocked peptides were found to be accumulated in the final product. The only fundamental solution to this problem may be the use of symmetrical anhydrides (see Section VI,A), as suggested by Bodanszky and Tolle (1977).

3. Scope and Limitations

The REMA procedure is a highly efficient and useful approach to the synthesis of peptides in solution, as evident from the list of successful applications compiled in Table VI. Reporting on a recent use of this procedure in syntheses of luliberin analogues, Beddell *et al.* (1975) describe it as "highly satisfactory, giving pure intermediates in high yields at each stage. The simple workup procedure after each coupling allowed the synthesis to proceed rapidly and without complications." An appealing aspect of the REMA procedure is the opportunity to stop at any given stage and purify intermediate products as often or as little as desired or required. This flexibility is a distinct advantage of REMA compared to solid phase synthesis. Since Beyerman and collaborators (see, e.g., van Zon and Beyerman, 1976) like to consider REMA as a direct competitor to the solid phase method, it must also be acknowledged that the speed and ease of operation of the latter cannot be fully matched by REMA.

The successful synthesis of secretin is certainly the most spectacular accomplishment, yet it has also served to indicate the constraints of REMA. Van Zon and Beyerman (1976) state in their paper: "we feel that the size-limit of a purifiable peptide that can be synthesized by the REMA-method has been reached at present, a limit which is of about thirty amino acid residues.* Further development, especially for avoidance of anomalous coupling ... is needed."

C. Peptide Segment Condensation

The mixed carbonic anhydride method has been successfully used in early insulin syntheses (Zahn *et al.*, 1965a) to condense peptide segments. To circumvent the possibility of racemization, carboxyl components with COOH-terminal glycine were selected for activation (see Table VII,A), and proline residues have been utilized similarly. In these mixed anhydride segment condensations triethylamine has commonly been used as a base. Instead *N*-methylmorpholine, as recommended by Anderson *et al.* (1967b), should be used with COOH-terminal glycine as well, because 5(4H)-oxazolone formation may cause racemization at the penultimate residue (Dzieduszycka *et al.*, 1973).

Solvent impurity, e.g., trace contamination of dimethylformamide or hexamethylphosphoramide with moisture or, much worse, alkylamines, is another concern in segment condensation of larger peptides. For example,

* The size limitation of ca. 30 residues applies rather generally to peptides synthesized by stepwise strategies and seems independent of the method used. It is dictated by the sensitivity limits of present analytical procedures (see Gross and Meienhofer, this volume, Chapter 1).

Table VII Peptide Segment Condensation by the Mixed Carbonic Anhydride Procedure[a]

Segments		Activation[b]			Coupling				
Carboxyl component	+ Amine component	Solvent	Time (min)	Temperature (°C)	Solvent	Time (hr)	Peptide bond[c] formed	Yield (%)	References
A. Via activation of COOH-terminal glycine									
Insulin A 1–9	+ A10–21	HMP	15	−20	HMP	24	-Ala-Gly → Val-	70	Zahn *et al.* (1965b)
Insulin B21–23	+ B24–30	THF	15	−10	CH_3CN	10	-Arg-Gly → Phe-	73	Schnabel (1964)
Insulin B13–20	+ B21–30	DMF	30	−15	DMF	6	-Cys-Gly → Glu-[d]	25	Schnabel (1963)
Insulin B 1–8	+ B 9–30	HMP	20	−15	HMP	72[e]	-Cys-Gly → Ser-[d]	51	Meienhofer and Schnabel (1965)
Eledoisin analogue 6–9	+ 10–11	DMF	8	−15	DMF	2	-Ile-Gly → Leu-	90	Neubert *et al.* (1977)
B. Via activation of chiral carboxyl terminus using Wieland–Anderson Conditions[f]									
Antamanide	5 + 5	THF	6	−4	THF	2–3	-Phe-Phe → Val-	81	Wieland *et al.* (1969b)
Antamanide	5 + 5	THF	6	−4	THF	2–3	-Phe-Phe → Phe-	80	Wieland *et al.* (1971c)
Antamanide	2 + 3	THF	6	−4	THF	2–3	-Pro-Phe → Pro-	88	Wieland *et al.* (1974)
Norphalloine	2 + 2	THF	6	−4	THF	2–3	-*a*Hyp-Ala → Trp-	79	Wieland *et al.* (1971a)
Norphalloine analogues	2 + 2	DMF–THF	6	−4	THF	2–3	-Hyp-Ala → Trp-	64	Nebelin *et al.* (1973)
	3 + 1	THF	6	−4	DMF	2–3	-Thr-Cys → Pro-[d]	77	
	3 + 1	THF	6	−4	DMF	2–3	-Thr-Cys → *a*Hyp-[d]	83	
Bungarotoxin 20–28	+ 29–37[g]	—	—	—	—	—	-Met-Trp → Cys-[g]	84	Ivanov *et al.* (1976)

[a] Abbreviations: *a*Hyp, *allo*-4-hydroxyproline; DMF, dimethylformamide; HMP, hexamethylphosphoramide; Hyp, 4-hydroxyproline; THF, tetrahydrofuran.

[b] Isobutyl chlorocarbonate was used in all experiments, triethylamine in Part A, *N*-methylmorpholine in Part B of the table.

[c] The amino acid residue penultimate to the COOH-terminal one of the carboxyl component is shown because of the potential danger of its racemization during oxazolone formation (Bergmann and Zervas, 1929; Neuberger, 1948; Weygand *et al.*, 1966b; Dzieduszycka *et al.*, 1973).

[d] Cys was protected as Cys(Bzl).

[e] A second and third coupling with freshly prepared mixed anhydride was conducted, for 24 hr each.

[f] For background, see Wieland *et al.* (1968a); for reaction conditions, see Wieland *et al.* (1968b, 1969b; Heber *et al.*, 1974).

[g] Experimental details not available; Cys was *S*-protected by *p*-methoxybenzyl.

0.02% dimethylamine in 100 ml dimethylformamide could cause about 9% side product formation in a coupling of 1.8 g of the dipeptide Z-Gly-Phe-OH. However as much as 31% of a mixed anhydride of 1.8 g insulin A(1–9) (1.4 mmol) could be blocked by the same amount of amine. Thus, amounts of solvent impurities that may well be tolerable in syntheses of small peptides could have disastrous effects in the condensation of larger segments. Considerable efforts (e.g., by gas chromatographic tests) must be made to assure the absence of impurities and moisture from solvents, reagents, peptide segments to be condensed, glassware to be used—in addition to that of excess tertiary base, discussed before—during mixed anhydride condensations.

Although the "Anderson conditions" for mixed anhydride coupling of peptides at low risk of racemization have been published more than 10 years ago, little or no effort has been made to apply this easy and efficient method to segment condensation in actual synthesis, except for one case (bungarotoxin, see Table VII,B) described below. However, Wieland and collaborators who developed a similar set of conditions in their own work (1968b) and adopted Anderson's proposed use of *N*-methylmorpholine (1968a), have taken frequent advantage of the refined mixed anhydride procedure for peptide segment condensation (see Table VII,B, for select examples). Thus, in numerous syntheses of the cyclic decapeptide antamamide (Wieland *et al.*, 1968a) and analogues and approaches to syntheses of the complex bicyclic phallotoxins, the mixed anhydride procedure served not only in peptide coupling but also in cyclization of linear peptides (see Section V,D). In fact, the method has become the "preferred" one for Wieland *et al.* (1974).

In efforts to synthesize α-bungarotoxin, Ivanov *et al.* (1976) assembled a protected 74-peptide, ("the longest peptide as yet prepared in solution by the method of maximum hydrophobic protection of the side chains"), in the course of which mixed anhydride activation of the N^{α}-*tert*-butyloxycarbonyl-(20–28)-peptide was used for condensation with the (29–37)-peptide in 84% yield by forming a Trp → Cys(Mob) bond, Scheme 11. Experimental details are not yet available.

Evidently, the refined mixed anhydride procedure is a very versatile and effective method. As such it is at present unduly underutilized and more frequent use should be encouraged.

D. Special Applications

Special applications of the mixed carbonic anhydride method in peptide synthesis include: (1) ester bond formation, (2) preparation of active esters of N^{α}-alkyloxycarbonylamino acids, (3) preparation of active esters of N^{α}-alkyloxycarbonyl peptides by "backing off," and (4) peptide cyclization.

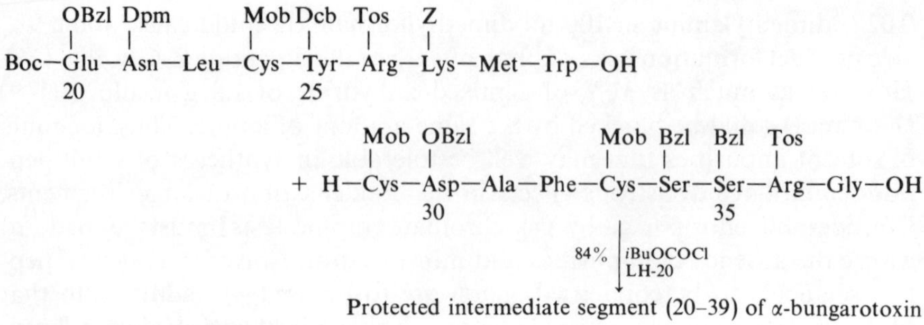

Mob = *p*-methyloxybenzyl

Scheme 11

1. The mixed anhydride procedure permits formation of certain ester bonds. For example, treatment of *N*-benzyloxycarbonyl-L-*N*-methylvaline isobutyloxycarbonic acid anhydride **(37)** with *tert*-butyloxycarbonyl-L-threonine triethylammonium salt **(38)** in tetrahydrofuran at −10°C provided 65–85% of *O*-(benzyloxycarbonyl-L-*N*-methylvalyl)-*N*-*tert*-butyloxycarbonyl-L-threonine **(39)** (Meienhofer, 1970). Similarly, *O*-(*tert*-butyl-

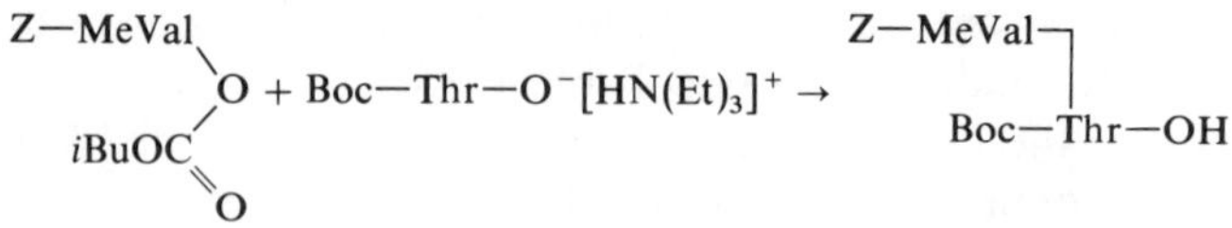

(37) **(38)** **(39)**

oxycarbonyl-L-phenylglycyl)-*N*-benzyloxycarbonyl-L-threonine was prepared in 54% recovery (Ondetti and Thomas, 1965).

2. Prior to the universal use of dicyclohexylcarbodiimide for the preparation of N^{α}-alkyloxycarbonylamino acid active esters (Elliott and Russell, 1957; Rothe and Kunitz, 1957; Bodanszky and du Vigneaud, 1959) the mixed anhydride method provided one of the means to prepare these compounds. Thus, Wieland *et al.* (1951) obtained crystalline thiophenyl esters of benzyloxycarbonyl amino acids, e.g., Z-Gly-SPh, in 85–95% yields (ethyl chlorocarbonate, tetrahydrofuran, 10–15 min activation time at 0°C). Benzyloxycarbonylglycine 4-nitrothiophenyl ester was similarly prepared (65% yield, ethyl chlorocarbonate, dimethylformamide, 5 min activation time at −5°C) (Farrington *et al.*, 1957).

The synthesis of *N*-hydroxysuccinimide esters of *N*-protected amino acids and peptides has been described by Anderson *et al.* (1967a). By applying the same conditions of optimum suppression of racemization, as elaborated for peptide coupling, [i.e., equivalent amounts of acyl peptide, *N*-methylmorpholine, and isobutyl chlorocarbonate in EtOAc, 30 sec activation time at −15°C, followed by addition of 1.1 equiv of HOSu] acyl peptide

N-hydroxysuccinimide esters were prepared with preservation of chiral integrity. Unfortunately, the procedure seems to be restricted to *N*-hydroxysuccinimide esters, since *p*-nitrophenyl, 2,4,5-trichlorophenyl and related esters could not be obtained in this manner.

Active ester preparation via mixed carbonic anhydrides is not used very much at present. It may, however, offer advantages in cases where urea by-products and/or side products of the commonly used dicyclohexylcarbodiimide method should be difficult to remove.

3. The "backing off" procedure (Goodman and Stueben, 1959) has been designed to provide for the preparation of chirally pure acyl peptide active esters. Since they may be condensed with amine components in high yield and with apparently low risk of racemization (see, for example, Kovacs *et al.*, 1969), acyl peptide active esters are useful and desirable intermediates in peptide synthesis and in preparation of sequential polyamino acids (see Wünsch and Wendlberger, 1974). However, all known procedures for the direct esterification of optically active carboxyl-terminal residues of acyl peptides are liable to cause partial racemization. For example, diastereoisomer contamination was found in many benzyloxycarbonyl dipeptide *p*-nitrophenyl esters, whether these were prepared by the carbodiimide procedure (Lübke and Schröder, 1961), by di-*p*-nitrophenyl sulfite (Iselin and Schwyzer, 1960), or by tri-*p*-nitrophenyl phosphite (Guttmann, 1960).

In the backing off procedure, Scheme 12, an alkyloxycarbonylamino acid such as compound **43** is coupled by the mixed anhydride procedure with a

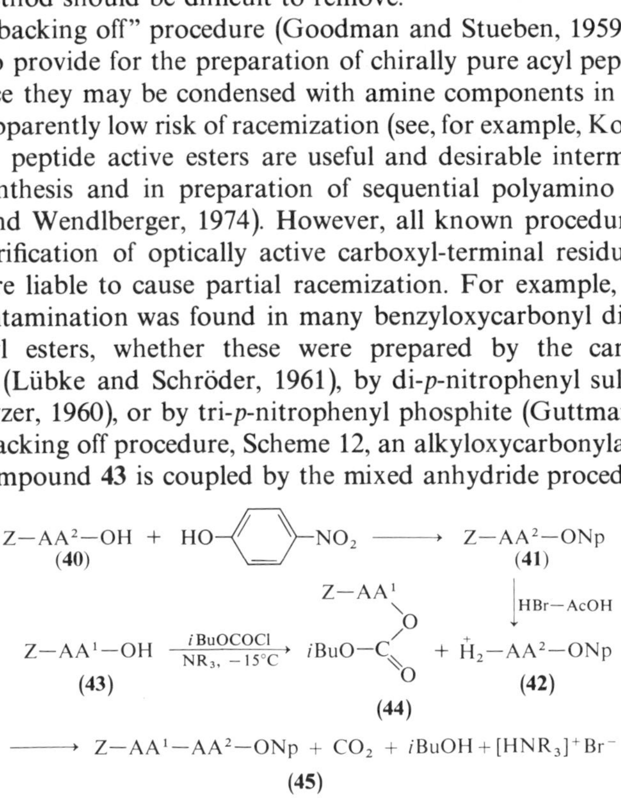

Scheme 12

chirally pure amino acid active ester, for example a *p*-nitrophenyl ester (**42**). Compound **42**, in turn was prepared from the N^{α}-alkyloxycarbonyl protected form (**40**) by reaction with *p*-nitrophenol and carbodiimide, purification of the ensuing product (**41**), and N^{α}-protective group cleavage. The *p*-nitrophenyl ester acts as a protecting group in the backing off reaction since the mixed anhydride acylation proceeds at a much faster rate than *p*-nitrophenyl ester aminolysis. High yields of pure compounds are generally obtained. The mixed anhydride is particularly suited, as compared to the frequently used carbodiimide, because of the volatility of its by-products. The backing off procedure has been applied to the preparation of a wide range of different *N*-protected peptide active esters (see, e.g., Iselin and Schwyzer, 1960; Lübke and Schröder, 1961; Goodman *et al.*, 1962; Stewart, 1965; Kapoor and Gerencser, 1969; Kovacs *et al.*, 1969, 1972).

4. Cyclization of linear peptides by the mixed anhydride method (Boissonnas and Schumann, 1952) has been seldom utilized. Anhydride formation (a) and cyclization (b) are carried out in separate successive stages which permits to conduct the former at high concentration and then add solvent, before liberating the amino group by tertiary base, to obtain optimal cyclization at high dilution (Scheme 13). This procedure was applied by Wieland

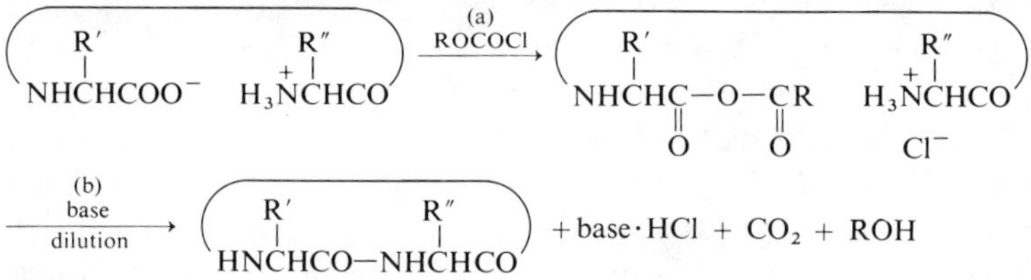

Scheme 13

and Sangl (1964) to the cyclization of *seco*-ketophalloidine (ethyl chlorocarbonate, hexamethylphosphoramide—tetrahydrofuran, 15 min activation at −15°C, dilution and triethylamine addition, 24-hr reaction at 2–3°C). A 6% recovery of toxic ketophalloidine was obtained, whereas cyclization attempts with carbodiimide failed. To improve the procedure, pyridinium chloride was added to cyclization mixtures for more complete protonation of the α-amino group (Wieland *et al.*, 1968b). Increased cyclization yields (> 20%) were indeed obtained in a single-stage operation.

In a further recent modification (Munekata *et al.*, 1977) resynthesis of phalloidine and phallisine from the seco compounds was achieved. After Sephadex LH-20 chromatography of secophalloidin to provide pure anion, mixed anhydride was generated from isobutyl chlorocarbonate in the presence of *N*-methylmorpholine and trifluoroacetic acid in dimethylformamide at −10°C, followed by dilution and base addition. Workup after 14 hr provided an 8.7% yield of phalloidine.

VI. RELATED METHODS

A large number of anhydride methods have been proposed for use in peptide synthesis* but a rather smaller group could measure up to the

* For a detailed description of the entire range of anhydride procedures in peptide synthesis, see Wünsch (1974). The relative merits of the more frequently used anhydride methods from the present vantage point have been surveyed by Jones (this volume, Chapter 2).

simplicity and efficiency of the mixed carbonic anhydride method. Among chemically more closely related procedures the carboxylic anhydrides, the *N*-carboxy (Leuchs) anhydrides and the symmetrical anhydrides of *N*-protected amino acids and peptides will be discussed, the latter because of their increasing use in solid phase synthesis.

A. Symmetrical Anhydrides of Alkyloxycarbonylamino Acids

Formation of symmetrical anhydrides of *N*-protected amino acids or peptides from mixed anhydrides was reported by Wieland *et al.* (1950), Sheehan and Frank (1950) and von Brunn-Leube and Schramm (1956). During studies on acylurea formation in carbodiimide reaction with benzyloxycarbonyl amino acids, two symmetrical anhydrides, i.e., $[Z\text{-}\beta Ala]_2O$ and $[Z\text{-}Glu(OMe)]_2O$, were identified (Muramatsu and Hagitani, 1959; Muramatsu, 1961; see Schnabel, 1965, 1966). Schüssler and Zahn (1962) obtained $[Z\text{-}Gly]_2O$ and $[Z\text{-}Val]_2O$ by treatment of the benzyloxycarbonylamino acids in CH_3CN with dicyclohexylcarbodiimide as stable crystalline compounds, recrystallized from hot benzene. Reaction with H-Phe-OEt in CH_3CN was rapid and complete within 5 min and provided crystalline Z-Gly-Phe-OEt and Z-Val-Phe-OEt, both in 88% yield.

Weygand and collaborators prepared symmetrical anhydrides by treatment of *tert*-butyloxycarbonylamino acids with inamine (methylethinyldiethylamine) (Weygand *et al.*, 1967; Weygand and Di Bello, 1969). Again, it was observed that coupling with esters of amino acids or peptides was completed within a few minutes. Quantitative yields of products were obtained by using excess (1.4 equiv) symmetrical anhydride which was removed after completion of the coupling reaction by treatment with *N*-(2-aminoethyl)piperazine. The efficacy of the procedure was demonstrated by a repetitive sequential synthesis of Boc-Leu-Ile-Ile-Leu-Val-Pro-Pro-Phe-Phe-OBzl, and a stepwise synthesis of Ser-Ser-Thr-Ser-Ala-Ala. On the basis of several comparative syntheses Weygand and Di Bello concluded that the symmetrical anhydride procedure was superior to the carbodiimide, *N*-hydroxysuccinimide, and mixed anhydride procedures.

Wieland, Birr, and Flor employed *tert*-butyloxycarbonylamino acid anhydrides with excellent results in both solution and solid phase synthesis (Wieland *et al.*, 1971b, 1973; Flor *et al.*, 1973). An economic preparation of symmetrical anhydrides with the use of phosgene was developed. Several isolated crystalline products were not stable enough to be of preparative value. In practice, the anhydrides are used *in situ*. Overall, Wieland *et al.* consider the method to be at least as effective as the mixed anhydride procedure.

Hagenmaier and Frank (1972) used symmetrical anhydrides, prepared *in situ* by carbodiimide, in solid phase synthesis with excellent results. Compared with standard carbodiimide coupling considerably better yields were obtained. The use of this procedure is increasing (see, e.g., Bayer and Mutter, 1974; Hemmasi and Bayer, 1974; Yamashiro and Li, 1974; Wünsch and Stelzel, 1974, pp. 259–270). The change from the regular carbodiimide to the symmetrical anhydride procedure along with several other improvements effected remarkable increases in yields in solid phase syntheses of human ACTH and analogues, i.e., from previously 3% overall yield (Yamashiro and Li, 1973) to 20–30% yields (Lemaire *et al.*, 1977).

The *in situ* procedure of Hagenmaier and Frank (1972) is identical with that of Schüssler and Zahn (1962) who first isolated *stable symmetrical anhydrides.* Recently, Benoiton *et al.* (1978) reported that "symmetrical anhydrides are stable to washing by dilute acid or aqueous sodium bicarbonate at 23°C and consequently can readily be obtained pure using water-soluble carbodiimide. Furthermore, when pure, most of the crystalline products are stable at 23°C."*

In summary, symmetrical anhydrides of *N*-protected amino acids and peptides are very effective tools in peptide synthesis with good potential for increasing use. They might even be stored for ready use, but their principal advantage lies in fast reaction at low temperature leading to homogeneous products in high yields. Careful control of reaction conditions as elaborated for mixed anhydrides by Anderson *et al.* (1967b) might serve to minimize the danger of racemization and permit the use of symmetrical anhydrides for peptide segment condensation. The argument that one-half of these reagents is wasted appears trivial in view (a) of the routine use of nonrecoverable 2- to 3-fold excess of *tert*-butyloxycarbonylamino acids in solid phase synthesis and (b) of the substantial price reductions for amino acids over the years.

B. Other Related Methods

N-Carboxy anhydrides (oxazolidinedione-2,5) of α-amino acids (structure (**46**); Leuchs, 1906) have been widely used for the preparation of polyamino acids by polymerization (see Wünsch and Thamm, 1974). This intriguing class of compounds combines amino group protection with carboxyl group activation by *N*-carbonic acid. The single by-product is carbon dioxide. The

* Using 0.5 equiv of 1-ethyl-3-(dimethylaminopropyl)-carbodiimide · HCl in CH_2Cl_2, 2 hr at 0°C for *tert*-butyloxycarbonylamino acids, 1 hr at 23°C for N^{α}-benzyloxycarbonylamino acids, followed by washing with acid and bicarbonate, the following pure anhydrides (L-isomers) have been obtained in 60–80% yields (Benoiton *et al.*, 1978): $[Boc—Ala]_2O$, $[Boc—Phe]_2O$, $[Boc—Val]_2O$, $[Boc—Ile]_2O$, $[Boc—Met]_2O$, $[Boc—Tyr(Bzl)]_2O$, $[Boc—Asp(OBzl)]_2O$, $[Boc—Glu(OBzl)]_2O$, $[Z—Ala]_2O$, $[Z—Val]_2O$, $[Z—Phe]_2O$, $[Z—MeLeu]_2O$, $[Z—Leu]_2O$.

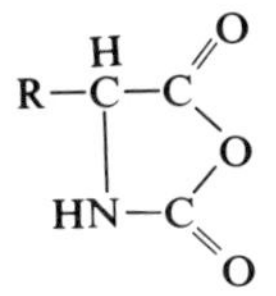

(46)

mechanism of the frequently studied polymerization reaction has recently been reinvestigated (Kricheldorf, 1977).

In sequential peptide synthesis *N*-carboxy anhydrides have been used by Hirschmann *et al.* (1971; see also Hirschmann, 1971) in ribonuclease A synthesis. Several N-substituted *N*-carboxy anhydrides, such as the *o*-nitrophenylsulfenyl derivatives (structure **47**) proved to be useful in stepwise

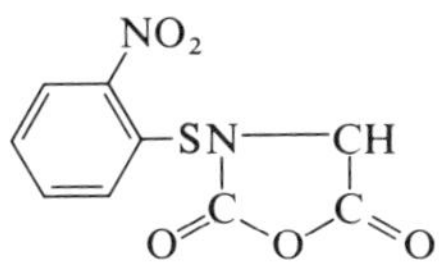

(47)

synthesis, see, for example, Halstrøm *et al.* (1974) and Katakai (1975; see also Katakai and Oya, 1975).

Mixed carboxylic acid anhydrides of N^{α}-alkyloxycarbonylamino acids (structure **49**) have been successfully employed in peptide synthesis (Wieland

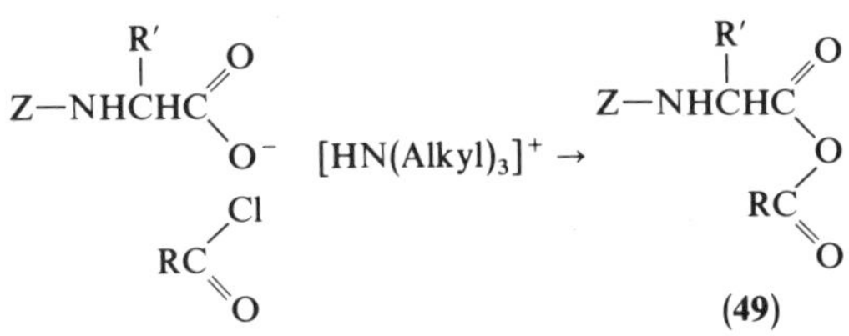

(48) **(49)**

R = Alkyl, aryl

et al., 1950; Vaughan and Osato, 1951). The pivalic mixed anhydride [**49**, R = $C(CH_3)_3$] (Zaoral, 1962) deserves special mention because of its remarkable capability of forming peptide bonds with good results in cases where other carboxylic as well as carbonic mixed anhydrides consistently fail. For example, mixed carbonic anhydrides of N^{α}-alkyloxycarbonyl-asparagine undergo undesired side reactions (see Wünsch and Stelzel, 1974, also see pp. 214–221; Rydon and Smith, 1956) whereas the use of pivalic anhydride provided Z-Asn-Cys(Bzl)-OMe (Zaoral, 1962) and Boc-Asn-Pro-OBzl (Wolman *et al.*, 1972) in 78% and 85% yields, respectively. Similarly, pivalic acid anhydrides were formed smoothly with N^{α}-tosylamino acids (Zaoral, 1962) and hindered compounds such as benzyloxycarbonyl-β-aminoisovaleric acid where other anhydride procedures failed or produced mixtures [see Zaoral and Rudinger (1961) and Drey *et al.* (1973), respectively]. Because of its excellent performance, the pivalic mixed anhydride procedure has found routine use in G. W. Kenner's laboratory (Galpin *et al.*, 1976).

VII. OPPORTUNITIES AND CONSTRAINTS

A. Comparative Evaluation

The mixed carbonic anhydride method combines several unique features. Its principal strength lies in very rapid reaction at low temperature. Consequently, coupling yields are frequently very high. The by-products are volatile which greatly facilitates work-up. Side reactions are few and their extent is often small. It is by far the most economic method of peptide synthesis.

The overall effectiveness of the mixed anhydride procedure in comparison with those of the active ester (AE), azide (AZ), and carbodiimide (DCC) methods, is (a) superior to all in rapid reaction rate at low temperature, (b) superior to AZ and DCC, but inferior to AE in giving rise to side reactions, (c) superior to AE, but inferior to AZ and DCC—HOBt in large segment condensation, and (d) at par with DCC but inferior to AE and AZ in maintaining chiral integrity.

Among the constraints, the predominant one is the high liability of racemization which makes it necessary to observe certain reaction conditions such as absence of excess base with extraordinary care. Furthermore, the competition of the azide and DCC—HOBt procedures in peptide segment condensation has effectively diminished the use of mixed anhydrides. Thus the efficiency of racemization control under the Anderson conditions in actual practice remains to be established.

B. Prognosis

It is clear that at present the mixed anhydride method is under-utilized. The long standing cautious attitude against using the procedure for anything else than coupling of single amino acid residues has not been overcome since the conditions for racemization suppression became known. However, if the recent promise of stable symmetrical anhydrides should hold, an increase in the use of these anhydrides would be predictable.

ACKNOWLEDGMENT

I want to thank Mrs. Debra Dabaghian for her excellent work in the preparation of the manuscript including the arrangement of the tables, and Mr. R. C. Makofske for help with the literature references.

REFERENCES

Abe, H., Sato, K., Kato, T., and Izumiya, N. (1976). *Bull. Chem. Soc. Jpn.* **49**, 3113–3118.

Albertson, N. F. (1962). *Org. React.* **12**, 157–355.

Amiard, G., Heymes, R., and Velluz, L. (1955). *Bull. Soc. Chim. Fr.* pp. 191–193.

Anantharamaiah, G. M., and Sivanandaiah, K. M. (1974). *Int. J. Pept. Protein Res.* **6**, 51–54.

Anderson, G. W. (1970). *In* "Peptides: Chemistry and Biochemistry" (B. Weinstein and S. Lande, eds.), pp. 255–266. Dekker, New York.

Anderson, G. W. (1972). *In* "Progress in Peptide Research" (S. Lande, ed.), pp. 343–345. Gordon & Breach, New York.

Anderson, G. W., and Callahan, F. M. (1958). *J. Am. Chem. Soc.* **80**, 2902–2903.

Anderson, G. W., Callahan, F. M., and Zimmerman, J. E. (1965). *Acta Chim. Acad. Sci. Hung.* **44**, 51–59.

Anderson, G. W., Zimmerman, J. E., and Callahan, F. M. (1966). *J. Am. Chem. Soc.* **88**, 1338–1339.

Anderson, G. W., Callahan, F. M., and Zimmerman, J. E. (1967a). *J. Am. Chem. Soc.* **89**, 178.

Anderson, G. W., Zimmerman, J. E., and Callahan, F. M. (1967b). *J. Am. Chem. Soc.* **89**, 5012–5017.

Antonovics, I., and Young, G. T. (1965). *Chem. Commun.* pp. 398–399.

Antonovics, I., and Young, G. T. (1967). *J. Chem. Soc. C* pp. 595–601.

Antonovics, I., Heard, A. L., Hugo, J., Williams, M. W., and Young, G. T. (1966). *In* "Peptides, Proceedings of the Sixth European Peptide Symposium" (L. Zervas, ed.), pp. 121–130. Pergamon Press, New York.

Applewhite, T. H., and Nelson, J. S. (1964). *Tetrahedron Lett.* pp. 819–825.

Barton, M. A., Lemieux, R. U., and Savoie, J. Y. (1973). *J. Am. Chem. Soc.* **95**, 4501–4506.

Battersby, A. R., and Robinson, J. C. (1955). *J. Chem. Soc.* pp. 259–269.

Bayer, E., and Mutter, M. (1974). *Chem. Ber.* **107**, 1344–1352.

Beddell, C. R., Fraser, P. J., Gilbert, D., Goodford, P. J., Lowe, L. A., and Wilkinson, S. (1975). *J. Med. Chem.* **18**, 417–423.

Beecham, A. F. (1957). *J. Am. Chem. Soc.* **79**, 3257–3261.

Belleau, B., Martel, R., Lacasse, G., Menard, M., Weinberg, N. L., and Perron, Y. G. (1968). *J. Am. Chem. Soc.* **90**, 823–824.

Belleau, D., and Malek, G. (1968). *J. Am. Chem. Soc.* **90**, 1651–1652.

Benoiton, N. L., Chen, F. M. F., and Kuroda, K. (1978). *Abstr. Mid Atl. Reg. Am. Chem. Soc. Soc. Meet.*, Hunt Valley, D, April 5–7, 1978.

Benson, J. R., and Hare, P. E. (1975). *Proc. Natl. Acad. Sci. U.S.A.* **72**, 619–622.

Bergmann, M., and Zervas, L. (1929). *Biochem. Z.* **203**, 280–292.

Berse, C., Massiah, T., and Piche, L. (1961). *J. Org. Chem.* **26**, 4514–4516.

Beyerman, H. C. (1972). *In* "Chemistry and Biology of Peptides" (J. Meienhofer, ed.), pp. 351–357. Ann Arbor Sci. Publ., Ann Arbor, Michigan.

Beyerman, H. C., de Leer, E. W. B., and Floor, J. (1973). *Recl. Trav. Chim. Pays-Bas* **92**, 481–492.

Beyerman, H. C., Hirt, J., Kranenburg, P., Syrier, J. L. M., and van Zon, A. (1974). *Recl. Trav. Chim. Pays-Bas* **93**, 256–257.

Beyerman, H. C., Rammeloo, T., Renirie, J. F. M., Syrier, J. L. M., and van Zon, A. (1976). *Recl. Trav. Chim. Pays-Bas* **95**, 143–148.

Bodanszky, M. (1960). *Ann. N.Y. Acad. Sci.* **88**, 655–664.

Bodanszky, M. (1971). *In* "Prebiotic and Biochemical Evolution" (A. P. Kimball and J. Oro, eds.), pp. 217–222. North-Holland Publ., Amsterdam.

Bodanszky, M. (1977). *In* "Peptides, Proceedings of the Fifth American Peptide Symposium" (M. Goodman and J. Meienhofer, eds.), pp. 1–13. Wiley, New York.

Bodanszky, M., and du Vigneaud, V. (1959). *J. Am. Chem. Soc.* **81**, 5688–5691.

Bodanszky, M., and Ondetti, M. A. (1966). "Peptide Synthesis." Wiley, New York.

Bodanszky, M., and Tolle, J. C. (1977). *Int. J. Pept. Protein Res.* **10**, 380–384.

Bodanszky, M., Klausner, Y. S., and Ondetti, M. A. (1976). "Peptide Synthesis," 2nd ed. Academic Press, New York.

Bodanszky, M., Fink, M. L., Klausner, Y. S., Natarajan, S., Tatemoto, K., Yiotakis, A. E., and Bodanszky, A. (1977). *J. Org. Chem.* **42**, 149–152.

Boissonnas, R. A. (1951). *Helv. Chim. Acta* **34**, 874–879.

Boissonnas, R. A., and Schumann, I. (1952). *Helv. Chim. Acta* **35**, 2229–2235.

Brown, J., Lauren, D. R., and Williams, R. E. (1972). *In* "Chemistry and Biology of Peptides" (J. Meienhofer, ed.), pp. 217–219. Ann Arbor Sci. Publ., Ann Arbor, Michigan.

Chen, F. M. F., and Benoiton, N. L. (1978). *Can. J. Biochem.* **56**, 38–42.

Determann, H. (1967). *In* "Peptides, Proceedings of the Eighth European Peptide Symposium" (H. C. Beyerman, A. van de Linde, and W. Massen van den Brink, eds.), pp. 73–78. North-Holland Publ., Amsterdam.

Determann, H., and Kahle, I. (1969). *Justus Liebig's Ann. Chem.* **725**, 203–211.

Determann, H., and Wieland, T. (1963). *Justus Liebig's Ann. Chem.* **670**, 136–140.

Drey, C. N. C., Lowbridge, J., and Ridge, R. J. (1973). *J. Chem. Soc., Perkins Trans. 1* pp. 2001–2006.

Dzieduszycka, M., Smulkowski, M., and Taschner, E. (1973). *In* "Peptides 1972" (H. Hanson and H.-D. Jakubke, eds.), pp. 103–107. North-Holland Publ., Amsterdam.

Elliott, D. F., and Russell, D. W. (1957). *Biochem. J.* **66**, 49 p.

Erlanger, B. F., Sachs, H., and Brand, E. (1954). *J. Am. Chem. Soc.* **76**, 1806–1810.

Farrington, J. A., Hextall, P. J., Kenner, G. W., and Turner, J. M. (1957). *J. Chem. Soc.* pp. 1407–1413.

Feurer, M. (1977). *In* "Peptides, Proceedings of the Fifth American Peptide Symposium" (M. Goodman and J. Meienhofer, eds.), pp. 448–457. Wiley, New York.

Finn, F. M., and Hofmann, K. (1976). *Proteins, 3rd Ed.* **2**, 105–253.

Flor, F., Birr, C., and Wieland, T. (1973). *Justus Liebig's Ann. Chem.* **1973**, 1601–1605.

Gabriel, T. F., Michalewsky, J., and Meienhofer, J. (1976). *J. Chromatogr.* **129**, 287–293.

Gabriel, T. F., Jimenez, M. H., Felix, A. M., Michalewsky, J., and Meienhofer, J. (1977). *Int. J. Pept. Protein Res.* **9**, 129–136.

Galpin, I. J., Handa, B. K., Hudson, D., Jackson, A. G., Kenner, G. W., Ohlsen, S. R., Ramage, R., Singh, B., and Tyson, R. G. (1976). *In* "Peptides 1976" (A. Loffet, ed.), pp. 247–256. Editions de l'Université de Bruxelles, Belgium.

Gillessen, D., Schnabel, E., and Meienhofer, J. (1963). *Justus Liebig's Ann. Chem.* **667**, 164–171.

Goodman, M., and Glaser, C. (1970). *In* "Peptides: Chemistry and Biochemistry" (B. Weinstein and S. Lande, eds.), pp. 267–335. Dekker, New York.

Goodman, M., and Levine, L. (1964). *J. Am. Chem. Soc.* **86**, 2918–2922.

Goodman, M., and McGahren, W. J. (1965). *J. Am. Chem. Soc.* **87**, 3028–3029.

Goodman, M., and Stueben, K. C. (1959). *J. Am. Chem. Soc.* **81**, 3980–3983.

Goodman, M., and Stueben, K. C. (1962). *J. Org. Chem.* **27**, 3409–3416.
Goodman, M., Schmitt, E. E., and Yphantis, D. A. (1962). *J. Am. Chem. Soc.* **84**, 1283–1303.
Goodman, M., Keogh, P., and Anderson, H. (1977). *Bioorg. Chem.* **6**, 239–247.
Greenstein, J. P., and Winitz, M. (1961). "Chemistry of the Amino Acids," Vol. 2, Chapter 10, pp. 763–1295. Wiley, New York.
Guttmann, S. (1960). *Chimia* **14**, 368.
Hagenmeier, H., and Frank, H. (1972). *Hoppe-Seyler's Z. Physiol. Chem.* **353**, 1973–1976.
Halstrøm, J., Brunfeldt, K., and Kovacs, K. (1974). *Hoppe-Seyler's Z. Physiol. Chem.* **355**, 82–84.
Heard, A. L., and Young, G. T. (1963). *J. Chem. Soc.* pp. 5807–5810.
Heber, H., Faulstich, H., and Wieland, T. (1974). *Int. J. Pept. Protein Res.* **6**, 381–389.
Hemmasi, B., and Bayer, E. (1974). *Hoppe-Seyler's Z. Physiol. Chem.* **355**, 481–489.
Hillmann-Elies, A., Hillmann, G., and Jatzkewitz, H. (1953). *Z. Naturforsch., Teil B* **8**, 445–446.
Hirschmann, R. (1971). *Intra-Sci. Chem. Rep.* **5**, 203–228.
Hirschmann, R., Schwam, H., Strachan, R. G., Schoenewaldt, E. F., Barkemeyer, H., Miller, S. M., Conn, J. B., Garsky, V., Veber, D. F., and Denkewalter, R. G. (1971). *J. Am. Chem. Soc.* **93**, 2746–2754.
Hoffmann, J. A., and Tilak, M. A. (1975). *Org. Prep. Proced. Int.*, Vol. 7, pp. 215–219.
Iselin, B., and Schwyzer, R. (1960). *Helv. Chim. Acta* **43**, 1760–1766.
Ivanov, V. T., Mikhaleva, I. I., Volpina, O. M., Myagkova, M. A., and Deigin, V. I. (1976). *In* "Peptides 1976" (A. Loffet, ed.), pp. 219–229. Editions de l'Université de Bruxelles, Belgium.
Izeboud, E., and Beyerman, H. C. (1978). *Recl. Trav. Chim. Pays-Bas* **97**, 1–6.
Izumiya, N., and Muraoka, M. (1969). *J. Am. Chem. Soc.* **91**, 2391–2392.
Izumiya, N., Muraoka, M., and Aoyagi, H. (1971). *Bull. Chem. Soc. Jpn.* **44**, 3391–3395.
Jakubke, H.-D., and Baumert, A. (1974). *J. Prakt. Chem.* **316**, 67–74.
Kapoor, A., and Gerencser, L. W. (1969). *J. Pharm. Sci.* **58**, 976–980.
Katakai, R. (1975). *J. Org. Chem.* **40**, 2697–2702.
Katakai, R., and Oya, M. (1975). *Biopolymers* **14**, 2507–2515.
Kemp, D. S. (1973). *In* "Peptides 1971" (H. Nesvadba, ed.), pp. 1–19. North-Holland Publ., Amsterdam.
Kemp, D. S., and Chien, S. W. (1967). *J. Am. Chem. Soc.* **89**, 2743–2745 and 2745–2746.
Kemp, D. S., and Rebek, J., Jr. (1970). *J. Am. Chem. Soc.* **92**, 5792–5793.
Kemp, D. S., Wang, S. W., Busby, G., III, and Hugel, G. (1970a). *J. Am. Chem. Soc.* **92**, 1043–1055.
Kemp, D. S., Bernstein, Z., and Rebek, J., Jr. (1970b). *J. Am. Chem. Soc.* **92**, 4756–4757.
King, F. E., Clark-Lewis, J. W., Wade, R., and Swindin, W. A. (1957). *J. Chem. Soc.* pp. 873–880.
Kiso, Y., and Yajima, H. (1972). *J. Chem. Soc., Chem. Commun.* pp. 942–943.
Kiso, Y., Kai, Y., and Yajima, H. (1973). *Chem. Pharm. Bull.* **21**, 2507–2510.
Klausner, Y. S., and Bodanszky, M. (1972). *Synthesis* pp. 453–463.
Klausner, Y. S., Lin, C. Y., Mutt, V., and Bodanszky, M. (1973). *Bioorg. Chem.* **2**, 345–353.
König, W., and Geiger, R. (1970). *Chem. Ber.* **103**, 788–798.
Kopple, K. D., and Renick, R. J. (1958). *J. Org. Chem.* **23**, 1565–1567.
Kovacs, J. (1979). *In* "The Peptides: Synthesis" (E. Gross and J. Meienhofer, eds.), Vol. 2, Chapter 8. Academic Press, New York.
Kovacs, J., Kisfaludy, L., Ceprini, M. Q., and Johnson, R. H. (1969). *Tetrahedron* **25**, 2555–2573.
Kovacs, J., Kalita, C., and Ghatak, U. R. (1972). *J. Org. Chem.* **37**, 30–34.
Kovacs, J., Cover, R., Jham, G., Hsieh, Y., and Kalas, T. (1975). *In* "Peptides: Chemistry, Structure and Biology" (R. Walter and J. Meienhofer, eds.), pp. 317–324. Ann Arbor Sci. Publ., Ann Arbor, Michigan.
Kricheldorf, H. R. (1977). *Makromol. Chem.* **178**, 905–939 and 1959–1970.

Krumdieck, C. L., and Baugh, C. M. (1969). *Biochemistry* **8**, 1568–1572.
Lauren, D. R., and Williams, R. E. (1972). *Tetrahedron Lett.* pp. 2665–2668.
Lemaire, S., Yamashiro, D., Behrens, C., and Li, C. H. (1977). *J. Am. Chem. Soc.* **99**, 1577–1580.
Leuchs, H. (1906). *Ber. Dtsch. Chem. Ges.* **39**, 857–861.
Longosz, E. J., and Tarbell, D. S. (1961). *J. Org. Chem.* **26**, 2161–2169.
Lübke, K., and Schröder, E. (1961). *Z. Naturforsch. Teil B* **16**, 765–767.
McDermott, J. R., and Benoiton, N. L. (1973). *Can. J. Chem.* **51**, 2562–2570.
Meienhofer, J. (1970). *J. Am. Chem. Soc.* **92**, 3771–3777.
Meienhofer, J., and Schnabel, E. (1965). *Z. Naturforsch., Teil B* **20**, 661–665.
Meienhofer, J., Jacobs, P. M., Godwin, H. A., and Rosenberg, I. H. (1970). *J. Org. Chem.* **35**, 4137–4140.
Merrifield, R. B. (1963). *J. Am. Chem. Soc.* **85**, 2149–2154.
Merrifield, R. B. (1969). *Adv. Enzymol.* **32**, 221–296.
Merrifield, R. B., Mitchell, A. R., and Clarke, J. E. (1974). *J. Org. Chem.* **39**, 660–668.
Mühlemann, M., Titov, M. I., Schwyzer, R., and Rudinger, J. (1972). *Helv. Chim. Acta* **55**, 2854–2860.
Munekata, E., Faulstich, H., and Wieland, T. (1977). *Angew. Chem., Int. Ed. Engl.* **16**, 267–268.
Muramatsu, I. (1961). *Nippon Kagaku Zasshi* **82**, 83–87, *Chem. Abstr.* **56**, 10273 (1962).
Muramatsu, I., and Hagitani, A. (1959). *Nippon Kagaku Zasshi* **80**, 1497–5001, *Chem. Abstr.* **55**, 6394 (1961).
Nebelin, E., Faulstich, H., and Wieland, T. (1973). *Justus Liebig's Ann. Chem.* **1973**, pp. 45–49.
Neuberger, A. (1948). *Adv. Protein Chem.* **4**, 297–383 (see p. 359).
Neubert, K., Mansfeld, H.-W., Mansfeld, J., and Jakubke, H.-D. (1977). *J. Prakt. Chem.* **319**, 34–38.
North, M. B., and Young, G. T. (1955). *Chem. Ind.* (*London*) p. 1597.
Ondetti, M. A., and Thomas, P. L. (1965). *J. Am. Chem. Soc.* **87**, 4373–4380.
Perrin, D. D., Armarego, W. L. F., and Perrin, D. R. (1966). "Purification of Laboratory Chemicals." Pergamon, Oxford.
Pettit, G. R. (1976). "Synthetic Peptides," Vol. 4, pp. 22–32. Elsevier, Amsterdam.
Prox, A., Weygand, F., König, W., and Schmidhammer, L. (1966). *In* "Peptides, Proceedings of the Sixth European Peptide Symposium" (L. Zervas, ed.), pp. 139–154. Pergamon Press, New York.
Prox, A., Schmid, J., and Ottenheym, H. (1969). *Justus Liebig's Ann. Chem.* **722**, 179–196.
Rothe, M., and Kunitz, F. W. (1957). *Justus Liebig's Ann. Chem.* **609**, 88–102.
Rydon, H. N., and Smith, P. W. G. (1956). *J. Chem. Soc.* pp. 3642–3650.
Rzeszotarska, B., and Kmiecik-Chmura, H. (1977). *Rocz. Chem.* **51**, 1523–1526.
Rzeszotarska, B., Taschner, E., and Kuziel, A. (1966). *In* "Peptides, Proceedings of the Sixth European Peptide Symposium" (L. Zervas, ed.), pp. 131–133. Pergamon Press, New York.
Sakakibara, S., and Itoh, M. (1967). *Bull. Chem. Soc. Jpn.* **40**, 656–661.
Sarges, R., and Witkop, B. (1965). *J. Am. Chem. Soc.* **87**, 2020–2027.
Schellenberg, P., and Ullrich, J. (1959). *Chem. Ber.* **92**, 1276–1287.
Schnabel, E. (1963). *Justus Liebig's Ann. Chem.* **667**, 171–179.
Schnabel, E. (1964). *Z. Naturforsch., Teil B* **19**, 120–124.
Schnabel, E. (1965). *Justus Liebig's Ann. Chem.* **688**, 238–249.
Schnabel, E. (1966). *In* "Peptides," Proceedings of the Sixth European Peptide Symposium (L. Zervas, ed.), pp. 71–78. Pergamon Press, New York.
Schröder, E., and Lübke, K. (1965). "The Peptides," Vols. 1 and 2. Academic Press, New York.
Schüssler, H., and Zahn, H. (1962). *Chem. Ber.* **95**, 1076–1080.
Schwarz, H., and Bumpus, F. M. (1959). *J. Am. Chem. Soc.* **81**, 890–897.
Sheehan, J. C., and Frank, V. S. (1950). *J. Am. Chem. Soc.* **72**, 1312–1316.

Sheehan, J. C., and Yang, D.-D. H. (1958). *J. Am. Chem. Soc.* **80**, 1154–1158.
Sipos, F., and Gaston, D. W. (1971). *Synthesis* p. 321.
Smart, N. A., Young, G. T., and Williams, M. W. (1960). *J. Chem. Soc.* pp. 3902–3912.
Steglich, W., Mayer, D., Barocio de la Lama, X., Tanner, H., and Weygand, F. (1967). *In* "Peptides, Proceedings of the Eighth European Peptide Symposium" (H. C. Beyerman, A. Van De Linde, and W. Massen Van Den Brink, eds.), pp. 67–72. North-Holland Publ., Amsterdam.
Stewart, F. H. C. (1965). *Aust. J. Chem.* **18**, 887–901.
Stewart, F. H. C. (1966). *Aust. J. Chem.* **19**, 489–501.
Tarbell, D. S. (1969). *Acc. Chem. Res.* **2**, 296–300.
Taschner, E., Lubiewska, L., Smulkowski, M., and Wojciechowska, H. (1968). *Experientia* **24**, 521–523.
Tilak, M. A. (1970). *Tetrahedron Lett.* pp. 849–854.
Tilak, M. A., and Hollinden, C. S. (1968a). *Tetrahedron Lett.* pp. 1297–1300.
Tilak, M. A., and Hollinden, C. S. (1968b). *In* "Peptides 1968" (E. Bricas, ed.), pp. 173–178. North-Holland Publ., Amsterdam.
Tilak, M. A., Hendricks, M. L., and Wedel, D. S. (1972). *In* "Progress in Peptide Research" (S. Lande, ed.), Vol. 2, pp. 351–359. Gordon and Breach, New York.
Udenfriend, S., Stein, S., Böhlen, P., Dairman, W., Leimgruber, W., and Weigele, M. (1972). *Science* **178**, 871–872.
van Zon, A., and Beyerman, H. C. (1973). *Helv. Chim. Acta* **56**, 1729–1740.
van Zon, A., and Beyerman, H. C. (1976). *Helv. Chim. Acta* **59**, 1112–1126.
Vaughan, J. R., Jr. (1951). *J. Am. Chem. Soc.* **73**, 3547.
Vaughan, J. R., Jr. (1952). *J. Am. Chem. Soc.* **74**, 6137–6139.
Vaughan, J. R., Jr., and Eichler, J. A. (1953). *J. Am. Chem. Soc.* **75**, 5556–5560.
Vaughan, J. R., Jr., and Osato, R. L. (1951). *J. Am. Chem. Soc.* **73**, 5553–5555.
Vaughan, J. R., Jr., and Osato, R. L. (1952). *J. Am. Chem. Soc.* **74**, 676–678.
Vaughn, J. W. (1967). *In* "The Chemistry of Non-aqueous Solvents" (J. J. Lagowski, ed.), Vol. 2. Academic Press, New York.
von Brunn-Leube, I., and Schramm, G. (1956). *Chem. Ber.* **89**, 2045–2055.
Wang, C. T., Kulesha, I. D., Stefko, P. L., and Wang, S. S. (1974). *Int. J. Pept. Protein Res.* **6**, 59–64.
Weigele, M., De Bernardo, S., and Leimgruber, W. (1973). *Biochem. Biophys. Res. Commun.* **50**, 352–356.
Weygand, F., and Di Bello, C. (1969). *Z. Naturforsch., Teil B* **24**, 314–317.
Weygand, F., Prox, A., Schmidhammer, L., and König, W. (1963a). *Angew. Chem., Int. Ed. Engl.* **2**, 183–188.
Weygand, F., Prox, A., Schmidhammer, L., and König, W. (1963b). *In* "Peptides, Proceedings of the Fifth European Peptide Symposium" (G. T. Young, ed.), pp. 97–107. Pergamon Press, Oxford.
Weygand, F., Steglich, W., Mayer, D., and von Philipsborn, W. (1964). *Chem. Ber.* **97**, 2023–2028.
Weygand, F., Prox, A., and König, W. (1966a). *Chem. Ber.* **99**, 1451–1460.
Weygand, F., Prox, A., and König, W. (1966b). *Chem. Ber.* **99**, 1446–1450.
Weygand, F., Huber, P., and Weiss, K. (1967). *Z. Naturforsch., Teil B* **22**, 1084–1085.
Wieland, H. (1912). *Ber. Dsch. Chem. Ges.* **45**, 484–493.
Wieland, T., and Bernhard, H. (1951). *Justus Liebig's Ann. Chem.* **572**, 190–194.
Wieland, T., and Sangl, I. (1964). *Justus Liebig's Ann. Chem.* **671**, 160–164.
Wieland, T., Kern, W., and Sehring, R. (1950). *Justus Liebig's Ann. Chem.* **569**, 117–121.
Wieland, T., Schäfer, W., and Bokelmann, E. (1951). *Justus Liebig's Ann. Chem.* **573**, 99–104.

Wieland, T., Heinke, B., Vogeler, K., and Morimoto, H. (1962). *Justus Liebig's Ann. Chem.* **655**, 189–194.

Wieland, T., Lüben, G., Ottenheym, H., Faesel, J., de Vries, J. X., Prox, A., and Schmid, J. (1968a). *Angew. Chem. Int. Ed. Engl.* **7**, 204–208.

Wieland, T., Faesel, J., and Faulstich, H. (1968b). *Justus Liebig's Ann. Chem.* **713**, 201–208.

Wieland, T., Lüben, G., and Ottenheym, H. (1969a). *Justus Liebig's Ann. Chem.* **722**, 173–178.

Wieland, T., Faesel, J., and Konz, W. (1969b). *Justus Liebig's Ann. Chem.* **722**, 197–209.

Wieland, T., Faulstich, H., and Fahrenholz, F. (1971a). *Justus Liebig's Ann. Chem.* **743**, 77–82.

Wieland, T., Birr, C., and Flor, F. (1971b). *Angew. Chem., Int. Ed. Engl.* **10**, 336.

Wieland, T., Lapatsanis, L., Faesel, J., and Konz, W. (1971c). *Justus Liebig's Ann. Chem.* **747**, 194–206.

Wieland, T., Flor, F., and Birr, C. (1973). *Justus Liebig's Ann. Chem.* **1973**, 1595–1600.

Wieland, T., Birr, C., Burgermeister, W., Trietsch, P., and Rohr, G. (1974). *Justus Liebig's Ann. Chem.* **1974**, 24–36.

Williams, M. W., and Young, G. T. (1963). *J. Chem. Soc.* pp. 881–889.

Williams, M. W., and Young, G. T. (1964). *J. Chem. Soc.* pp. 3701–3708.

Willstätter, H., and Waldschmidt-Leitz, E. (1921). *Ber. Dsch. Chem. Ges.* **54B**, 113–138.

Wolman, Y., Schejter, A., and Sokolovsky, M. (1972). *J. Am. Chem. Soc.* **94**, 1720–1723.

Wünsch, E. (1974). *In* "Houben-Weyl, Methoden der organischen Chemie" (E. Müller, ed.), Vol. 15, Part II, pp. 169–364. Thieme, Stuttgart.

Wünsch, E., and Stelzel, P. (1974). *In* "Houben-Weyl, Methoden der organischen Chemie" (E. Müller, ed.), Vol. 15, Part II, pp. 171–183. Thieme, Stuttgart.

Wünsch, E., and Thamm, P. (1974). *In* "Houben-Weyl, Methoden der organischen Chemie" (E. Müller, ed.), Vol. 15, Part II, pp. 397–428. Thieme, Stuttgart.

Wünsch, E., and Wendlberger, G. (1974). *In* "Houben-Weyl, Methoden der organischen Chemie" (E. Müller, ed.), Vol. 15, Part II, pp. 23–26 and 397–428. Thieme, Stuttgart.

Yajima, H., and Kawatani, H. (1971). *Chem. Pharm. Bull.* **19**, 1905–1913.

Yajima, H., Kawasaki, K., Okada, Y., Minami, H., Kubo, K., and Yamashita, I. (1968). *Chem. Pharm. Bull.* **16**, 919–928.

Yajima, H., Okada, Y., Kawatani, H., and Mizokami, N. (1969a). *Chem. Pharm. Bull.* **17**, 1229–1236.

Yajima, H., Mizokami, N., Okada, Y., and Kawasaki, K. (1969b). *Chem. Pharm. Bull.* **17**, 1958–1962.

Yamashiro, D., and Li, C. H. (1973). *J. Am. Chem. Soc.* **95**, 1310–1315.

Yamashiro, D., and Li, C. H. (1974). *Proc. Natl. Acad. Sci. U.S.A.* **71**, 4945–4949.

Young, G. T. (1967). *In* "Peptides, Proceedings of the Eight European Peptide Symposium" (H. C. Beyerman, A. van de Linde, and W. Massen van den Brink, eds.), pp. 55–66. North-Holland Publ., Amsterdam.

Zahn, H., Meienhofer, J., and Schnabel, E. (1965a). *Acta Chim. Acad. Sci. Hung.* **44**, 109–128.

Zahn, H., Bremer, H., and Zabel, R. (1965b). *Z. Naturforsch., Teil B* **20**, 653–660.

Zaoral, M. (1959). *Angew. Chem.* **71**, 743.

Zaoral, M. (1962). *Collect. Czech. Chem. Commun.* **27**, 1273–1277.

Zaoral, M., and Rudinger, J. (1961). *Collect. Czech. Chem. Commun.* **26**, 2316–2332.

Zervas, L., and Theodoropoulos, D. M. (1956). *J. Am. Chem. Soc.* **78**, 1359–1363.

Zervas, L., Borovas, D., and Gazis, E. (1963). *J. Am. Chem. Soc.* **85**, 3660–3666.

Chapter 7

Racemization in Peptide Synthesis

D. S. KEMP

I. Introduction 317
  A. General Considerations 317
  B. Reviews of the Racemization Problem 320
  C. The Numerical Problem Posed by Epimerization . . . 320
  D. Problems of Diastereomer Separation 323
  E. Experimental Methods—A Cautionary Note 324
II. Racemization of Amino Acids, Peptides, and their Unactivated Derivatives 325
  A. Racemization of Amino Acids 325
    1. Racemization in Dilute Aqueous Solution 325
    2. Racemization in Strongly Acidic Aqueous Solution . 326
    3. Racemization in Acetic Acid Solution 327
    4. Racemization in Aqueous Alkaline Solution . . . 328
  B. Epimerization of Peptides and Other Unactivated Amino Acid Derivatives 329
    1. Racemization during Acidic Hydrolysis or Methanolysis of Peptides 329
    2. Base-Catalyzed Epimerization of Peptides 329
    3. The Influence of N^{α}-Protective Groups on Rates of Racemization 331
    4. Racemization during the Acidic or Basic Deblocking Steps of Peptide Synthesis 332
    5. Epimerization of Diketopiperazines 332
    6. Racemization of N-Alkylamino Acid Derivatives . . 332
  C. Mechanism and Summary 333
    1. Mechanism 333
    2. Summary of Observations 335

THE PEPTIDES, VOL. 1

ISBN 0-12-304201-1

III. Racemization of Activated Acyl Derivatives of Amino Acids and Peptides 336
A. General Considerations 336
1. Primary Features of Racemization of Activated Acyl Derivatives 336
2. The Problem of Amine Basicity 338
B. Racemizing Tendencies of Activated, Urethane-Blocked Amino Acid Derivatives 339
1. Mechanistic Significance of the *N*-Acyl Protecting Group 339
2. Effects of Amine Leaving Group, Protecting Group, Side-Chain Substituent, and Solvent on Tertiary Amine-Catalyzed Racemization by the Enolization Mechanism 340
3. Racemization Expected during Coupling Reactions of Activated Urethane-Protected Amino Acids 341
C. Racemization of Activated Acyl Aminoacyl Derivatives—The 5(4H)-Oxazolone Mechanism 342
1. General Considerations 342
2. Evidence for the Intermediacy of 5(4H)-Oxazolones in Peptide Coupling Reactions 343
3. Catalysis of 5(4H)-Oxazolone Formation 345
4. Racemization and Aminolysis of 5(4H)-Oxazolones . 347
5. The Effect of Substituents on the Ease of 5(4H)-Oxazolone Formation 349
D. Comparison of the Two Racemization Mechanisms . . 350
IV. Control of Racemization during Peptide Synthesis 351
A. General Considerations 351
B. Racemization during Protective Group Transformation or Removal 352
C. Important Variables for Amide-Forming Reactions . . 352
1. Temperature 352
2. Amine Bases and Salts 353
3. Concentration 354
4. Solvent 354
D. Special Amino Acids 355
1. Cysteine 355
2. Histidine 355
3. COOH-Terminal Proline or Glycine 356
E. Racemization during Coupling Reactions of Solid-Phase Synthesis 357
F. Epimerization during Fragment-Condensation Coupling Reactions 357
1. Mixed Anhydrides and EEDQ (2-Ethyloxy-1-Ethyloxycarbonyl-1,2,-Dihydroquinoline) 357
2. Acyl Azides 359
3. The Dicyclohexylcarbodiimide–*N*-Hydroxysuccinimide and Dicyclohexylcarbodiimide–1-Hydroxybenzotriazole Procedures 360
V. Strategies for Minimizing Racemization during Amide-Forming Steps 362

A. The Key Issues: Properties and Accessibility of Acylating Agent 362
B. Design of Minimally Racemizing Activated Acyl Derivatives 362
1. Anchimeric Assistance of Aminolysis 362
2. Nature of Aminolytic Anchimeric Assistance . . . 363
C. Nonracemizing Routes to Acylating Agents 364
1. Trapping Agents 364
2. The Intramolecular Energy Leak 365
3. Backing-Off Synthesis and Safety-Catch Activation . 366
VI. Assays for Chiral Purity 366
A. Types of Assays 366
B. Assay Procedures 369
C. Selection of an Epimerization Assay 369
D. Chiral Integrity of Peptide Coupling Procedures . . . 372
E. Unresolved Problems 372
References 378

I. INTRODUCTION

A. General Considerations

Epimerization at the amino acid α-carbon has plagued the field of polypeptide synthesis throughout its history, placing severe constraints on the development of satisfactory synthetic routes to these molecules. This chapter is primarily concerned with aspects of the epimerization problem that are relevant to synthesis. Nevertheless, it seems appropriate to enlarge its scope to include the racemization behavior of simple amino acids and their derivatives, since similar factors influence the racemization of both amino acids and peptides derived from them.

In considering the terminology of the epimerization problem, it seemed to both the author and the editors that current usage is in need of some refinement to avoid awkward ambiguities. We have no desire to increase the complexity of the already abundant vocabulary for describing chiral molecules. Moreover, it is clear that any changes should be modest in aim, conforming as far as possible with precise present usage.

It seems appropriate to distinguish between terms applied to specific substances or their reactions and general reactions, and the following rules have been followed in this review:

1. Reference to specific molecules is exact. Alanine racemizes, as does any amino acid or peptide that bears a single chiral center. Isoleucine epimerizes, as does any peptide that bears several chiral centers but undergoes loss of chiral integrity at only one of them.

A rate constant for racemization, k_{rac}, describes the irreversible process, L → DL; a rate constant for epimerization, $k_{L \to D}$, describes the forward step of a reversible process, L ⇌ D.

When comparing tests of chiral integrity, it is important to recall that racemization measures the fraction of racemate that is formed, while epimerization measures only the fraction of D-epimer.

Enantiomeric case		Epimeric case
k_{rac}	Should be compared with	$2(k_{L \to D})$
%DL	Should be compared with	2(%D)

2. In accord with common usage, the term racemization is also applied to general processes or operations that result in loss of chiral integrity and that may occur with substances bearing either one or several chiral centers, provided the chirality at only one of these centers is affected. Thus, taken as a class, the enolization and 5(4H)-oxazolone-forming mechanisms (p. 39) of activated acyl derivatives are described as racemization mechanisms.

3. Chiral integrity is the term here used to describe the degree of contamination of a peptide by enantiomers or diastereomers. The mechanism of loss of chiral integrity at an amino acyl site can be described as α-inversion. These terms have been applied where they lend clarity or precision.

Perhaps the best understood α-inversion processes are the very slow racemization reactions of simple amino acids and their unactivated derivatives. With occasional exceptions to be noted, these appear to occur via reaction (1), simple enolization of the carbonyl, a mechanism that is strongly

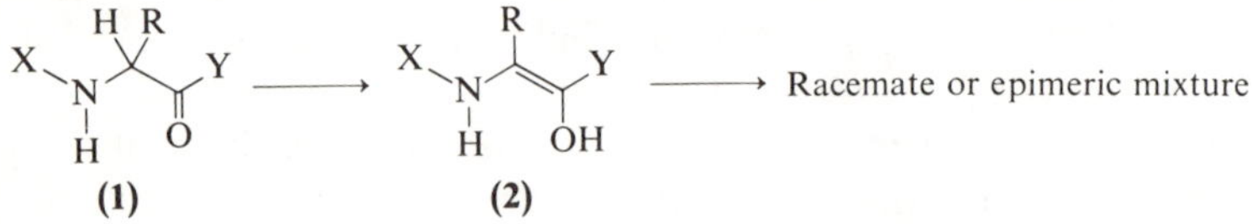

(1)

influenced by the nature of the substituents X, R, and Y, the temperature, and the basicity of the medium. These reactions usually involve a reaction condition and a single substance and are describable by a single rate constant or a set of parallel rate constants. The ubiquity of these reactions, as well as their simplicity, argues for their being considered first, even though they are extremely slow and contribute to the problems of maintaining chiral integrity in synthesis only under special circumstances.

The α-inversion reactions of acylating agents derived from amino acids and peptides occur much more rapidly, and their control is therefore central to the problem of stereospecific peptide synthesis. For this reason these reactions have received considerable careful attention, and their general mechanistic features are now clear, although detailed, predictive rationalizations still elude investigators in this complex area. Two features create complexity. First, the extent of epimerization of the product of a peptide coupling reaction usually depends on a rate ratio, since it depends on the rate of loss of chiral integrity of the acylating agent that generates it as well as on the lifetime of this species under coupling conditions. Second, two mechanisms, simple enolization, reaction (1), and 5(4H)-oxazolone formation, reaction (2), can result in configurational inversion at the α-site. For

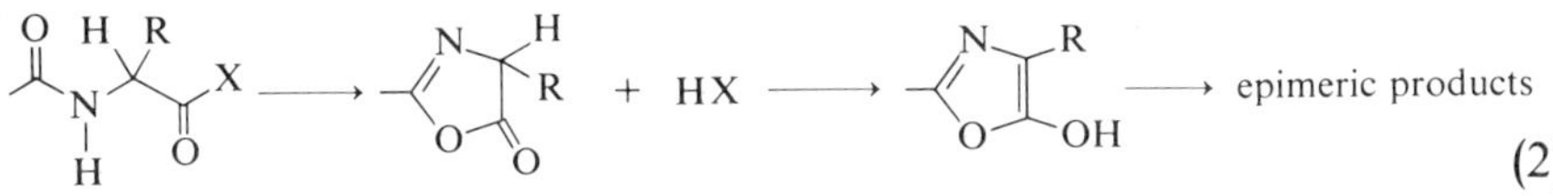

(2)

(3) **(4)** **(5)**

several important cases in which both mechanisms can contribute, the relative importance of each has been defined, and information is available concerning the reaction conditions and substituent effects that favor each mechanism. Such information provides a necessary background for the design of reagents that minimize contributions by either mechanism, and even though the available data form an incomplete picture, they can provide very useful guidelines for planning peptide synthesis with available reagents. Discussion of this topic must therefore combine mechanistic generalizations with empirical summaries of the observed effects of structure and reaction conditions on the racemization level.

For many of the most widely used assays, the error limits of the analysis are close to the upper limits of epimerization that are tolerable in practical synthesis. For this reason and because variables that may influence epimerization have yet to be defined, one expects and indeed often finds for the same experiment, discordances between results obtained by different assay procedures and from different laboratories. Because of the difficulties of monitoring chiral integrity in most practical synthetic situations, the results of model experiments must be viewed with a skeptical eye. Accordingly, in cases of unexplained inconsistency, high levels of racemization or epimerization are given greater weight in this chapter.

Along with the limits of tolerable epimeric content (considered in Section I,C), the available analytical procedures for monitoring this quantity define the practicality of studies of the phenomenon. The past decade has seen a proliferation of assays for chiral integrity of peptides and amino acid derivatives. Few of these have been extensively used or subjected to comparisons with other assays. It therefore seems inappropriate to attempt more than a listing of the available methods, grouped according to their procedural similarities, and accompanied by an overview of the problem of detection.

Through a combination of advances in synthesis and an increased knowledge of pitfalls that must be avoided, presently available methods appear to be marginally capable of generating polypeptides that meet reasonable standards for chiral homogeneity. A hoped for improvement in that art would result from an increased capacity to predict situations in which unacceptably high levels of epimer are generated, and further advances in our understanding of mechanisms should lead to that improvement. Realization of the ideal of synthesis in which the problem of racemization can be ignored rests with the development of reagents and procedures that are truly racemization-free. It seems appropriate to speculate briefly on strategies that may achieve that ideal, and these are considered in Section V,C, which is primarily devoted to existing reagents that have been constructed to minimize racemization.

B. Reviews of the Racemization Problem

Since 1968, critical yearly reviews of the synthesis-related aspects of the racemization problem have appeared in "Amino-acids, Peptides, and Proteins," the Specialist Periodical Reports of the Chemical Society. The Proceedings of Peptide Symposia, European and American, have contained summaries and overviews of the problem by Young, Goodman, and others (Antonovics *et al.*, 1966; Young, 1967; Goodman and Glaser, 1970a; Kemp, 1973; Kovacs *et al.*, 1975).

Older reviews of synthesis or of the properties of amino acids contain valuable summaries, along with information that has been superceded: (Neuberger, 1948; Greenstein and Winitz, 1961; Schröder and Lübke, 1965). A short, practical summary is available in Bodanszky *et al.* (1976). The magisterial treatise by Wünsch and his collaborators (Wünsch, 1974) provides much useful information, including a convenient summary by Jaeger, with experimental examples, of the important racemization tests.

C. The Numerical Problem Posed by Epimerization

How much racemization at a single amino acid residue can be tolerated in a peptide synthesis? The answer clearly must vary with the size of the ultimate synthetic peptide, the strategy used to prepare it, the reliability of purification procedures and the number of stages at which they can be used, as well as the use for which the final product is intended. Underlying these variables and for a large peptide dominating them, is the grim arithmetic by which mere traces of chiral impurities at each site combine to result in a product which, overall, is diastereomerically inhomogeneous.

Although the analysis has been discussed previously (see, for example, Greenstein and Winitz, 1961, p. 945), it is so central to understanding the

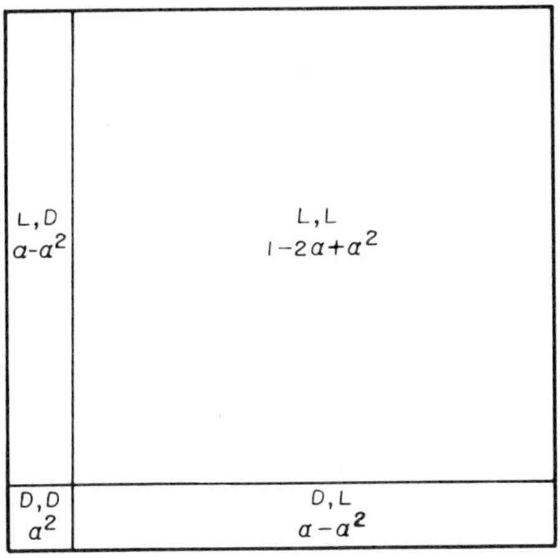

Figure 1. Stereoisomers of a dipeptide containing a fraction, α, of D-epimer at each site.

problems of synthesis that it deserves brief attention. Consider a dipeptide, H—A—B—OH, that contains a fraction, α, of D-epimer at each amino acid site. As shown in Fig. 1, four stereoisomers are possible, and their ratios are conveniently represented as areas of sections of a square. For small values of α, the important stereoisomers are the LL, LD, and DL in relative abundances, $1\text{–}2\alpha$, α, and α, respectively.

For a polypeptide composed of n amino acids, there are in principle 2^n stereoisomers. In practice, for low values of α ($\alpha < 0.01$), only $(n + 1)$ of these are expected in significant amounts, these being the all L-peptide and the n peptides that have the D-configuration at only one amino acid site. The ratios are, roughly, α for each of these and $(1 - n\alpha)$ for the all L-peptide.

The reduction of impurities from 2^n to n for small α is a welcome simplification, but unfortunately, the n impurities that can be expected are not to be taken lightly. For example, a peptide containing one hundred amino acids, each with 0.2% D-enantiomer contaminant, consists of 82% all L-peptide and 16% of an impurity consisting of 100 diastereomeric peptides. At least for some purposes, such a mixture is unacceptably impure.

If one assumes no separation of impurities during intermediary steps of a synthesis, 0.05–0.2% D-enantiomer at each amino acid site is a likely figure for "trouble-free" steps of many synthetic schemes. It is also likely that at a few "vulnerable" sites, the epimeric content is much higher, perhaps as high as 5–10%. Figure 2 is provided as an exercise in visualizing the consequences of constructing a 50-residue peptide with 0.2% D-enantiomer at 44 sites and 5% D-enantiomer at 5 sites. The primary consequence of the latter inclusion is an increase in the number of diastereomeric impurities that contribute significantly to product inhomogeneity.

All L	70.8% (1 isomer)
One D at any of 5 sites	18.6% (5 isomers)
One D at any of 44 sites	6.3% (44 isomers)
One D at any of 49 sites	24.9% (49 isomers)
Two D at any of 5 sites	2.0% (10 isomers)
One D at any of 5, one D at any of 44 sites	1.6% (220 isomers)
Two D at any of 49 sites	3.6% (230 isomers)

A total of 280 isomers, 60 major and 220 minor, account for 99.3% of product

For a given type of isomer, number of isomers = $\dfrac{n!}{k!\,(n-k)!}$

n = number of sites where D occurs;

k = number of D-epimers per isomer

$$0.708 = 1(0.95)^5(0.998)^{44} \qquad 0.020 = 10(0.95)^3(0.05)^2(0.998)^{44}$$

Figure 2. Stereoisomers of a pentacontapeptide for which α equals 0.002 for 44 sites and α equals 0.05 for 5 sites.

In principle, there are three strategies for resolving this problem. One might rely on sensitive methods for assay of diastereomeric content to indicate when the problem of chiral inhomogeneity exists. In the absence of such assays, one might employ exhaustive purification procedures to remove diastereomers. In the absence of reliable purification procedures one might seek synthetic conditions that provide a large safety margin, yielding lower racemization levels in model experiments than the chiral requirements of the synthesis at hand.

Although much recent progress has been made in assay procedures, these still appear to fall short of the most exacting needs of homogeneity control. Purification procedures face a formidable challenge with complex mixtures of diastereomers, and at present the reliability of existing procedures and the potential for their extension are equally difficult to judge. Aspects of these problems are considered in Section I,D.

The third approach, development of reagents and reaction conditions that provide a safety margin for control of epimerization, has motivated much recent work in the peptide field. It is clear that an important requirement of this approach is an analytical method that can detect chiral inhomogeneity at levels at least an order of magnitude below those required in the most stringent practical synthesis. Among several convenient assays that approach this limit (Section V), the isotopic dilution assays appear to be the most suited to the needs of the problem (Kemp *et al.*, 1970a).

If one adopts the safety margin approach, then chirally safe procedures should routinely yield only 0.005–0.02% loss of chirality in a variety of model experiments, run under the reaction conditions of synthesis.

D. Problems of Diastereomer Separation

How reliably can one assume that careful purification will result in removal of unwanted diastereomers from a polypeptide? For relatively small peptides, the numbers of possible impurities are also small, and it seems likely that existing purification techniques are adequately matched to the dimensions of the problem. For peptides of size, the problem assumes a correspondingly formidable aspect, illustrated in the arithmetic of the preceding section.

One can consider two limiting cases, a Merrifield synthesis that allows purification only after all synthetic operations have been completed, and a fragment condensation synthesis in which crystallization of every intermediate has been possible. The problems of the two are different.

In the case of Merrifield synthesis, unseparated trace impurities in the final product are expected to exhibit nearly a continuum of similar properties. For example, with a given pair of immiscible solvent phases, the distribution coefficient of a single isomer is likely to be shared with many others. The achievement of clean separations is likely to depend on the availability of several different separation procedures, each of which is suited to removing different components of the mixture. Fortunately, the problem is ameliorated by the low levels of racemization expected during most amide bond-forming steps in this type of synthesis.

A fragment condensation synthesis with the possibility of purification by crystallization after each synthesis operation has the greatest intrinsic capacity for removing undesired diastereomeric contaminants. However, it must be remembered that the efficiency of crystallization as a separation technique may be greatly reduced as the concentration of impurities is reduced to low levels. Surface adsorption effects or solid solution formation have been demonstrated to reduce the efficiency of crystallization in inorganic systems. Their roles in the analogous purification of organic substances has not been widely investigated, although a probable example arises with the tripeptide Z—Gly—LPhe—Gly—OEt, for which repeated recrystallization from ethanol fails to remove 0.2% racemate (Kemp *et al.*, 1970a).

The contrary case can be made for likelihood of separability of a mixture containing a few diastereomers, even with peptides of appreciable size. The CIBA–Geigy group has reported two such cases that arose during their synthesis of human calcitonin M (Sieber *et al.*, 1970). Epimerization at His^{20} of this 32 residue peptide was inferred and while separation of diastereomers was not observed by thin layer chromatography, countercurrent distribution proved capable of this task. At an earlier stage in the synthesis a similar epimerization at Phe^{16} was observed, and separation by countercurrent distribution was again possible. Still more striking is the separation of diastereomers observed during the insulin synthesis by this group (Sieber *et al.*, 1977). Rivier *et al.* (1977) and Rivier and Burgus (1976) have successfully separated a series of diastereomers by high pressure liquid chromatography, and Li and co-workers (Yamashiro *et al.*, 1977) have reported separation of β-endorphin from three out of four of its diastereomers by partition chromatography.

These demonstrations are impressive, and it would be unwise to discount the resolving power of modern chromatographic techniques when applied to these problems. At the same time it is probably also unwise to suppose that resolutions of complex mixtures can ever be achieved casually or automatically, without the application of experience, effort, and skill.

E. Experimental Methods—A Cautionary Note

A variety of different measurements can be related to chiral integrity. Although the point is familiar to workers in the field, it is worth noting that unless large changes in chiral integrity are studied, optical rotation is an unsatisfactory assay, since it is a highly condition-sensitive property that measures the concentration of the major, not the minor chiral species. A low rotation of a species isolated from a reaction is not a readily interpretable observation, and purification by any process that can alter chiral purity must be avoided.

Percentages or rates of exchange of hydrogen isotopes at the α-carbon of an aminoacyl derivative can be equated with racemization rates or percentages provided it can be shown that loss of chirality at this site occurs at the same rate as exchange. If the α-hydrogen is not isotopically homogeneous, kinetic isotope effects on enolization rate are expected and can lead to a misleading rate discrimination. Moreover, several peptide derivatives have been demonstrated to exhibit the phenomenon of isoracemization, racemization rates exceeding those for exchange by as much as twentyfold (Kovacs *et al.*, 1971). Clearly, the equivalence of racemization and exchange rates must be established, not assumed, in each new system to be studied.

A final point arises with assays for chiral impurities. It is a common practice to report a product whose enantiomeric or diastereomeric content falls below the limits of the detection method at hand as "racemization free." Many examples so reported have subsequently been shown to contain racemate or diastereomer levels just at or below the detection limit. In this review, the term "racemization-free" has been reserved for levels such that no likely variation of conditions could be expected to result in diastereomer formation that complicates purification. A level of 0.005 to 0.02% D-isomer is suggested as providing a satisfactory margin of safety, and this definition is followed throughout.

II. RACEMIZATION OF AMINO ACIDS, PEPTIDES, AND THEIR UNACTIVATED DERIVATIVES

A. Racemization of Amino Acids

1. Racemization in Dilute Aqueous Solution

Studies of the behavior of amino acids under conditions of acidic hydrolysis of peptides, industrial requirements for conversion of D- to L-amino acids, and studies of paleobiological dating have independently added to our information concerning the rates of racemization of amino acid derivatives in aqueous media. The data of Bada and co-workers provide a convenient starting point (Bada, 1972, 1974; Bada *et al.*, 1970).

With the objective of establishing a dating procedure for paleobiological samples based on amino acid racemization, these workers examined the rate constants for racemization of several amino acids in water as functions of pH and temperature. For valine at 135°C, the rate constant is pH independent over the range of 3.5 to 8.3 and has the value, $4 \times 10^{-7}\ \text{sec}^{-1}$. Lowering the pH to 0.9 reduces the rate constant by sixfold. Rate constants at pH values above 8.3 were not reported. As the series (3) shows, there is roughly a

Relative rates of exchange:	Val <	Ile <	Ala <	Phe <	Asp	(3)
pH 7.6, 135°C	0.8	1.0	2.4	4.4	8.6	

tenfold increase in racemization rate as the side chain is varied from isopropyl to methylenecarboxylate. The racemization of aspartic acid, which of course differs from the other amino acids in the above series in having an additional ionizable group, was studied over a wider pH range. After correction for the ammonia elimination process unique to aspartic acid, a pH–rate profile for racemization of aspartic acid shows an overall variation in rate constant of approximately thirtyfold over the pH range −1 to 12. A maximum is observed at pH 3, and minima at pH 6 and 0. No change in rate constant is observed over the pH range 9–12, and the rate constant at pH 7 is roughly half that observed in 6 *N* HCl. The pH–rate profile can be rationalized by assuming that exchange occurs by hydroxide-catalyzed enolization of the three ammonium species (**6**), (**7**), and (**8**). A large reactivity difference

α-Hydrogen reactivity toward hydroxide ion:

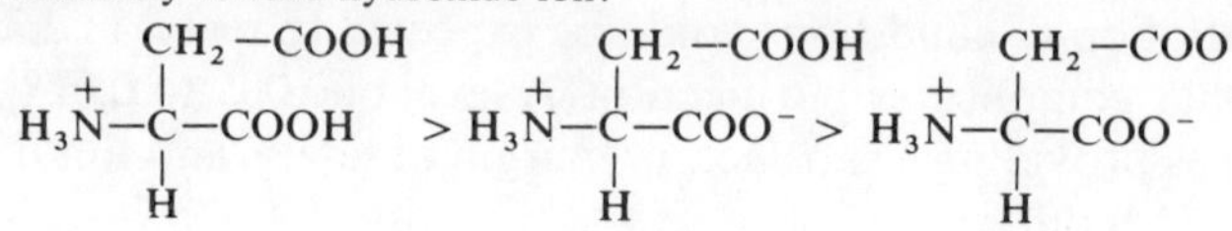

(4)

(**6**) pH 0–5 (**7**) pH 5–8 (**8**) pH 8–12

between these species must be postulated, with the reactivity order as shown above. All these data are thus consistent with inductive and resonance stabilization of a developing enolate anion at C-α as the major effect governing substituent effects on racemization of amino acids under aqueous conditions.

The effect of temperature on the rate constants for the racemization or epimerization was also investigated, and Eq. (5) was found to represent data

$$\log k(\mathrm{yr}^{-1}) = 17.98 - 6853/T \quad k_{\mathrm{Ile} \rightarrow a\mathrm{Ile}} \tag{5}$$

observed for the Ile → *a*Ile conversion over the temperature range of 90°–135°C, pH 7.6 (Bada *et al*., 1970). This equation corresponds to an activation energy of −31.2 kcal/mole. Similar E_a values are claimed for other amino acids. At 20°C, pH 7, epimerization thus proceeds on a geological time scale, with an estimated halftime of 180,000 years. On the other hand at 100°C, the halftime is ca. 600 days, and at 200°C the estimated halftime is 2 hr. This estimate is in reasonable agreement with patent claims of complete racemization of Lys and Glu by heating under a variety of aqueous conditions to temperatures of 170°–200°C for periods of 5 to 12 hr (Tatsumi and Ono, 1965; Yoshitake and Kanaka, 1965).

Two effects have been noted here, acceleration of racemization by electron-withdrawing groups in the side chain and an unusual, large activation energy for racemization with a consequent large increase in racemization rate with temperature. These are features that are also characteristic of most α-inversion processes of peptide derivatives.

2. *Racemization in Strongly Acidic Aqueous Solution*

Racemization of amino acids under strongly acidic aqueous conditions is particularly pertinent to the behavior of amino acids under the conditions used for protein and peptide hydrolysis, prior to amino acid analysis. In a particularly thorough study, Manning (1970) has studied both the tritium α-exchange and the racemization of amino acids in 6 *N* HCl at 110°C and has established that under these conditions, for simple amino acids, the rates are equal. In 22 hr, between 0.4 and 3.0% conversion to D-isomer was observed in a series of 10 amino acids. The relative order of ease of racemization is shown in (6). Ile is converted to *a*Ile to the extent of 1.0% under these

Relative order of racemization rate:

$$\text{Lys} > \text{Met,Pro} > \text{Arg} > \text{Leu} > \text{Ile,Ala} > \text{Val} > \text{Thr,Ser} \qquad (6)$$

6 *N* HCl, 110°C

conditions, implying an approximate rate constant of $8 \times 10^{-6}\ \text{min}^{-1}$, close to the value obtained by Bada of $2 \times 10^{-6}\ \text{min}^{-1}$ at neutral pH, and supporting the observation of Bada that 6 *N* acid causes relatively little change in rate over that observed at pH 7.

One striking feature of Manning's data is the relatively slow racemization of Ser and Thr. This feature is not shared with cystine. Rapoport's group (Jacobson *et al.*, 1974) observed that in refluxing 6 *N* hydrochloric acid, cystine epimerizes forty times more rapidly than the epimerization of Ile. Similar, rapid racemization is observed under these conditions for α,β-diaminopropionic acid (~ 100 × Ile) and phenylglycine (Phg) (~ 10 × Ile). Because of decomposition, it was not possible to study the high temperature epimerization of cystine in less acidic solutions.

As a final point, it may be noted from early reports (Rittenberg *et al.*, 1938) that amino acids racemize moderately rapidly in very strongly acidic media. For example, when heated to 100°C for 48 hr in sulfuric acid containing 8% water, Leu is completely racemized. This process must occur roughly a thousand times faster than the equivalent racemization processes in dilute aqueous solution, not a surprising result, given the vastly greater protonating power of the 92% acid.

3. Racemization in Acetic Acid Solution

Studies by several groups have established that vulnerable amino acids racemize unusually easily if heated in acetic acid as solvent. The first observation of this type was by Liberek and co-workers (1966) who noted that when H—LCys(Bzl)—OH is refluxed for 2 hr in acetic acid, 80% racemate is recovered. Moreover, it was noted that N-blocked Cys(Bzl) derivatives are much more resistant to racemization. In a subsequent study, Matsuo *et al.* (1970) compared the ease of racemization of derivatives of propionic acid, Phg, and Ala in hot acetic acid. Both H-Phg-OH and H-Phg-OEt were significantly racemized after 1 hr at 60°C, but N-blocked derivatives required heating to much higher temperatures to generate detectable racemate. Both H-Ala-OH and H-Phe-OH racemized significantly in an hour at 120–150°C. The rate of racemization of H-Phg-OH was slowed by dilution of acetic acid with water; in a 1:1 D_2O—HOAc mixture, heating to 120°C for an hour was required to generate major amounts of racemate. The racemization observed in 10% D_2SO_4—D_2O was comparable only at the much higher temperature of 180°C.

Since addition of sodium acetate to acetic acid is observed to greatly accelerate the racemization rate of Ac-Phg-OH, these workers propose that rapid racemization in this solvent results when acetate ion as basic catalyst encounters a carboxylic acid derivative bearing an α-ammonium ion. The structure of salts in acetic acid, and particularly the very feeble dissociation of ion pairs are doubtless relevant considerations (Kolthoff and Bruckenstein, 1956).

4. *Racemization in Aqueous Alkaline Solution*

Racemization of amino acids under basic conditions has long been noted to be a relatively rapid process, particularly with amino acids that bear electron-withdrawing or α-conjugating substituents (Neuberger, 1948). Thus hot aqueous barium hydroxide has been the reagent of choice for preparing most DL-amino acids by racemization (Greenstein and Winitz, 1961, Vol. 3).

A study by Hill and Leach (1964) provides an estimate of the exchange rates of α-hydrogens of several amino acids at pH 12.5, 25°C. In ten days the susceptible amino acids Ser and Thr exchanged 0.01 to 0.02 gram atoms of tritium per mole, while Cys incorporated a ca. tenfold greater amount. The exchange observed for Ser at pH 12.5 roughly corresponds to a pseudo first-order rate constant of 1×10^{-6} min^{-1}, more than 50,000 times faster than Bada's estimate of the rate of epimerization of Ile at 25°C, pH 7.6, and comparable to the rate at 100°C at this pH (Section II,A,1).

It should be noted that with the exception of cystine, which is unusually vulnerable to epimerization under either acidic or basic conditions, the relative sensitivity of amino acids to base-catalyzed racemization differs from that observed with acid. Thus Ser and Thr are *less* susceptible to acid-catalyzed exchange than amino acids with aliphatic side chains. Under base-catalyzed conditions, cystine, Ser, and Thr are the most vulnerable amino acids. Gly and Asp also undergo rapid exchange, but the former cannot racemize, and the latter appears to undergo most rapid base-catalyzed exchange at its β-position. Some indication of the unusual vulnerability of the above amino acids is given by the observations that Trp is only slowly racemized by boiling 5 *N* sodium hydroxide, and Tyr requires 30 hr for complete racemization under these conditions (Neuberger, 1948).

B. Epimerization of Peptides and Other Unactivated Amino Acid Derivatives

1. Racemization during Acidic Hydrolysis or Methanolysis of Peptides

Considerable variations are reported for the amount of racemate observed during the acidic hydrolysis of peptides and proteins, and this has been interpreted (Manning, 1970, and earlier references therein) to result from the greater ease of acid-catalyzed racemization of N-acylated versus free amino acids, together with a large variation in hydrolysis rate of peptide-derived amide bonds. Because of concomitant hydrolysis, systematic study of acid-catalyzed racemization of amino acid components of peptides is difficult. Manning (1970) noted that hydrolysis of bradykinin for 22 hr at 110°C with 6 *N* HCl resulted in Ser, Pro, and Arg with the same racemization levels as were observed with free amino acids under these conditions. By contrast, bacitracin yields extensively epimerized Ile (15.7% *a*Ile) and Leu (6.8%) under the same conditions.

Weygand *et al.* (1966a) examined the acidic hydrolysis of di- and tripeptides containing Ile and detected epimerization (0.5–1.0% *a*Ile) only for tripeptides in which the Ile residue occupies the 2-position. In a related study, methanolysis of tri- and tetrapeptides, catalyzed by hydrogen chloride (8.5 *N*, 24 hr, 70°C), resulted in the formation of dipeptides for which only the dipeptide formed from the COOH-terminal fragment was free of detectable diastereomer. Other dipeptides were found to contain 1.5–5.0% epimer; an example is shown in Eq. (7). A novel acid-catalyzed formation of 5(4H)-oxazolones was proposed to rationalize these findings.

$$\underset{\substack{\textbf{(9)}\\ \text{(all L)}}}{\text{H—Leu—Phe—Val—Val—OH}} \xrightarrow[\substack{\text{MeOH}\\ 70^\circ\text{C}\\ 24\ \text{hr}}]{\substack{8.5\ N\\ \text{HCl}}} \underset{\substack{\textbf{(10)}\\ 4.7\%\ \text{epimer}}}{\text{H}_2^+\text{—Leu—Phe—OMe}} + \underset{\substack{\textbf{(11)}\\ 3.6\%\ \text{epimer}}}{\text{H}_2^+\text{—Phe—Val—OMe}} + \underset{\substack{\textbf{(12)}\\ 0\%\ \text{epimer}}}{\text{H}_2^+\text{—Val—Val—OMe}} \qquad (7)$$

2. Base-Catalyzed Epimerization of Peptides

It is not surprising that the reactivity order observed for base-catalyzed racemization of amino acids is also observed for peptides if one compares amino acids in identical peptide sites. Racemization of amino acids at internal sites of peptides was noted in early work by Levene and co-workers

(Neuberger, 1948). In a study of the base catalyzed epimerization of gliadin, Nyilasi (1957) established a vulnerability order of Ser,His > Thr,Arg,Lys. Hill and Leach (1964) studied tritium exchange of the α-hydrogens of ribonuclease (pH 12.5, 25°C, 4 days) and noted the following exchange order (8).

$$\text{Gly} > \text{Ser,Thr,Cys} > \text{Ala,Arg,Pro,Phe} > \text{Ile,Leu} \quad (8)$$

The facile epimerization of serine derivatives has received considerable attention. For example, Schnabel (1959) has noted that Z-LSer-Gly-LAla-OEt and its DSer diastereomer interconvert in wet methanol containing 6% triethylamine with a half life of around a day.

In an important study, Bohak and Katchalski (1963) determined rate constants for the epimerization of Ser residues of poly-LSer. The model, Ac-LSer-NH_2, exhibited a second order rate constant of 4.6×10^{-2} M^{-1} min^{-1} for L → D conversion, catalyzed by hydroxide in water at 24°C. The epimerization of Ser residues of polyserines showed no cooperative effects and a rate constant per residue that is nearly identical to that of the model. In the presence of 7 *M* lithium bromide, a slightly smaller rate constant was observed. The solubility properties of poly-Ser depend strikingly on its chiral identity. The effect of epimerization on polymer shape has been analyzed theoretically by Flory's group (Miller *et al.*, 1967).

The relative susceptibility of blocked amino acids has been examined by Sato *et al.* (1970) who studied the racemization of benzoylamino acyl anilides at 50°C in 1:1 ethanol-DMSO containing 0.1 *M* sodium ethoxide. For Phg (α-phenylglycine), Ala, Leu, and Val, respective racemization halftimes of 6.3, 27, 70, 128, and 909 min were observed. The 34-fold rate difference between Phe and Val may be compared with a ca. 6-fold rate increase reported for racemization of the free amino acids at pH 7.6 in water, 135°C.

A qualitative study by Fridkin *et al.* (1970) defines the importance of amino acid position in a peptide chain. Using pmr-monitored deuterium exchange, this group noted amino acids that are in terminal positions and therefore bear free amino or anionic carboxylate groups exchange α-hydrogen for deuterium at much slower rates than amino acids in internal positions. The Ser of Ac—Ser—Gly—OH was found to exchange ten times faster than the Ala of Ac—Ala—Gly—OH, Gly^2 of tetraGly exchanged faster than Gly^3, and in H—Gly—Ala—Gly_2—OH, Ala exchange was roughly 10 times slower than Gly^3 exchange. At pH 13.5, $t = 25°C$, the latter had an approximate rate constant of 5×10^{-3} min^{-1} (50% exchanged in 2.2 hr).

Matsuo *et al.* (1967), had previously established for a series of derivatives of Phg in deuteriomethanol containing 1 *M* sodium methoxide that the exchange rate of α-hydrogen is equal to the rate of racemization and the exchange rate follows the following order (9).

Order of exchange in $CH_3OD/CH_3O^-Na^+$:

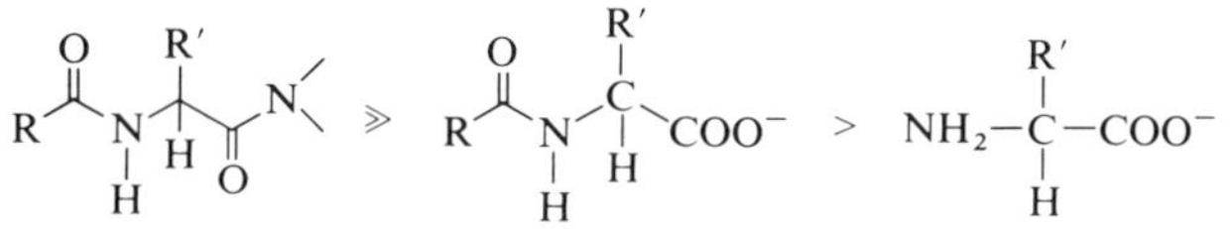

(13) (14) (15) (9)

The above studies have usually involved strong oxygen bases in protic media at elevated temperatures. To put these results in the perspective of synthesis, it is useful to compare them with the behavior of Z-Gly-Phe-Gly-OEt in ethanol at 30°C in the presence of 0.3 *M* triethylamine. Kemp *et al.* (1970a) observed 0.015% racemate after 28 days ($k_{L \rightarrow D} = 2 \times 10^{-9}$ min^{-1}). In DMSO, at 22°C, in the presence of 0.1 *M* of the stronger base, tetramethylguanidine, 0.01% racemate was observed in two days (Kemp *et al.*, 1974c) ($k_{L \rightarrow D} = 3 \times 10^{-8}$ min^{-1}). It may also be noted that Cheung and Benoiton (1977a,b) have prepared Boc-*N*-methylamino acids by methylation of the urethane conjugate base, observing less than 0.1% racemization during reaction conditions involving contact with excess sodium hydride at 0°C in THF as solvent for 24 hr.

3. The Influence of N^{α}-Protective Group on Rates of Racemization

An increase in electron-withdrawing character of the protective group attached to an NH_2-terminal amino acid greatly increases its susceptibility to epimerization. The phthalimido group induces more racemization than the carbobenzoxy (Liberek, 1963a,b; Liberek *et al.*, 1963; Liberek and Grzonka, 1964a,b; Matsuo *et al.*, 1967). An *N*-benzoylaminoacylanilide is somewhat more vulnerable to base-catalyzed racemization than its *N*-acetyl analogue (Neuberger, 1948). *N*-alkyl-*N*-nitrosoamino acid derivatives are very easily racemized (Quitt *et al.*, 1964).

Schiff bases of amino acids and peptides can undergo relatively rapid base-catalyzed racemization, particularly if the imine nitrogen is coordinated with a metal, as in the case of salicylaldimine copper complexes (Toi *et al.*, 1963; O'Connor *et al.*, 1968; Auld and Davison, 1968). Rates of triethylamine-catalyzed racemization of *p*-nitrobenzaldehyde imines of aminoacyl esters have been measured (Siemion and Wilschowitz, 1971). Formaldehyde Schiff bases lead to partial racemization under basic conditions (Cheung and Benoiton, 1977b). The racemization of amino acids in the presence of α-ketoacids has been patented (Gosei, 1967).

4. Racemization during the Acidic or Basic Deblocking Steps of Peptide Synthesis

In a study primarily aimed at comparisons with the anomalous behavior of *N*-alkylamino acid derivatives, McDermott and Benoiton (1973a) studied the epimerization of peptide derivatives under conditions of ester saponification. In accord with earlier reports, less than 0.1% of epimeric acid was formed during saponification of Z-Ala-Leu-OMe or Boc-Ala-Pro-OMe. Saponification of Z-Ile-OMe generated 1% of Z—*a*Ile—OH. Barton *et al.* (1973) reported 0.3% epimerization when a Merrifield-type COOH-terminal benzyl ester was transesterified with dimethylethanolamine and the resulting amino ester was hydrolyzed in aqueous DMF.

Kenner and Seely (1972) observed that the hydrolysis of Z-Ala-Phe-OMe and Z-Leu-Ala-OMe (1 equiv 0.25 *N* sodium hydroxide, 1:4 water–acetone 1 hr, 20°C) gave, respectively, 2.8% and 0.8% diastereomer. No diastereomers could be detected (< 0.1%) when the corresponding phenyl esters were subjected to peroxide-catalyzed hydrolysis.

It is generally believed that treatment with HBr/HOAc at room temperature for short periods does not affect chiral integrity. In accord with this view, McDermott and Benoiton (1973a) found that Boc-Ala-Pro-OH and Z-Ala-Leu-OH in 6 *N* HBr/HOAc for 4 hr formed less than 0.1% diastereomer. In contrast, they observed 0.5% *a*Ile from Ile under the same conditions. Epimerization had been noted during deblocking steps in an earlier study by Taschner *et al.* (1963a).

5. Epimerization of Diketopiperazines

The ease of base-catalyzed epimerization of diketopiperazines has long been recognized (Neuberger, 1948), although rationalizations for this phenomenon remain obscure. Conformational activation similar to that responsible for the enhanced carbonyl and α-hydrogen reactivity of certain lactones may well be involved.

Among several recent studies (Eguchi and Kakuta, 1974; Schmidt and Nikiforov, 1975), we draw attention to the synthetic work of Ott *et al.* (1963), in which cyclo(LPro–LPhe) is rapidly converted to its DL diastereomer by treatment with 0.5 *N* sodium hydroxide (> 80% in 15 min, 25°C). Remarkably, the proline residue and not the more activated phenylalanine is epimerized.

6. Racemization of *N*-Alkylamino Acid Derivatives

From early work of Bovarnik and Clarke (1938) and from the important recent study of McDermott and Benoiton (1973a), it appears that *N*-

methylamino acid derivatives are unusually susceptible to racemization. Thus, saponification of peptide esters bearing COOH-terminal MeLeu or MeIle residues resulted in 9–12% epimerization. Part but not all of this anomalously high α-hydrogen exchange can be explained as resulting from prolonged exposure of the more slowly hydrolyzing, hindered ester to hydroxide. However, a comparison of *tert*-butyl esters of Z-Ala-Leu-OH and Z-Ala-MeLeu-OH revealed that the latter epimerizes ca. 10 times more rapidly. Exposure of Z-Ala-MeLeu-OH to 5.6 *M* HBr in acetic acid gave 17% of LD epimer, and 9% of epimer was formed from H-Ala-MeLeu-OH under the same conditions. Decarbobenzoxylation of Z-MeIle-OH with HBr resulted in formation of 34% epimer. Hydrogenation of Z-Ala-MeLeu-OH generated 0.4% LD epimer and is clearly the method of choice for removing the carbobenzoxy protective group from substances containing these sensitive amino acids.

In considering rationalizations for these observations it is significant that proline peptides, which are also tertiary amides, do not usually share these racemizing tendencies (McDermott and Benoiton, 1973a). These workers have suggested that the acidity of the amide hydrogen of a normal peptide plays a protective role in saponification, inhibiting racemization. This proposal requires that > 90% of peptide amides be converted to conjugate bases under saponification conditions, inconsistent with a normal secondary amide pK_a of 14–15.

An alternative possibility is that unlike secondary amides which exist as *s-trans* rotamers, tertiary amides of *N*-alkylamino acids have rotamers of nearly equal energy. Conceivably, the racemizing tendency of the *s-cis* conformation is greater than that of the *s-trans*. It is interesting to note that among other structural peculiarities, the amide functions of diketopiperazines are obliged to assume the *s-cis* conformation.

C. Mechanism and Summary

1. Mechanism

Bada's model (1972) for the aqueous racemization of aspartic acid probably holds generally for other amino acids. In its general form the model states that racemization depends on enolate formation, the rate of which is proportional to the attack of base or acid on the amino acid species in solution. Four processes are probably involved, as seen in Fig. 3. Near neutrality, hydroxide is presumed to react primarily with the amino acid conjugate acid (path b), and at higher pH, with its zwitterion (path c). At very low pH the mechanism is expected to involve protonation of the conjugate acid (path a), and at very high pH, one expects the mechanism to

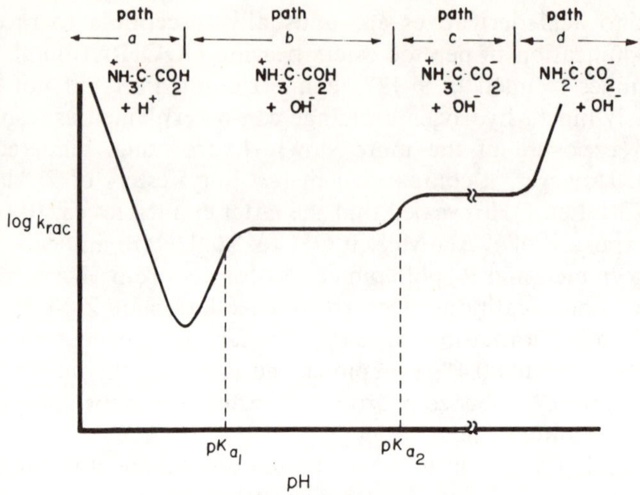

Figure 3. Generalized pH–rate profile for racemization of amino acids, assuming that four mechanisms (a, b, c, and d) contribute to the observed racemization (see text). pK_{a_1} and pK_{a_2} are the first and second dissociation constant for the amino acid. Note that, depending on the amino acid, the rate in region C can be either smaller or larger (as shown) than that observed in region B.

involve hydroxide-catalyzed enolization of the amino acid conjugate base (path d). Figure 3 depicts these four separate processes that can contribute to racemization of an amino acid. The general behavior that can be expected is shown in Fig. 3, although a change in substituent may change the appearance of the graph by altering the relative contributions of these four processes.

Considering the expected effects of electronic changes at three sites of attachment of substituent, X, R, and Y, one finds that the above model can give generally consistent and simple rationalization for many empirical exchange results. Electron-withdrawing substituents at sites X, R, or Y are expected to accelerate the three base-catalyzed processes, as observed (10).

Since acetylalanine amide undergoes base catalyzed exchange more rapidly than propionamide, the α-amido function has an accelerating effect on exchange, relative to hydrogen (Matsuo *et al.*, 1967).

Results of exchange under acidic conditions are more difficult to rationalize using the above model. Thus we have seen that electron-withdrawing substituents at R inhibit exchange in the cases of Ser or Thr, but accelerate it for α,β-diaminopropionic acid. The exchange order, isobutyric acid > βAla > Ala (Manning, 1970) is consistent with inhibition of acid catalysis by an α-ammonium ion, yet results in acetic acid clearly establish the very marked accelerating effect of an α-ammonium ion on exchange rate.

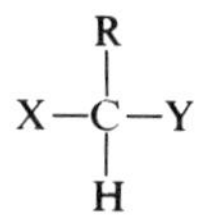

(16)

Base-catalyzed rate orders:

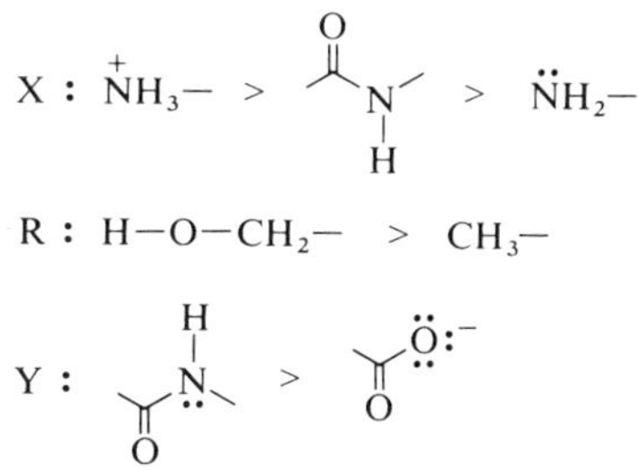

(10)

Probably the largest riddle is posed by the rapid exchange of cystine under acidic conditions and Cys derivatives under basic conditions. At least part of the effect is inductive, and it is interesting to note that a β-thioalkyl substituent changes the basicity of ethylamine more than a β-methoxy substituent (Perrin, 1965, pp. 22–23; also, see Table II, Section III,B,2). Although β-elimination has been established as operating for Cys derivatives under vigorous conditions (MacLaren *et al.*, 1958; Bodanszky and Bodanszky, 1967) it appears not to be the important mechanism operative at 25°C. Details relevant to the cystine exchange are discussed by Jacobson *et al.* (1974).

Given the form of the pH–rate profile observed by Bada for Asp, it is likely that paths a and b each contribute significantly to enolization at pH 0 for amino acids with aliphatic side chains. Variation of substituent or solvent is expected to change the relative importance of these two pathways, and since substituent effects may reverse for the two, this variation may be an explanation for some of the contradictory data on substituent effects for the acid-catalyzed racemization processes. Complete pH–rate profiles are needed to resolve this point.

Finally, it may be noted that an enol derived from an α-amino acid derivative has a double bond substituted with three heteroatoms. Rich chemistry might be expected for such a system, and it is striking that only reprotonation at the α-carbon seems to be observed.

2. Summary of Observations

In Table I, a number of results taken from the preceding sections have been assembled to permit comparison of the vulnerability of a variety of amino acids under different conditions. The important generalization that is

Table I Conditions Required To Induce Formation of 0.1% Enantiomer or Diastereomer

Substance	Conditions	Time for 0.1% L → D
Amino acids		
H-Val-OH	6 *N* HCl, 110°C	ca. 2 hr
H-Val-OH	H_2O, pH 8, 135°C	ca. 1 hr
H-Val-OH	H_2O, pH 8, 25°C	ca. 100 yr
H-Cys(Bzl)OH	HOAc, reflux	< 0.5 min
H-Ser-OH	H_2O, pH 12.5, 25°C	ca. 15 hr
Peptides		
-Ser-*Ser*-Ser-	H_2O, pH 12.5, 25°C	ca. 1 min
H-Gly-*Ala*-Gly_2-O^-	H_2O, pH 13.5, 25°C	ca. 3 min
Z-Gly-*Phe*-Gly-OEt	EtOH, 0.3*M* Et_3N, 25°C	300 days

not reflected in this brief summary is the additivity of effects. The effects of temperature, catalyst, and the three substituents operate largely independently of each other, and their total effect can be approximated by multiplying together independent rate factors for each effort. In designing experiments it is always important to remember the very large increase of racemization rate with temperature.

III. RACEMIZATION OF ACTIVATED ACYL DERIVATIVES OF AMINO ACIDS AND PEPTIDES

A. General Considerations

1. Primary Features of Racemization of Activated Acyl Derivatives

We have observed that the rate of racemization of an α-amino acid derivative is sensitive to the temperature, the reaction conditions, and the three substituents, X, R, and Y. The principal difference that must be noted in examining the racemization behavior of acylating agents derived from peptides and amino acids is the considerable increase in ease of racemization conferred by an activating acyl substituent, Y. An increase in acylating efficiency (measured in terms of rate constant) is usually achieved by an increase in thermodynamic activation of the acyl function (Gerstein and Jencks, 1964; Johnson, 1967). The latter increase almost invariably results in increased vulnerability to racemization.

As a result, we must distinguish between three stages of an acylation process.

Stage 1. Conversion of an α-NH_2-protected amino acid or peptide acid, with its low racemizing tendency, to an activated acylating agent, with a much higher racemizing tendency. During stage one, reactive intermediates may be formed whose racemizing tendencies are very much greater than those of the activated acyl product. In these cases, successful retention of chiral integrity is possible only if these highly racemizing intermediates are very short lived.

Stage 2. Isolation, purification, and storage of the activated acyl derivative prepared in stage 1. In many cases, this step can be avoided.

Stage 3. Aminolysis of the activated acyl derivative. The product of this step is an unactivated amide that has a very small tendency to racemize.

Epimerization at the α-carbon thus may be incurred as an activated acyl derivative is formed, as it is manipulated, or as it is aminolyzed. Although there are certain rare exceptions (to be noted later), vulnerability to epimerization is confined to the α-carbon of the acylating component in the case of activated amino acid derivatives. In the case of activated peptide derivatives, vulnerability occurs simultaneously at the last *two* amino acid residues of the peptide chain of the acylating agent. We will see that the vulnerability of the penultimate amino acid is a feature of the 5(4H)-oxazolone-mediated racemization mechanism.

Epimerization that occurs in stage 2 resembles the racemization processes of Section II; it is a function of conditions and duration of exposure and is described by a single rate constant. Epimerization during stages one or three depends on a *ratio* of rate constants and is proportional to the rates of racemization of the vulnerable species times their average lifetimes. Thus, an acylating agent that is epimerized exceedingly rapidly may lead to chirally integral product if it is generated in the presence of amine nucleophile and aminolyzes orders of magnitude more rapidly than it racemizes.

For peptide acids, epimerization during stage 1 poses particularly serious problems, since in many cases, acyl derivatives with low tendencies toward racemization, such as substituted phenyl esters, can only be formed from carboxylic acids through the intermediacy of highly activated intermediates, such as anhydrides or acid chlorides, that spontaneously and rapidly form 5(4H)-oxazolones. Many otherwise useful acyl derivatives of H_2N-blocked peptide acids are not available in chirally pure form from the acids themselves. This problem is one of the primary reasons for the utility of acylating agents attainable by special or indirect routes, such as acyl azides (from acyl hydrazides) or certain enol or phenyl esters derived from isoxazolium salts. The problem has motivated the development of highly efficient acyl trapping agents, such as *N*-hydroxysuccinimide and 1-hydroxybenzotriazole, and it provides a primary impetus for Young's safety-catch activation principle (Jones and Young, 1968).

2. The Problem of Amine Basicity

During the coupling reaction of stage three, an activated acyl derivative is exposed to an amine nucleophile that has basic properties. Racemization is expected to be catalyzed by the nucleophile itself and to depend on its base strength. Moreover, since peptide amines are commonly generated *in situ* by reaction of their ammonium halide salts with a stronger, nonnucleophilic tertiary amine base, deviations from exact stoichiometry can result in catalysis of racemization by the latter. The important issue is the relative efficiency of amine bases in the aprotic solvents commonly used for aminolysis.

In Table II are listed pK_a values reported for amine salts in water. It may be noted that α-amino acid esters or amides are relatively weak bases that are only half protonated at pH 8 in water. As a result, the basicity of a peptide coupling reaction mixture is much lower than would be the case if the nucleophile were a simple aliphatic amine.

The change from water to a dipolar aprotic solvent used in a coupling reaction is expected to have relatively little effect on the relative strengths of bases within a common structural class, but is expected to have very large effects on the relative strengths of bases that belong to different classes (Coetzee, 1967). Moreover, several types of salts form in the aprotic solvents (Kolthoff and Chantooni, 1963). Because hydrogen bonding is an important determinant of these effects it is likely that primary and tertiary amines of the same basicity in water may differ in strength in dipolar aprotic solvents.

A first step toward unraveling these complex issues for bases of interest to the peptide chemist was taken by Williams and Young (1972), who

Table II Basicities of Amines in Water at 25°C

Amine	pK_a of conjugate acid[a]	Amino acid	pK_a[a]
		H_2^+-Gly-O^-	9.8
Triethylamine	10.8	H_2^+-Ala-O^-	9.9
N-Methylmorpholine	7.4	H_2^+-Leu-O^-	9.7
H-Gly-OEt	7.6	H_2^+-Val-O^-	9.7
H-Leu-OEt	7.6	H_2^+-Phe-O^-	9.2
H-Phe-OMe	7.0	H_2^+-Ser-O^-	9.2
H-Ser-OMe	7.1	H_2^+-Cys(Et)-O^-	8.7
H-Phg-OEt[b]	2.1	H_2^+-Phg-O^-[b]	4.4
H-Gly_2-OEt	7.8	H_2^+-Gly_2-O^-	8.3
H-Gly_5-OEt	7.9	H_2^+-Gly_4-O^-	8.1

[a] From Perrin (1965).
[b] Phg, *C*-Phenylglycine.

examined the reactions of the indicator, 2,4-dinitrophenol, with a series of amines, including *N*-methylmorpholine, triethylamine, and H—Gly—OEt. In dioxane, ethyl acetate, tetrahydrofuran, and chloroform, triethylamine is between ten and several thousandfold more basic than ethyl glycinate. *N*-methylmorpholine is comparable in strength to H—Gly—OEt in dioxane and ethyl acetate, more basic in chloroform, and less basic in tetrahydrofuran (THF). It is somewhat more basic than H—Gly—OEt in acetonitrile.

B. Racemizing Tendencies of Activated Urethane-Blocked Amino Acid Derivatives

1. *Mechanistic Significance of the N-Acyl Protecting Group*

In the course of their study of neighboring group effects at alkyl carbon, Winstein and co-workers (Scott *et al.*, 1957) noted a reactivity order that finds an exact parallel in the behavior of acyl derivatives of peptide synthesis, (11). As neighboring nucleophiles, the oxygen of amide conjugate base (**17**)

Reactivity order for cyclization:

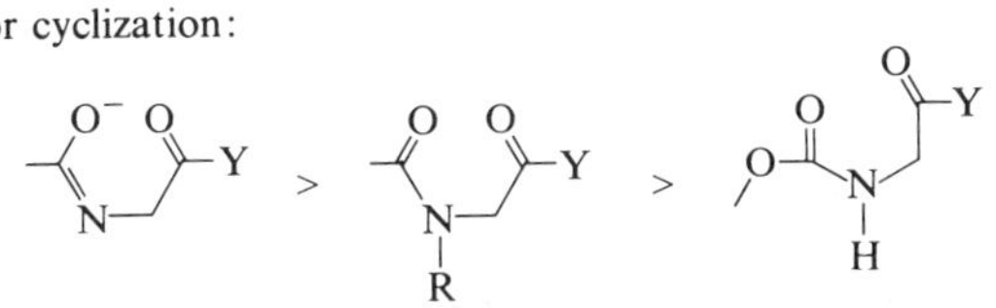

(11)

(**17**) (**18**) (**19**)

exceeds that of the neutral amide (**18**), which in turn is much more reactive than the oxygen of a urethane, (**19**). For this reason, cyclization at the acyl carbon to form a 5(4H)-oxazolone is observed with urethane protecting groups only in the presence of exceedingly good leaving groups, such as chloride, and under relatively vigorous conditions (cf. the synthesis of Leuchs anhydrides). Accordingly, the 5(4H)-oxazolone mechanism is relevant to the epimerizations of acylamino acids but not to urethane-protected amino acids. For these, the less rapid enolization mechanism is well-established. Evidence consistent with this mechanism includes first order kinetic dependence on each base that is present (general base catalysis), and rate retardation with an increase in steric bulk of base.

Most racemization studies of this type have been conducted with stable esters of the phenyl, thiophenyl, or *N*-succinimidoyl types, all of which can be prepared conveniently in several ways, for example, from urethane-blocked amino acids by reaction with the corresponding H—Y compound and dicyclohexylcarbodiimide (12). No detectible racemization attends this procedure, presumably because racemization via enolization requires an efficient base, which is absent.

$$X{-}AA{-}OH + H{-}Y + DCC \rightarrow X{-}AA{-}OY + DCU \qquad (12)$$

X = Z or Boc
Y = ONp,SPh,OSu

An interesting complication to the behavioral disjunction between *N*-acyl and *N*-alkoxycarbonyl amino acids has been provided by the observation of Jones and Witty (J. H. Jones, private communication) that the species reported by Miyoshi (1973) and observed to result from the reaction of benzyloxycarbonylamino acids and phosphorus pentachloride are in fact 2-benzyloxy-5(4H)-oxazolones. These species are observed to aminolyze to form optically active products. There are thus two levels to the protection offered by a urethane group. Urethane-blocked amino acids form 5(4H)-oxazolones reluctantly, and once formed, these species racemize slowly.

2. Effects of Amine Leaving Group, Protecting Group, Side Chain Substituent, and Solvent on Tertiary Amine-Catalyzed Racemization by the Enolization Mechanism

Following an initial observation of Bodanszky and Bodanszky (1967), Williams and Young (1972) observed that hindered, highly basic tertiary amines such as diisopropylethylamine catalyze the racemization of esters of Z-Phe-OH or Z-Phg-OH from five to forty times less efficiently than triethylamine; these workers, as well as Sakakibara and Itoh (1967), have noticed the expected increase of catalytic efficiency with base strength (triethylamine > *N*-methylmorpholine ~ *N,N*-dimethylglycine ethyl ester). Williams and Young (1972) note that the amine structure determines whether DMSO or MeCN induces faster racemization. Toward Z-Phe-ONp in acetonitrile, unhindered tertiary amines are more reactive than hindered secondary amines, but in DMSO the order is reversed.

For a given phenyl ester and amine, an increase in polarity of aprotic solvent is observed to increase the rate constant for racemization. Bodanszky and Bodanszky (1967) and Kovacs *et al.* (1975) have observed a solvent order of DMSO > DMF > $CHCl_3$ > EtOAc > THF > dioxane, with an overall range of fiftyfold for a pentachlorophenyl ester.

At any of the three sites, X, R, and Y (cf. structure **16**), an increase in electron-withdrawing character accelerates the rate of racemization. Thus Pht—AA—Y is a racemization-prone function, even with as weak a base as pyridine (Liberek, 1963b; Liberek and Grzonka, 1964a,b), and Pht—Phe—ONp racemizes more than ten times faster than Z—Phe—ONp (CH_2Cl_2, 20°C, Et_3N) (Young, 1967). Kovacs *et al.* (1975) report that the racemizing tendency of the *tert*-butyloxycarbonyl group is four times lower than that of the benzyloxycarbonyl group.

Considerable attention has been given to the correlations between structure of a phenol and the racemizing tendency of its aminoacyl active esters (Liberek and Grzonka, 1964a; Martynor and Samartesev, 1969; Kovacs *et al.*, 1970b; Morawiec *et al.*, 1971). For example, within a series of phenyl esters of Z-Cys(Bzl)-OH, triethylamine-catalyzed racemization varies three hundredfold with a change from phenyl to pentafluorophenyl esters. There is an approximate correlation between the acidity of a phenol and the ease of racemization of its esters.

The influence of side chain on racemizing tendency of Z-AA-Y esters parallels those on base-catalyzed racemization of unactivated amino acids and on pK_a of amino acids (Table II, Section III,B,2). The unnatural amino acids phenylglycine (Bodanszky and Bodanszky, 1967) and β-cyanoalanine (Liberek and Michalik, 1965, 1966a,b; Liberek and Grzonka, 1964a) show unusually high sensitivity. From these data and results of Kovacs *et al.* (1975) the following order of approximate relative reactivities can be set up: Trp, 0.9; Ala, 1.0; Glu(OMe), 2.5; Phe, 7.1; Asp(OMe), 20; Ser, 29; Cys(Bzl), 280; Phg, ca. 700; β-cyanoAla, ca. 2500.

Kovacs *et al.* (1975) have made the important observation that the influence of solvent and substituents X, R, and Y (cf. structure **16**) contribute roughly independently to racemization rate, which can therefore be approximated as a product of partial rate factors.

This analysis is expected to apply if two conditions are met. The reactions must proceed by the same mechanism, and the pertinent reaction variables must contribute independently to the activation energy. Given these limitations, the model is a good first approximation to experiment.

3. Racemization Expected during Coupling Reactions of Activated Urethane-Protected Amino Acids

A strong tertiary amine base such as triethylamine is not a normal constituent of peptide coupling mixtures. It is therefore difficult to relate the results of the preceding section to racemization levels expected for practical amide-forming reactions.

Phenolic esters of urethane-protected amino acids show no measurable tendency to racemize under neutral or mildly acidic conditions (Goodman and Stueben, 1962; Anderson *et al.*, 1965) at ordinary temperatures, and it can be presumed that the weak bases that are the nucleophiles of peptide synthesis are considerably less effective than triethylamine at catalyzing racemization. However, if one assumes that the side-chain effects on racemization induced by a peptide amine parallel that observed for triethylamine, then a ratio of rate constants for racemization by triethylamine and coupling by peptide amine must be proportional to the actual percent racemate formed in the coupling. (If the assumption fails, a nonadditivity of steric effects of base and amino acid side chain is the likely cause.) Using this analysis, Kovacs *et al.* (1975) have pointed out that Cys(Bzl) derivatives are likely to yield more than two orders of magnitude more racemate than amino acids with aliphatic side chains. (Comparable figures for Ser and Asp(OR), the other vulnerable natural amino acids, are about ten times smaller.)

In isolated cases, data are actually available on racemization of urethane-protected aminoacyl species under conditions of aminolysis. Weygand *et al.* (1963) have observed 29% racemate from the coupling of Z—Val—SPh with H—Val—OH in acetic acid at 130°C. Since this reaction involves a leaving group, a solvent, and a temperature that are atypical, it is a difficult result to interpret. A more representative result is likely to be that of Rebek and Feitler (1975). Using a sensitive isotope dilution assay, these workers established that for the coupling of Boc—LPhe—OH with H—Gly—OMe in CH_2Cl_2 by the DCC procedure, a concentration-dependent racemate formation of 0.01–0.06% obtains. Under the average conditions of Merrifield peptide synthesis, 0.03% racemate is expected.

C. Racemization of Activated Acyl Aminoacyl Derivatives—The 5(4H)-Oxazolone Mechanism

1. General Considerations

Long recognized (Bergmann and Zervas, 1928), the 5(4H)-oxazolone-mediated racemization mechanism provides a constant hazard for amide-forming processes involving activated acyl derivatives of *N*-blocked peptide acids or acylamino acids. Under most circumstances, these derivatives form 5(4H)-oxazolones far more rapidly than they enolize, and this is in part the explanation for the extensively documented finding that most of the simple reagents for converting carboxylic acids to activated derivatives do so with extensive to complete epimerization at the α-carbon, when applied to *N*-blocked peptide acids or acylamino acids (Iselin and Schwyzer, 1960; Lübke and Schröder, 1961; Goodman and Stueben, 1962; Theodoropoulus and Gazopoulos, 1962; Liberek, 1963a; Anderson *et al.*, 1965; Kovacs *et al.*, 1967, 1969; Bator-Sawicka, 1968; Wieland and Seelinger, 1971). (The three solutions to this problem are discussed in Section V,C.)

In its usual form, the 5(4H)-oxazolone-mediated racemization mechanism involves a rate-determining cyclization that is base catalyzed. The 5(4H)-oxazolone is then proposed to racemize via a resonance-stabilized tautomer and rapidly react with nucleophiles at its acyl carbon to form a mixture of enantiomers or epimers [Eq. (13)].

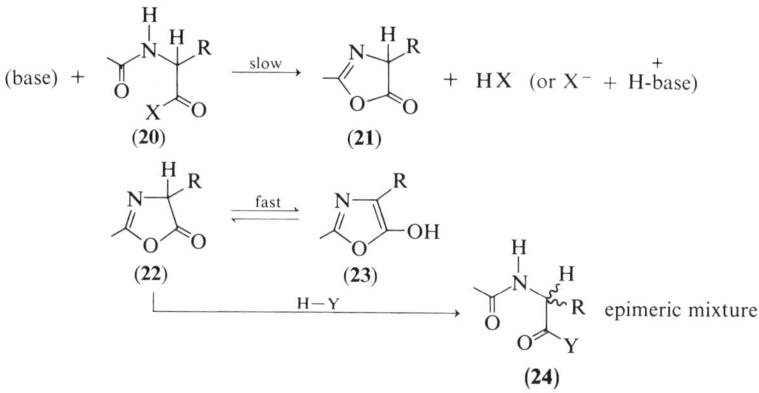

(13)

In the sections that follow, we first consider the evidence for 5(4H)-oxazoline-mediated racemization (Section III,C,2) and then consider three issues: (a) the nature of the catalysis for the rate-determining ring closure (Section III,C,3), (b) the chemical behavior of 5(4H)-oxazolones (Section III,C,4), and (c) the effect of substituents on the ease of 5(4H)-oxazolone formation (Section III,C,5).

2. *Evidence for the Intermediacy of 5(4H)-Oxazolones in Peptide Coupling Reactions*

In many circumstances 5(4H)-oxazolones can be detected directly when an acylamino acid or peptide acid is converted to an activated aminoacyl derivative. Treatment with carbodiimides or conversion to mixed anhydrides in the absence of nucleophiles is a well-defined procedure for preparing 5(4H)-oxazolones (reaction 15). The characteristic infrared absorption at

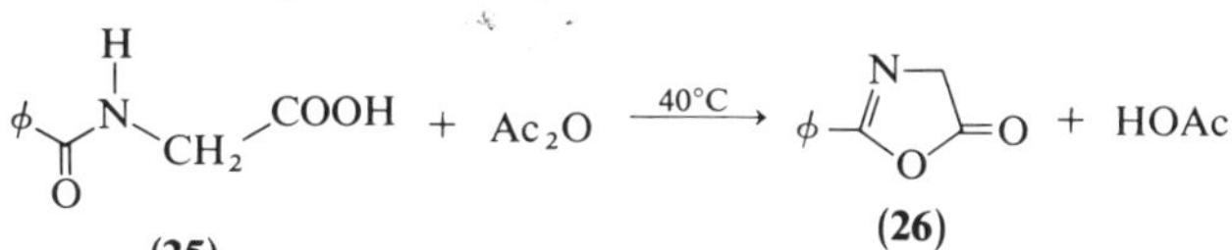

(14)

mp 90°C ir: 1835 cm^{-1}

1830 cm^{-1} allows sensitive detection of these species, and presence of trace absorption in this region has been used as an indicator of the operation of an 5(4H)-oxazolone-mediated mechanism. Unfortunately, absence of this absorption does not exclude the mechanism, nor, in the limiting argument, need presence of 1830 cm^{-1} absorption establish that all observed racemization is 5(4H)-oxazolone-mediated.

This latter point has been resolved elegantly by a trapping experiment of Antonovics and Young (1967). A dichloromethane solution containing a 1:10 ratio of Bz—Gly—LPhe—ONp and the 5(4H)-oxazolone derived from

Z-Gly-DLPhe-OH was treated with triethylamine. After 51% loss of optical rotation in the solution, workup was carried out, resulting in the recovery of 46% Bz-Gly-Phe-ONp which was optically pure (> 91% L). Reversing the roles of Z and Bz, these workers found that Z-Gly-LPhe-ONp and the 5(4H)-oxazolone from Bz-Gly-Phe-OH react in the presence of triethylamine to form Bz-Gly-LPhe-ONp. Clearly, an equilibrium is established in which

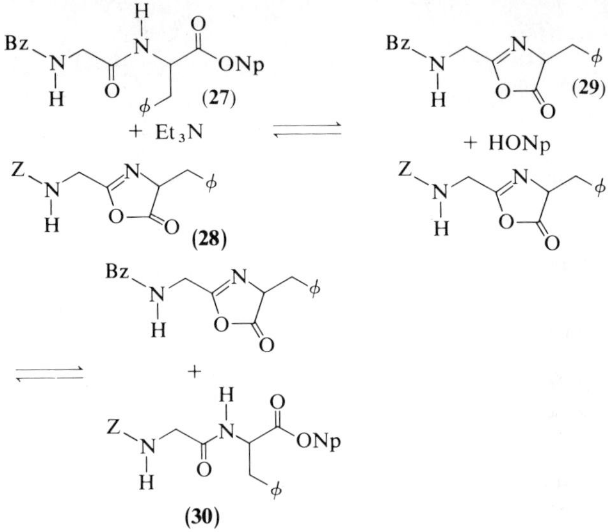

(15)

the nitrophenoxy group is exchanged between the two dipeptide moieties. The prior experiment establishes that when the equilibrium is shifted to one side by use of a large excess of 5(4H)-oxazolone, rate of loss of nitrophenoxide must equal rate of racemization. The reasonable conclusion is that under the experimental conditions all racemization proceeds via an 5(4H)-oxazolone pathway.

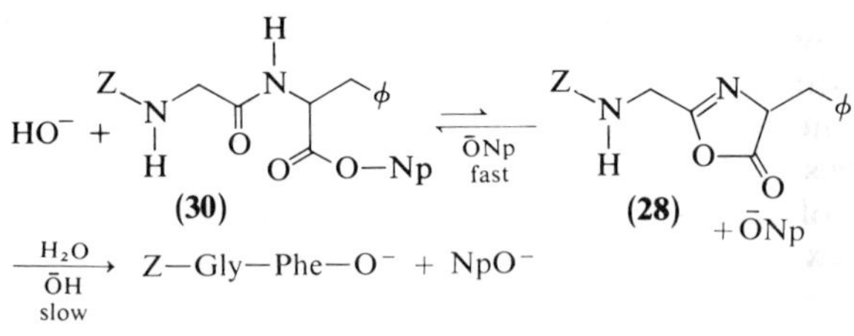

(16)

Goodman and Stueben (1962) had earlier noted that the alkaline hydrolysis of Z—Gly—Phe—ONp is accompanied by racemization that occurs ten times more rapidly than hydrolysis. Formation of 5(4H)-oxazolone, which reacts reversibly with *p*-nitrophenolate more rapidly than it hydrolyzes, is proposed to explain this result. Similar equilibrium experiments have been demonstrated by Kemp *et al.* (1974c) for tetramethylguanidine-catalyzed racemization of catechol half esters, and by Veber (1975) for acyl azides.

It can be argued that because the above studies involved unusually basic conditions or special systems, they need not generalize to routine peptide coupling conditions. A kinetic isotope effect can be used as a general criterion for distinguishing between 5(4H)-oxazolone and enolization mechanisms of racemization. Because the rate-determining step of the 5(4H)-oxazolone mechanism precedes the α—C—H bond cleavage step, while the two steps must coincide for the enolization mechanism, substitution of an α-hydrogen by deuterium is expected to slow the latter but not the former. An isotope effect of ca. 1.0 is strong evidence for operation of the 5(4H)-oxazolone mechanism.

Kemp and Rebek (1970) observed that in addition to phenolic esters, for which one would expect dominance of the 5(4H)-oxazolone mechanism by analogy with the above results, the mixed anhydride and acyl azide methods, applied to the synthesis of Bz-Leu-Gly-OEt, also give kinetic isotope effects of 1.0, implying a dominance of the 5(4H)-oxazolone-mediated racemization mechanism under the normal conditions of synthesis with these reagents. With the acyl azide method the synthesis of Z-Gly-Phe-Gly-OEt in the presence of triethylamine was observed to give an isotope effect which varied from 1.3 to 2.6 with experimental conditions, implying that when treated in ether with a tertiary amine base, this acyl azide racemizes in part by direct enolization. Clearly, therefore, the two mechanisms can operate at similar efficiencies under some circumstances.

3. Catalysis of 5(4H)-Oxazolone Formation

There are two questions concerning catalysis: does 5(4H)-oxazolone formation occur from the neutral amide without base under some circumstances, and can a distinction be made between general and specific base catalysis? In both cases, the answer is yes.

Because the neutral amide is expected to be a much less nucleophilic neighboring group than the amide anion or its kinetic equivalent, base catalysis of 5(4H)-oxazolone formation is a likely requirement unless the acyl leaving group is very good or the conditions are very vigorous. Although 5(4H)-oxazolones can be formed under neutral or acidic conditions, e.g., (**25**) → (**26**), it is difficult to rule out catalysis by traces of bases, especially in solvents of low polarity in which acids and bases can easily coexist. Tertiary amides cannot form neutral 5(4H)-oxazolones, and it was believed by many that the reluctance of acyl proline derivatives to racemize when converted to activated acyl derivatives was convincing evidence against the possibility of neighboring group attack at acyl carbon by a neutral amide. (For a summary of these arguments, pro and con, see Goodman and Glaser, 1970a.)

The recent observations of McDermott and Benoiton (1973b; see also, Boyd and Wright, 1972) appear to have settled the issue decisively in favor of neutral amide attack. Using *N*-methylamino acid derivatives, these workers observed large amounts of racemization during normal coupling reactions and succeeded in trapping an intermediary 5(4H)-oxazolone by reaction with a 1,3-dipolarophile.

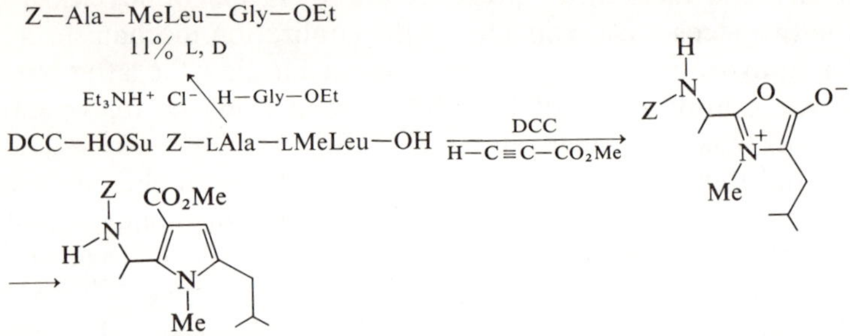

(17)

For processes that are base-catalyzed, N–H proton transfer can be envisaged as occurring concertedly with attack of the amide at the acyl carbon (general base catalysis), or in a prior step generating an equilibrium amount of amidate anion (specific base catalysis). In aprotic solvents, specific cat-

General catalysis: 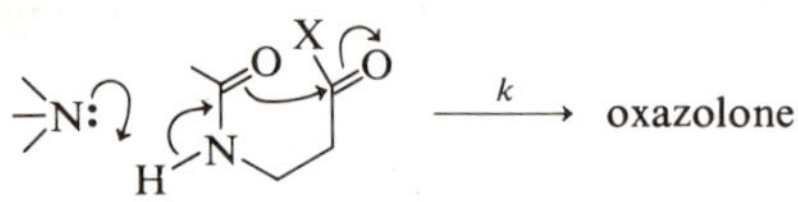

(18)

$$\text{Rate} = k[\text{amine}]\,[\text{ester}]$$

Specific catalysis: 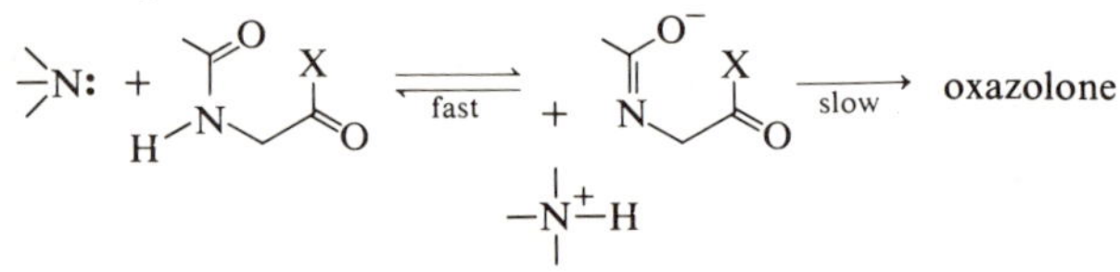

$$\text{Rate} = \frac{[\text{amine}]}{[\text{amine salt}]}\,[\text{ester}] \qquad (19)$$

alysis can be established by showing that the reaction rate is proportional to the *ratio* of amine to amine conjugate acid concentrations. This behavior has been demonstrated for a variety of phenyl esters under strongly basic conditions in dipolar aprotic solvents (Kemp and Chien, 1967b; Kemp *et al.*, 1974b). It is likely that in less polar solvents, the general catalysis mechanism, which appears to involve less charge separation, is favored.

Specific base catalysis carries the implication that racemization can be minimized by adding a conjugate acid of the amine nucleophile. Addition of one equivalent of H_2^+-Gly-OEtCl$^-$ to a coupling reaction of Z-Gly-LPhe-ONp and H-Gly-OEt in DMF lowered the percentage of racemate formed from 0.5 to 0.06% (Kemp, 1973).

A further testable implication of the specific base catalysis mechanism is the expectation that steric bulk of the amine catalyst should not be an important factor in catalysis of 5(4H)-oxazolone formation. Williams and Young (1972) observed that triethylamine and ethyldiisopropylamine catalyze racemization of Bz-Leu-ONp and Z-Gly-Phe-ONp with nearly identical efficiency, the more hindered amine being the better catalyst. As expected, the actual racemization of the 5(4H)-oxazolone from Bz-Leu-OH is catalyzed ten times more rapidly by triethylamine (Grahl-Nielsen, 1971).

4. Racemization and Aminolysis of 5(4H)-Oxazolones

This topic has recently been thoroughly reviewed by Goodman and Glaser (1970a), and only the highlights are summarized here. Working with the crystalline 5(4H)-oxazolone derived from Bz-LPhe-OH (**33**), Goodman and Levine (1964) measured the rate constants of ring opening and racemization for reaction with a variety of nucleophiles. For both reactions, in water, oxygen nucleophiles such as *p*-nitrophenoxide anion are more reactive by at least an order of magnitude than peptide amines of comparable basicity. Buffer catalysis of hydrolysis is observed.

Under these conditions, racemization is a much faster process than ring opening. Studies in dioxane showed the expected dependence of racemization rate on amine basicity. For example, with 2-phenyl-4-benzyl-5(4H)-oxazolone in dioxane at 25°C, racemization half-lives of 230 min for pyridine and 2 min for H-Phe-OME were observed (0.1 *M* catalyst). For a 0.1 *M* 1:1 mixture of *p*-nitrophenol and tributylamine, the half life was 3.6 sec.

With peptide-derived 5(4H)-oxazolones, such as (**31**), the rates of racemization and aminolysis by peptide amines are similar in many solvents (Goodman and McGahren, 1966, 1967a,b; Goodman and Glaser, 1969). Chirality is preserved to the greatest degree in the least polar solvent, toluene, with the least hindered nucleophile, H-Gly-OEt, (reaction 20). A

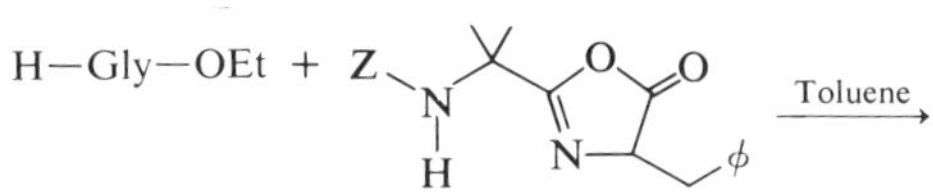

(**31**)

Z—Aib—Phe—Gly—OEt (20)

78% L (**32**)

22% DL

more hindered nucleophile, H—Ala—OMe, leads to a larger percentage of racemized product, and complete racemization is observed with

H-Aib-OMe,* an exceedingly hindered nucleophile. Product composition depends on the ratio of racemization and aminolysis rate constants. Both rate constants are significantly reduced by increased steric bulk of nucleophile, but as expected, the ring opening process has the larger sensitivity to bulk, and thus determines the trend for the rate ratio.

In studies with very important implications for suppression of 5(4H)-oxazolone-derived racemization, Siemion, Goodman and co-workers studied the reactions of 5(4H)-oxazolones with hydroxylamine, hydrazine, and their derivatives (Siemion and Morawiec, 1964; Goodman and McGahren, 1966; Siemion, 1968; Goodman and Glaser, 1970b). With these nucleophiles, the aminolysis rate exceeds the rate of racemization, and products are formed with a high degree of chiral integrity. The unusual reactivity of this

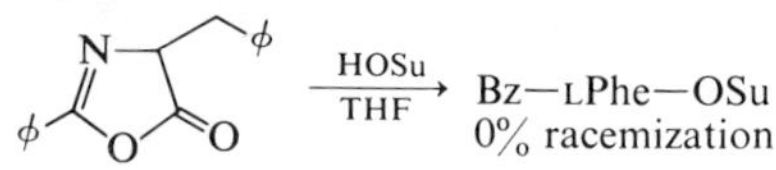

(33)

(21)

class of nucleophiles with acyl carbon is well known (Johnson, 1967), and whatever its origin it allows trapping of 5(4H)-oxazolones by reagents such as HOSu or HOBt before significant racemization can occur.

A further point of interest is the pronounced asymmetric induction that is observed when racemic 5(4H)-oxazolones are aminolyzed with chiral amino acid esters (Steglich *et al.*, 1967, 1971; Tomida and Kayahara, 1976).

A final point concerning the intermediacy of 5(4H)-oxazolones is their potential for effecting epimerization at two chiral sites of a peptide acylating agent [Eq. (22)] (Neuberger, 1948). First recognized by Bergmann and

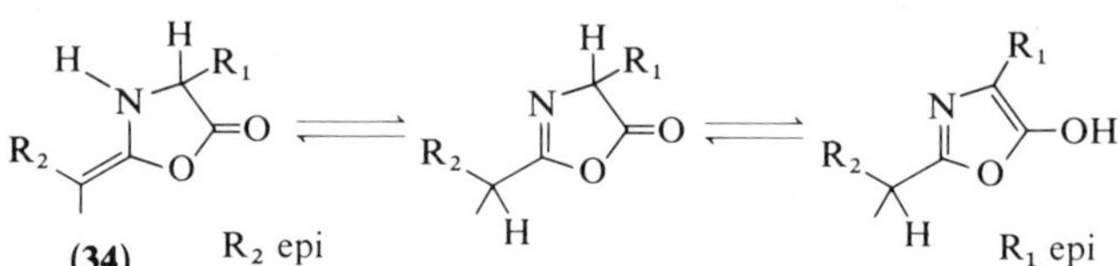

(34) R_2 epi **(35)** **(36)** R_1 epi

(22)

Zervas (1928), this phenomenon has been examined by Weygand *et al.* (1966b), using the model coupling **(37)** → **(38)**. In an extensive study by

$$\text{Z—Ile—Phe—OH} + \text{H—Val—O}t\text{Bu} \xrightarrow{\text{DCC}} \text{Z—Ile—Phe—Val—O}t\text{Bu} \qquad (23)$$

(37) **(38)**

6.3% αIle

Taschner's group (Dzieduszycka *et al.*, 1973) epimerization of the penultimate COOH-terminal amino acid was detected with the tripeptides

* Aib, α-aminoisobutyric acid.

Z-Phe-Val-Phe-OH, Z-Phe_3-OH, Z-Phe-Phe-Gly-OH, Z-Phe-Val_2-OH, and Z-Phe-Val-Gly-OH. With DCC, the mixed anhydride procedure, or with acyl azide in the presence of triethylamine, the amounts of epimerization lay in the range, 2–5%. At the present writing, the effects of conditions on the relative ease of epimerization of the two labile C—H sites of an 5(4H)-oxazolone are unknown.

5. The Effect of Substituents on the Ease of 5(4H)-Oxazolone Formation

Given a series of activated acyl aminoacyl derivatives (structure **39**), what can be said about the effect of substituents R_1 and R_2 on the ease of 5(4H)-

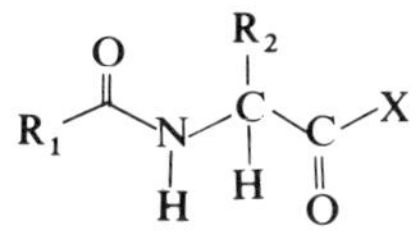

(39)

oxazolone formation? Young (1967) has reported data for the triethylamine-catalyzed racemization rates of species X-A-Phe-ONp, in which X = Z or Bz and A = Gly or Sar. These are all likely to be 5(4H)-oxazolone-mediated processes, and an overall rate variation of ca. tenfold was observed, establishing that R_1 has a strong influence on ease of 5(4H)-oxazolone formation. For simple acyl derivatives under the mildly basic conditions of peptide coupling reactions, an order Tfa > Bz > Ac > For, may be inferred from the racemization observed in coupling reactions involving amino acids protected by these groups (Heard and Young, 1963; Weygand *et al.*, 1963; Halpern *et al.*, 1967).

Unusually large amounts of epimerization are observed for couplings of Z-Pro-Val-OH with Pro-O*t*Bu (König and Geiger, 1970b). It is not clear how much of this effect is a Z—Pro-induced acceleration of 5(4H)-oxazolone formation or the indirect result of a Pro-Val coupling that may occur unusually slowly (Kemp *et al.*, 1974f).

Data for triethylamine-catalyzed racemization of a series of derivatives, Z-Gly-B-X, in which X = ONp or OPcp have been reported by Kovacs *et al.* (1970a, 1975). Variation of B among Ala, Phe, Glu(OMe), Cys(Bzl), Ser, Asp(OMe), and Trp results in an overall 15-fold rate change, markedly smaller than the rate change observed for the same series of amino acids for derivatives Z-B-OTcp that racemize by enolization. There is a poor correlation between the effects of R_2 substituents for the two processes. Although For-Gly-OH is observed to form a 5(4H)-oxazolone only slowly (Schnabel, 1965), it will be recalled that Gly derivatives undergo base-catalyzed α-hydrogen exchange unusually readily.

In a coupling reaction, a substituent R_2 can augment racemization in three ways: (a) it can slow the coupling process, increasing the lifetime of the racemization-prone activated species; (b) it can accelerate 5(4H)-oxazolone formation; and (c) it can change the disposition of the 5(4H)-oxazolone to epimerize before it couples (Goodman and McGahren, 1967a,b; Schnabel, 1965). Substituent effects are expected to influence these three processes differently. Conformational effects are likely to dominate (b). Increased steric bulk of R_2 is expected to increase racemization by either (a) or (c). Increased electronegativity of R_2 should decrease net racemization by (a), increase it to a lesser degree by (b) and probably by (c). Given the complexity of the situation, it is not surprising that the available data for the effect of R_2 on racemization by the azlactone route are difficult to rationalize.

If one assumes that 5(4H)-oxazolone-formation occurs entirely by a base-catalyzed path, then one expects the acidity of the amide at R_1 to be the major determinant of racemization induced by this substituent, in accord with the simple acyl order, Tfa > Bz > Ac. The observation by Young (1967) that Z-Sar-Phe-ONp racemizes less rapidly than Z-Gly-Phe-ONp, X = Z or Bz, is not easily explained in this way, and probably must be attributed to conformational or hydrogen bonding effects. Until more data are available, it seems inappropriate to propose a detailed rationalization.

D. Comparison of the Two Racemization Mechanisms

Ignoring for the moment the more complex behavior of a peptide coupling reaction, one can summarize the factors that influence rates of the two racemization mechanisms as shown in Table III.

It is clear that under most circumstances, most derivatives that can racemize by both pathways are at least a hundred times more vulnerable to the 5(4H)-oxazolone-mediated pathway than to simple enolization. Unusual vulnerability to racemization by enolization is furthered by presence of an unhindered base (Table III, point 6) and vulnerable amino acids, such as Cys, or to a lesser degree, Phe or Ser (Kovacs *et al.*, 1975). Ethyl acetate, chloroform, or dioxane appear to be the solvents that are most likely to lead to competition by the enolization pathway, if one examines the data of Williams and Young (1972) and assumes that rate of aminolysis is only slightly affected by the change from Z—A—ONp to Z—Gly—A—ONp. [This assumption appears to be valid in DMF or DMSO (Kemp *et al.*, 1974a); it is not valid in THF (Kovacs *et al.*, 1975).]

Excluding these special cases, one may conclude that efficient racemization control can be achieved by processes that suppress 5(4H)-oxazolone formation or trap 5(4H)-oxazolone before they can racemize. The intriguing

Table III Effect of Structure and Conditions on Base-Catalyzed Racemization by Enolization (E) or 5-(4H)-Oxazolone Formation (O)

```
          A               B
          |               |
X—NH—CH—CONH—CH—CO—OY + base → epimer or racemate
```

Variable	Mechanism	Effect of variable
1. Temperature increase 2. Solvent polarity increase 3. Increase in strength of base 4. Increase in activating capacity of O—Y		Marked rate acceleration for both mechanisms
5. Site of epimerization	E	Only at Site B
	O	Sites A and B (vulnerability: B > A; relative ratio uncertain)
6. Increase in steric bulk of base	E	Slows rate
	O	Small effect, may accelerate or slow
7. Substituent at B	E	Steric bulk slows, electronegativity accelerates
	O	Uncertain; reaction favored by substituents that favor geometry of intramolecular cyclization
8. Substituent at A	E	Electronegativity accelerates; other effects probably small
	O	A—B bond must belong to a secondary amide for base-catalysis; electronegativity accelerates; other effects also important
9. Substituent at X	E	Effects probably unimportant
		Effects important but undefined

proposal (J. Rebek, private communication) to introduce selective reagents that trap 5(4H)-oxazolone and efficiently convert them to products that are not acylating agents has not been realized as yet, but may be very productive, given the versatile chemistry of 5(4H)-oxazolones.

IV. CONTROL OF RACEMIZATION DURING PEPTIDE SYNTHESIS

A. General Considerations

With the information of the foregoing sections as background, it is now possible to consider available evidence concerning the racemization levels expected during the operations of peptide synthesis. In the sections that follow, primary attention is given the amide-forming steps and the types of coupling reagents that have been most thoroughly investigated by racemization assay. Other coupling reagents are discussed in Section V, devoted to the design of "nonracemizing" reagents, or in Section VI, in which racemization assays are enumerated.

B. Racemization during Protective Group Transformation or Removal

The racemization levels detected for saponification of esters or removal of protective groups with HBr/HOAc have been discussed in Section II,B,4. Though apparently trouble-free by other methods, racemization has been reported during the preparation of *tert*-butyl esters by the $POCl_3$/HO*t*Bu-pyridine procedure (Schröder and Lübke, 1965, p. 58).

The same source (p. 208) reports racemization of Z-LSer-OMe on hydrazinolysis, and this problem has also been observed for Cys esters (MacLaren *et al*., 1958), as well as for Z-Gly-Ala-OMe, by a more sensitive assay (Mitin and Maksimov, 1973). These results appear to argue for use of the safety-catch Boc-hydrazide procedure whenever possible.

Given the relatively facile racemization of certain amino acid derivatives with free amino termini in warm acetic acid (Section II,A,3), it would be of interest to have comparable data for trifluoroacetic acid, the solvent widely used for cleaving Boc or O*t*Bu protective groups. From upper bounds that can be set on racemization by solid phase synthesis (Section IV,E), one can estimate that on the average, less than 0.1% epimerization must occur in 30 min at room temperature.

Racemization has been reported by several groups during cleavage of benzyl groups by sodium in liquid ammonia (Brenner and Pfister, 1951; Hope, 1968). Hydrogenolysis of benzyl groups appears to cause little or no epimerization (McDermott and Benoiton, 1973a).

Probably as a result of the temperatures required, fusion of amino acids with phthalic anhydride results in extensive racemization. Formation of Pht derivatives is more satisfactorily achieved by other methods (Nefkens *et al*., 1960; Reinhoudt *et al*., 1971).

Deprotection of Tfa-Val-Val-O*t*Bu with aqueous base (2 *N* NaOH, MeOH—H_2O, 40°C) has been found to give 3.5% diastereomer (W. König, unpublished results).

C. Important Variables for Amide-Forming Reactions

1. Temperature

The factor determining the fraction of racemized product in a coupling reaction is a ratio of racemization rate to coupling rate. If both processes were to show a similar rate increase with temperature, the fraction of racemate or diastereomer that is formed would be temperature-independent. In no documented case has this behavior been observed. Instead, chiral integrity always decreases with increase in temperature of coupling reaction, implying a much larger temperature coefficient for the racemization rate constant (see, for example, Weygand *et al.*, 1963, 1968; Kemp *et al.*, 1974c; Kemp, 1973).

Evidently the large temperature coefficient noted for the racemization of amino acids (Section II,A,1) is shared with other amino acid derivatives, regardless whether they racemize by enolization or 5(4H)-oxazolone-mediated mechanisms. This is probably the explanation for racemization observed in slow reactions that must be conducted at elevated temperatures such as formation of cyanomethyl esters (Iselin *et al.*, 1955; Taschner *et al.*, 1965). A cardinal rule of peptide chemistry seems to be, whenever possible, to use a minimal reaction temperature.

2. Amine Bases and Salts

Three rules that have received general attention are use as weak a tertiary amine base as possible, use as hindered a base as possible, and avoid salts if possible. The first of these rules appears to be general. The second two are subject to strictures and are less reliable.

Results that argue for the use of a weak tertiary amine base such as *N*-methylmorpholine, rather than a stronger base, such as triethylamine, for liberating amine nucleophiles from their hydrochloride salts have been documented for major coupling methods, particularly the azide and mixed anhydride procedures. These are described in sections devoted to these methods.

As we have seen, hindered amines are less effective promoters of racemization by the enolization pathway (Bodanszky and Bodanszky, 1967) but not by the more prevalent 5(4H)-oxazolone pathway (Williams and Young, 1972; Bodanszky and Conklin, 1967).

Amine salts, such as the triethylamine hydrochloride formed by the liberation of a peptide amine, have long been recognized as inducers of racemization for DCC or active ester mediated couplings (Williams and Young, 1964; Weygand *et al.*, 1963). These studies primarily involved relatively nonpolar solvents, such as chloroform, dichloromethane, tetrahydrofuran, or ethyl acetate. More recent studies have revealed a more complicated picture. Earlier (Section II,C,3) we have noted that in DMF, addition of the hydrochloride salt of the amine nucleophile can suppress racemization. Studies of DCC—HOSu and DCC—HOBt mediated couplings in dipolar aprotic solvents like DMF can show either a small effect of amine salts or even an inhibitory effect of salts on racemization (Weygand *et al.*, 1966c; König and Geiger, 1970b). The problem of the origin of salt effects has been discussed by Williams and Young (1972).

It appears likely that salts exert both enhancing and inhibitory effects on racemization processes. In planning synthesis the chemist is advised to consult the available examples to identify which result is likely to be relevant to his experimental conditions.

3. Concentration

Specific base catalysis of 5(4H)-oxazolone formation by the amine nucleophile in an aprotic solvent that dissociates ion pairs results in a racemization process that must show less than first-order dependence on amine concen-

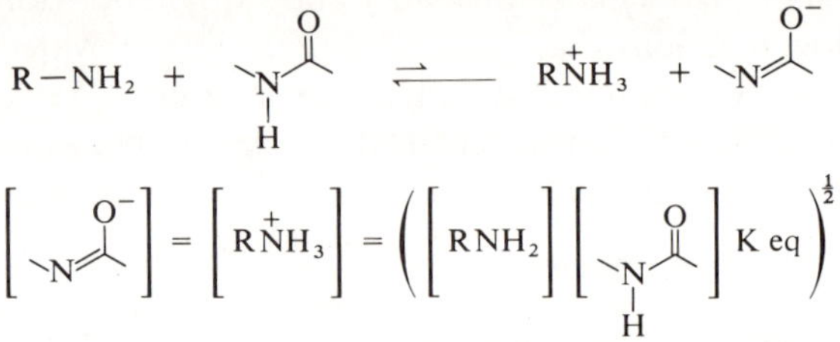

(24)

tration. If the coupling and racemization processes show different kinetic dependences on reactant concentration, then the amount of racemate formed will depend on concentration. Solubility often necessitates low concentrations of reagents in amide-forming steps, and consequently, this effect deserves more scrutiny than it has received. With one exception (Rebek and Feitler, 1975; see Section II,B,3), the available evidence implies an increase in racemate level with reagent dilution (Liberek and Michalik, 1965; Kemp *et al.*, 1974c,e; Kemp, 1973). The exception, a DCC coupling, probably involves a coupling rate that is zero-order in amine concentration.

4. Solvent

All studies of solvent effects on rates of racemization have demonstrated an increase in rate with solvent polarity (e.g., Kovacs *et al.*, 1975; Williams and Young, 1972). Since dipolar solvents such as DMF or DMSO are potent catalysts for the aminolysis of some but by no means all activated acyl derivatives (see Section V,A), the effects of solvent on the chiral integrity of peptide coupling products are difficult to predict.

For ONp esters, Kemp reported (1973) a range of 0.03 to 2% racemate, with a racemate-inducing order of DMSO ~ HMP > DMF > MeCN > toluene > pyridine > chloroform > THF ~ EtOAc. For DCC couplings, Weygand *et al.* (1963) observed a general order, DMF > CH_2Cl_2 ~ THF. For DCC—HOBt mediated couplings, Kemp *et al.* (1974c) observed an overall range of tenfold and a solvent order of HMP ~ DMSO > DMF > MeCN ~ CH_2Cl_2 ~ THF. König and Geiger (1970a) report a twofold change in epimeric content for this procedure, with the order, pyridine ≥ DMSO ≥ DMF ≥ THF.

D. Special Amino Acids

1. Cysteine

The special racemizing tendencies of Cys derivatives have been commented on in earlier sections. The need for careful pH control during preparation of Z-Cys(Bzl)-OH has been noted by Schröder and Lübke, 1965 (1965, p. 23). A reversible *β*-elimination mechanism has been repeatedly proposed for these racemization processes (Schröder and Lübke, 1965, p. 227; Bodanszky and Bodanszky, 1967), and there is no doubt that in the presence of strong bases and good leaving groups (Photaki, 1963) or at high temperatures (Bodanszky and Bodanszky, 1967; MacLaren *et al.*, 1958) these eliminations can occur.

However, it seems likely from the observations of Kovacs *et al.* (1968, 1970b), who noted that Z-Cys(Bzl)-OH activated esters racemize in the presence of triethylamine without exchanging sulfur with radiolabeled benzylmercaptan, that at room temperature or below, cysteine derivatives racemize by enolization, complicated by the peculiarity of isoracemization (Kovacs *et al.*, 1971). Reversible *β*-elimination must occur more slowly under these conditions.

Barber and Jones (1976) have reported the remarkable observation that active esters of N^{α}-benzyloxycarbonyl-L-thiazolidine-2-carboxylic acid (compound **40**) are at least fifty times less susceptible to triethylamine-

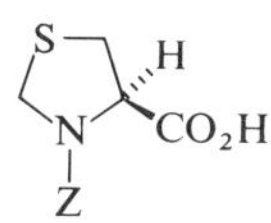

(40)

catalyzed racemization than the corresponding *S*-alkylcysteine derivatives. The interpretation of this result and its relevance to the low racemizing tendency of the isostructural proline cases remain unclear at this time.

2. Histidine

Since the recognition that acyl imidazoles formed from peptide acids undergo unusually facile racemization (Schröder and Lübke, 1965, p. 116; Anderson *et al.*, 1965), it has been clear that the presence of a free imidazole function (or an *N*-alkylimidazole) in a peptide coupling reaction creates a risk of racemization. Weygand *et al.* (1963) noted that the addition of imidazole or benzylimidazole to a DCC-mediated coupling increased racemization levels, and similar effects have been noted to result from DCC-mediated couplings in the presence of a His residue (Hofmann *et al.*, 1958; Riniker and Schwyzer, 1961). More recently, epimerizations during azide couplings have occurred at a COOH-terminal His function during a calcitonin M synthesis (Sieber *et al.*, 1970). By careful control of base, epimer was reduced to a minimum.

Two important recent studies have concerned cases in which a His residue lies at the terminus of the activated carboxyl fragment. Veber (1975) has demonstrated reversible cyclization of a His azide (compound **41**) to an imidazolide (**42**) which undergoes racemization, and has reviewed the racemization problem for COOH-terminal His.

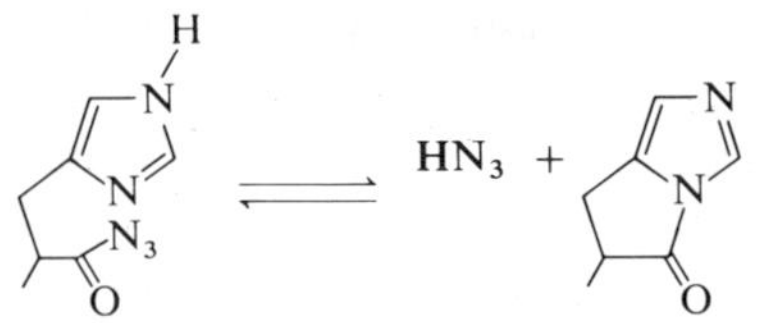

(25)

(41) **(42)**

Jorgensen and associates have observed as much as 22% epimer formation during a DCC-mediated coupling of Boc-His(Bzl)-OH to a resin-bound peptide amine. Addition of HOBt reduced this remarkable racemization of a urethane-blocked amino acid derivative to 0.1% (Jorgensen *et al.*, 1970; Windridge and Jorgensen, 1971).

Regardless of mechanism, ease of racemization is expected to depend on the basicity of the imidazole group. One might therefore expect that racemization by imidazole participation would be slow for His residues protected by the strongly electron-withdrawing N^{im}-2,4-dinitrophenyl group. The available data (Syrier and Beyerman, 1974) imply, however, that considerable racemization can result in this instance as well.

3. COOH-Terminal Proline or Glycine

Many peptide syntheses by fragment condensation have been planned to permit final couplings to COOH-terminal Pro or Gly residues, on the premise that 5(4H)-oxazolone-mediated racemization cannot occur with these residues. It is now clear that this premise is based on relative, not absolute strictures. We have noted earlier (Section III,C,4) Taschner's evidence for 5(4H)-oxazolone-induced epimerization of the penultimate Phe residue of Z-Phe_2-Gly-OH, and the apparent reluctance of COOH-terminal Gly acids to form 5(4H)-oxazolones. The latter would appear to be the important barrier to epimerization at the penultimate site in cases of COOH-terminal Gly.

Acylproline derivatives cannot form 5(4H)-oxazolone by a base-catalyzed mechanism. To the degree that 5(4H)-oxazolone ions can be formed from these species, one expects a rate order, Ac > Bz ≫ Tfa, a fact which explains the stability of Tfa-Pro-Cl (Westley, 1970). The results of McDermott and Benoiton (1973b) demonstrate 0.1% D-epimer with a mixed anhydride-mediated coupling of Boc-Ala-Pro-OH and 6.5% D-enantiomer for a similar coupling with Bz-Pro-OH. Earlier results of Williams and Young (1964) showed expected optical activity in couplings of Z-Gly-Pro-OH, using a variety of reagents, but a partial racemization for a DCC-mediated ONp ester preparation from *p*-nitrobenzoyl-Pro-OH. The issue of proline racemization would seem to deserve the general scrutiny now possible with highly sensitive racemization tests.

E. Racemization during Coupling Reactions of Solid-Phase Synthesis

With the exception of the His(Bzl) problem (Section IV,D,2), which is controllable by HOBt, solid-phase synthesis or other procedures based on step-wise elongation would appear to encounter little racemization during coupling procedures (Merrifield and Marglin, 1967). An attempt to measure racemization directly encountered less racemization than that incurred during the acidic hydrolysis of the assay. The value was set at less than 0.1% D-enantiomer (Bayer *et al.*, 1970). The data of Rebek and Feitler (1975) are consistent with this estimate.

F. Epimerization during Fragment-Condensation Coupling Reactions

1. *Mixed Anhydrides and EEDQ (2-Ethyloxy-1-Ethyloxycarbonyl-1,2-Dihydroquinoline)*

Given its convenience and its disposition toward racemization, the intensive study given the control of racemization for the mixed carbonic-carboxylic anhydride procedure is not surprising. These acyl intermediates vie with the imino anhydride intermediates of DCC synthesis for the most activated acyl derivatives in common use in peptide synthesis. Although most observed racemization reactions of mixed anhydrides (MA) occur with peptides and are undoubtedly 5(4H)-oxazolone-induced, an isolated report exists of racemization during a MA coupling of Z-Trp-OH with H-Gly-OEt that must occur by enolization (Geiger *et al.*, 1963).

The MA procedure is executed in two stages. First, an alkyl chloroformate (usually ethyl or isobutyl) is added to a chilled solution of the peptide acid and an equivalent of a tertiary amine. Second, this solution is treated with the peptide amine. Determann and Wieland (1963) observed that extensive epimerization occurred if the first step is prolonged and that MA formation is complete within a very short time.

Anderson *et al.* (1966) noted that a 4-min activation time at −15°C in THF is more than adequate and that trimethylamine or *N*-methylmorpholine give higher yields and lower racemization levels than triethylamine. In a later paper (Anderson *et al.*, 1967b) these workers noted that an unhindered tertiary amine that possesses at least one *N*-methyl group is required and proposed that the amine reacts as a nucleophile with the chloroformate, generating an alkoxycarbonylammonium ion that reacts with the peptide carboxylate in a second step to form the MA. The optimal procedure generates the MA in THF at −15°C for 30 sec, using exact equivalents of carboxylic acid and *N*-methylmorpholine. Couplings could then be carried out by adding this solution to the peptide amine in DMF. No racemate could be detected from Z-Gly-Phe-OH + H-Gly-OEt, following this procedure. This result was confirmed by Sakakibara and Itoh (1967) who noted that the weakly basic *N*,*N*-diethylglycine ethyl ester can be substituted for *N*-alkylmorpholine. Kemp *et al.* (1970b) noted 0.2% racemate for a normally executed coupling of Z-Gly-Phe-OH and H-Gly-OEt, using *N*-methylmorpholine as base, but observed only 0.01% racemate if very carefully weighed equivalents of acid and amine were used. Using the coupling of Z-Gly-Ala-OH with H-Leu-OBzl, Izumiya *et al.* (1971) observed 2.4% D-epimer for the *N*-methylmorpholine procedure (Izumiya and Muraoka, 1969).

There is a large solvent effect on the racemization observed with the MA procedure. Anderson *et al.* (1967b) report an approximate order, DMF or DMAc > CH_2Cl_2 > THF or other oxygen bases. The base used influences the solvent order, and since the stage at which solvent-induced racemization is occurring is not established, it is difficult to compare these results with other solvent effects.

The EEDQ reagent developed by Belleau and Malek (1968) provides an alternative route to MA derivatives of peptide acids, and was not observed to give detectable racemate by the Young test. Modified EEDQ derivatives have been prepared and studied by Yajima's group (Kiso *et al.*, 1973). Later

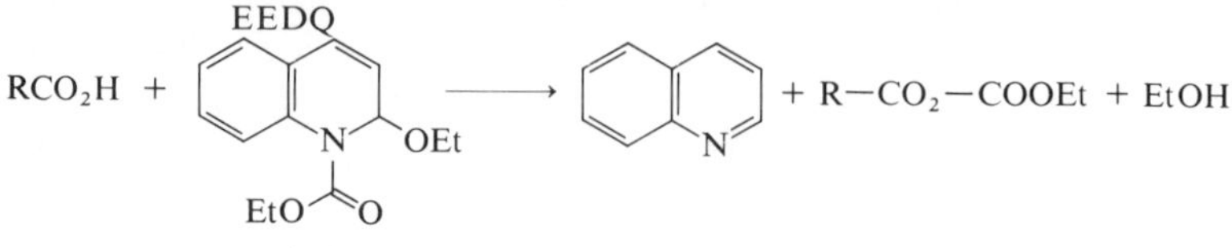

(43)

(26)

reports involving X-Gly-Ala-OH and H-Leu-Y couplings have claimed significant epimer formation. Lauren and Williams (1972) observe 6% D-epimer for a polymer based study of EEDQ, while Barton *et al.* (1973) report EEDQ-mediated couplings of Boc-Gly-Ala-OH with H-Leu-functionalized polymer that give 8.5% D-epimer in DMF and 0.9% D-epimer in dioxane. In their study of *N*-alkylamino acid derivatives, McDermott and Benoiton (1973b) report less than 0.1% epimer in EEDQ couplings, and for a Z-Gly-Ala-OH + H.Leu-OBzl coupling in THF, Izumiya and Muraoka report 0.2% D-epimer (1969).

From these variations, it appears that all of the MA-derived coupling procedures share a sensitivity to reaction conditions and can be expected to deviate from chiral integrity with changes in reaction variables.

2. *Acyl Azides*

Despite its evident drawbacks (Klausner and Bodanszky, 1974) the acyl azide coupling procedure has until recently been the primary cornerstone of peptide synthesis by fragment condensation. In large part, this role has been based on the widespread conviction, confirmed by all early studies, that the procedure is truly racemization-free under all useful coupling conditions (Smart *et al.*, 1960; Heard and Young, 1963; Williams and Young, 1963; Taschner *et al.*, 1968).

Two reports, Weygand *et al.* (1966d) and Anderson *et al.* (1966), established that if exposed to tertiary amines in excess, substantially racemized products are formed from acyl azides. Using an isotope dilution assay, Kemp *et al.* (1970a) established that under normal coupling conditions, the formation of the Phe–Gly bond of Z-Gly-Phe-Gly-OEt by the azide procedure is accompanied by only 0.01 to 0.04% racemate formation. In a subsequent study, the sensitivity to excesses of tertiary bases was confirmed, and repeated bicarbonate extraction during manipulation of the acyl azide was found to increase racemate content of product formed from it (Kemp *et al.*, 1970b).

Demonstration of the importance of the control of base used in azide couplings was provided by a detailed study of epimerization incurred during practical synthesis involving azide couplings (Sieber and Riniker, 1973). A variety of Phe—Val couplings were carried out by the Honzl–Rudinger procedure, in which excess hydrogen chloride used in the hydrazide diazotization step is neutralized with tertiary amine at −20°C in the presence of the hydrochloride of the amine nucleophile, and triethylamine was found to induce formation of between 4 and 16% D-epimer. Substitution of *N*-methyl- or *N*-ethylmorpholine reduced the epimerization level to 0.2–1.4%. (Interestingly, the hindered base, ethyldiisopropylamine, also significantly reduced the level of epimerization, a result that is more consistent with epimerization by direct enolization than by 5(4H)-oxazolone formation.) It was pointed out that use of *N*-methylmorpholine results in slower coupling rates than stronger bases, an undesirable outcome under many circumstances. The limits of the chiral assay were close to the diastereomeric levels seen for *N*-ethylmorpholine or ethyldiisopropylamine, and it was suggested that circumstances would best dictate the choice between these two.

Subsequent studies by Mitin and Maksimov (1973) and by Kisfaludy and Nyéki (1972) have confirmed the occurrence of azide racemization with the Honzl–Rudinger procedure. The latter study has recommended diethyl-2-hydroxypropylamine as superior to other amines in reducing racemization. Its effect on coupling rate is not clear.

3. The Dicyclohexylcarbodiimide—N-Hydroxysuccinimide and Dicyclohexylcarbodiimide—1-Hydroxybenzotriazole Procedures

In 1966 Weygand *et al.* (1966c) reported that the extensive racemization incurred in peptide couplings by the dicyclohexylcarbodiimide procedure could be reduced to a low level by the addition of *N*-hydroxysuccinimide, and in accord with an earlier observation of Wünsch and Drees, the coupling yields were found to be higher in the presence of this additive. Thus for coupling of Z-Leu-Phe-OH with H-Val-O*t*Bu 9.4% of epimer, observed at 0°C in the absence of HOSu, was reduced to 2.4% in its presence. If the initial reaction was carried out at −20°C, less than 1% of epimer was observed. Similar results were obtained by Zimmerman and Anderson (1967), who used the procedure to prepare OSu esters, but noted that Bz-Leu-OSu is easily racemized in workup. It is clear (Kemp, 1973, p. 16) that racemization of HOSu esters can occur on standing in solvents such as DMF. The "one-pot" procedure of Weygand–Wünsch has been widely adopted.

A variety of racemization tests have been applied to the reaction of acid, amine, and DCC, in the presence of HOSu. No epimer could be detected with the coupling of Z-Gly-Ala-OH with H-Leu-OBzl (Izumiya and Muraoka, 1969). Isotope dilution assay of the coupling, Z-Gly-Phe-OH with H-Gly-OEt, gave 0.02 to 0.6% racemate, depending on solvent (DMF > THF) and concentration (Kemp, 1973). Other tests revealed substantial racemization. For the coupling, Ac-Ile-OH with H-Gly-OEt, Itoh observed 2.7% epimer (1973), and the difficult system, Z-Pro-Val-OH + Pro-O*t*Bu, yielded 19–46% epimer (König and Geiger, 1970b). For Ala-Ala couplings, Bosshard *et al.* (1973) observed 0.5–1.0% epimer formation in DMF, but 5% epimer for a coupling with the dipeptide H-Ala_2-OH in dioxane–water containing bicarbonate. In an early study with the coupling Tfa-Pro-Val-OH + Pro-OMe, Weygand *et al.* (1968) noted a temperature-dependent variation of from 0.5 to 2.0% epimer.

Although many successful syntheses testify to the practicality of the DCC—HOSu procedure, racemization has also been observed during practical coupling conditions (Kaurov *et al.*, 1973), and β-alanine-derived impurities can accompany products (Gross and Bilk, 1968).

In 1970, König and Geiger (1970a,b,c) reported the successful outcome of an exhaustive search for a more satisfactory racemization-suppressing additive. Their primary candidate, 1-hydroxybenzotriazole (**44**), was found to give cleaner reaction products than HOSu as well as lower levels of epimerization by several criteria. [Although (**45**) and (**46**) give still lower epimer levels with the most stringent tests, and (**46**) is useful for preactivation (W.

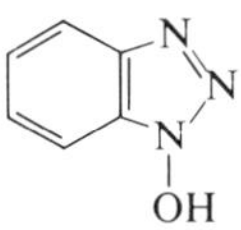

1-Hydroxybenzotriazole

(**44**)

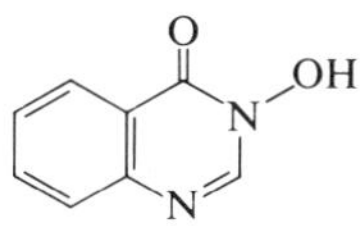

3-Hydroxy-4-oxo-3,4-dihydroquinazoline

(**45**)

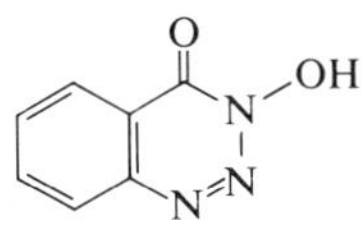

3-Hydroxy-4-oxo-3,4-dihydro-1,2,3-benzotriazine

(**46**)

König, private communication), they had other disadvantages, including lower yields and generation of impurities.]

With 1.0 equiv of HOBt in THF or DMF, and with the DCC-mediated coupling of Boc-Leu-Phe-OH with H-Val-O*t*Bu, less than 1% of epimer was detected; in DMSO or pyridine, this level rose to 1.7–2.0%. Substitution of HOBt for HOSu in the coupling of Tfa-Pro-Val-OH with Pro-O*t*Bu reduced the level of epimer from 17.1% to 3.4%, and for the exceedingly stringent coupling of Z-Pro-Val-OH with Pro-O*t*Bu in DMF in the absence of salts, HOBt yielded 9.0% epimer versus 46.0% with HOSu.

Using the isotope dilution procedure, Kemp *et al.* (1974e) measured racemate formation for the DCC—HOBt coupling of Z-Gly-Phe-OH with H-Gly-OEt and observed a concentration-dependent variation of 0.2–0.5% in DMF, 0.7–3.0% in DMSO, HMP, or HMP—DMF, and 0.08–0.2% in THF, CH_2Cl_2, MeCN. Young (1973) has noted 10–14% racemate formation for the DCC—HOBt procedure, applied to the Bz-Leu-OH + H-Gly-OEt coupling. In the synthesis of insulin by the CIBA–Geigy group, 30% of the B16-Tyr epimer was formed during a coupling of B-chain fragments (Sieber *et al.*, 1977).

The DCC–HOBt procedure gives very rapid reactions (König and Geiger, 1970a) which are free from the normal DCC-derived by-products, and though not racemization-free, for a majority of coupling procedures it appears to give epimerization levels that are insensitive to minor variations in reaction conditions, a considerable advantage in a practical reagent.

V. STRATEGIES FOR MINIMIZING RACEMIZATION DURING AMIDE-FORMING STEPS

A. The Key Issues: Properties and Accessibility of Acylating Agent

A successful coupling procedure for fragment condensation must solve two independent problems. The acylating agent must react rapidly with peptide amines, yielding chirally integral products, and the acylating agent must also be obtainable directly and easily from peptide carboxylic acids. Rational planning that addresses the first problem must seek to accelerate the aminolysis process selectively, leaving unchanged or inhibiting the processes that lead to racemization.

The second problem confronts the fact that most reagents that react quantitatively and rapidly with carboxylic acids generate "overactivated," unselective acyl derivatives that rapidly form 5(4H)-oxazolones. For this reason, the average new route devised by the organic chemist for converting carboxylic acids to amides is of no use to the peptide chemist. A correlary of this problem is that many useful, stable acyl derivatives of the active or phenolic ester class that can be shown to have satisfactory capacity for minimizing racemization are difficult to generate from peptide acids by routes that preserve chiral integrity. Three strategies have emerged to solve this problem, the trapping strategy, the intramolecular energy leak, and the backing-off and safety-catch strategies, which are closely related. In the sections that follow, examples of each are briefly considered.

B. Design of Minimally Racemizing Activated Acyl Derivatives

1. *Anchimeric Assistance of Aminolysis*

In 1965, Young and co-workers (Beaumont *et al.*, 1965; Handford *et al.*, 1965) proposed that the low racemization observed for amide formation from 1-hydroxpiperidine esters (Weygand and König, 1965) could be explained if ester aminolysis occurred by intramolecular anchimeric assistance by the piperidine nitrogen (see structure **47**). (For geometrical reasons, equivalent catalysis of 5(4H)-oxazolone formation is interdicted.)

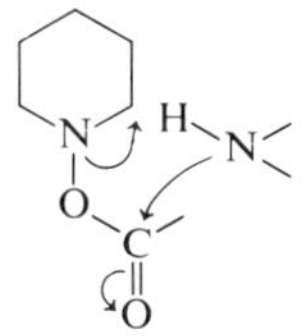

(47)

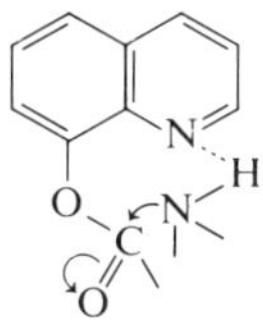

(48)

A similar tendency toward low racemization was recognized for esters of 8-hydroxyquinoline (**48**), and following Young's suggestion, anchimeric assistance of aminolysis was proposed to be operative in this case by Jakubke *et al.* (Jakubke, 1965; Jakubke *et al.*, 1967).

Unusually rapid hydrolysis of catechol half-esters had been observed some years previously, and two groups independently recognized the potential of these esters for peptide synthesis (Jones and Young, 1968; see also, Jones and Young, 1967; Young, 1967; Kemp and Chien, 1967a). Besides these esters (**49**), which appear to aminolyze via their conjugate bases (structure **50**), esters (**51**) of 2-pyridine thiol (Lloyd and Young, 1968, 1971) and of 2-hydroxypyridine (**52**) (Dutta and Morley, 1971) are further examples. All

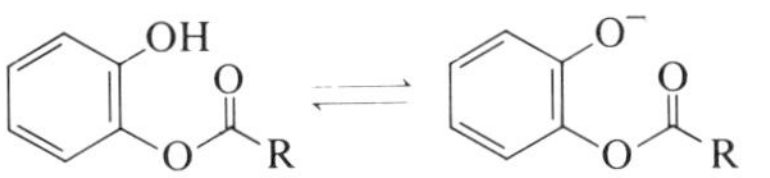

(49) **(50)**

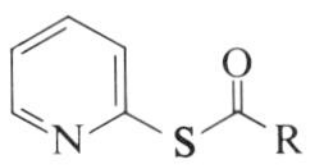

(51)

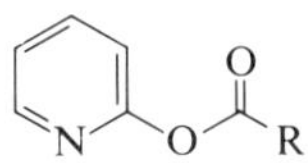

(52)

these esters show very favorable rate ratios, with rate constants for aminolysis greatly exceeding those for racemization. Where studied, racemization levels appear to lie in the range of 0.01–0.2% DL.

2. *Nature of Aminolytic Anchimeric Assistance*

Although the process depicted in (**47**) can be envisaged as involving a full N—H proton transfer, it is disquieting to note that relatively nonbasic lone pair atoms that are adjacent to the acyl carbon appear to be able to accelerate aminolysis, as in (**53**) (Bankowski and Drabarek, 1972) and (**54**) (Guarneri *et al.*, 1973). Catalysis of aminolysis of phenyl esters in nonaqueous

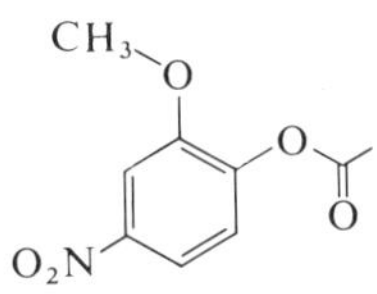

(53)

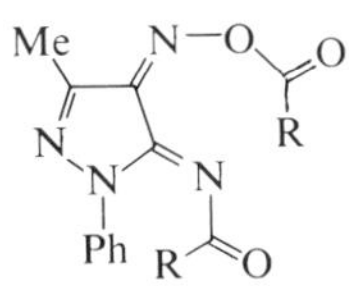

(54)

solvents is a property of species that form strong hydrogen bonds (Su and Watson, 1974), and hydrogen bonding that facilitates intramolecular proton transfer has been proposed as the likely form of anchimeric assistance of aminolysis (Kemp, 1973). Many of the successful activated species that are used in peptide synthesis can be viewed as providing anchimeric aminolytic assistance, see structures **(55)**, **(56)**, **(57)**, **(58)**.

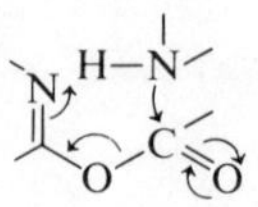

(55)

2-Hydroxypyridyl

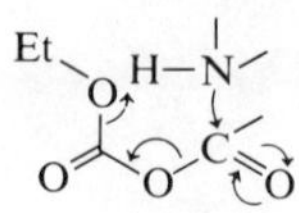

(56)

MA

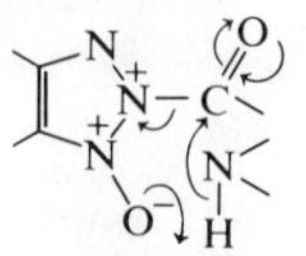

(57)

HOBt ester

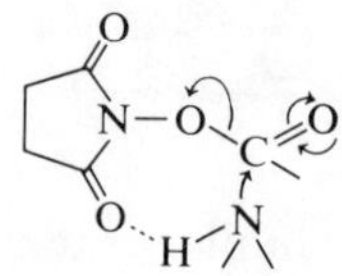

(58)

HOSu ester

C. Nonracemizing Routes to Acylating Agents

1. Trapping Agents

The successful trapping agent strategy is best exemplified by HOSu and HOBt. These species appear to react with highly activated acylating agents, including 5(4H)-oxazolones, more rapidly than they can racemize. It may be noted that the most successful trapping agents are all derivatives of hydroxylamine.

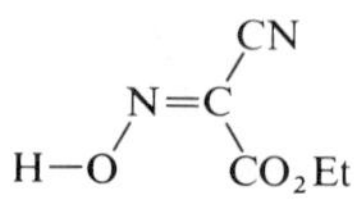

(59)

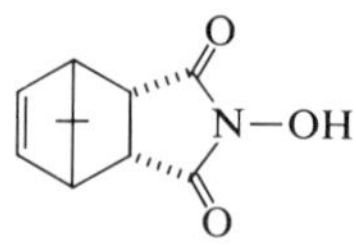

(60)

The reagent **(59)** devised by Itoh (1973), is an interesting addition to the list of effective trapping agents and appears to be more efficient than either HOBt or HOSu in controlling racemization or epimerization (1.8% for the formation of Ac-Ile-Gly-OEt). Yajima has reported trapping studies with reagents of different types (Yajima *et al.*, 1973a,b). Fujino and co-workers have reported that the hydroxysuccinimide derivative **(60)** yields active esters that lack the instability of HOSu esters. Only 0.5% epimer was detected by an assay of the Bodanszky type (Fujino *et al.*, 1974).

Anderson has introduced HOSu as an additive for controlling racemization in mixed anhydride synthesis (Anderson *et al.*, 1967a), and Kovacs and co-workers (1967, 1969) have reported that acidic phenols can trap 5(4H)-oxazolones and that the complex formed between pentachloro- or pentafluorophenol and carbodiimide lacks the tendency of the latter to promote 5(4H)-oxazolone racemization. The procedure is not able to preserve complete chiral integrity, a result confirmed by studies of König (W. König, unpublished observations).

2. *The Intramolecular Energy Leak*

The reaction of isoxazolium salts with carboxylate anions generates two sequentially formed acylating agents. The first, an imino anhydride (structure **62**), is highly activated and readily forms 5(4H)-oxazolones; the second is an enol or phenyl ester (**63**), formed from the imino anhydride by an

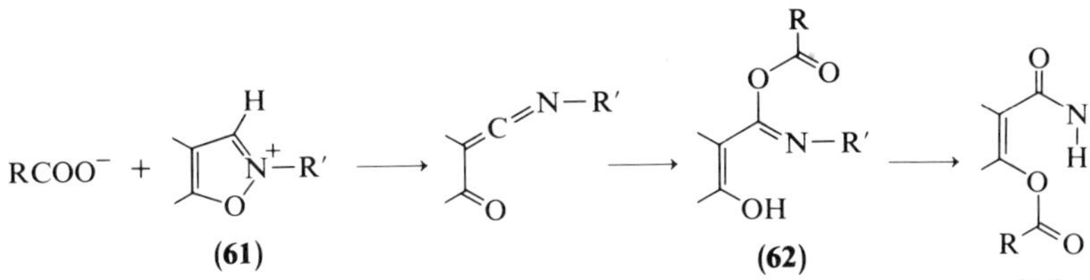

(**61**) (**62**) (**63**)

irreversible intramolecular acyl transfer. Provided the acyl transfer occurs highly efficiently, without leakage of imino anhydride to 5(4H)-oxazolones, the isoxazolium salt pathway may provide a racemization-free route from carboxylic acid to activated esters of a variety of structural types. The overactivated nature of the intermediate formed directly from the carboxylate is tempered in the final activated ester by irreversible leakage of acyl activation at the intramolecular transfer step.

A further refinement of this principle is provided by the 3-acyloxy-2-hydroxy-*N*-ethylbenzamides (**65**), available in high yield from the 7-hydroxy-2-ethylbenzisoxazolium cation (**64**) by a mechanism that involves two intramolecular acyl transfer steps. The esters (**65**) are of the catechol

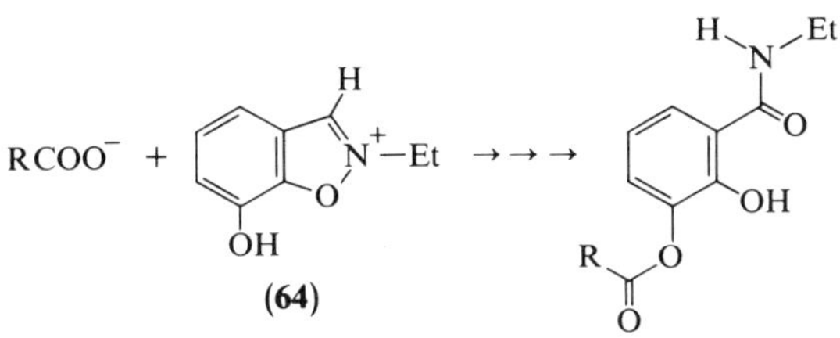

(**64**) (**65**)

half-ester type and for their degree of activation, display unusually rapid aminolytic cleavage. Very low levels of racemate formation are demonstrated during coupling reactions with these esters (0.03–0.2% under various conditions for Z-Gly-Phe-Gly-OEt), even under coupling conditions involving strong bases. A special buffering mechanism has been shown to protect these esters from racemization at high pH. For discussions of these issues, see Kemp *et al.* (1974b,c,d).

3. Backing-Off Synthesis and Safety-Catch Activation

In both of these procedures, an ester or other carboxyl derivative of a urethane-blocked amino acid derivative is prepared and deblocked to obtain an amino ester, protected as its salt. No racemization is expected for either step. Release of free amine and condensation with a second *N*-blocked amino acid or peptide acid can then be carried out, provided a coupling method is chosen that allows sufficiently rapid amide formation so that insignificant aminolysis of the original ester function occurs (Iselin and Schwyzer, 1960; Goodman and Stueben, 1962). The method is illustrated below (27).

$$\begin{array}{l} \text{Z—Tyr(Z)—OH} + \text{HONp} + \text{DCC} \rightarrow \text{Z—Tyr(Z)—ONp} + \text{DCU} \\ \text{Z—Tyr(Z)—ONp} \xrightarrow[\text{HOAc}]{\text{HBr}} \text{H}_2^+\text{—Tyr—ONpBr}^- \\ \text{Z—Val—OH} + \text{Et}_3\text{N} + \text{H}_2^+\text{—Tyr—ONpBr}^- + \text{DCC} \rightarrow \\ \text{Z—Val—Tyr—ONp} + \text{DCU} + \text{Et}_3\text{N}^+\text{HBr}^- \end{array} \tag{27}$$

Safety-catch activation is a phrase coined by Rudinger and used by Young to describe procedures in which a COOH-terminal protective group can be converted into an active acyl function without racemization and without intermediacy of the carboxylic acid. The *N*-blocked acyl hydrazides (Eq. 28) are early examples, Jones and Young's *o*-benzyloxyphenyl esters (Eq. 29)

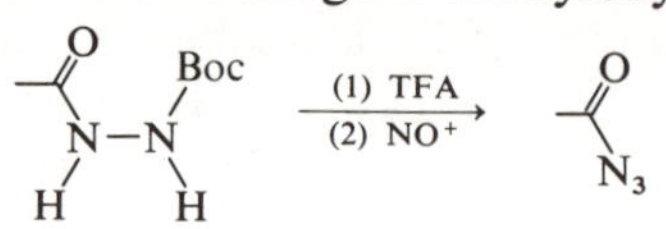

(28)

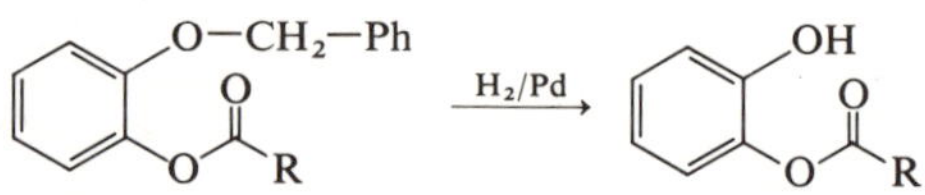

(29)

provide another (1968), and Trudelle's *o*-phenacyloxyphenyl esters are a third case (1971).

The four-component condensation procedure of Ugi and the prior amine capture principle of Kemp represent coupling strategies that confront the racemization problem in new ways. These are beyond the scope of this review, but are discussed elsewhere in this series.

VI. ASSAYS FOR CHIRAL PURITY

A. Types of Assays

The early literature on assay procedures has been reviewed by Weygand *et al.* (1963). A recent review (Jones, 1971) lists five principal assays for racemization, as shown in Table IV. These illustrate all but three of the principles on which such tests have been based. As indicated in the Table, the Young tests (Smart *et al.*, 1960; Heard and Young, 1963; Williams and Young, 1963; Anderson *et al.*, 1966) depend on measurement of the specific rotation of the amide product. When dealing with samples of nearly full optical purity, this measurement has a high intrinsic inaccuracy which is partly compensated by the vulnerability of Bz-Leu-OH to 5(4H)-oxazolone formation and consequent racemization.

The Anderson test (see Anderson *et al.*, 1966, and many earlier references) measures the weight of racemate obtained by fractional crystallization. From solubility data, the limit of detection of the Anderson test has been shown to be 1.5% racemate, and this percentage should therefore be added to the percentage obtained by isolation to obtain the true yield of racemate, as pointed out by Kemp *et al.* (1970a). This paper also demonstrated that by application of an experimentally simple multiple isotope dilution procedure, the sensitivity of the Young and Anderson tests can be extended by several orders of magnitude.

Weygand and his collaborators have developed a number of tests based on gas chromatographic (glc) separations of diastereomers. The simplest of these involves coupling of Tfa-Val-OH with H-Val-OMe, followed by glc assay (Weygand *et al.*, 1963); it is open to the criticism that the Tfa amido function is likely to show reactivity atypical of peptide-derived amides. The coupling of Z-Leu-Phe-OH with H-Val-O*t*Bu, followed by treatment with 9 *N* HCl in methanol, then by Tfa_2O, yields Tfa-Phe-Val-OMe (Weygand *et al.*, 1966d). A simpler procedure involves glc assay of the products of coupling of Tfa-Pro-Val-OH with Pro-OMe (Weygand *et al.*, 1968). A glc procedure in which Tfa-aminoacyl esters are separated as enantiomers on a chiral column has been developed for assay of optical purity of amino acids (Gil-Av and Feibush, 1967; Bayer *et al.*, 1970).

The first of the assays based on ion exchange separations was the Bodanszky test which involves a coupling of Ac-Ile-OH with H-Gly-OEt, followed by acidic hydrolysis and measurement of the ratio of *a*Ile to Ile (Bodanszky and Conklin, 1967). Perhaps because it involves study of a peptide acid rather than an acylamino acid, the Izumiya test has been more widely used. This test (Izumiya and Muraoka, 1969) employs the ion exchange resin of an amino acid analyzer to effect separation of diastereomeric free peptides, H-Gly-Ala-Leu-OH, formed by coupling and hydrogenolysis of Z-Gly-Ala-OH and H-Leu-OBz.

Tritium exchange from a protic coupling medium was employed by the Merck group as a racemization assay of their NCA-based synthetic procedure (Dewey *et al.*, 1968). Utilizing pmr peak area ratios of diastereomers, Weinstein and others have employed magnetic resonance as an

Table IV The Principal Racemization Tests[a]

Test	Model reaction product	Method of determining racemization	Sensitivity limit
1. Anderson	Z-Gly-Phe—┼—Gly-OEt	Specific rotation of crude material and isolation of racemate by fractional crystallization	ca. 1–2% racemization
2. Izumiya	Z-Gly-Ala—┼—Leu-OBzl	Deprotection and determination of Gly-DAla-Leu on an amino acid analyzer	ca. 0.1–1.0% racemization
3. Kemp	As in the Anderson or Young tests, but using ^{14}C-labeled carboxy components	Isolation of radioactive racemate after dilution with unlabeled racemate	ca. 0.001–0.01% racemization
4. Weygand	Z-Leu-Phe—┼—Val-O*t*Bu[b]	Deprotection and partial hydrolysis followed by determination of DPhe-Val by glc	ca. 0.1–1.0% racemization
5. Young	Bz-Leu—┼—Gly-OEt	Specific rotation of crude material: presence of racemate can be confirmed by isolation after saponification	ca. 1–2% racemization[c]

[a] From Jones (1971). Quoted by permission of publisher and author.
[b] And others.
[c] The model reaction is some 10 times as susceptible to racemization as most typical peptide coupling reactions.

assay (Halpern *et al.*, 1967; Weinstein and Pritchard, 1972). Bosshard *et al.* (1973) have demonstrated a novel technique in which Z—LAla—DAla—OH is first coupled by the procedure to be assayed with H—LAla—LAla—ONb. Hydrogenolysis and treatment with leucine aminopeptidase liberates NH_2-terminal LAla, assayed by conventional amino acid analysis, only from product in which the original DAla residue was epimerized.

B. Assay Procedures

Assay procedures for racemization or epimerization of amino acids that have not been discussed above are listed in Table V. Procedures 1d, 1e, 4a, and 4b in Table V should be noted as providing special information, not available with comparable accuracy by other means.

An important recent development is the application of ion exchange and HPLC chromatographic techniques to the determination of trace diastereomers in samples of peptides of the medium size range. Work by Li (Yamashiro *et al.*, 1977) and by Rivier *et al.* (1977; Rivier and Burgus, 1976) clearly demonstrates that in the hands of the proper investigators, most if not all contaminating epimers can be separated from a major component and assayed. These methods have the demonstrated potential for proving product homogeneity in many troublesome situations. However, it must not be assumed that they can be applied casually, without the patient application of experience and skill.

C. Selection of an Epimerization Assay

Clearly the sensitivity of the test employed must be tailored to the likely application of the coupling method under scrutiny. A method that is only to be used to prepare small peptides may be suitable even though it is anticipated to generate 1% levels of epimer.

However, for the most exacting needs of synthesis, 0.1% epimer is an upper limit for the tolerable amount of epimeric impurity. Moreover, the careful researches into levels of chiral purity encountered for couplings involving the most widely used coupling procedures reveal that changes in peptide structure or unavoidable changes in reaction conditions for practical couplings (cf. the acyl azide–amine problem) can frequently change racemization levels by one or two orders of magnitude. For this reason, it is most important to explore the "chirally pure" range of 0.01–0.001% racemate. Other factors being equal, a reagent that routinely generates model peptides at the lower limit of this range is likely to be safer than one that routinely yields 0.1–0.2% racemate in model systems.

Table V Additional Assay Procedures for Chiral Purity

Reference	System	Sensitivity	Use[a]
1. Separation of diastereomeric product by ion exchange (amino acid analyzer)			
a. Stewart (1970)	(1) Z-Gly-Ala-OH + H-Leu-OTmb[b] (2) HBr	~0.1%	PC (variant on Izumiya test)
b. Barton *et al.* (1973)	(1) Boc-Gly-Ala-OH + H-Leu-OResin (2) Ester exchange (3) Hydrolysis (4) TFA	~0.3%	(Solid-phase var. of Izumiya)
c. Manning and Moore (1968), Manning (1970), Manning *et al.* (1972), Barooshian *et al.* (1972)	Amino acid + LLeu-NCA	~0.1%	AA
d. McDermott and Benoiton (1973b)	(1) Z-Ala-MeLeu-OH + H-Gly-OBzl (2) Pd/H_2	~0.1%	*N*-alkylamino acid
e. Windridge and Jorgensen (1971)	(1) Boc-His(Bzl)-OH + H-Glu(OBzl)-OResin (2) HBr	~0.1%	Boc-His(Bzl)-OH rac.
2. Separation of diastereomeric product by glc			
a. König and Geiger (1970b)	(1) Z-Pro-Val-OH + H-Pro-O*t*Bu (2) H_2/Pd (3) TFA_2O-TFA (4) CH_2N_2	0.1%	PC
b. König and Geiger (1970b) (see paper for similar methods)	(1) Tfa-Pro-Val-OH + H-Pro-O*t*Bu (2) TFA (3) CH_2N_2	0.1%	PC

c. Tomida *et al.* (1973)	(1) Z-Pro-Val-OH + H-Pro-OMe (2) HBr-HOAc (3) $CF_3CO_2CH_3$ Also: Z-Pro-Ile-OH, Z-Pro-Ala-OH, Z-Sar-Val-OH, Z-Ala-Val-OH with H-Pro-OMe	0.1%	PC
d. Lande and Landowne (1966)	(1) hydrolysis (2) α-chloropropionyl chloride (3) CH_2N_2	—	AA
e. Parr and Howard (1972)	(1) TFA_2O (2) esterification Tfa-X-O-(1-C_3H_7)	0.1%	AA
f. Barooshian *et al.* (1971)	(1) Tfa-Pro-Cl (2) esterification Tfa-Pro-X-OMe	0.1%	AA
3. Tlc or column chromatographic separation of diastereomers			
a. Taschner *et al.* (1963b)	16 diastereomeric dipeptides	ca. 0.1%	
b. Taschner *et al.* (1978)	Z-Val-Phe-OH + H-Phe-O*t*Bu		
c. Pravda *et al.* (1964)	Z-X-Y-OMe diastereomers		
4. Enzymatic assay			
a. Bosshard *et al.* (1973)	Assay for LAla		PC
b. Syrier and Beyerman (1974)	Application to TRH		PC
5. Isotopic dilution assay			
a. Rebek and Feitler (1975)	Boc-Phe-OH + H-Gly-OMe	<0.001%	Racemization in coupling to Boc-Phe-OH

[a] PC, Model for epimerization during peptide coupling; AA, assay for racemate content in individual amino acids.
[b] Tmb, 2,4,6-trimethylbenzyl.

D. Chiral Integrity of Peptide Coupling Procedures

In Table VI are summarized a variety of results obtained with racemization tests applied to amide-forming procedures. In Table VI, early observations of high racemization or epimerization under conditions now seldom in use are usually grouped.

A variety of other methods have been found to give extensive racemization under general conditions. Carbonyldiimidazole is observed by Weygand *et al.* (1963, 1966d) and Anderson *et al.* (1965) to result in considerable racemization, a result in accord with general observations. On the other hand, Wieland and Determann (1963) and Weinstein and Pritchard (1972) report low levels of racemate with this reagent.

The phosphorazo procedure, procedures using dialkyl phosphochloridates, and the carbon tetrachloride–triphenylphosphine procedure are observed by several groups to racemize strongly (Williams and Young, 1963; Weygand *et al.*, 1966d; Wieland and Seelinger, 1971; Shioiri *et al.*, 1972; Weinstein and Pritchard, 1972), although Takeuchi and Yamada (1974) note that low temperatures and choice of a more suitable phosphine can minimize racemate formation. Condition-dependent racemization that can be minimized appears to characterize the phosphine-bis-2-pyridyl disulfide reduction–oxidation procedure of Mukaiyama (1972).

Jones and Fairweather (1969) have noted that extensive racemization characterizes the coupling of Z-Gly-Phe-OH with an isocyanate. The use of ynamines as coupling reagents is accompanied by extensive racemization (Weygand *et al.*, 1965). In contrast to the chiral stability of *N*-carboxy anhydrides, the *N*-(*o*-nitrophenylsulfenyl)-*N*–carboxy anhydrides described by Kricheldorf and Fehrle (1974) appear to racemize easily.

E. Unresolved Problems

It can be readily recognized from this review that existing synthetic procedures fall short of the most stringent needs of peptide synthesis, and despite much ingenious planning and effort this problem remains most acute for the amide-forming steps of synthesis by fragment condensation. Ultimately, the resolution of this problem may rest on the construction of improved reagents. Nevertheless, it seems appropriate to conclude by listing several simple studies which could aid the planning of synthesis in its present form.

1. A more complete survey of the extent of epimerization incurred at various centers during protective group removal.
2. A clarification of the acid–base chemistry of peptide-related species in the most convenient solvents of synthesis. Particularly important are the

Table VI Levels of Chiral Purity for Coupling Reactions

Test	Conditions	Result[a]	References
A. Mixed anhydride procedure			
1. Triethylamine and ethyl chloroformate			
a. Z-Gly-Phe+Gly-OEt	Variable activation time	2% rac. (1–3 min) Up to 80% rac. (prolonged)	Determann and Wieland (1963) Anderson *et al.* (1966)
b. Z-Leu-Phe+Val-O*t*Bu	THF	5% LDL	Weygand *et al.* (1966d)
c. Z-Gly-Phe+Phe-O*t*Bu		3–13% DL	Taschner *et al.* (1968)
d. Z-Gly-Ala+Leu-OBzl		9.5% DL	Izumiya *et al.* (1971)
e. Boc-Leu-Phe+Val-O*t*Bu	DMF or THF, −10°C	13–55% DPhe	W. König and M. Walter (unpublished)
2. Ethyl chloroformate and *N*-methylmorpholine or other bases			
a. Z-Gly-Phe+Gly-OEt	THF, 12 min, −15°C	"0% rac." also seen with other *N*-methylamines	Anderson *et al.* (1967b)
b. Z-Gly-Phe+Gly-OEt	1–2 min	"0% rac."	Sakakibara and Itoh (1967)
c. Z-Gle-Phe+Gly-OEt	−15°C, THF	0.2–0.01% rac.	Kemp *et al.* (1970b)
d. Z-Gly-Ala+Leu-OBzl		2.4% DL	Izumiya *et al.* (1971)
e. Tfa-Pro-Val+Pro-O*t*Bu	3 min, −30°C, *N*-ethylmorpholine, THF	0.5% LDL	W. König and M. Walter (unpublished)
f. Tfa-Pro-Val+Pro-O*t*Bu	3 min, −30°C, *N*-ethylmorpholine, DMF	2% LDL	W. König and M. Walter (unpublished)
g. Tfa-Pro-Val+Pro-O*t*Bu	3 min, −30°C, DMF, 1,1,3,3-tetramethylguanidine	17–21% LDL	W. König and M. Walter (unpublished)
3. Ethyl chloroformate, *N*-methylmorpholine, THF, −15°C, HOSu added			
a. Z-Gly-Phe+Gly-OEt		"no rac."	Anderson *et al.* (1967a)
b. Z-Gly-Ala+Leu-OBzl		0.2% DL	Izumiya *et al.* (1971)

(*continued*)

Table VI (*Continued*)

Test	Conditions	Result[a]	References
4. EEDQ, 2-ethyloxy-1-ethyloxycarbonyl-1,2-dihydroquinoline			
a. Bz-Leu—┼—Gly-OEt		"no rac."	Belleau and Malek (1968)
b. Z-Gly-Ala—┼—Leu-OBzl	THF	0.2% DL	Izumiya *et al.* (1971)
c. Z-Gly-Ala—┼—Leu-OPolymer		6% DL	Lauren and Williams (1972)
	DMF	8.5%	Barton *et al.* (1973)
	THF	0.9%	
d. Z-Ala-MeLeu—┼—Gly-OBzl		7–15% LD	McDermott and Benoiton (1973b)
B. Dicyclohexylcarbodiimide-mediated couplings			
1. DCC without additives			
Most tests	20°C	4–80% rac.	Most references cited below
	0°C	2–5%	
2. DCC with HOSu, "one-pot" procedure			
a. Z-Leu-Phe—┼—Val-O*t*Bu		<1.0–2.8% LDL	Weygand *et al.* (1966c)
b. Z-Gly-Phe—┼—Gly-OEt		"0% rac."	Zimmerman and Anderson (1967)
	THF	0.02–0.01% rac.	Kemp (1973)
	DMF	0.2 –0.6% rac.	
c. Z-Ala$_2$—┼—Ala$_2$-ONb	DMF	0.5–1.0% epi	Bosshard *et al.* (1973)
d. Tfa-Pro-Val—┼—Pro-OMe	CH_2Cl_2	0.5–1.5%	Weygand *et al.* (1968)
e. Z-Gly-Ala—┼—Leu-O*t*Bu	THF	0.0% DL	Izumiya *et al.* (1971)
		0–3.5% (Temperature-dependent)	Mitin and Maksimov (1973)
f. Ac-Ile—┼—Gly-OEt	DMF	2.7% *a*Ile	Itoh (1973)
g. Tfa-Pro-Val—┼—Pro-O*t*Bu	DMF	17.4% LDL	König and Geiger (1970b)
h. Z-Pro-Val—┼—Pro-O*t*Bu	DMF	19–46% LDL	
i. Z-Ala-MeLeu—┼—Gly-OEt	THF	3–11% LD	McDermott and Benoiton (1973b)
3. DCC with HOSu, preactivation in DMF			
a. Boc-Leu-Phe—┼—Val-O*t*Bu		1–10% LDL	König and Geiger (1970b)
b. Z-Gly-Phe—┼—Gly-OEt		3–10% rac.	Kemp (1973)

4. DCC with HOBt, "one-pot" procedure			
a. Boc-Leu-Phe—┼—Val-O*t*Bu		≤1 % LDL	König and Geiger (1970a,b)
b. Tfa-Pro-Val—┼—Pro-O*t*Bu		3.4 % LDL	König and Geiger (1970a,b)
c. Z-Pro-Val—┼—Pro-O*t*Bu		8–9 % LDL	König and Geiger (1970a,b)
d. Ac-Ile—┼—Gly-OEt		9 % *a*Ile	Itoh (1973)
e. Z-Gly-Phe—┼—Gly-OEt		0.1–3.0 % rac. (solvent-dependent)	Kemp *et al.* (1974e)
f. Bz-Leu—┼—Gly-OEt		10–14 % rac.	Young (1973)
5. DCC with HOBt, preactivation			
Boc-Leu-Phe—┼—Val-O*t*Bu	DMF	4–9 % LDL	König and Geiger (1970a,b)
6. DCC with other Additives			
a. 8-hydroxyquinoline derivatives			
Ac-Ile—┼—Gly-OEt		5–21 % *a*Ile	Yajima *et al.* (1973a)
b. *N*-hydroxybenzenesulfonamide			
Ac-Ile—┼—Gly-OEt		8 % *a*Ile	Yajima *et al.* (1973b)
c. Ethyl 2-oximinocyanoacetate			
Ac-Ile—┼—Gly-OEt		1.8 % *a*Ile	Itoh (1973)
d. 4-keto-3-hydroxy-3,4-dihydro-1,2,3-benzotriazene			
Z-Pro-Val—┼—Pro-O*t*Bu		1.3 % LDL	König and Geiger (1970b)
e. For results with 18 HOBt derivatives, see König and Geiger (1970a)			
C. Acyl azide couplings (many workers prior to 1966 report no detectable racemization for azide couplings)			
1. No excess base			
a. Z-Gly-Phe—┼—Gly-OEt		0.01–0.04 % rac.	Kemp *et al.* (1970a)
b. Bz-Leu—┼—Gly-OEt		0.15–0.50 % rac.	Kemp *et al.* (1970a)
2. Acyl azides in presence of tertiary amines		1–20 % rac.	See Section IV,F,2

(*continued*)

Table VI (*Continued*)

Test	Conditions	Result[a]	References
D. Isoxazolium salt couplings			
1. Woodward's reagent K			
a. Ac-Phe—+—Ala-OMe		15% DL	Weinstein and Pritchard (1972)
b. Z-Leu-Phe—+—Val-O*t*Bu		3% LDL	Weygand *et al.* (1966d)
c. Tfa-Val—+—Val-OMe		69% DL	Weygand *et al.* (1963)
d. Z-Gly-Ala—+—Leu-OBzl		1.7% DL	Izumiya *et al.* (1971)
e. Ac-Ile—+—Gly-OEt		4–20%[a]	Bodanszky and Conklin (1967)
f. Z-Gly-Ala—+—Phe-OMe		5% DL	Halpern *et al.* (1967)
g. Z-Gly-Phe—+—Gly-OEt		2–4% rac.	Kemp *et al.* (1970a)
h. Bz-Leu—+—Gly-OEt		8–10% rac.	Kemp *et al.* (1970a)
2. *O*-Acyl-*N*-ethylsalicylamides, from 2-ethylbenzisoxazolium salts			
a. Z-Gly-Phe—+—Gly-OEt		1–1.5% rac.	Kemp *et al.* (1970a)
b. Bz-Leu—+—Gly-OEt		7–14% rac.	Kemp *et al.* (1970*a*)
3. 3-Acyloxy-2-hydroxy-*N*-ethylbenzamides, from 7-hydroxy-2-ethylbenzisoxazolium salt			
a. Z-Gly-Phe—+—Gly-OEt	0°C	0.01% rac.	Kemp *et al.* (1970a, 1974c)
	20°C	0.15% rac.	Kemp *et al.* (1970a, 1974c)
b. Bz-Leu—+—Gly-OEt	0°C	0.2% rac.	Kemp *et al.* (1970a, 1974c)
	20°C	0.8% rac.	Kemp *et al.* (1970a, 1974c)
c. Z-Gly-Phe—+—Gly-O^-	0°C	0.01–0.03% rac.	Kemp *et al.* (1970a, 1974c)
d. Bz-Leu—+—Gly-OEt	0°C	0.1–0.3% rac.	Kemp *et al.* (1970a, 1974c)
4. 5-Nitro-3-acyloxy-2-hydroxy-*N*-ethylbenzamides			
Z-Gly-Phe—+—Gly-OEt	0°C	0.007% rac.	Kemp *et al.* (1974b)
E. *p*-Nitrophenyl Esters, prepared by back-off procedure, Section V,C,3			
1. Salt Effects in $CHCl_3$ and EtOAc			
Bz-Leu—+—Gly-OEt		0.30% rac.	Williams and Young (1964)
2. Solvent Effects (see reference)			
Z-Gly-Phe—+—Gly-OEt		0.03–2% rac.	Kemp (1973)

3. Z-Leu-Phe—┼—Val-O*t*Bu	"0% DL"	Weygand *et al.* (1966d)
F. Other active esters		
1. Thioesters		
a. No racemization has been reported by the listed groups for thiophenyl ester couplings with amino esters, temperature < 40°C		Weygand *et al.* (1966d) Determann and Wieland (1963) Liberek and Michalik (1966b)
b. racemization in prep of Z-Ala(β-CN)—┼—Gly-OMe from thiophenyl ester		
c. Tfa-Val—┼—Val-OMe via thiophenyl ester 130°C, HOAc	39% DL	Weygand *et al.* (1963)
2. *N*-hydroxypiperidine esters No racemization observed (Bz-Leu—┼—Gly-OEt, Z-Leu-Phe—┼—Val-O*t*Bu)		Weygand and König (1965) Beaumont *et al.* (1965)
3. Cyanomethyl esters		
a. Bz-Leu—┼—Gly-OEt	"no rac."	Williams and Young (1963)
b. Tfa-Val—┼—Val-OMe (110°C)	16–20% DL	Weygand *et al.* (1963)
4. Vinyl esters		
a. Tfa-Val—┼—Val-OMe	<2% DL	Weygand *et al.* (1963)
b. Z-Leu-Phe—┼—Val-O*t*Bu	"0% DL"	Weygand *et al.* (1966d)
5. 8-Hydroxyquinoline esters		
a. Z-Leu-Phe—┼—Val-O*t*Bu	"0% DL"	Weygand *et al.* (1966d)
b. Z-Gly-Phe—┼—Gly-OEt	0.16% rac.	Kemp *et al.* (1970b)
G. *N*-Carboxy anhydride procedures		
1. Tyr-NCA + H-Ser-O^-	0.004% DL	Denkewalter *et al.* (1966)
2. 4-methyl-2,5-thiazolidinedione + H-Phe-O^-	1.3% DL	Dewey *et al.* (1968)

[a] rac, racemate.
DL, D,L diastereomer.
LDL, L,D,L diastereomer.

relative strengths and catalytic efficiencies of tertiary and peptide amines and the relevance, if any, of association phenomena.

3. A determination, for the commonly used activating acyl functions, of their degree of thermodynamic activation (Gerstein and Jencks, 1964). The primary requirement of such a study, an acyl-derived species to which all other acyl derivatives can be quantitatively related, appears to have been met in the form of acyl imidazoles (Veber, 1975) or 5(4H)-oxazolones (Goodman and Glaser, 1970a; Antonovics and Young, 1967; Kemp *et al.*, 1974c). With a scale of thermodynamic activation, identification of anomalies in racemization and coupling rates would be greatly facilitated.

4. A determination of combinations of substituent, acyl derivative, and amine for which peptide derivatives racemize more rapidly by enolization than 5(4H)-oxazolone formation.

5. Development of a procedure for predicting the chiral purity expected for a coupling reaction, carried out with a given pair of peptide fragments. A first step in developing such a procedure involves determining whether racemization level can be analyzed in terms of separate contributions from substituents, coupling agent, and conditions. A second step involves determining the degree to which the entire substituent contribution is controlled by the penultimate and COOH-terminal acyl substituents, together with the amine substituent. Implicit in a prediction procedure is an understanding of the scope of 5(4H)-oxazolone-induced epimerization at the penultimate site.

REFERENCES

Anderson, G. W., Callahan, F. M., and Zimmerman, J. E. (1965). *Acta Chim. Acad. Sci. Hung.* **44**, 51.

Anderson, G. W., Zimmerman, J. E., and Callahan, F. M. (1966). *J. Am. Chem. Soc.* **88**, 1338–1339.

Anderson, G. W., Callahan, F. M., and Zimmerman, J. E. (1967a). *J. Am. Chem. Soc.* **89**, 178.

Anderson, G. W., Zimmerman, J. E., and Callahan, F. M. (1967b). *J. Am. Chem. Soc.* **89**, 5012–5017.

Antonovics, I., and Young, G. T. (1967). *J. Chem. Soc. C* pp. 595–601.

Antonovics, I., Heard, A. L., Hugo, J., Williams, M. W., and Young, G. T. (1966). *In* "Peptides, Proceedings of the Sixth European Symposium" (L. Zervas, ed.), pp. 121–130. Pergamon Press, Oxford.

Auld, G. A., and Davison, A. (1968). *Inorg. Chem.* **7**, 306–309.

Bada, J. L. (1972). *J. Am. Chem. Soc.* **94**, 1371–1373.

Bada, J. L. (1974). *Nature (London)* **252**, 379–380.

Bada, J. L., Luyendyk, B. P., and Maynard, J. B. (1970). *Science* **170**, 730–732.

Bankowski, K., and Drabarek, S. (1972). *Rocz. Chem.* **46**, 607–612.

Barber, M., and Jones, J. H. (1976). *In* "Peptides, Proceedings of the Fourteenth European Symposium" (A. Loffet, ed.), pp. 109–116. North-Holland Publ., Amsterdam.

Barooshian, A. V., Lautenschleger, M. J., and Harris, W. G. (1971). *Anal. Biochem.* **44**, 543–547.

Barooshian, A. V., Lautenschleger, M. J., Greenwood, J. M., and Harris, W. G. (1972). *Anal. Biochem.* **49**, 602–606.

Barton, M. A., Lemieux, R. U., and Savoie, J. Y. (1973). *J. Am. Chem. Soc.* **95**, 4501–4506.

Bator-Sawicka, B. (1968). *Rocz. Chem.* **42**, 47–56.

Bayer, E., Gil-Av, E., König, W. A., Nakaparksin, S., Oró, J., and Parr, W. (1970). *J. Am. Chem. Soc.* **92**, 1738–1740.

Beaumont, S. M., Handford, B. O., and Young, G. T. (1965). *Acta Chim. Acad. Sci. Hung.* **44**, 37–43.

Belleau, B., and Malek, G. (1968). *J. Am. Chem. Soc.* **90**, 1651–1652.

Bergmann, M., and Zervas, L. (1928). *Biochem. Z.* **203**, 280–292.

Bodanszky, M., and Bodanszky, A. (1967). *Chem. Commun.* pp. 591–593.

Bodanszky, M., and Conklin, L. (1967). *Chem. Commun.* pp. 773–774.

Bodanszky, M., Klausner, Y., and Ondetti, M. A. (1976). "Peptide Synthesis," 2nd ed., pp. 136–157. Wiley, New York.

Bohak, Z., and Katchalski, E. (1963). *Biochemistry* **2**, 228–237.

Bosshard, H. R., Schechter, I., and Berger, A. (1973). *Helv. Chim. Acta* **56**, 717–723.

Bovarnik, M., and Clarke, H. T. (1938). *J. Am. Chem. Soc.* **60**, 2426–2430.

Boyd, G. V., and Wright, P. H. (1972). *J. Chem. Soc., Perkin Trans.* **1**, pp. 909–913.

Brenner, M., and Pfister, R. W. (1951). *Helv. Chim. Acta* **34**, 2085–2096.

Cheung, S. T., and Benoiton, N. L. (1977a). *Can. J. Chem.* **55**, 906–910.

Cheung, S. T., and Benoiton, N. L. (1977b). *Can. J. Chem.* **55**, 916–921.

Coetzee, J. F. (1967). *Prog. Phys. Org. Chem.* **4**, 45–92.

Denkewalter, R. G., Schwam, H., Strachan, R. G., Beesley, T. E., Veber, D. F., Schoenewaldt, E. F., Barkemeyer, H., Paleveda, W. J., Jacob, T. A., and Hirschmann, R. (1966). *J. Am. Chem. Soc.* **88**, 3163–3164.

Determann, H., and Wieland, T. (1963). *Justus Liebig's Ann. Chem.* **670**, 136–140.

Dewey, R. R., Schoenewaldt, E. F., Joshua, H., Palveda, W. J., Schwam, H., Barkemeyer, H., Arison, B. H., Veber, D. F., Denkewalter, R. G., and Hirschmann, R. (1968). *J. Am. Chem. Soc.* **90**, 3254–3255.

Dutta, A. S., and Morley, J. S. (1971). *J. Chem. Soc. C* pp. 2896–2902.

Dzieduszycka, M., Smulkowski, M., and Taschner, E. (1973). *In* "Peptides 1972" (H. Hanson and H.-D. Jakubke, eds.), pp. 103–107. North-Holland Publ., Amsterdam.

Eguchi, C., and Kakuta, A. (1974). *J. Am. Chem. Soc.* **96**, 3985–3989.

Fridkin, M., Wilchek, M., and Syheinblatt, M. (1970). *Biochem. Biophys. Res. Commun.* **38**, 458–464.

Fujino, M., Kobayashi, S., Obayashi, M., Fukuda, T., Shinagawa, S., and Nishimura, O. (1974). *Chem. Pharm. Bull.* **22**, 1857–1863.

Geiger, R., Sturm, K., and Siedel, W. (1963). *Chem. Ber.* **96**, 1080.

Gerstein, J., and Jencks, W. P. (1964). *J. Am. Chem. Soc.* **86**, 4655–4663.

Gil-Av, E., and Feibush, B. (1967). *Tetrahedron Lett.* pp. 3345–3347.

Goodman, M., and Glaser, C. (1969). *Tetrahedron Lett.* pp. 3473–3475.

Goodman, M., and Glaser, C. (1970a). *In* "Peptides, Proceedings of the First American Peptide Symposium" (B. Weinstein and S. Lande, eds.), pp. 267–335. Dekker, New York.

Goodman, M., and Glaser, C. (1970b). *J. Org. Chem.* **35**, 1954–1962.

Goodman, M., and Levine, L. (1964). *J. Am. Chem. Soc.* **86**, 2918–2922.

Goodman, M., and McGahren, W. J. (1966). *J. Am. Chem. Soc.* **88**, 3887–3888.

Goodman, M., and McGahren, W. J. (1967a). *Tetrahedron* **23**, 2017–2030.

Goodman, M., and McGahren, W. J. (1967b). *Tetrahedron* **23**, 2031–2050.

Goodman, M., and Stueben, K. C. (1962). *J. Org. Chem.* **27**, 3409–3416.

Gosei, T. (1967). Chemical Industrial Co. Ltd., Netherlands Patent 6,602,194; *Chem. Abstr.* **66**, 2769uP.

Grahl-Nielsen, O. (1971). *Chem. Commun.* pp. 1588–1589.

Greenstein, J. P., and Winitz, M. (1961). "The Chemistry of the Amino Acids," 3 vols. Wiley, New York.

Gross, H., and Bilk, L. (1968). *In* "Peptides, Proceedings of the Ninth European Symposium" (E. Bricas, ed.), pp. 156–158. North-Holland Publ., Amsterdam.

Guarneri, M., Giori, P., Ferroni, R., Guggi, A., Menegatti, E., Tomatis, R., and Benassi, C. A. (1973). *In* "Peptides 1971" (H. Nesvadba, ed.), pp. 22–27. North-Holland Publ., Amsterdam.

Halpern, B., Chew, L. F., and Weinstein, B. (1967). *J. Am. Chem. Soc.* **89**, 5051–5052.

Handford, B. O., Jones, J. H., Young, G. T., and Johnson, T. F. N. (1965). *J. Chem. Soc.* pp. 6814–6827.

Heard, A. L., and Young, G. T. (1963). *J. Chem. Soc.* pp. 5807–5810.

Hill, J., and Leach, S. J. (1964). *Biochemistry* **3**, 1814–1818.

Hofmann, K., Woolner, M. E., Spühler, G., and Schwartz, E. T. (1958). *J. Am. Chem. Soc.* **80**, 1486–1489.

Hope, D. B. (1968). *In* "Peptides, Proceedings of the Ninth European Peptide Symposium" (E. Bricas, ed.), pp. 111–117. North-Holland Publ., Amsterdam.

Iselin, B., and Schwyzer, R. (1960). *Helv. Chim. Acta* **43**, 1760–1766.

Iselin, B., Feurer, M., and Schwyzer, R. (1955). *Helv. Chim. Acta* **38**, 1508–1516.

Itoh, M. (1973). *Bull. Chem. Soc. Jpn.* **46**, 2219–2221.

Izumiya, N., and Muraoka, M. (1969). *J. Am. Chem. Soc.* **91**, 2391–2392.

Izumiya, N., Muraoka, M., and Aoyagi, H. (1971). *Bull. Chem. Soc. Jpn.* **44**, 3391–3395.

Jacobson, S. J., Willson, C. G., and Rapoport, H. (1974). *J. Org. Chem.* **39**, 1074–1077.

Jakubke, H.-D. (1965). *Z. Naturforsch., Teil B* **20**, 273–274.

Jakubke, H.-D., Voigt, A., and Burkhardt, S. (1967). *Chem. Ber.* **100**, 2367–2372.

Johnson, S. L. (1967). *Adv. Phys. Org. Chem.* **5**, 237–330 (see, esp. p. 291).

Jones, J. H. (1971). "Amino-acids, Peptides, and Proteins," Vol. 3, p. 239. Chem. Soc., London.

Jones, J. H., and Fairweather, R. (1969). *Makromol. Chem.* **128**, 279–283.

Jones, J. H., and Young, G. T. (1967). *Chem. Commun.* pp. 35–36.

Jones, J. H., and Young, G. T. (1968). *J. Chem. Soc. C* pp. 436–441.

Jorgensen, E. C., Windridge, G. C., and Lee, T. C. (1970). *J. Med. Chem.* **13**, 352–356.

Kaurov, O. A., Martynov, V. F., and Popernatskii, O. A. (1973). *J. Gen. Chem. USSR* (*Engl. Transl.*) **43**, 907–913.

Kemp, D. S. (1973). *In* "Peptides 1971" (H. Nesvadba, ed.), p. 1–20. North-Holland Publ., Amsterdam.

Kemp, D. S., and Chien, S. W. (1967a). *J. Am. Chem. Soc.* **89**, 2743–2745.

Kemp, D. S., and Chien, S. W. (1967b). *J. Am. Chem. Soc.* **89**, 2745–2746.

Kemp, D. S., and Rebek, J. (1970). *J. Am. Chem. Soc.* **92**, 5792–5793.

Kemp, D. S., Wang, S. W., Busby, G., and Hugel, G. (1970a). *J. Am. Chem. Soc.* **92**, 1043–1055.

Kemp, D. S., Bernstein, Z., and Rebek, J. (1970b). *J. Am. Chem. Soc.* **92**, 4756–4757.

Kemp, D. S., Bernstein, Z. W., and McNeil, G. N. (1974a). *J. Org. Chem.* **39**, 2831–3835.

Kemp, D. S., Wang, S.-W., Mollan, R. C., Hsia, S. L., and Confalone, P. N. (1974b). *Tetrahedron* **30**, 3677–3688.

Kemp, D. S., Wang, S.-W., Rebek, J., Mollan, R. C., Banquer, C., and Subramanyam, G. (1974c). *Tetrahedron* **30**, 3955–3967.

Kemp, D. S., Wrobel, S. J., Wang, S.-W., Bernstein, Z., and Rebek, J. (1974d). *Tetrahedron* **30**, 3969–3980.

Kemp, D. S., Trangle, M., and Trangle, K. (1974e). *Tetrahedron Lett.* pp. 2695–2696.

Kemp, D. S., Choong, S. L. H., and Pekaar, J. (1974f). *J. Org. Chem.* **39**, 3841–3847.

Kenner, G. W., and Seely, J. H. (1972). *J. Am. Chem. Soc.* **94**, 3259–3260.

Kisfaludy, L., and Nyéki, O. (1972). *Acta Chim. Acad. Sci. Hung.* **72**, 75–80.
Kiso, Y., Kai, Y., and Yajima, H. (1973). *Chem. Pharm. Bull.* **21**, 2507–2510.
Klausner, Y., and Bodanszky, M. (1974). *Synthesis* pp. 549–559.
Kolthoff, I. M., and Bruckenstein, S. (1956). *J. Am. Chem. Soc.* **78**, 1–9.
Kolthoff, I. M., and Chantooni, M. K. (1963). *J. Am. Chem. Soc.* **85**, 426–430.
König, W., and Geiger, R. (1970a). *Chem. Ber.* **103**, 788–798.
König, W., and Geiger, R. (1970b). *Chem. Ber.* **103**, 2024–2033.
König, W., and Geiger, R. (1970c). *Chem. Ber.* **103**, 2034–2040.
Kovacs, J., Kisfaludy, L., and Ceprini, M. Q. (1967). *J. Am. Chem. Soc.* **89**, 183–184.
Kovacs, J., Mayers, G. L., Johnson, R. H., and Ghatak, U. R. (1968). *Chem. Commun.* pp. 1066–1067.
Kovacs, J., Kisfaludy, L., Ceprini, M. Q., and Johnson, R. H. (1969). *Tetrahedron* **25**, 2555–2573.
Kovacs, J., Mayers, G. L., Johnson, R. H., Cover, R. E., and Ghatak, U. R. (1970a). *J. Chem. Soc. D* pp. 53–54.
Kovacs, J., Mayers, G. L., Johnson, R. H., Cover, R. E., and Ghatak, U. R. (1970b). *J. Org. Chem.* **35**, 1810–1815.
Kovacs, J., Cortegiano, H., Cover, R. E., and Mayers, G. L. (1971). *J. Am. Chem. Soc.* **93**, 1541–1543.
Kovacs, J., Cover, R., Jham, G., Hsieh, Y., and Kalas, T. (1975). *In* "Peptides: Chemistry, Structure and Biology" (R. Walter and J. Meienhofer, eds.), pp. 317–324. Ann Arbor Science Publ., Ann Arbor.
Kricheldorf, H. R., and Fehrle, M. (1974). *Chem. Ber.* **107**, 3533–3547.
Lande, S., and Landowne, R. A. (1966). *Tetrahedron* **22**, 3085–3091.
Lauren, D. R., and Williams, R. E. (1972). *Tetrahedron Lett.* pp. 2665–2668.
Liberek, B. (1963a). *Tetrahedron Lett.* pp. 925–928.
Liberek, B. (1963b). *Tetrahedron Lett.* pp. 1103–1108.
Liberek, B., and Grzonka, Z. (1964a). *Tetrahedron Lett.* pp. 159–165.
Liberek, B., and Grzonka, Z. (1964b). *Bull. Acad. Pol. Sci., Ser. Sci. Chim.* **12**, 367–371.
Liberek, B., and Michalik, A. (1965). *Acta Chim. Acad. Sci. Hung.* **44**, 71–75.
Liberek, B., and Michalik, A. (1966a). *Rocz. Chem.* **40**, 781–784.
Liberek, B., and Michalik, A. (1966b). *Rocz. Chem.* **40**, 941–946.
Liberek, B., Nowicka, A., and Grzonka, Z. (1963). *Tetrahedron Lett.* pp. 1479–1483.
Liberek, B., Grzonka, Z., and Michalik, A. (1966). *Rocz. Chem.* **40**, 683–687.
Lloyd, K., and Young, G. T. (1968). *Chem. Commun.* pp. 1400–1401.
Lloyd, K., and Young, G. T. (1971). *J. Chem. Soc. C* pp. 2890–2896.
Lübke, K., and Schröder, E. (1961). *Z. Naturforsch., Teil B* **16**, 765–767.
McDermott, J. R., and Benoiton, N. L. (1973a). *Can. J. Chem.* **51**, 2555–2561.
McDermott, J. R., and Benoiton, N. L. (1973b). *Can. J. Chem.* **51**, 2562–2570.
MacLaren, J. A., Savige, W. E., and Swan, J. M. (1958). *Aust. J. Chem.* **11**, 345–350.
Manning, J. M. (1970). *J. Am. Chem. Soc.* **92**, 7449–7454.
Manning, J. M., and Moore, S. (1968). *J. Biol. Chem.* **243**, 5591–5597.
Manning, J. M., Marglin, A., and Moore, S. (1972). *In* "Progress in Peptide Research" (S. Lande, ed.), Vol. II, pp. 173–183. Gordon & Breach, New York.
Martynov, V. F., and Samartsev, M. A. (1969). *J. Gen. Chem. USSR (Engl. Transl.)* **39**, 911–912.
Matsuo, H., Kawazoe, Y., Sato, M., Ohnishi, M., and Tatsuno, T. (1967). *Chem. Pharm. Bull.* **15**, 391–398.
Matsuo, H., Kawazoe, Y., Sato, M., Ohnishi, M., and Tatsuno, T. (1970). *Chem. Pharm. Bull.* **18**, 1788–1794.
Merrifield, R. B., and Marglin, A. (1967). *In* "Peptides 1966" (H. C. Beyerman, A. van de Linde, and W. Maassen van den Brink, eds.), pp. 85–90. North-Holland Publ., Amsterdam.

Miller, W. G., Brant, D. A., and Flory, P. J. (1967). *J. Mol. Biol.* **23**, 67–80.
Mitin, Y. V., and Maksimov, E. E. (1973). *Zh. Obshch. Khim.* **43**, 203–207.
Miyoshi, M. (1973). *Bull. Chem. Soc. Jpn.* **46**, 1489–1493.
Morawiec, J., Konopinska, D., and Siemion, I. Z. (1971). *Rocz. Chem.* **45**, 771–775.
Mukaiyama, T. (1972). *Synth. Commun.* **2**, 243–265.
Nefkens, G. H. L., Tesser, G. I., and Nivard, R. J. F. (1960). *Recl. Trav. Chim. Pays-Bas* **79**, 688–698.
Neuberger, A. (1948). *Adv. Protein Chem.* **4**, 344–380.
Nyilasi, J. (1957). *Acta Chim. Acad. Sci. Hung.* **10**, 353–365.
O'Connor, M. J., Ernst, R. E., Schoenborn, J. E., and Holm, R. H. (1968). *J. Am. Chem. Soc.* **90**, 1744–1752.
Ott, H., Frey, A. J., and Hofmann, A. (1963). *Tetrahedron* **19**, 1675–1684.
Parr, W., and Howard, P. Y. (1972). *Angew. Chem., Int. Ed. Engl.* **11**, 529–530.
Perrin, D. D. (1965). "Dissociation Constants of Organic Bases in Aqueous Solution." Butterworth, London.
Photaki, I. (1963). *J. Am. Chem. Soc.* **85**, 1123–1126.
Pravda, Z., Poduska, K., and Blaha, K. (1964). *Collect. Czech. Chem. Commun.* **29**, 2626–2632.
Quitt, P., Studer, R. O., and Vogler, K. (1964). *Helv. Chim. Acta* **47**, 166–173.
Rebek, J., and Feitler, D. (1975). *Int. J. Pept. Protein Res.* **7**, 167–169.
Reinhoudt, D. N., Hoogwater, D. A., and Beyerman, H. C. (1971). *In* "Peptides 1969" (E. Scoffone, ed.), pp. 7–11. North-Holland Publ., Amsterdam.
Riniker, B., and Schwyzer, R. (1961). *Helv. Chim. Acta* **44**, 658–667.
Rittenberg, D., Keston, A. S., Schoenheimer, R., and Foster, G. L. (1938). *J. Biol. Chem.* **125**, 1–12.
Rivier, J., and Burgus, R. (1976). *In* "Peptides, Proceedings of the Fourteenth European Peptide Symposium," pp. 235–240. North-Holland Publ., Amsterdam.
Rivier, J., Wolbers, R., and Burgus, R. (1977). *In* "Peptides, Proceedings of the Fifth American Peptide Symposium," pp. 52–55.
Sakakibara, S., and Itoh, M. (1967). *Bull. Chem. Soc. Jpn.* **40**, 656–661.
Sato, M., Tatsuno, T., and Matsuo, H. (1970). *Chem. Pharm. Bull.* **18**, 1794–1798.
Schmidt, U., and Nikiforov, A. (1975). *Monatsh. Chem.* **106**, 313–320.
Schnabel, E. (1959). *Hoppe-Zeyler's Z. Physiol. Chem.* **314**, 114–117.
Schnabel, E. (1965). *Justus Liebig's Ann. Chem.* **688**, 238–249.
Schröder, E., and Lübke, K. (1965). "The Peptides," 2 vols. Academic Press, New York.
Scott, F. L., Glick, R. E., and Winstein, S. (1957). *Experientia* **13**, 183–185.
Shioiri, T., Ninomiya, K., and Yamada, S. (1972). *J. Am. Chem. Soc.* **94**, 6203–6205.
Sieber, P., and Riniker, B. (1973). *In* "Peptides 1971" (H. Nesvadba, ed.), pp. 49–53. North-Holland Publ., Amsterdam.
Sieber, P., Riniker, B., Brugger, M., Kamber, B., and Rittel, W. (1970). *Helv. Chim. Acta* **53**, 2135–2150.
Sieber, P., Kamber, B., Hartmann, A., Jöhl, A., Riniker, B., and Rittel, W. (1977). *Helv. Chim. Acta* **60**, 27–37.
Siemion, I. Z. (1968). *Rocz. Chem.* **42**, 237–241.
Siemion, I. Z., and Morawiec, J. (1964). *Bull. Acad. Pol. Sci., Ser. Sci. Chim.* **12**, 295–298.
Siemion, I. Z., and Wilschowitz, L. (1971). *Z. Naturforsch., Teil B* **26**, 762–763.
Smart, N. A., Young, G. T., and Williams, M. W. (1960). *J. Chem. Soc.* pp. 3902–3912.
Steglich, W., Mayer, D., Barocio de la Lama, X., Tanner, H., and Weygand, F. (1967). *In* "Peptides, Proceedings of the Eighth European Peptide Symposium" (H. C. Beyerman, A. van de Linde, and W. Maassen van den Brink, eds.), pp. 67–72. North-Holland Publ., Amsterdam.
Steglich, W., Frauendorfer, E., and Weygand, F. (1971). *Chem. Ber.* **104**, 687–690.

Stewart, F. H. C. (1970). *Aust. J. Chem.* **23**, 1073–1075.
Su, C., and Watson, J. W. (1974). *J. Am. Chem. Soc.* **96**, 1854–1857.
Syrier, J., and Beyerman, H. C. (1974). *Recl. Trav. Chim. Pays-Bas* **93**, 117–120.
Takeuchi, Y., and Yamada, S. (1974). *Chem. Pharm. Bull.* **22**, 832–855.
Taschner, E., Chimiak, A., Biernat, J. F., Sokolowska, T., Wasielewski, C., and Rzeszotarska, B. (1963a). *In* "Peptides, Proceedings of the Fifth European Symposium" (G. T. Young, ed.), pp. 109–111. Pergamon Press, Oxford.
Taschner, E., Sokolowska, T., Biernat, J. F., Chimiak, A., Wasielewski, C., and Rzeszotarska, B. (1963b). *Justus Liebig's Ann. Chem.* **663**, 197–202.
Taschner, E., Rzeszotarska, B., and Kuziel, A. (1965). *Acta Chem. Acad. Sci. Hung.* **44**, 67–71.
Taschner, E., Lubiewska, L., Smulkowski, M., and Wojciechowska, H. (1968). *Experientia* **24**, 521–523.
Tatsumi, S., and Ono, H. (1965). Ajinomoto Co., Japanese Patent 21,231., *Chem. Abstr.* **62**, 11912eP.
Theodoropoulus, D., and Gazopoulos, J. (1962). *J. Org. Chem.* **27**, 2091–2093.
Toi, K., Izumi, Y., and Akabori, S. (1963). *Bull. Chem. Soc. Jpn.* **36**, 734–738.
Tomida, I., and Kayahara, H. (1976). *Agric. Biol. Chem.* **40**, 2033–2036.
Tomida, I., Kayahara, H., and Iriye, R. (1973). *Agric. Biol. Chem.* **37**, 2557–2563.
Trudelle, Y. (1971). Chem. Commun. pp. 639–641.
Veber, D. F. (1975). *In* "Peptides: Chemistry, Structure and Biology" (R. Walter and J. Meienhofer, eds.), pp. 307–315. Ann Arbor Science Publ., Ann Arbor.
Weinstein, B., and Pritchard, A. E. (1972). *J. Chem. Soc., Perkin Trans. 1* pp. 1015–1020.
Westley, J. (1970). *In* "Peptides: Chemistry and Biochemistry" (B. Weinstein and S. Lande, eds.), pp. 359–370. Dekker, New York.
Weygand, F., and König, W. (1965). *Z. Naturforsch., Teil B* **20**, 710–711.
Weygand, F., Prox, A., Schmidhammer, L., and König, W. (1963). *Angew. Chem.* **75**, 282–287.
Weygand, F., König, W., Buyle, R., and Viehe, H. G. (1965). *Chem. Ber.* **98**, 3632–3636.
Weygand, F., König, W., Prox, A., and Burger, K. (1966a). *Chem. Ber.* **99**, 1443–1445.
Weygand, F., Prox, A., and König, W. (1966b). *Chem. Ber.* **99**, 1446–1450.
Weygand, F., Hoffmann, D., and Wünsch, E. (1966c). *Z. Naturforsch., Teil B* **21**, 426–428.
Weygand, F., Prox, A., and König, W. (1966d). *Chem. Ber.* **99**, 1451–1460.
Weygand, F., Hoffmann, D., and Prox, A. (1968). *Z. Naturforsch., Teil B* **23**, 279–281.
Wieland, T., and Determann, H. (1963). *Angew. Chem.* **75**, 539–551.
Wieland, T., and Seelinger, A. (1971). *Chem. Ber.* **104**, 3992–3994.
Williams, M. W., and Young, G. T. (1963). *J. Chem. Soc.* pp. 881–889.
Williams, M. W., and Young, G. T. (1964). *J. Chem. Soc.* pp. 3701–3708.
Williams, M. W., and Young, G. T. (1972). *J. Chem. Soc., Perkin Trans. 1*, pp. 1194–1200.
Windridge, G. C., and Jorgensen, E. C. (1971). *J. Am. Chem. Soc.* **93**, 6318–6319.
Wünsch, E. (1974). *In* "Houben-Weyl's Methoden der organischen Chemie" (E. Mueller, ed.), Vol. 15, Nos. 1 und 2. Thieme, Stuttgart.
Yajima, H., Kurobe, M., and Koyama, K. (1973a). *Chem. Pharm. Bull.* **21**, 1612.
Yajima, H., Kitagawa, K., and Kurobe, M. (1973b). *Chem. Pharm. Bull.* **21**, 2566–2567.
Yamashiro, D., Tseng, L., Doneen, D., Loh, H., and Li, C. H. (1977). *Int. J. Pept. Protein Res.* **10**, 159–166.
Yoshitake, H., and Kanaka, Y. (1965). Ube Industries Ltd., Japanese Patent 26,386; *Chem. Abstr.* **62**, 10508gP.
Young, G. T. (1967). *In* "Peptides, Proceedings of the Eighth European Peptide Symposium" (L. Zervas, ed.), p. 55. North-Holland Publ., Amsterdam.
Young, G. T. (1973). *In* "Peptides 1972" (H. Hanson and H. Jakubke, ed.), p. 132. North-Holland Publ., Amsterdam.
Zimmerman, J. E., and Anderson, G. W. (1967). *J. Am. Chem. Soc.* **89**, 7151–7152.

Author Index

Numbers in italics refer to the pages on which the complete references are listed.

A

Abe, H., 294, *309*
Abe, K., 228, *239*
Acher, R., 7, *56*
Acosta, M., 51, *61*
Adachi, Y., 73, *100*
Adamiak, D. A., 8, *62*
Adams, J. B., 205, *235*
Adams, M. J., 8, *56*
Agarwal, K. L., 214, *233*
Akabori, S., 331, *383*
Akerkar, A. S., 133, 146, *194*
Albertson, N. F., 71, *100*, 265, 269, 270, 272, 285, 286, 288, 291, *309*
Allewell, N. M., 20, *64*
Amiard, G., 271, *309*
Anantharamaiah, G. M., 291, *309*
Anderson, G. W., 74, 76, 84, *100, 102*, 123, 147, 154, 155, 160, 166, 184, 185, *187, 194*, 207, 219, 220, 221, 222, *233, 237*, 242, 249, 254, *260, 261*, 265, 268, 269, 270, 271, 272, 273, 274, 275, 276, 277, 278, 279, 280, 281, 282, 283, 284, 290, 291, 298, 299, 301, 302, 306, *309*, 341, 342, 355, 358, 359, 360, 364, 367, 368, 372, 373, 374, *378, 383*
Anderson, H., 282, 283, *311*
Anderson, J. C., 180, *187*
Ando, T., 15, *59*
Anfinsen, C. B., 15, *56*, 149, *189*, 259, *261*
Antonovics, I., 276, 279, *309*, 320, 343, 378, *378*
Aoyagi, H., 41, *59*, 222, 225, *235*, 283, 284, *311*, 358, 373, 374, 376, *380*
Apelt, H., 167, *193*
Applewhite, T. H., 276, 278, *309*
Arens, J. F., 71, 78, *100, 103*, 119, 134, *187, 193*
Argay, G., 33, *60*
Arguilla, E. R., 10, *61*
Ariely, S., 165, *189*
Arimura, A., 51, *56*
Arison, B. H., 86, *101*, 367, 377, *379*
Ariyoshi, Y., 73, *100*
Armarego, W. L. F., 272, *312*
Arnold, Z., 68, *104*
Arora, A., 45, 49, 55, *63*, 95, *103*
Aswanikumar, S., 21, *56*
Aten, B., 18, *63*
Atherton, E., 10, 46, 55, *56, 61*, 150, *193*
Audrieth, L. F., 37, *58*
Auld, G. A., 331, *378*
Axelrod, A. E., 176, *193*

B

Babusiaux, P., 78, *100*
Bach, A. N., 258, *261*
Bach, H., 164, *189*
Bada, J. L., 325, 326, 327, 328, 333, 335, *378*
Bailey, W. J., 229, 230, *233*
Bajusz, S., 214, *236*
Baker, E. N., 8, *56*
Balaspiri, L., 51, *61*
Balasubramanian, R., 10, 11, *62*
Ballardin, A., 5, 50, *63*
Balog, A., 175, *187*
Bankowski, K., 125, 143, 154, 155, *187*, 363, *378*
Banquer, C., 40, *59*, 331, 345, 353, 354, 365, 375, 376, 378, *380*
Barany, G., 50, *61*
Baratti, J., 16, *61*
Barber, M., 355, *378*
Barkemeyer, B., 85, *101*
Barkemeyer, H., 85, 86, *101*, 203, 227, *235*, 307, *311*, 367, 377, *379*
Barocio de la Lama, X., 276, *313*, 348, *382*

Barooshian, A. V., 370, 371, *378, 379*
Barrett, G. C., 257, *260*
Barth, A., 116, 125, *187, 192*
Barton, M. A., 293, *309*, 332, 359, 370, 374, *379*
Bates, A. J., 89, 90, *100*
Bath, R. J., 32, *56*, 114, 174, *187, 188*
Bator, B., 141, *195*
Bator-Sawicka, B., 342, *379*
Battersby, A. R., 285, *309*
Bauer, K., 29, *56*
Bauer, W., 51, 53, *62*
Baugh, C. M., 294, *311*
Baumert, A., 174, *190*, 295, *311*
Bauschke, S., 201, 203, 204, *236*
Bayer, E., 168, *195*, 306, *309, 311*, 357, 367, *379*
Beacham, J., 44, 45, *56*, 184, 185, *187, 195*, 209, 228, *237*
Beaumont, S. M., 125, 155, *187*, 253, *260*, 362, 377, *379*
Becker, J. M., 168, *187*
Beckwith, A. J., 66, *100*
Beddell, C. R., 297, 299, *309*
Beecham, A. F., 271, *309*
Beesley, T. E., 85, *101*, 377, *379*
Beglinger, U., 95, *100*, 123, *189*
Behrens, C., 46, *60*, 306, *312*
Belleau, B., 75, *100*, 289, *309*, 358, 374, *379*
Belleau, D., 267, 289, 290, 292, *309*
Bello, J., 20, *59*
Benassi, C. A., 127, *190*, 363, *380*
Bender, M. L., 30, *56*
Benedetti, E., 4, *56*
Benedict, R. G., 25, *56, 62*
Benkovic, S. J., 34, *57*
Benoiton, N. L., 290, 292, 306, *309, 310, 312*, 331, 332, 333, 346, 352, 357, 359, 370, 374, *379, 381*
Benson, J. R., 275, *309*
Bentley, P. H., 180, *187*
Benuck, M., 52, *56, 61*
Benz, E., 177, *192*
Berger, A., 201, 204, *238*, 360, 369, 371, 374, *379*
Bergmann, M., 300, *309*, 342, 348, *379*
Bernal, I., 4, *63*
Berndt, H., 177, *194*
Bernhard, H., 73, *103*, 112, 164, *195*, 264, 270, 273, *313*
Bernstein, Z. W., 40, *59*, 83, *102*, 214, 220, 221, 222, *236*, 283, 284, *311*, 350, 358, 359, 365, 373, 377, *380*
Berse, C., 271, *309*
Bev, L., 175, *187*
Beyerman, H. C., 74, 84, *100, 103*, 142, 177, *187*, 231, *238*, 265, 266, 268, 272, 273, 286, 287, 295, 296, 297, 298, 299, *309, 310, 311, 313*, 352, 356, 371, *382, 383*
Biedrzycka, D., 209, *234*
Biernat, J. F., 209, *234*, 332, 371, *383*
Bilk, L., 249, *260*, 361, *380*
Birkhimer, C. A., 135, 153, 155, 176, *187, 188*
Birktoft, J. J., 17, *56*
Birr, C., 46, *63*, 109, 118, 126, 171, *195, 196*, 205, 231, *234*, 254, *261*, 268, 294, 295, 300, 301, 305, *310, 314*
Bittner, S., 125, *187*, 253, *260*
Blaha, K., 371, *382*
Blake, J., 46, 47, *56, 64*
Blobel, G., 18, *56, 57*
Blodinger, J., 76, *100*, 160, *187*
Bloom, S. M., 165, *187*
Blotny, G., 141, *195*
Blout, E. R., 126, 165, 185, *187, 192*, 213, 229, *234*
Blow, D. M., 17, *56*
Blundell, T. L., 8, 9, 10, 50, 54, *56, 58, 62*
Bodanszky, A., 32, 45, *56*, 92, *100*, 150, 153, 154, 157, 171, *188*, 259, *260*, 296, *310*, 335, 340, 341, 353, 355, *379*
Bodanszky, M., 31, 32, 40, 43, 45, 46, 49, *56, 60, 61*, 66, 68, 69, 91, 92, *100, 102*, 108, 111, 113, 114, 129, 131, 134, 135, 140, 143, 144, 145, 146, 147, 148, 149, 150, 151, 152, 153, 154, 155, 157, 159, 161, 163, 167, 168, 171, 172, 174, 176, 177, 178, 180, *187, 188, 189, 191, 193*, 198, 199, 207, 214, 216, 220, 229, *234, 236*, 242, 247, 259, *260*, 264, 265, 286, 287, 289, 290, 292, 296, 298, 302, *310, 311*, 320, 335, 340, 341, 353, 355, 359, 364, 367, 376, *379, 381*
Bode, W., 10, 16, 17, *58*
Böhlen, P., 219, *234*, 275, 296, *313*
Boers-Boonekamp, C. A. M., 177, *187*
Bohak, Z., 330, *379*
Boissonnas, R. A., 32, 34, *62*, 69, 73, *100*, 108, 115, 147, 176, 177, 179, 180, 181, *190, 194*, 209, 214, *234, 235*, 264, 270, 271, 279, 304, *310*

Bokelmann, E., 107, 108, 112, 129, *195*, *196*, 302, *314*
Bolivar, F., 55, *59*
Bolton, A. E., 167, *188*
Borin, G., 49, *56*, 209, 218, *234, 236*
Borovas, D., 205, *239*, 271, 279, *314*
Borsook, H., 26, *56*
Boshart, G. L., 81, *103*
Bosshard, H. R., 360, 369, 371, 374, *379*
Bovarnik, M., 332, *379*
Bovey, F. A., 5, 10, *56*, *57*, 229, *234*
Bower, J. D., 92, *100*
Boyd, G. V., 346, *379*
Boyer, H. W., 55, *59*
Brack, A., 164, 165, *188*
Bradbury, A. F., 19, *57*
Brady, S. F., 44, *63*
Brand, E., 276, *310*
Brandenburg, D., 8, 45, 50, *57*, *58*, *60*, *62*, 166, 170, *188*, *192*, *195*, 214, *239*
Brant, D. A., 330, *382*
Brazeau, P., 45, *63*
Breazu, D., 175, *187*
Brechbühler, H., 231, *234*
Bremer, H., 150, *196*, 200, *314*
Brenner, M., 37, *57*, 95, *100*, 116, 123, 152, *188*, *189*, 204, 213, *234*, 352, *379*
Brewster, A. I., 5, *57*
Bricas, E., 52, *57*, 185, *192*
Brinigar, W. S., 164, *189*
Brink, L., 48, *58*
Broadbent, M., 166, 180, *189*
Brown, J., 94, *100*, 293, *310*
Brown, J. H., 25, *58*
Brownlee, G. G., 18, *61*
Bruckenstein, S., 328, *381*
Bruckner, V., 214, *236*
Brugger, M., 45, *63*, 69, *103*, 199, 200, 219, 220, 221, 222, *237*, 249, *261*, 324, 356, *382*
Bruice, T. C., 31, 34, *57*, 243, *260*
Brundish, D. E., 70, *100*
Brunfeldt, K., 256, *260*, 307, *311*
Büchi, H., 231, *234*
Buescher, H. H., 51, 53, *62*
Buetzer, P., 33, *62*
Buijle, E., 78, *100*, 121, *189*
Bumpus, F. M., 51, *59*, 276, *312*
Burckhardt, S., 32, 39, *59*
Burger, K., 329, *383*
Burgermeister, W., 268, 294, 300, 301, *314*
Burghard, H., 45, 49, 55, *63*, 95, *103*
Burgus, R., 45, *63*, 99, *100*, 324, 369, *382*
Burkhardt, S., 363, *380*
Burmeister, P., 53, *63*
Busby, G., III., 41, *59*, 216, 219, 220, 221, 224, 251, *236*, *261*, 283, *311*, 323, 331, 359, 367, 375, 376, *380*
Busch, D. H., 109, 141, *196*
Busse, W. D., 8, *57*, 169, *189*
Buyle, R., 372, *383*

C

Callahan, F. M., 74, *100*, 123, 147, 154, 155, 184, 185, *187*, *194*, 207, 219, 220, 221, *233*, *237*, 242, 254, *260*, 265, 268, 269, 270, 271, 272, 274, 275, 276, 277, 278, 279, 280, 281, 282, 283, 284, 291, 298, 299, 302, 306, *309*, 341, 342, 355, 358, 359, 364, 367, 372, 373, *378*
Camble, R., 171, *189*, 202, 215, 228, *234*
Capozza, R. C., 230, *233*
Cardinaux, F., 51, 53, *62*
Carpenter, F. H., 166, 169, *189*, *192*
Carpino. L. A., 15, *57*, 205, 231, *234*
Cary, P., 44, *62*
Cash, W. D., 161, *189*
Castro, B., 89, *100*, 216, *234*
Cavaggion, F., 218, *234*
Caviezel, M., 46, *56*
Ceprini, M. Q., 115, 134, 148, 156, 181, 183, *191*, *192*, 243, 247, 248, *261*, 303, 304, *311*, 342, 364, *381*
Cernosek, S. F., 185, *195*, 209, 228, *237*
Chan, S. J., 18, *57*
Chan, T. H., 78, *100*
Chang, A., 114, 174, *188*
Chang, J. K., 204, *234*
Chantooni, M. K., 338, *381*
Chapman, T. M., 126, *192*
Chaturvedi, N., 140, 161, *188*
Chauvet, J., 7, *56*
Chauvet, M. T., 7, *56*
Chen, F. M. F., 290, 306, *309*, *310*
Cheung, H. T., 213, *234*
Cheung, S. T., 331, *379*
Chew, L. F., 349, 369, 376, *380*
Chick, W. L., 18, *57*
Chien, S. W., 40, *59*, 83, *102*, 122, *191*, 276, *311*, 346, 363, *380*
Chimiak, A., 332, 371, *383*
Choong, S. L. H., 341, 349, *380*

Chorev, M., 150, *191*
Choudhury, A. M., 49, *57*
Christensen, T., 256, *260*
Chung, D., 19, *60*, 82, *102*, 176, *192*
Clark, J. L., 18, *63*
Clarke, H. T., 332, *379*
Clarke, J. E., 255, *261*, 288, 295, *312*
Clark-Lewis, J. W., 271, *311*
Clayton, D. W., 78, *100*
Cleeland, R., 275, 296, *313*
Clement, B. A., 68, *100*
Closse, A., 51, 53, *62*
Coetzee, J. F., 338, *379*
Cohen, L. A., 23, *62*
Collier, H. O. J., 147, *193*
Confalone, P. N., 40, *59*, 346, 365, 376, *380*
Conklin, L., 353, 367, 376, *379*
Conn, J. B., 307, *311*
Cook, R. M., 208, 209, *238*
Corcoran, B. A., 21, *56*
Corey, R. B., 4, 10, *57*
Corley, L., 15, *56*, 149, *189*
Cortegiano, H., 32, 35, *60*, 154, *192*, 324, 355, *381*
Cosand, W. L., 50, *61*
Cosmatos, A., 76, *100*
Costopanagiotis, A. A., 179, *189*
Cotton, R., 150, *193*, 202, *234*
Cover, R. E., 31, 32, 33, 34, 35, 36, 37, 42, *60*, 154, 184, *192*, 279, *311*, 320, 324, 340, 341, 342, 349, 350, 354, 355, *381*
Cowell, R. D., 93, 94, *100*
Coy, D. H., 51, *56*
Coy, E. J., 51, *56*
Crea, R., 55, *59*
Creighton, T., 88, *101*, 110, *189*
Crick, F. H. C., 27, *57*
Cruickshank, P. A., 81, *103*
Curtius, T., 68, 72, *100*, 107, *189*, 198, 200, 213, 215, 218, 225, 226, *234*
Czugler, M., 33, *60*

D

Dairman, W., 275, 296, *313*
Dane, E., 119, 175, *189*
Danho, W., 184, 185, *189*, *196*, 207, *234*
Darvas, F., 33, *60*
Dasgupta, S. K., 165, *187*
Davie, E. W., 17, *57*
Davies, J. S., 95, *101*
Davis, E. J., 32, 35, *60*
Davison, A., 331, *378*
Day, A. R., 108, 113, 118, *190*, 202, *234*
Deak, J. E., 202, 203, 207, 227, *238*
Deber, C. M., 229, *234*
DeBernardo, S., 219, *234*, 275, 296, *313*
Deer, A., 207, *237*
Degani, Y., 11, *63*
Deigin, V. I., 45, *59*, 300, 301, *311*
Deimer, K.-H., 80, *104*, 201, *238*
Deisenhofer, J., 10, *57*
de Leer, E. W. B., 297, 298, *309*
Del Pozo, E., 53, *63*
Demuth, F., 78, *102*
Denkewalter, R. G., 44, 49, *58*, 85, 86, *101*, 198, 227, 228, *235*, 307, *311*, 367, 377, *379*
Denning, G. S., Jr., 149, 176, *188*
Desnuelle, P., 16, 17, *57*, *61*
DeTar, D. F., 164, 165, *189*, 244, 246, 247, 248, 255, 258, *260*
Determann, H., 142, 160, *196*, 217, *234*, 276, 277, 278, 288, *310*, 358, 372, 373, 377, *379*, *383*
Devillers-Thiery, A., 18, *57*
de Vries, J. X., 268, 294, 301, *314*
De Wald, H. A., 147, *193*
Dewey, R. R., 367, 377, *379*
Dewey, R. S., 86, *101*, 203, 227, *235*
Diaconescu, C., 8, 50, *57*, *58*
Di Bello, C., 273, 296, 305, *313*
Dickinson, M. J., 202, 203, 207, 227, *235*, *238*
Dietrich, H., 214, *237*
Dmitrieva, M. G, 32, *59*
Dobberstein, B., 18, *56*
Dockerill, S., 8, *58*, *62*
Dockner, T., 119, 175, *189*
Dodson, E. J., 8, *56*
Dodson, G. G., 8, 9, 50, 54, *56*
Doleschall, G., 243, *260*
Doneen, B. A., 47, 48, 50, *60*
Doneen, D., 324, 369, *383*
Donohue, J., 4, *61*
Dorman, L. C., 256, *260*
Dormoy, J.-R., 89, *100*, 216, *234*
Douraghi-Zadeh, K., 243, 258, *261*
Dourtoglon, B., 89, *100*
Dowd, S. R., 70, *103*, 209, 228, *237*
Drabarek, S., 70, *103*, 125, 143, 154, 155, *187*, 209, 228, *237*, 363, *378*

Drees, F., 80, *104*, 119, 155, 175, *189*, *196*, 212, *238*, 248, 253, *261*
Dreux, J., 78, *100*
Drey, C. N. C., 88, *101*, 308, *310*
Duguid, J. R., 18, *57*
Dunitz, J. D., 4, 11, *64*
Dunn, M. S., 203, *237*
Dupraz, C. A., 156, 181, *192*
Dupuis, G., 44, 45, *56*, 184, *187*, 215, 228, *234*
Dutta, A. S., 40, *57*, 115, 139, *189*, 202, *234*, 363, *379*
duVigneaud, V., 49, 52, 55, *57*, 91, *100*, 108, 109, 110, 114, 134, 143, 144, 149, 157, 176, *188*, *189*, 242, *260*, 296, 302, *310*
Dvonch, W., 25, *56*
Dvortsak, P., 244, *261*
Dyckes, D. F., 88, *101*, 110, *189*
Dzieduszycka, M., 222, *234*, 299, 300, *310*, 348, *379*

E

Eastlake, A., 15, *56*
Eberle, A., 151, *189*
Eberle, C., 95, *103*
Eberle, G., 45, 49, 55, *63*, 95, *103*
Eckert, H., 45, 49, 55, *63*, 95, *103*
Eguchi, C., 332, *379*
Eichler, J. A., 276, 278, 285, *313*
Eipper, B. A., 19, *61*
Eisler, K., *103*
Elliott, D. F., 133, 155, *189*, 247, *260*, 302, *310*
Elmore, D. T., 135, *189*
Endo, S., 21, *57*
Engelhard, M., 50, *61*
Erickson, B. W., 43, 46, 47, 48, *57*, *60*, *61*, 204, *234*, 243, 246, *260*
Erin, G., 89, *100*
Erlanger, B. F., 276, *310*
Ernst, R. E., 331, *382*
Eschenmoser, A., 231, *234*
Esko, K., 247, *260*

F

Faesel, J., 265, 268, 294, 300, 301, 304, *314*
Fagan, D., 168, *188*
Fahrenholz, F., 265, 268, 294, 300, *314*
Fairweather, R., 72, 74, 78, 91, 93, *101*, 372, *380*
Falkenburg, H. R., 166, *196*
Farrington, J. A., 32, *57*, 78, *100*, 114, 129, 160, *189*, 302, *310*
Fasman, G. D., 85, *101*
Fauchere, J. L., 151, *189*
Faulstich, H., 265, 268, 294, 300, 301, 304, *311*, *312*, *314*
Fehrle, M., 76, 84, 86, *102*, 372, *381*
Feibush, B., 367, *379*
Feinberg, R. S., 44, 47, *57*
Feitler, D., 246, 247, 255, *261*, 342, 354, 357, 371, *382*
Felix, A. M., 44, *57*, 70, 99, *101*, 219, 230, *234*, *235*, 275, *310*
Fenichel, R. L., 99, *103*
Fenwick, G. R., 259, *260*
Ferren, R. A., 202, *234*
Ferroni, R., 127, *190*, 363, *380*
Feurer, M., 37, 44, 53, *57*, *62*, 107, 111, 132, 139, 141, *190*, *194*, 265, *310*, 353, *380*
Fieser, L. F., 75, *101*
Fieser, M., 75, *101*
Filippi, B., 49, *56*, 218, *234*
Fink, M. L., 32, 45, *56*, 91, 92, *100*, 114, 144, 150, 171, 172, 174, 178, *188*, *189*, 296, *310*
Finn, F. M., 20, 43, 44, 45, 49, *56*, *57*, 66, *101*, 184, 185, *187*, *195*, 199, 209, 217, 220, 228, *234*, *235*, *237*, 243, *260*, 265, 296, *310*
Fischer, E., 2, *57*, 90, *101*, 107, *189*
Fischer, R. F., 219, *234*
Flanigan, E., 163, *189*
Fleckenstein, P., 118, *196*
Fletcher, G. A., 80, 97, 98, *101*, 131, 150, 180, *189*, 251, *260*
Flippen, J. L., 10, *57*
Floor, J., 297, 298, *309*
Flor, F., 46, *63*, 254, *261*, 295, 305, *310*, *314*
Flory, P. J., 330, *382*
Föhles, J., 173, *191*
Foley, C., 25, *62*
Foster, G. L., 327, *382*
Fothergill, L. A., 19, *59*
Fourneau, E., 90, *101*
Frank, H., 46, *58*, 254, 255, 257, *260*, 306, *311*
Frank, V. S., 67, *103*, 305, *313*
Frankel, M., 125, 166, *187*, *189*, 253, *260*
Fraser, P. J., 297, 299, *309*
Frauendorfer, E., 348, *382*

Frenkel, A., 165, *189*
Fretzdorff, M., 177, *194*
Frey, A. J., 332, *382*
Fridkin, M., 94, *102*, 109, 126, 127, 163, 165, 184, *189, 191, 193, 194*, 204, *235*, 330, *379*
Fried, J., 176, *188*
Friesen, H. J., 8, 50, *57, 58*
Fromageot, C., 52, *57*
Fruton, J. S., 198, 226, *235*
Fujii, N., 43, *64*, 209, 228, *238*
Fujii, S., 13, *61*
Fujino, M., 125, 159, 166, *189*, 205, *235*, 250, 253, *260*, 364, *379*
Fukuda, K., 226, *236*
Fukuda, T., 205, 231, *235, 237*, 364, *379*
Funakoshi, S., 43, *64*
Funk, K. W., 32, *56*, 91, *100*, 114, 144, 171, 174, 178, *188*

G

Gabriel, T. F., 70, 99, *101, 235*, 275, *310*
Gallop, P. M., 84, *104*, 201, 204, *238*
Galpin, I. J., 49, *58*, 89, *100*, 308, *310*
Gammeltoft, S., 8, *62*
Gardner, J., 180, *188*
Garner, R., 150, 171, *189*
Garsky, V., 202, 203, 207, 227, *238*, 307, *311*
Gaston, D. W., 293, *313*
Gattner, H. G., 8, 50, *57, 58*
Gazis, E., 205, *239*, 271, 279, *314*
Gazopoulos, J., 78, *103*, 342, *383*
Geiger, R., 32, 37, 38, 45, *60*, 80, *102*, 109, 110, 125, 143, 146, 148, 150, 155, 166, *189, 190, 191*, 201, 205, 212, *235, 236*, 250, 251, 253, 256, *260, 261*, 265, *311*, 349, 353, 354, 357, 360, 361, 370, 374, 375, *379, 381*
George, G., 45, 49, 55, *63*, 95, *103*
Gerencser, L. W., 304, *311*
Gerstein, J., 336, 378, *379*
Ghatak, U. R., 32, *60*, 154, 165, 184, *192*, 304, *311*, 341, 349, 355, *381*
Gianotti, R., 181, *192*
Gibbons, W. A., 5, 50, *63, 64*
Gibian, H., 206, *236*
Giger, U., 37, *57*
Gil-Av, E., 357, 367, *379*
Gilbert, D., 297, 299, *309*
Gilchrist, M., 244, *260*
Gillessen, D., 202, 214, *235*, 285, *310*
Gilon, C., 166, *189*
Giori, P., 127, *190*, 363, *380*
Gish, D. T., 148, *190*, 242, *260*
Gisin, B. F., 201, *238*, 258, *261*
Glaser, C., 41, *58*, 206- *235*, 242, *260*, 276, *310*, 320, 346, 347, 348, 378, *379*
Glasoe, P. K., 37, *58*
Glatthard, R., 130, 131, *190*
Glick, R. E., 339, *382*
Gliemann, J., 8, *57, 62*, 170, *188*
Glinskaya, O. V., 76, 84, *102*, 159, *193*
Godwin, H. A., 295, *312*
Gödicke, W., 147, *190*
Gönczy, F., 175, *187*
Gokel, G., 45, 49, 55, *63*
Goldman, L., 177, *192*
Goldschmidt, S., 76, 78, *101*
Golubow, J., 176, *193*
Goodford, P. J., 297, 299, *309*
Goodman, M., 41, *58*, 111, 145, 156, 160, 161, 176, *190, 194*, 206, 229, 230, *234, 235*, 242, 247, *260*, 276, 282, 288, 303, 304, *310, 311*, 320, 341, 342, 344, 346, 347, 348, 350, 366, 378, *379*
Gordon, M., 108, 113, 118, *190*
Gordon, S., 49, 52, *57*, 109, 110, *189*, 242, *260*
Goren, H. J., 204, *235*
Gormley, J. J., 202, *234*
Gorup, B., 82, *102*
Gosei, T., 331, *379*
Goto, K., 181, *193*
Gouge, M., 164, *189*
Govindachari, T. R., 123, *190*, 253, *260*
Grahl-Nielsen, O., 347, *380*
Grant, G., 45, *63*
Grathwohl, C., 10, *64*
Grattan, J., 16, 44, 45, 49, 55, *59*, 95, *102*, 205, *236*
Greenstein, J. P., 199, 203, 213, *235*, 265, 273, *311*, 320, 328, *380*
Greenwald, S. M., 114, 174, *188*
Greenwood, J. M., 370, *379*
Gregory, H., 180, *187*
Gresser, M. J., 34, *58*
Grieser, G., 173, *195*
Griffiths, E. C., 52, *58*
Gross, E., 21, 23, 24, 25, *58, 60, 61*, 110, 169, *190*
Gross, H., 68, *102*, 249, *260*, 361, *380*

Gruber, K. A., 48, *58*
Grunberg, E., 275, 296, *313*
Grzonka, Z., 154, *192*, 327, 331, 340, 341, *381*
Guarneri, M., 127, *190*, 363, *380*
Güntelberg, A. V., 19, *58*
Guest, K. P., 92, *100*
Guggi, A., 70, *103*, 363, *380*
Guillemin, R., 19, 45, *60*, *63*
Gukel, G., 95, *103*
Gurd, F. R. N., 141, *191*
Gut, V., 32, *58*, 88, *101*, 174, *194*
Gutfreund, H., 16, *58*
Gutte, B., 15, 46, 47, *58*, 184, *196*
Guttmann, S., 69, *100*, 147, 176, 179, *190*, 209, 214, *234*, *235*, 303, *311*

H

Haas, W., 184, *190*
Hagenmaier, H., 46, *58*, 174, *190*, 254, 255, 257, *260*, 306, *311*
Hagitani, A., 305, *312*
Hagiwara, D., 166, *190*
Hagler, A. T., 5, *58*
Hahne, W., 180, *188*
Hall, M. M., 51, *59*
Hall, S. R., 11, *61*
Hallett, A., 49, *64*, 89, *100*, 231, *237*
Halpern, B., 349, 369, 376, *380*
Halstrøm, J., 173, *191*, 307, *311*
Hamada, Y., 76, 77, *101*, 216, *235*
Hamalidis, C., 205, *239*
Hambrook, J. M., 52, *58*
Han, G. Y., 205, 231, *234*
Hancock, W. S., 15, 46, *61*
Handa, B. K., 49, *58*, 308, *310*
Handford, B. O., 125, 155, 179, 180, 186, *187*, *190*, 253, *260*, 362, 377, *379*, *380*
Hansen, S. R., 169, *189*
Hanser, D., 51, 53, *62*
Harding, M. M., 8, *56*
Hardman, K., 20, *64*
Hardy, P. M., 257, *260*
Hare, P. E., 275, *309*
Harker, D., 20, *59*
Harris, J. I., 226, *235*
Harris, W. G., 370, 371, *378*, *379*
Harrison, T. M., 18, *61*
Harrow, T. A., 257, *260*
Harryvan, E., 71, *103*
Hartley, B. S., 17, *56*
Hartmann, A., 45, 48, 49, *63*, 81, 95, 99, *100*, *103*, 123, *189*, 225, *237*, 324, 361, *382*
Hase, S., 184, *194*
Hashimoto, T., 228, *239*
Hassall, C. H., 95, *101*
Hasspacher, K., 88, *103*, 110, *195*, 252, *261*
Hatanaka, C., 159, 166, *189*
Hatz, E., 231, *234*
Havinga, E., 181, 182, *195*
Hayakawa, S., 201, *238*
Hayashi, N., 249, *261*
Hayward, C. F., 202, *234*
Heard, A. L., 199, *235*, 271, 276, 279, *306*, *311*, 320, 349, 359, 367, *378*, *380*
Heber, H., 294, 300, *311*
Heffernan, J. T., 208, 209, *238*
Hegedüs, B., 244, *261*
Hegerty, A. F., 243, *260*
Heikens, D., 132, *190*
Heinke, B., 76, *103*, 129, 130, *195*, *196*, *238*, 272, 273, 275, 288, *314*
Hemmasi, B., 306, *311*
Hendricks, M. L., 297, *313*
Henes, J. B., 163, 167, 180, *188*, *194*
Herbert, E., 19, *62*
Herling, P. A., 88, *101*
Herlinger, H., 45, 49, 55, *63*, 95, *103*
Hermann, P., 72, *103*
Hermans, P. H., 132, *190*
Herzog, H., 177, 185, *194*
Herzog, K. H., 173, *195*
Hess, G. P., 15, 44, *60*, *62*, *63*, 79, *103*, 120, 133, 145, 148, *190*, *194*, 242, *260*, *261*
Hetzer, G., 117, 125, *192*
Hextall, G. W., 114, 160, *189*
Hextall, P. J., 302, *310*
Heymes, R., 271, *309*
Heyneker, H. L., 55, *59*
Heynemann, H., 113, 128, *191*
Higashi, F., 76, *104*, 131, *196*
Hill, J., 328, 330, *380*
Hill, R. C., 51, 53, *62*
Hill, R. L., 11, 13, *58*
Hillmann, G., 271, *311*
Hillmann-Elies, A., 271, *311*
Hirohata, R., 13, *61*

Hirose, T., 55, *59*
Hirotsu, Y., 51, *56*
Hirs, C. H. W., 11, 19, *58*
Hirschmann, R., 44, 49, 55, *58, 63*, 85, 86, *101*, 198, 202, 203, 205, 207, 217, 220, 221, 222, 227, 228, *35, 238*, 307, *311*, 367, 377, *379*
Hirt, J., 298, *310*
Hiskey, R. G., 205, 231, *235*
Hladar, J., 7, *61*
Hlavka, J. J., 242, *261*
Hodges, R. S., 46, 47, *58, 61*
Hodgkin, D. C., 8, 9, 50, 54, *56*
Hörnle, S., 173, 177, *190, 195*
Hofer, W., 204, 213, *234*
Hoffman, P. L., 50, 52, *63*
Hoffmann, D., 80, *103*, 155, *195*, 212, *238*, 249, 250, 253, *261*, 353, 360, 361, 367, 374, *383*
Hoffmann, J. A., 218, 231, *238*, 297, *311*
Hoffmann, K. H., 117, 125, *192*
Hoffmann, P., 45, 49, 55, *63*, 95, *103*, 177, 185, *194*
Hofmann, K., 20, 43, 44, 45, 49, *56, 57*, 66, 70, *101, 103*, 184, 185, *187, 190, 195*, 199, 200, 204, 208, 209, 213, 215, 217, 220, 226, 228, 229, *234, 235, 236, 237*, 242, 243, *260*, 265, 296, *310*, 332, 356, *380, 382*
Hofmeister, F., 2, *58*
Holley, R. W., 219, *235*
Hollinden, C. S., 294, *313*
Holly, F. W., 44, 49, *58*, 198, 202, 203, 207, 227, 228, *235, 238*
Holm, R. H., 331, *382*
Honda, I., 231, *235, 237*
Honsberg, U., 164, *189*
Honsberg, W., 164, *189*
Honzl, J., 69, *101*, 213, 214, 215, 221, 222, 230, 232, *235, 236*
Hoogwater, D. A., 352, *382*
Hope, D. B., 352, *380*
Hopkins, K. H., 95, *101*
Howard, P. Y., 371, *382*
Hoyng, C., 16, 44, 45, 49, 55, *59*, 205, *236*
Hruby, V. J., 5, *58*
Hsia, S. L., 31, 33, 34, 37, 40, *59*, 346, 365, 376, *380*
Hsieh, Y., 31, 33, 34, 35, 36, 37, 42, *60*, 279, *311*, 320, 340, 341, 342, 349, 350, 354, *381*
Huber, P., *195*, 273, 296, 305, *313*
Huber, R., 10, 16, 17, *58*
Hudson, D., 49, *58*, 89, *100*, 140, 161, *188*, 308, *310*
Hugel, G., 41, *59*, 216, 219, 220, 221, 224, *236*, 251, *261*, 283, *311*, 323, 331, 359, 367, 375, 376, *380*
Hughes, J., 19, *59*
Hugo, J., 276, 279, *309*, 320, *378*
Huguenin, R. L., 51, 53, *62*, 179, *190*
Huhn, M., 244, *261*
Hunger, K., 166, *195*, 231, *238*
Hunter, W. M., 167, *188*
Hylton, T. A., 176, 179, 180, *190*

I

Ichikizaki, I., 94, *101*
Ikota, N., 216, *239*
Inagami, T., 20, *64*
Inouye, H., 226, *235*
Inukai, N., 131, 168, *194*
Iriye, R., 371, *383*
Iselin, B., 37, *62*, 107, 111, 116, 117, 130, 131, 132, 139, 141, 161, 167, *190, 194, 195*, 210, 214, 231, *237*, 303, 304, *311*, 342, 353, 366, *380*
Itakura, K., 55, *59*
Ito, H., 94, *101*
Itoh, M., 125, 136, 140, 161, 166, *188, 190*, 254, *260*, 283, 284, *312*, 340, 358, 360, 364, 373, 374, 375, *380, 382*
Ivanov, V. T., 5, 45, *59, 61*, 229, *237*, 300, 301, *311*
Iwai, K., 15, *59*
Iwakura, Y., 86, *102*
Iyer, V. S., 124, *194*, 253, *261*
Izeboud, E., 272, 297, *311*
Izumi, Y., 331, *383*
Izumiya, N., 41, *59*, 151, 163, *191, 193*, 204, 207, 215, 222, 225, 226, *235, 236, 237, 238*, 283, 284, 292, 294, *309, 311*, 358, 359, 360, 367, 368, 373, 374, 376, *380*

J

Jackson, A. G., 49, *58*, 76, *101*, 308, *310*
Jackson, R. W., 25, *62*
Jacob, T. A., 44, 49, *58*, 85, *101*, 198, 227, 228, *235*, 377, *379*
Jacobs, P. M., 117, 163, *190*, 295, *312*

Jacobson, S. J., 327, 335, *380*
Jaeger, E., 30, 49, *64*, 73, *104*
Jäger, G., 45, *58*, 148, *190*, 205, *235*, 251, 256, *260*
Jaenicke, F., 113, 129, *195*
Jakubke, H.-D., 31, 32, 39, *59*, 91, 93, *101*, 114, 174, 186, *190*, 295, 300, *311*, *312*, 363, *380*
Jaquenoud, P.-A., 69, *100*, 209, *234*
Jatzkewitz, H., 271, *311*
Jencks, W. P., 31, 34, *58*, 244, *260*, 336, 378, *379*
Jenkins, J. A., 10, *62*
Jenkins, S. R., 202, 203, 207, 227, *235*, *238*
Jeschkeit, H., 123, *190*, 253, *260*
Jham, G., 31, 33, 34, 35, 36, 37, 42, *60*, 279, *311*, 320, 340, 341, 342, 349, 350, 354, *381*
Jimenez, M. H., 44, *57*, 70, 99, *101*, 219, *234*, *235*, 275, *310*
Jöhl, A., 45, 48, 49, *63*, 81, 99, *103*, 225, *237*, 324, 361, *382*
John, W. D., 73, *101*
Johnson, B. J., 93, *101*, 117, 163, 165, 173, 181, 182, *190*
Johnson, L. F., 4, *63*
Johnson, L. N., 20, *64*
Johnson, P. L., 4, 11, *62*
Johnson, R. H., 32, 35, *60*, 115, 134, 154, 183, 184, *191*, *192*, 303, 304, *311*, 341, 342, 349, 355, 364, *381*
Johnson, S. L., 31, *59*, 336, 348, *380*
Johnson, T. F. N., 125, 186, *190*, 362, *380*
Jones, D. S., 68, 72, 87, *101*, *102*, 179, *191*, 256, *260*, *261*
Jones, J. H., 30, 40, *59*, 66, 68, 70, 72, 74, 78, 87, 91, 93, 94, 97, 98, *100*, *101*, *102*, 117, 125, 142, 155, 166, 180, 186, *187*, *189*, *190*, *191*, 253, *260*, 337, 340, 355, 362, 363, 366, 368, 372, *378*, *380*
Jorgensen, E. C., 252, 256, *260*, *261*, 356, 370, *380*, *383*
Joshua H., 85, 86, *101*, 367, 377, *379*
Jukes, T. H., 21, *61*

K

Kaczmarek, M., 115, 116, *192*
Kägi, H., 111, 132, 141, *194*
Kahle, I., 288, *310*
Kahle, W., 142, 143, 160, *195*, *196*
Kai, Y., 75, *102*, 289, 290, 292, 293, *311*, 358, *381*
Kaiser, E. T., 49, *63*
Kajtar, M., 214, *236*
Kakuta, A., 332, *379*
Kalas, T., 31, 32, 33, 34, 35, 36, 37, 42, *60*, 154, *192*, 279, *311*, 320, 340, 341, 342, 349, 350, 354, *381*
Kalir, R., 94, *102*, 126, 127, *189*, *191*
Kalita, C., 304, *311*
Kaloy, K., 244, *261*
Kamber, B., 45, 48, 49, *63*, 69, 70, 80, 81, 99, *102*, *103*, 182, *191*, 199, 200, 215, 219, 220, 221, 222, 225, *236*, *237*, 249, *261*, 324, 356, 361, *382*
Kamen, M. D., 15, 46, 50, *62*
Kamiya, T., 166, *190*
Kanaka, Y., 326, *383*
Kanda, Y., 13, *61*
Kapoor, A., 165, 181, *191*, *192*, 304, *311*
Kappeler, H., 214, 218, 225, 226, *237*
Karle, I. L., 4, 8, 10, *57*, *59*
Karlsson, S. M., 166, 173, *191*, *194*, 247, *260*
Karrer, P., 113, 128, *191*
Kartha, G., 20, *59*
Kaschig, J., 76, 84, *102*
Kashelikar, D. V., 148, *194*
Katakai, R., 86, 87, *102*, 307, *311*
Katchalski, E., 85, 94, *102*, 109, 126, 163, 168, *187*, *189*, *193*, 330, *379*
Kato, T., 163, *191*, 207, 215, *236*, *238*, 294, *309*
Katsoyannis, P. G., 49, 52, *57*, 109, 110, 148, *189*, *190*, 226, *236*, 242, *260*
Katz, J. J., 15, *59*, *60*
Kaufmann, K.-D., 201, 203, 204, *236*
Kaurov, O. A., 183, *191*, 361, *380*
Kawabata, N., 230, *233*
Kawakami, T., 249, *261*
Kawasaki, K., 70, *103*, 205, 209, 215, 228, *234*, *237*, *239*, 285, 286, *314*
Kawatani, H., 286, 290, 293, *314*
Kawazoe, Y., 327, 330, 331, 334, *381*
Kayahara, H., 348, 371, *383*
Kazi, H., 37, *62*
Keikichi, U., 86, *102*
Keim, P., 18, *57*
Keller, O., 231, *237*
Kemmler, W., 18, *63*

Kemp, D. S., 16, 31, 33, 34, 37, 40, 41, 44, 45, 49, 55, *59*, 83, 95, *102*, 122, *191*, 199, 205, 214, 216, 219, 220, 221, 222, 224, *236*, 251, *261*, 276, 281, 282, 283, 284, *311*, 320, 323, 331, 341, 345, 346, 347, 349, 350, 353, 354, 358, 359, 360, 361, 363, 364, 365, 366, 367, 368, 373, 374, 375, 376, 377, 378, *380*
Kenner, G. W., 32, 49, *57, 58*, 68, 72, 76, 77, 78, 87, 89, *100, 101, 102*, 113, 114, 117, 129, 160, 180, *187, 189, 191*, 214, *233*, 256, *260, 261*, 302, 308, *310*, 332, *380*
Keogh, P., 282, *311*
Keresztessy, P., 49, *60*
Kerling, K. E. T., 181, 182, *195*
Kern, C. W., 10, *62*
Kern, W., 288, 305, 307, *313*
Kerp, L., 53, *63*
Keston, A. S., 327, *382*
Khairallah, P. A., 51, *59*
Khan, N. H., 200, 208, 209, *235*
Khan, N. M., *195*
Khan, S. A., 172, *191*
Khorana, H. G., 120, 133, *191*, 242, *261*
Khosla, M. C., 51, *59*
Khurgin, Y. I., 32, *59*
Kilday, W., 84, *102*
Kiltz, H. H., 21, 25, *58, 60*
Kim, M., 75, *102*
Kindt, T., 18, *57*
King, F. E., 271, *311*
Kinzle, F., 124, *195*
Kisfaludy, L., 33, 49, *60*, 91, *102*, 115, 116, 134, 156, 183, 184, *191, 192*, 214, 220, 221, 225, *236*, 243, 247, 248, 249, *261*, 303, 304, *311*, 342, 360, 364, *381*
Kishida, Y., 231, *237*
Kiso, Y., 75, *102*, 208, 209, *239*, 289, 290, 292, 293, *311*, 358, *381*
Kitagawa, K., 364, 375, *383*
Klauke, E., 177, 185, *194*
Klausner, A., 259, *260*
Klausner, Y. S., 31, 43, 45, *56, 60*, 66, 68, 69, 92, *100, 102*, 108, 114, 116, 134, 150, 157, 174, *188*, 199, 207, 214, 216, 220, 229, *234, 236*, 265, 289, 290, 292, 296, *310, 311*, 320, 359, *379, 381*
Kleinberg, J., 37, *58*
Kleinkauf, H., 29, *56, 60*
Kleimann, H., 45, 49, 55, *63*, 95, *103*
Klieger, E., 152, *191*, 206, *236*
Klostermeyer, H., 173, 177, 184, *191, 193, 194*
Klusacek, H., 45, 49, 55, *63*, 95, *103*
Kmiecik-Chmura, H., 291, *312*
Knobler, Y., 125, *187*, 253, *260*
Knof, S., 49, *64*
Kobayashi, A., 250, 253, *260*, 364, *379*
Koch, A. L., 15, *60*
Kocy, O., 207, *237*
Kömives, T., 33, *60*
König, W. A., 32, 37, 38, 45, *58, 60*, 80, *102*, 109, 110, 125, 143, 146, 150, 155, *187, 189, 191, 195*, 201, 212, 218, 219, 220, 221, 225, *236, 238*, 250, 251, 253, 256, *260, 261*, 265, 271, 277, 278, 279, 300, *311, 312, 313*, 329, 342, 348, 349, 352, 353, 354, 355, 357, 359, 360, 361, 362, 364, 366, 367, 370, 372, 373, 374, 375, 376, 377, 379, *381, 383*
Kohl, U., 10, *58*
Kolaskar, A. S., 11, *62*
Kolthoff, I. M., 328, 338, *381*
Koltun, W. L., 141, *191*
Kondo, M., 32, *56*, 144, 171, 174, 178, *188*
Konopinska, D., 341, *382*
Konrad, P., 119, 175, *189*
Konz, W., 265, 268, 294, 300, 301, *314*
Kopple, K. D., 229, *236*, 288, *311*
Korenczki, F., 49, *60*
Kosterlitz, H. W., 19, *59*
Kotake, H., 71, *102*
Kovacs, J., 31, 32, 33, 34, 35, 36, 37, 42, *60*, 115, 134, 148, 154, 156, 165, 181, 183, 184, *191, 192*, 243, 247, 248, *261*, 279, 303, 304, *311*, 320, 324, 340, 341, 342, 349, 350, 354, 355, 364, *381*
Kovács, K., 32, *60*, 174, *192*, 307, *311*
Kovacs, L., 115, 183, *191*
Koyama, K., 254, *261*, 364, 375, *383*
Krail, G., 8, 45, *60, 62*, 166, *192*
Kranenburg, P., 298, *310*
Krebs, E., 53, *63*
Kricheldorf, H. R., 33, *60*, 76, 84, 86, *102*, 307, *311*, 372, *381*
Krieger, D. E., 48, *60*
Krumdieck, C. L., 294, *311*
Kubo, K., 205, *239*, 286, *314*
Kubota, M., 228, *239*
Kühne, W. F., 16, *60*
Kukla, D., 10, *58*
Kulesha, I. D., 201, 204, 210, *238*, 293, *313*

Kunitz, F. W., 133, 155, *194*, 302, *312*
Kupryszewski, G., 115, 116, 119, *192*
Kurath, P., 70, *102*
Kurobe, M., 254, *261*, 364, 375, *383*
Kuroda, K., 306, *309*
Kuromizu, K., 43, *60*, 204, *237*
Kurzer, F., 243, 258, *261*
Kusch, P., 173, *191*
Kutny, R., 201, 210, *238*
Kuziel, A., 115, *195*, 276, 279, *312*, 353, *383*

L

Lacasse, G., 289, *309*
Lackner, H., 10, 55, *60*
Ladkany, D., 166, *189*
Laird, A. H., 180, *187*
Lakshminarayanan, A. V., 11, *62*
Lam, H.-L., 45, 49, 55, *63*
Lamont, W. A., 15, *60*
Lande, S., 371, *381*
Landowne, R. A., 371, *379*
Lang, P., 205, 231, *234*
Langeray, R., 78, *100*
Lanza, T., Jr., 203, 227, *235*
Lanzilotti, A. E., 177, *192*
Lapatsanis, L., 268, 294, 300, *314*
Lapiccirella, A., 5, *58*
Larkin, J. B., 203, *237*
Larn, H.-L., 95, *103*
Laufer, D. A., 126, 185, *192, 193*
Lauren, D. R., 293, *310, 312*, 359, 374, *381*
Lautenschlager, H., 76, *101*
Lautenschleger, M. J., 370, 371, *378, 379*
Law, H. D., 135, *192*
Lawson, W. B., 23, *61*
Lazarus, L. H., 19, *60*
Leach, S. J., 328, 330, *380*
Lee, T. C., 252, *260*, 356, *380*
Lee, Y. C., 44, *63*
Leemann, H. G., 34, *63*, 139, *195*
Lefrancier, R., 185, *192*
Leimgruber, W., 219, *234*, 275, 296, *313*
Lemaire, S., 46, 47, 48, 50, *60*, 306, *312*
Lemieux, R. U., 293, 309, 332, 359, 370, 374, *379*
Lempert, K., 243, *260*
Lenard, J., 15, 44, *60*
Leplawy, M. T., 72, 87, *102*, 256, *261*
Le Quesne, W. J., 73, *102*
Leuchs, H., 306, *312*
Levine, L., 276, *310*, 347, *379*
Levine, S. D., 157, *188*
Levitt, T. E., 78, *102*
Levy, D., 166, *192*
Levy, L., 213, *234*
Lewalter, J., 118, *196*
Li, C. H., 15, 19, 46, 47, 48, 49, 50, *56, 60, 64*, 82, *102*, 147, 150, 176, 185, *189, 192, 194*, 207, *234*, 255, *261*, 306, *312, 314*, 324, 369, *383*
Liberek, B., 142, 154, 186, *191, 192*, 206, *236*, 327, 331, 340, 341, 342, 354, 377, *381*
Lien, E. L., 99, *103*
Liener, I. E., 163, *193*
Lillycrop, J. E., 13, *61*
Lin, C. Y., 32, 91, *56, 100*, 144, 171, 174, 178, *188*, 290, 292, *311*
Lindeberg, G., 173, *191*
Lindenfelser, L. A., 25, *56*
Lindenfelser, A. L., 25, *62*
Lindenmann, A., 200, 208, 209, *235*
Linderstrom-Lang, K. U., 16, *60*
Lindsay, D. G., 8, *62*
Ling, N., 19, 45, *60, 61, 63*
Lipmann, F., 29, *60*
Lloyd, K., 40, *60*, 114, *192*, 363, *381*
Lo, T. B., 176, *192*
Lochinger, W., 205, 231, *234*
Löw, M., 33, 49, *60*, 116, 131, 150, 183, 184, *189, 191, 192*, 214, *236*, 249, *261*
Loh, H., 324, 369, *383*
Long, D. A., 13, 14, *61*
Longosz, E. J., 273, 288, *312*
Losse, G., 78, *102*, 116, 117, 125, 147, *187, 190, 192*
Lossen, W., 250, 250, 252, *261*
Lotan, N., 229, *234*
Lovell, F. M., 10, *62*
Low, B. W., 10, *62*
Low, M., 251, *260*
Lowbridge, J., 7, *61*, 88, *101*, 308, *310*
Lowe, L. A., 297, 299, *309*
Lubiewska-Nakonieczna, L., 123, 124, 132, 156, *192, 195*, 253, *261*, 284, *313*, 359, 371, 373, *383*
Ludescher, U., 92, *102*
Lüben, G., 268, 294, 301, *314*
Lübke, K., 43, *62*, 72, *103*, 173, *193*, 199,

201, 206, *237*, 243, 254, *261*, 265, 303, 304, *312*, 320, 342, 352, 355, *381*, *382*
Lunkenheimer, W., 173, *191*
Luyendyk, B. P., 325, 326, *378*
Lynen, F., 112, *193*

M

Maassen van den Brink, W., 84, *100*, 142, *187*
Maassen van den Brink-Zimmermannova, H., *187*
McCormick, D. B., 168, *193*
McDermott, J. R., 290, 292, *312*, 332, 333, 346, 352, 357, 359, 370, 374, *381*
McGahren, W. J., 206, *235*, 276, *310*, 347, 348, 350, *379*
McGregor, A. C., 166, *187*
McGuire, R. F., 11, *61*
McKay, A. F., 207, *236*
McKerrow, J. H., 44, *62*
McKillop, A., 124, *195*
MacLaren, J. A., 201, 202, 207, 208, *236*, 335, 352, 355, *381*
MacLeod, J. K., 180, *187*
McNeil, G. N., 350, *380*
Macrea, R., 205, 209, *236*
Madison, V., 229, *234*
Magee, M. Z., 200, 208, 209, 229, *235*, *236*
Maia, H. L., 97, *102*
Mains, R. E., 19, *61*
Makisumi, S., 215, 226, *236*, *238*
Makofske, R., 201, 210, *238*
Maksimov, E. E., 352, 360, 374, *382*
Malek, G., 75, *100*, 267, 289, 290, 292, *309*, 358, 374, *379*
Manhart, K., 219, *234*
Manning, J. M., 210, *236*, 327, 329, 335, 370, *381*
Manning, M., 7, 51, *61*
Mansfeld, H.-W., 300, *312*
Mansfeld, J., 300, *312*
Marchiori, F., 49, *56*, 145, 167, 176, *193*, *194*, 203, 204, 208, 209, 218, *234*, *236*, *237*
Marchiori, I., 176, 177, *194*
Marglin, A., 357, 370, *381*
Margoliash, E., 18, *63*
Marks, N., 52, *56*, *61*
Marlborough, D. I., 126, *192*
Maroux, S., 16, *61*
Marquarding, D., 45, 49, 55, *63*, 95, *103*
Marsh, R. E., 4, *61*
Marshall, G. R., 15, 46, *61*, 163, *189*, *193*
Martel, R., 289, *309*
Martin, G. J., 176, *193*
Martinez, J., 137, *193*
Martynov, V. F., 183, *191*, 341, 361, *380*, *381*
Maruyama, H., 90, *103*, 137, 149, *193*
Marx, A. F., 119, 134, *193*
Marzotto, A., 203, 204, *237*
Massiah, T., 271, *309*
Mathews, M. B., 18, *61*
Mathur, R. B., 184, *193*
Matsubara, H., 21, *61*
Matsueda, R., 137, 138, 149, 181, *193*
Matsuo, H., 224, *237*, 327, 330, 331, 334, *381*, *382*
Matter, M., 131, *190*
Maximov, E. E., 206, *236*
Mayer, D., 271, 276, *313*, 348, *382*
Mayer, H., 82, *104*, 109, 122, *196*
Mayers, G. L., 32, *60*, 115, 154, 183, 184, *191*, *192*, 324, 341, 349, 355, *381*
Maynard, J. B., 325, 326, *378*
Mazur, R. H., 142, 160, *193*
Medzihradszky, K., 214, *236*
Medzihradszky-Schweiger, H., 201, 203, 204, *236*
Mehlis, B., 167, *193*
Meienhofer, J., 10, 43, 44, 45, 46, 49, 55, *57*, *60*, *61*, *63*, 68, 70, 73, 74, 99, *101*, *104*, 145, 146, 147,150, 176, 177, *188*, *192*, *193*, 201, 202, 204, 209, 210, 212, 214, 217, 233, *235*, *236*, *238*, *239*, 254, 275, 285, 295, 299, 300, *310*, *312*, *314*
Meiri, T. H., 116
Meister, A., 27, *61*
Menard, M., 289, *309*
Menegatti, E., 363, *380*
Menger, F. M., 91, *102*
Mercola, D. A., 8, 9, 10, 50, 54, *56*, *61*
Merrifield, R. B., 15, 43, 44, 46, 47, 48, 50, *57*, *58*, *60*, *61*, *114*, *125*, *172*, *193*, *204*, *210*, *211*, *219*, *230*, *234*, *236*, *238*, *234*, *246*, *255*, *258*, *260*, *261*, *288*, *295*, *312*, *323*, *332*, *342*, *357*, *381*
Michalewsky, J., 70, 99, *101 235*, *275*, *310*
Michalik, A., 206, *236*, *327*, *341*, *354*, *377*, *381*

Mikhaleva, I. I., 45, *59, 300, 301, 311*
Milkowski, J., 86, *101*
Miller, A. W., 206, *236*
Miller, J. G., 108, 113, 118, *190, 202, 234*
Miller, S. M., 307, *311*
Miller, W. G., 330, *382*
Milne, H. B., 84, *102*
Milstein, C., 18, *61*
Minami, H., 205, *239, 286, 314*
Mitaki, T., 203, *236*
Mitchell, A. R., 46, *61, 255, 251, 288, 295, 312*
Mitchell, J. W., 25, *62*
Mitin, Y, V., 76, 84, *102, 116, 129, 159, 192, 206, 236, 352, 360, 374, 382*
Mitsuyasu, N., 207, 215, 226, *235, 238*
Miyata, K., 209, 228, *238*
Miyoshi, M., 87, *102, 340, 382*
Mizoguchi, T., 231, *235*
Mizokami, N., 285, 286, *314*
Mochizuki, T., 228, *239*
Moffat, J. B., 12, *61*
Mozsov, S., 50, *61*
Mollan, R. C., 40, *59, 331, 345, 346, 353, 354, 365, 375, 376, 378, 380*
Momany, F. A. *61*
Moore, G. A., 76, *101*
Moore, S., 12, 19, 49, *57, 58, 61, 210, 236, 376, 381*
Morawiec, J., 341, 348, *382*
Morell, J. L., 21, 24, 25, *58*
Morgan, B. A., 19, 52, *58, 49, 92, 100*
Morimoto, H., 130, *195, 239, 261, 272, 273, 275, 288, 314*
Moritz, P., 247, *260*
Morley, J. S., 40, *47, 115, 139, 166, 180, 187, 189, 202, 234, 363, 379*
Moroder, L., 49, *56, 64, 203, 204, 209, 231, 236, 237*
Morozov, V. B., 183, *191*
Morris, H., 19, *59*
Morris, J. W. S., 10, *61*
Most, C. F., Jr., 84, *102*
Mowles, T., 44, *57*
Mühlemann, M., 75, *102*, 292, *312*
Mukaiyama, T., 90, *102, 103,* 127, 128, 139, 181, *192,* 372, *382*
Munekata, E., 294, 304, *312*
Munk, M. E., 78, *103,* 121, *195*
Munoz-Ramirez, H., 51, *59*
Muramatsu, I., 258, *261,* 305, *312*
Muramatu, M., 13, *61*
Murakoka, M., 41, *49,* 222, 224, *235,* 283, 284, 292, *311,* 358, 359, 360, 367, 373, 374, 376, *380*
Musich, J. A., 259, *261*
Mutt, V., 91, *100,* 290, 292, *311*
Mutter, M., 306, *309*
Muzalewski, F., 33, 38, *61,* 116, 141, *192, 193*
Myagkova, M.A., 45, *59*, 300, 301, *311*

N

Nadezhdina, L. B., 116, *193*
Nagamatsu, A., 13, *61*
Naganathan, P.S., 11, *61*
Nagarajan, K., 123, 124, *190, 194,* 253, *260, 261*, 352, *382*
Nah, H. S., 45, 49, 55, *63,* 95, *103*
Naithani, V. K., 8, 50, *57, 58,* 207, *237*
Nakamizo, N. *193*
Nakamura, M., 209, 228, *238*
Nakaparksin, S., 357, 367, *379*
Nakayama, Y., 87, *102*
Narayanan, V. L., 177, *193*
Natarajan, S., 45, *56, 61,* 92, *100,* 150, 151, 152, 177, *188, 193,* 296, 310
Nebelin, E., 21, *58, 61,* 300, *312*
Nefkens, G. H. L., 108, 123, 182, *192,* 253, *261,* 352, *382*
Nelson, J. S., 276, 278, *309*
Nelson, P., 78, *102*
Neuberger, A., 300, *312,* 320, 328, 330, 331, 332, 348, *382*
Neubert, K., 300, *312*
Neurath, H., 16, 17, *61*
Nicolaides, E. D., 147, *193*
Nikiforov, A., 332, *382*
Ninomiya, K., 213, *237, 372, 382*
Nintz, E., *187, 195*
Nishimura, O., 125, *189,* 205, *235,* 250, 253, *260,* 364, *379*
Nivard, R. J. F., 183, *193,* 251, *261,* 352, *382*
Niwano, M. *196*
Nolan, C., 18, *63*
North, M. B., 279, *312*
Nowak, J., 8, 50, *57, 58*
Nowicka, A., 331, *381*
Nulty, W. L., 15, 46, *61*

Nussey, A., 45, *63*
Nutt, R. F., 44, 49, *58*, 198, 202, 203, 207, 227, 228, *235*, *238*
Nyéki, O., 49, *60*, 91, *102*, 116, 182, 220, 221, 224, *236*, 360, *381*
Nyfeler, R., 37, *57*
Nyilasi, J., 330, *382*

O

Oakwood, T. S., 68, *102*
Obayashi, M., 250, 253, *260*, 364, *379*
Obermeier, R., 70, *102*, 209, 228, *237*
O'Connor, M. J., 331, *382*
Offermann, K., 45, 49,55, *63*, 95, *103*
Ogawa, H., 43, *64*, 204, 209, 228, *237*, *238*
Ohlsen, S. R., 49, *58*, 308, *310*
Ohnishi, M., 327, 330, 331, 334, *381*
Ohno, M., 15, *56*, 151, *193*, 204, *237*
Okada, Y., 205, *239*, 284, 285, *314*
Olofson, R. A., 82, *104*, 109, 122, *196*
Ondetti, M. A., 31, 43, 45, *56*, 66, *100*, 145, 148, 157, 176, 177, 185, *188*, *193*, 199, 207, 229, *234*, *237*, 265, 285, 302, *310*, *312*, 320, *379*
Ono, H., 326, *383*
Ono, T., 13, *61*
Ontjes, D., 15, *56*
Oró, J., 357, 367, *379*
Osato, R. L., 265, 270, 274, 277*313*
Ott, H., 332, *382*
Ottenheym, H., 268, 294, 301, *312*, *314*
Ottesen, M., 16, 19, *58*, *60*
Ovchinnikov, Yu. A., 5, *61*, 229, *237*
Over, H., 46, *56*
Oya, M., 86, *102*, 307, *311*
Oyer, P., 18, *63*
Ozawa, K., 77, *102*, 216, *237*

P

Pacsu, E., 107, *193*
Paleveda, W. J., Jr., 44, *63*, 85, 86, *101*, 202, 203, 207, 227, *235*, *238*, 367, 377, *379*
Pankstelis, J. V., 75, *102*
Panneman, H. J., 71, *103*, 119, 134, *193*
Panse, G. T., 126, *193*
Paquette, L. A., 124, *193*, 254, *261*
Parr, W., 357, 367, 371, *379*, *382*
Pataki, G., 204, *237*
Patchornik, A., 11, 23, *61*, *63*, 84, 94, *102*, *104*, 109, 126, 127, 163, *189*, *191*, *193*, 201, 204, *238*
Patel, D. J., 5, *7*
Patel, R. P., 145, 147, 154, 177, *187*, *193*
Pattabhi, V., 11, *61*
Paul, R., 84, *102*, 147, *194*, 207, *237*
Pauling, L., 4, 10, *57*
Peach, M. J., 51, *59*
Pekaar, J., 31, 33, 34, 37, *59*, 341, 349, *380*
Pelter, A., 78, *102*
Penke, B., 260, 174, *192*
Perrin, D. D., 272, *312*, 335, 338, *382*
Perrin, D. R., 272, *312*
Perron, Y. G., 289, *309*
Pestka, S., 27, 28, *63*
Peterson, J. D., 18, *63*
Pettit, G. R., 198, *237*, 264, *312*
Pfister, R. W., 352, *379*
Photaki, I., 76, *100*, 123, *188*, 355, *382*
Piala, J. P., 176, *188*
Piche, L., 271, *309*
Pless, J., 32, 34, 51, 53, *61*, *62*, 108, 115, 147, 176, 179, 180, 181, *190*, *194*
Pluscec, J., 148, 185, *193*, 207, *237*
Poduska, K., 68, *102*, 371, *382*
Pöldinger, W., 53
Polglose, W. J., 138, *195*
Popernatskii, O. A., 361, *380*
Pravda, Z., 371, *382*
Prelog, V., 214, *237*
Prescott, D. J., 15, 46, *61*
Preston, J., 176, 179, *189*, *190*, 209, *237*, 256, *260*
Preston, W. H., 25, *62*
Pridham, T. G., 25, *56*, *62*
Prinz, W., *195*
Pritchard, A. E., 369, 372, 376, *383*
Prox, A., *187*, 218, 219, 220, 221, 225, *238*, 249, 250, *261*, 268, 271, 277, 278, 279, 294, 300*312*, *313*, *314*, 329, 342, 348, 349, 353, 354, 355, 359, 360, 361, 366, 367, 372, 373, 374, 376, 377, *383*
Przybylski, J., 116, *192*
Pullen, R. A., 8, 10, *62*
Pullman, A., 10, 11, *62*
Pullman, B., 10, 11, *62*
Puls, W., 8, *57*

Q

Quitt, P., 95, *100,* 123, *188, 189,* 331, *382*

R

Raap, J., 181, 182, *195*
Racky, W., 171, *195*
Raddatz, F., 201, 203, 204, *236*
Radhakrishnan, A., 48, *58*
Ragnarsson, U., 155, 166, 172, *191, 194, 195*
Rajapda, S., 123, *190*
Rajappa, S., 124, 133, 146, *194,* 253, *260, 261*
Rakoczy, B., 115, 156, 183, *191*
Ramachandran, G. N., 4, 10, 11, *62*
Ramachandran, J., 82, *102,* 150, 176, *192, 194*
Ramage, R, 49, *57, 58,* 76, 89 *100, 101,* 308, *310*
Ramakrishnan, C., 10, 11, *62*
Rammeloo, T., 74, *100,* 265, 266, 272, 297, *310*
Rance, M. J., 52, *58*
Rapoport, H., 259, *261,* 327, 335, *380*
Ratnoff, O. D., 17, *57*
Ratzkin, H., 81, *102,* 148, *194*
Rebek, J., Jr., 40, 41, *59,* 83, *102,* 214, 220, 221, 222, *236,* 246, 247, 255, *261,* 276, 281, 283, 284, *311,* 331, 342, 345, 351, 353, 354, 357, 358, 359, 365, 371, 373, 375, 376, 377, 378, *380, 382*
Reczek, J., 16, 44, 45, 49, 55, *59,* 95, *102,* 204, *236*
Reed, L. L., 4, 22, *62*
Regel, W., 33, *60*
Rehn, D., 45, 49, 55, *63,* 95, *103*
Rehn, H., 95, *103*
Reich, E., 18, *62*
Reichert, E., 112, *193*
Rein, R., 11, *62*
Reinert, G. E., 230, *33*
Reinhoudt, D. N., 352, *382*
Renick, R. J., 288, *311*
Renirie, J. F. M., 74, *100,* 265, 266, 272, 297, *310*
Renugopalakrishnan, V., 11, *62*
Resofszki, G., 244, *261*
Ressler, C., 49, 52, *57,* 81, *102,* 109, 110, 148, *189, 194,* 242, *260, 261*
Richards, F. M., 19, 20, *62, 64*
Richards, P. M., 49, *57*
Ridge, B., 97, *102*
Ridge, R. J., 88, *101,* 308, *310*
Ried, W., 84, *102*
Rifkin, D. B., 18, *62*
Riggs, A. D., 55, *59*
Rimmer, B., 8, *56*
Riniker, B., 45, 48, 49, *63,* 69, 70, 80, 81, 99, *102, 103,* 199, 200, 214, 215, 219, 220, 221, 222, 223, 225, *236, 237,* 249, 258, *261,* 324, 356, 359, 361, *382*
Rink, H., 81, *103,* 258, *261*
Rishi, S., 76 *101,* 216, *235*
Rittel, W., 45, 48, 49, *63,* 69, 70, 80, 81, 99, *102, 103,* 111, 116, 117, 130, 131, 161, 176, *190, 194,* 199, 200, 214, 215, 219, 220, 221, 222, 225, *236, 237,* 249, *261,* 324, 356, 361, *382*
Rittenberg, D., 327, *382*
Rivier, J., 45, *63,* 99, *100,* 324, 369, *382*
Roberts, C. W., 49, 52, *52,* 109, 110, *189,* 242, *260*
Roberts, D., 16, 44, 45, 49, 55, *59,* 205, *236*
Roberts, J. E., 32, 35, *60,* 115, 154, 183, *191, 912*
Roberts, J. L., 19, *62*
Robinson, A. B., 15, 44, 46, 50, *62*
Robinson, J. C., 285, *309*
Rocchi, R., 70, *103,* 145, 176, 177, *192, 194,* 203, 204, 208, *237*
Rockland, L. B., 203, *237*
Rodin, R. L., 165, *192*
Roemer, D., 51, 53, *62*
Rogers, F. F., Jr., 164, *189*
Rohr, G., 268, 294, 300, 301, *314*
Romovacek, H., 70, *103,* 209, 215, 228, *234, 237*
Rosenberg, I. H., 295, *312*
Roskoski, R., Jr., 29, *56, 60*
Rothe, I., 95, *103*
Rothe, M., 95, *103,* 133, 155, *194,* 302, *312*
Rottenberg, M., 201, *237*
Roubicek, J., 53, *63*
Rouette, H. K., 171, *194*
Rubenstein, A. M., 18, *63*
Rubin, B., 176, *188*
Rudinger, J., 31, 32, 33, *58, 62,* 68, 69, 75, 88, 89, *101, 102, 103,* 174, *194,* 213, 214, 215, 221, 222, 120, 232, *235,* 271, 288, 292, 308, *312, 314*
Rudko, A. D., 10, *62*

Ruettinger, T. A., 117, 163, 90
Russell, D. W., 133, 155, *189*, 302, *310*
Ruttenberg, M. A., 184, *194*
Ryabtsev, M. N., 46, *61*
Ryan, C. A., 10, *58*
Rydon, H. N., 97, *102*, *103*, 229, 237, 257, *260*, 286, 307, *312*
Rzeszotarska, B., 115, 123, 124, 132, 156, 166, *192*, *194*, *195*, 253, *261*, 276, 279, 291, *312*, 332, 353, 371, *383*

S

Sabatini, D. D., 18, *56*
Sabo, E. F., 138, 177, 185, *193*
Sachs, D. H., 149, *189*
Sachs, H., 276, *310*
Said, S. I., 91, *100*, 180, *188*
Saito, T., 71, *102*
Sakagami, M., 228, *239*
Sakakibara, S., 15, 43, 44, *62*, *63*, 131, 168, 184, *194*, 231, *235*, *237*, 283, 284, *312*, 340, 358, 373, *382*
Salvadori, S., 70, *103*
Samartsev, M. A., 341, *381*
Sandberg, B. E., 166, *194*
Sandow, J., 150, *189*
Sandrin, E., 177, *194*
Sangl, I., 304, *313*
Sano, P., 145, 147, 177, *193*
Sarantakis, D., 99, *103*, 124, *194*
Sarges, R., 20, *62*, 294, *312*
Sasaki, K., 8, *62*
Sasaki, R., 21, *61*
Sasisekharan, V., 4, 10, *62*
Satchell, D. P. N., 12, 20, 21, *62*
Sato, H., 228, *239*
Sato, K., 294, *309*
Sato, M., 224, *237*, 327, 330, 331, 334, *381*, *382*
Savige, W. E., 202, 207, 208, *236*, 225, 352, 355, *381*
Savoie, J. Y., 293, *309*, 332, 359, 370, 374, *379*
Sawyer, W. H., 7, 51, *61*
Scatturin, V., 176, 177, 94
Schade, F., 169, *194*
Schäfer, W., 107, 108, 112, 113, 117, 129, *195*, *196*, 302, *314*
Schafer, D. J., 150, *189*
Schaich, E., 177, *194*
Schally, A. V., 15, 44, 51, *56*, *60*, 150, *189*
Scharf, R., 49, *64*
Schatz, K., 125, *192*
Schechter, I., 360, 369, 371, 374, *379*
Scheele, G., 18, *57*
Scheiner, S., 10, *62*
Schejter, A., 307, *314*
Schellenberg, P., 288, *312*
Schenk, P., 37, *57*
Scheraga, H. A., 11, *61*
Schermutzki, W., 170, *188*
Schiffman, E., 21 *56*
Schleimer, B., 84, *102*
Schmid, J., 268, 294, 301, *312*, *314*
Schmidhammer, L., *187*, 271, 277, 278, 279, *312*, *313*, 342, 349, 353, 354, 355, 366, 367, 372, 376, 377, *383*
Schmidt, G. N., 156, 165, 181, *191*
Schmidt, I., 49, *64*
Schmidt, U., 332, *382*
Schmiechen, R., 204, 213, 215, *235*
Schmir, G. L., 23, *62*
Schmitt, E. E., 304, *311*
Schnabel, 74, *104*, 147, 167, 175, 176, 177, *192*, *194*, 199, 202, 206, 212, 213, 214, 217, 225, 226, *235*, *236*, *237*, *239*, 265, 285, 299, 300, 305, *310*, *312*, *314*, 330, 349, 350, *382*
Schneider, F., 173, 177, *194*, *195*, 226, *237*
Schneider, G., 160, *196*
Schneider, W., 15, *63*, 95, *100*, 123, *189*
Schnyder, J., 201, *237*
Schön, I., 49, *60*, 116, 183, *191*
Schön, W., 23, *63*
Schoenborn, J. E., 331, *382*
Schoenewaldt, E. F., 85, 86, *101*, 203, 227, *235*, 307, *311*, 367, 377, *379*
Schoenheimer, R., 327, *382*
Schott, H. F., 203, *237*
Schramm, G., 288, 305, *313*
Schreiber, J., 231, *234*
Schreier, E., 72, *103*
Schröder, E., 43, *62*, 72, *103*, 152, *191*, 199, 201, 206, *237*, 243, 254, *261*, 265, 303, 304, *312*, 320, 342, 352, 355, *381*, *382*
Schüssler, H., 305, 306, *312*
Schumann, I., 304, *310*
Schwam, H., 85, 86, *101*, 307, *311*, 367, 377, *379*

Schwartz, E. T., 226, *235*, 242, *260*, 356, *380*
Schwartz, I. L., 5, 50, *63*, 88, *101*
Schwarz, H., 276, *312*
Schwyzer, R., 10, 37, 43, 54, *62*, *64*, 75, 92, *102*, 107, 111, 112, 116, 117, 130, 131, 132, 129, 141, 151, 157, 161, 162, 163, 176, *189*, *190*, *194*, 203, 209, 214, 218, 225, 226, *237*, 292, 303, 304, *311*, *312*, 342, 353, 356, 366, *380*, *382*
Scoffone, E., 145, 167, 176, 177, *193*, *194*, 203, 204, 208, 209, *236*, *237*
Scott, F. L., 339, *382*
Seelinger, A., 342, 372, *383*
Seely, J. H., 113, 117, *191*, 332, 380
Sehring, R., 288, 305, 307, *313*, *314*
Sela, M., 85, *102*
Selve, C., 89, *100*
Shaltiel, S., 184, *194*
Sharp, 15, 46, 50, *62*
Shaw, E., 18, *62*
Sheat, S., 8, *56*
Sheehan, J. C., 67, 79, 81, 88, *103*, 110, 120, 133, 145, *194*, *195*, 242, 252, *261*, 276, 305, *313*
Sheehan, J. T., 147, 148, 172, 176, 177, 185, *188*, *193*, 207, *237*
Sheppard, R. C., 46, *56*, 68, 72, 87, 88, *101*, *102*, 110, 180, *187*, *189*, 214, *233*, 256, *260*, *261*
Shibuya, S., 13, *61*
Shimizu, M., 204, *234*
Shimonishi, Y., 15, *62*, 185, *195*, 203, 231, *235*, *236*
Shin, K. H., 15, *62*, *63*
Shinagawa, S., 205, 250, 253, *235*, *260*, 364, *379*
Shiori, T., 76, 77, *101*, *102*, *103*, 213, 216, *235*, *237*, *239*, 372, *282*
Shorley, P. G., 147, *193*
Shotwell, O. L., 25, *56*, *62*
Sieber, P., 45, 48, 49, *63*, 69, 70, 80, 81, 99, *102*, *103*, 111, 116, 117, 130, 131, 157, 161, 162, 163, 167, *190*, *194*, *195*, 199, 200, 203, 210, 215, 219, 220, 221, 222, 223, 225, 231, *236*, *237*, 249, *261*, 324, 356, 359, 361, *382*
Siedel, W., 205, *235*, 357, *379*
Siemion, I. Z., 331, 341, 348, *382*
Sigler, G. F., 144, 171, 178, *188*
Silverstein, R., 244, 246, 247, 248, 255, 258, *260*
Singer, A., 21, *61*
Singh, B., 49, *58*, 308, *310*
Sipos, F., 293, *313*
Sivanandaiah, K. M., 172, *191*, 291, *309*
Smart, N. A., 276, *313*, 359, 367, *382*
Smeby, R. R., 51, *59*
Smith, C. F. C., 52, *58*
Smith, C. W., 7, 51, 55, *63*
Smith, E. L., 138, *195*
Smith, J. H., 91, *102*
Smith, P. A. S., 250, *261*
Smith, P. W. G., 229, 286, 307, *312*, *237*
Smith, T. W., 19, *59*
Smithers, M. J., 184, *190*, 202, *234*
Smithwick, E. L., 231, *235*
Smulkowski, M., 222, *234*, 284, 299, 300, *310*, *313*, 348, 359, 371, 373, *379*, *383*
Smyth, D. G., 19, *57*
Smyth, J., 135, *189*
Snell, C. R., 19, *57*
Sohar, P., 33, *60*
Sokolvsky, M., 307, *314*
Sokolowska, T., 332, 371, *383*
Sondey, J., 203, 227, *235*
Sondheimer, E., 219, *235*
Soulen, R. L., 68, *100*
Spach, G., 164, 165, *188*
Spackmann, D. H., 138, *195*
Spande, T. F., 11, *63*
Spühler, G., 242, *260*, 356, *380*
Staab, H. A., 135, *195*
Stahl, G. L., 7, *63*
Stahnke, G., 205, 231, *234*
Stedman, R. J., 148, *190*, 242, *260*
Steffen, K.-D., 95, *103*
Stefko, P. L., 293, *313*
Steglich, W., 118, *195*, 209, *238*, 271, 276, *313*, 348, *382*
Steigemann, W., 10, *57*
Stein, S., 48, *58*, *63*, 219, *234*, 275, 296, *313*
Stein, W. H., 19, 20, *58*, *63*
Steiner, D. F., 18, *57*, *63*
Stelzel, P., 30, *64*, 199, 201, 203, 207, 208, 210, 213, 214, 220, *238*, 265, 306, 307, *314*
Stengele, E., 33, *60*
Stern, F., 52, *61*
Stevens, C. L., 78, *103*, 121, *195*
Stewart, F. H. C., 115, 150, 165, *195*, 271, 272, 275, 279, 286, 291, 304, *313*, 370, *383*
Stich, K., 34, *63*, 139, *195*

Stirling, C. J. M., 206, *236*
Stodola, F. H., 25, *62*
Stone, B. E., 166, 180, *189*
Storey, H. T., 44, 45, *56*, 184, 185, *187*, *195*, 209, 228, *237*
Strachon, R. G., 85, 86, *101*, 202, 203, 207, 227, *235*, *238*, 307, *311*, 277, *379*
Studer, R. O., 331, *382*
Stueben, K. C., 111, 156, 160, 161, 176, *190*, 230, *235*, 247, *260*, 276, 288, 303, *310*, 341, 342, 344, 366, *379*
Stüber, S., 45, 49, 55, *63*, 95, *103*
Sturm, K., 357, *379*
Su, C., 363, *383*
Subramanyam, G., 40, *59*, 331, 345, 353, 354, 365, 375, 376, 378, *380*
Surbeck-Wegmann, E., 214, *237*
Sutherland, J. K., 124, *194*
Suzuki, M., 137, 138, *193*
Swan, J. M., 49, 52, *57*, 109, 110, *189*, 202, 207, 208, *236*, 242, *260*, 335, 352, 355, *381*
Swarc, M., 85, *103*
Swindin, W. A., 271, *311*
Syheinblatt, M., 330, *379*
Syrier, J. L. M., 74, *100*, 265, 266, 272, 297, 298, *310*, 356, 371, *381*
Szirmai, M., 49, *60*
Szirtes, R., 183, *191*
Szirtes, T., 49, *60*, 116, *191*, 220, 221, 225, *236*

T

Tachibana, S., 216, *239*
Tadema, G., 71, *103*
Tager, H. S., 18, *63*
Tahara, A., 164, *189*
Takamatsu, N., 94, *101*
Takayamagi, H., 201, *238*
Takeuchi, Y., 372, *383*
Tamarnasi, M., 95, *103*
Tamasi, M., 45, 49, 55, *63*
Tamburro, A. M., 203, 204, 208, *237*
Tamura, F., 209, 228, *238*
Tan, N. H., 49, *63*
Tanaka, A., 209, 228, *238*
Tanner, H., 276, *313*, 348, *382*
Tarbell, D. S., 265, 267, 269, 273, 285, 288, *312*, *313*
Taschner, E., 115, 123, 124, 132, 141, 156, *192*, *195*, 222, *234*, 253, *261*, 276, 279, 284, 299, 300, *310*, *312*, *313*, 332, 348, 353, 356, 359, 371, 373, *379*, *383*
Tatemoto, K., 45, *56*, 92, *100*, 150, *188*, 296, *310*
Tatsumi, S., 326, *383*
Tatsuno, T., 224, *237*, 327, 330, 331, 334, *381*, *382*
Taylor, E. C., 124, *195*
Teichman, J., 99, *103*
Terashima, S., 109, 141, *196*
Tesser, G. I., 108, 123, 151, 183, *189*, *193*, 251, 253, *261*, 352, *382*
Thamm, P., 30, 49, *64*, 201, *238*, 306, *314*
Theodoropoulos, D. M., 78, *103*, 271, *314*, 342, *383*
Thomas, P. L., 285, 302, *312*
Thompson, P. J., 114, 160, *191*
Thompson, T. A., 226, *235*
Thorpe, W. D., 49, *57*, 76, *101*
Tickle, I. J., 8, 10, *62*
Tilak, M. A., 218, 231, *238*, 297, *311*, 273, 287, 294, 295, 296, 297, *313*
Timasheff, S. N., 11, *58*
Titov, M. I., 75, *102*, 292, *312*
Toda, F., 86, *102*
Toi, K., 331, *383*
Tolle, J. C., 286, 287, 298, *310*
Tomatis, R., 70, *103*, 363, *380*
Tometsko, A. M., 226, *236*, 256, *261*
Tomida, I., 348, 371, *383*
Tonelli, E. A., 5, *57*
Toome, V., 219, *234*
Torchia, D. A., 5, *57*
Tortorella, C., 124, *194*
Tortorella, V., 124, *194*
Tith, T., 95, *103*
Trangle, K., 251, *261*, 354, 361, *380*
Trangle, M., 251, *261*, 354, 361, *380*
Trask, E. G., 93, *101*, 181, 182, *190*
Treiber, R. L., 118, *195*
Trietsch, P., 268, 294, 300, 301, *314*
Trudelle, Y., 117, *195*, 366, *383*
Truscott, T. G., 14, *61*
Tschopp, A., 37, *57*
Tseng, L., 324, 369, *383*
Tsuji, J., 201, *238*
Tsukamoto, S., 151, *193*
Tsunehiko, F., 250, 253, *260*
Turner, J. M., 32, *57*, 78, *100*, 114, 129, 160, *189*, *191*, 302, *310*

Tyson, R. G., 49, *58*, 308, *310*
Tzougraki, C., 201, *238*

U

Udenfriend, S., 48, *58*, *63*, 219, *234*, 275, 296, *313*
Ueki, M., 90, *103*, 137, 149, 181, *193*
Ugi, I., 45, 49, 55, *63*, 95, *103*, 177, 185, *194*
Ullrich, J., 288, *312*
Urban, R., 45, 49, 55, *63*, 95, *103*
Urry, D. W., 4, 6, 10, 20, 50, *63*

V

Vagelos, P. R., 15, 46, *61*
Vaidya, V. M., 126, *192*
Vajda, T., 165, *189*
Vale, W., 45, *63*
van Velden, F., 132, *190*
van Zon, A., 74, *100*, *103*, 231, *238*, 265, 266, 268, 272, 286, 295, 296, 297, 298, 299, *310*, *313*
Varga, S. L., 44, 49, *58*, 198, 227, 228, *235*
Vargha, E., 175, *187*
Vaughan, J. R., Jr., 73, 78, *103*, 264, 265, 270, 271, 272, 274, 276, 277, 278, 279, 281, 282, 285, 307, *313*
Vaughan, J. W., 272, *313*
Veber, D. F., 33, 37, 44, 49, *58*, *63*, 69, 85, 86, *101*, *103*, 198, 202, 203, 205, 207, 217, 219, 227, 228, *235*, *238*, 252, *261*, 307, *311*, 345, 356, 367, 377, 378, *379*, *383*
Vellaccio, F., 16, 44, 45, 49, 55, *59*, 95, *102*, 205, *236*
Velluz, L., 271, *309*
Venkatesan, K., 11, *61*
Vezer, C., 49, *60*
Vidali, G., 176, 177, *194*
Viehe, H. G., 78, *100*, 121, *189*, 372, *383*
Vijayan, M., 8, *56*
Vilchez-Martinez, J. A., 51, *56*
Villemoes, P., 256, *260*
Visser, J. P., 209, *235*
Visser, S., 181, 182, *195*
Viswanatha, R., 168, *195*
Vitali, R. A., 44, 49, *58*, 198, 202, 203, 207, 227, 228, *235*, *238*
Vithayathi, P. J., 19, *62*
Vlasov, G. P., 129, *193*
Vogeler, K., 84, *104*, 130, 142, 160, *196*, 272, 273, 275, 288, *314*
Vogler, K., 331, *382*
Vogt, H. P., 170, *188*
Voigt, A., 32, 39, *59*, 93, *101*, 186, *190*, 363, *380*
Volk, A., 45, *58*
Volpina, O. M., 45, *59*, 300, 301, *311*
von Brunn-Leube, I., 288, 305, *313*
von Graffenried, B., 53, *63*
von Hinrichs, E., 45, 49, 55, *63*, 95, *103*
von Philipsborn, W., 271, *313*
von Saltza, M., 177, *193*
von Zychlinski, H., 45, 49, 55, *63*, 95, *103*

W

Wackerle, L., 45, 49, 55, *63*, 95, *103*
Wade, R., 70, *100*, 247, *260*, 271, *311*
Wagatsuma, M., 109, 141, *196*
Wahl, S. M., 21, *56*
Wajcht, J., 116, *191*
Waki, M., 45, 49, *63*, 207, 215, 226, *236*, *238*
Waldschmidt-Leitz, E., 298, *314*
Waller, J. P., 69, *100*
Walter, R., 4, 5, 6, 7, 10, 50, 51, 52, 55, *63*, 88, *101*
Walton, E., 202, 203, 207, 227, *237*, *238*
Wang, C. T., 293, *313*
Wang, K. T., 180, *190*
Wang, S. S., 201, 204, 206, 210, 211, *234*, *238*, 293, *313*
Wang, S.-W., 40, 41, *59*, 83, *102*, 216, 219, 220, 221, 224, *236*, 251, *261*, 283, *311*, 323, 331, 345, 346, 353, 354, 359, 365, 367, 376, 378, *380*
Warshawsky, A., 94, *102*, 127, *189*
Wasielewski, C., 141, *195*, 332, 371, *383*
Watson, J. W., 363, *383*
Weber, U., 173, *195*
Wedel, D. S., 297, *313*
Wehrmüller, J., 95, *100*, 123, *188*, *189*
Weigele, M., 219, *234*, 275, 296, *313*
Weinberg, N. L., 289, *309*
Weinert, M., 166, *195*
Weinstein, B., 176, 179, 180, *189*, *190*, 208, 209, *237*, *238*, 349, 369, 372, 376, *380*, *383*
Weintraub, J., 15, 46, *61*
Weisgerber, C. A., 68, *102*
Weiss, K., 273, 296, 305, *313*
Weissbach, H., 27, 28, *63*

Weitzel, G., 173, *195*
Welcher, A. D., 76, *100*, 160, *187*
Wells, R. D., 204, 213, 215, *235*
Wendlberger, G., 30, 49, *64*, 71, 79, *104*, 209, *238*, 243, 254, *261*, 303, *314*
Wersin, G., 231, *237*
Westley, J., 357, *383*
Weygand, F., 80, *103*, 118, 155, 166, *187*, *195*, 209, 212, 218, 219, 220, 221, 225, 231, *238*, 249, 250, 253, *261*, 271, 273, 276, 277, 278, 279, 296, 300, 305, *312*, *313*, 329, 342, 348, 349, 353, 354, 355, 359, 360, 361, 362, 366, 367, 368, 372, 373, 374, 376, 377, *382*, *383*
Whetstone, R. R., 219, *234*
Wiejak, S., 166, *194*
Wieland, A., 164, *189*
Wieland, P., 214, *237*
Wieland, T., 23, 46, *63*, 73, 76, 84, *103*, *104*, 107, 108, 109, 112, 113, 117, 118, 126, 129, 130, 142, 143, 160, 164, 171, *195*, *196*, 217, *234*, *238*, 254, *261*, 264, 265, 168, 270, 272, 273, 275, 276, 277, 278, 288, 291, 294, 295, 298, 300, 301, 302, 304, 305, 307, *310*, *311*, *312*, *313*, *314*, 342, 358, 372, 373, 377, *379*, *383*
Wilchek, M., 168, 176, *187*, *195*, *196*, 330, *379*
Wilkinson, S., 297, 299, *309*
Williams, A. W., 224, *238*
Williams, M. W., 216, 220, *238*, 276, 277, 279, *309*, *313*, *314*, 320, 338, 340, 347, 350, 353, 354, 357, 359, 367, 372, 376, 377, *378*, *382*, *383*
Williams, N. J., 148, 157, 177, 185, *188*, *193*
Williams, R. E., 94, *100*, 293, *310*, *312*, 359, 374, *381*
Willson, C. G., 327, 335, *380*
Willstätter, H., 298, *314*
Wilschowitz, L., 49, *64*, 331, *382*
Wilson, E. J., 107, *193*
Windridge, G. C., 252, 256, *260*, *261*, 356, 370, *280*, *282*
Winitz, M., 199, 203, 213, *235*, 265, 273, *311*, 320, 328, *380*
Winkler, F. K., 4, 11, *64*
Winstein, S., 339, *382*
Winter, D. P., 201, 210, *238*
Winternitz, F., 137, *193*
Wissmann, H., 251, *260*
Witkop, B., 11, 20, 21, 23, 25, *58*, *61*, *62*, *63*, 110, 169, *190*, 294, *312*
Witty, M. J., 68, 87, 98, *101*
Wojciechowska, H., 284, *313*, 359, 371, 373, *383*
Wolbers, R., 324, 369, *382*
Wollmer, A., 8, *57*, *62*, 170, *188*
Wolman, Y., 84, *104*, 166, *189*, 201, 204, 213, 215, *235*, *238*, 307, *314*
Wolters, E. T. M., 251, *261*
Wong, L. T. L., 78, *100*
Wood, S. P., 8, 10, *58*, *62*
Woodman, D. J., 83, *104*
Woodward, R. B., 82, 83, *104*, 109, 122, *191*, *196*
Wooley, D. W., 207, 219, *236*, *238*
Woolner, M. E., 242, *260*, 356, *380*
Wright, P. H., 346, *379*
Wrobel, S. J., Jr., 40, *59*, 83, *102*, 365, *380*
Wu, Y., 109, 141, *196*
Wünsch, E., 30, 41, 43, 44, 49, *64*, 66, 67, 71, 73, 79, 80, 84, 87, 91, *104*, 155, 156, 177, 184, 185, *195*, *196*, 199, 201, 203, 207, 208, 209, 210, 212, 213, 214, 220, 231, *237*, *238*, 243, 248, 249, 253, 254, *261*, 265, 303, 304, 306, 307, *314*, 320, 353, 360, 374, *383*
Wüthrich, K., 10, *64*
Wyckoff, H. W., 20, *64*
Wyssbrod, H. R., 5, 50, *63*, *64*

Y

Yajima, H., 43, *64*, 75, *102*, 205, 208, 209, 226, 228, *235*, *238*, *239*, 254, *261*, 285, 286, 289, 290, 292, 293, *311*, *314*, 358, 364, 375, *381*, *383*
Yamada, S., 76, 77, *101*, *102*, *103*, 109, 141, *196*, 213, 216, *235*, *237*, *239*, 372, *382*, *383*
Yamaguchi, K., 228, *239*
Yamashiro, D., 15, 46, 47, 48, 49, 50, *60*, *64*, 255, *261*, 306, *312*, *314*, 324, 369, *383*
Yamashita, I., 205, *239*, 286, *314*
Yamatani, T., 73, *100*
Yamazaki, N., 76, *104*, 131, *196*
Yan, J. F., 11, *61*
Yanaihara, C., 44, 45, *56*, 184, 185, *187*, *195*, 209, 228, *237*, *239*
Yanaihara, N., 44, 45, *56*, 184, *187*, 204, 213, 215, 228, *234*, *235*, *239*
Yang, D.-D. H., 276, *313*

Yang Lin, C., 150, 177, *188*
Yeh, Y. L., 88, *103*, 110, *195*, 252, *261*
Yiotakis, A. E., 45, *56*, 91, 92, *100*, 150, 177, 180, *188*, 296, *310*
Yoshitake, H., 326, *383*
Young, G. T., 40, *59*, *60*, *64*, 70, 73, 80, 93, *101*, *102*, *104*, 114, 117, 125, 131, 142, 150, 155, 166, 171, 186, *187*, *189*, *190*, *191*, *192*, 199, 205, 209, 216, 219, 220, 224, *235*, *236*, *238*, *239*, 251, 253, *260*, 271, 276, 277, 279, 283, *309*, *311*, *312*, *313*, *314*, 320, 337, 338, 340, 343, 347, 349, 350, 353, 354, 357, 358, 359, 361, 362, 363, 366, 367, 368, 372, 375, 376, 377, 378, *378*, *379*, *380*, *381*, *382*, *383*
Young, R. W., 76, *100*, 160, *187*
Yphantis, D. A., 304, *311*

Z

Zabel, R., 150, *196*, 300, *314*
Zahn, H., 8, 45, 50, *57*, *58*, *60*, *62*, 74, *104*, 150, 166, 169, 171, 184, *188*, *192*, *193*, *194*, *195*, *196*, 206, 212, 213, 214, *237*, *239*, 265, 299, 300, 305, 306, *312*, *314*
Zahr, S., 45, 49, 55, *63*, 95, *103*
Zalut, C., 226, *236*
Zanetti, G., 184, *190*
Zaoral, M., 68, 72, *104*, 271, 288, 307, 308, *314*
Zaumeyer, W. J., 25, *62*
Zeiger, A. R., 259, *261*
Zervas, L., 76, *100*, 205, *239*, 271, 279, 300, *309*, *314*, 342, 348, *379*
Zeze, F., 228, *239*
Ziegler, J.-C., 89, *100*
Zimmerman, J. E., 74, *100*, 123, 154, 155, 184, 185, *187*, *196*, 219, 220, 221, *233*, 249, *261*, 265, 268, 269, 270, 271, 272, 274, 275, 276, 277, 279, 280, 281, 282, 283, 284, 291, 298, 299, 302, 306, *309*, 341, 342, 355, 358, 359, 360, 364, 367, 372, 373, *378*, *383*
Zimmermann, J. P., 95, *100*, 123, *188*, *189*
Zuber, H., 214, *237*
Zwick, A., 73, *104*

Subject Index

A

Ab initio techniques, in conformational calculations, 11
Acetal, in peptide synthesis, 30
Acetic acid, 141, 142, 327, 328
  catalysis of *p*-nitrophenyl ester aminolysis by, 141
  catalysis of phenyl ester aminolysis by, 141
  catalysis of 1-piperidyl ester aminolysis by, 142
  racemization of amino acids in, 327, 328
Acetoxime, as additive to dicyclohexylcarbodiimide coupling, 253
Acetylacetone, 84, 119
  enamine formation from , 119
Acetylation of peptides, by active esters, 167
N-Acetyl-S-benzyl-L-cysteine *p*-nitrophenyl ester, preparation by backing-off procedure of, 161
Acetylcoenzyme A, 112
Acid–base catalysis, of active ester reactions, 141–143
Acid catalysis
  of active ester aminolysis, 37, 38
  of dicyclohexylcarbodiimide activation, 243
Acidolysis
  of N^{α}-benzyloxycarbonyl group, in REMA procedure, 298
  of N^{α}-*tert*-butyloxycarbonyl group, in REMA procedure, 298
  graded, for selective protecting group cleavage, 43
Actinomycin D, 10, 54, 55
  crystal and solution structure identity in, 10
Activated acylaminoacyl derivatives
  5(4H)-oxazolone mechanism of racemization during coupling of, 342
  racemization during coupling of, 342–350
Activated alkyl esters, 111
Activated esters, *see* Active esters
Activation energy, of Z-Phe-OPcp aminolysis, 34
Activation time, in mixed carbonic anhydride method, 281
Active ester method, 30, 90–94, 105–196
  acid-base catalysis in, 141–143
  N-acylhydroxylamine esters in, 123–125
  O-acyloximes as active esters, 125
  alkyl esters in, 109–112
  amine component influence on ester aminolysis, 135
  aminoacylation of natural peptides, 165, 166
  aminolysis mechanism, 91
  aryl esters in, 112–118
  bifunctional catalysts in aminolysis, 142
  catalysis in, 138–143
  catalysis vs. solvent (dimethylformamide) effects in, 143
  classification of active esters, 109–128
  crosslinking by, 169–171
  cyclization by, 161–163
  derivatization of peptides by, 167–169
  enol esters in, 118, 119
  ester preparation, 128–138, 302
    through activated hydroxyl components, 132, 133
    through activated protected amino acids, 128–132
    by aryl carbonates, 130
    by aryl phosphites, 129, 130
    by aryl sulfites, 129–132
    by carbonyldiimidazole, 135
    by coupling reagents, 133–138
    by DCC, 133

by hexachlorocyclotriphosphatriazine, 137
by mixed anhydrides, 129–131, 302
by oxidation-reduction condensation, 137, 138
esters derived from coupling reagents in, 119–123
use of excess active ester in stepwise chain elongation, 159
history of development of, 106–109
insoluble active esters in, 125–128
kinetics of aminolysis of, 31–37, 139
metal catalysis in, 141
in situ peptide synthesis with, 171, 172
polymerization by, 163–165
protecting group introduction by, 166, 167
racemization during ester preparation, 152–156
reactivity of *para* vs. *ortho*-nitrophenyl esters in, 140
scope and limitations of, 174, 175
segment condensation by, 159–161
side reaction, *O*-acylation, 149, 150
side reaction, racemization by, 152–156
side reactions caused by arginine guanidine, 147, 148
side reactions caused by carboxamide group, 148, 149
side reactions by chain-terminating *N*-acylation, 150, 151
side reactions during coupling, 149–152
side reactions during ester preparation, 143–149
side reactions caused by hydroxyl groups, 145–147
in solid phase synthesis, 172–174
solvent effects in, 139, 140
stepwise chain elongation by, 157–159
steric effects on side reactions, 143, 144
strategy of peptide synthesis in, 156–165
in various techniques of peptide synthesis, 165–174
Active ester racemization, 34–36, 152–156, 376, 377
α-amine protecting group effect on, 154, 155
comparison with other methods, 153
during coupling, 152–156
of *N*-hydroxysuccinimde esters of peptides during coupling, 156
isoracemization mechanism of, 153, 154
in *p*-nitrophenyl ester formation of acylpeptides by DCC, 247
in pentachlorophenyl ester formation of acylpeptides by DCC, 248
of peptide *p*-nitrophenyl esters, 248
suppression by N^{α}-benzyloxycarbonyl groups of, 154, 155
suppression by 1-hydroxybenzotrizole of, 155
suppression by hydroxylamine derivatives of, 155
suppression by *N*-hydroxyphthalimide of, 155
suppression by *N*-hydroxypiperidine of, 155
suppression by *N*-hydroxysuccinimide of, 155
suppression by N^{α}-phthaloyl groups of, 154, 155
Active esters of protected peptides
preparation in optically pure form of, 159–161
racemization problem of, 156
Acyl azides, in peptide synthesis, 68–70
Acyl carrier protein, lability in liquid HF of, 15
Acyl chlorides, in peptide synthesis, 30, 67–68, 138
Acyl cyanides, in peptide synthesis, 30, 67–70, 77
Acyl imidazoles, facile racemization of, 355
O-Acyl oximes, as active esters, 125
Acyl peptide azides
racemization of, 69–70
racemization promotion by $NaHCO_3$ washing of, 220
Acyl transfer, 67
Acylating agents, reactivity of, 30
Acylation process, stages and racemization in, 337
O-Acylation
of serine, threonine residues,
side reaction of active esters, 150
suppression by 1-hydroxybenzotriazole, 150
side reactions by, in active ester method, 149, 150
1-Acyl-3,5-dimethylpyrazolides, 84
O-Acylhydroxylamine(s), reactivity of, 125
N-Acylhydroxylamine esters, 123–125
O-Acylhydroxylamine esters, 108, 123–125

O-Acylisourea, 79, 80, 243, 251
  intermediate in dicyclohexylcarbodiimide method, 243–247, 250, 252, 257
  as intermediate in solid phase dicyclohexylcarbodiimide coupling, 246, 247
3-Acyloxy-2-hydroxy-*N*-ethylbenzamides, rapid aminolysis, low racemization of, 365
Acyloxyphosphonium intermediates, 76
  in peptide synthesis, 89–90
Acyloxytriphenylphosphonium salts, 90
Acylphenyldiimine, 84
N-Acylurea, 80, 143, 144, 242–246, 254, 257, 259
  formation in carbodiimide method, 242, 243, 244, 245, 246
  major product of dicyclohexylcarbodiimide reaction with α-benzylphenylalanine, 257
  reaction with aziridine, 259
  side product formation by steric hindrance, 143
  side product in *o*-nitrophenyl ester preparation, 143, 144
  solvent influence on formation of, 245, 246
  suppression of formation of, 244–246
    by solvents of low dielectric constants, 245, 254
Additivity of effects, during racemization, 336
Additivity principle, in predicting active ester reactivity, 34
Affinity purification of synthetic peptides, 48
Alkali, aqueous, racemization of amino acids in, 328
Alkyl esters, 106, 107, 108, 109–112, 132
  hydrazine, hydroxylamine reaction with, 109
  preparation by alkyl halogenides of, 132
Alkyl nitrite, for azide formation, 215
N-Alkylamino acid derivatives, racemization and epimerization of, 332, 333
  during ester saponification, 333
  during HBr–AcOH treatment, 333
N-Alkylisoxazolium salts, 82
p-Alkyloxybenzyloxycarbonylhydrazide resin, 211
Alkyloxycarbonyl amino acids, symmetrical anhydrides of, 305–306
Alkyloxycarbonylammonium fluoroborate, 74, 75
tert-Alkyloxycarbonylhydrazide resin, 210, 211
Amide bonds, essential structural features of peptides, 2
N^{im}-Amidino derivatives, 81
Amine basicities in water, tabulation of, 338
Amine capture method, in peptide segment condensation, 45, 49, 55
Amine protective urethane groups, introduction by azides of, 231
Amino acids
  racemization of unactivated amino acids and derivatives, 325
  α,β-unsaturated, 21
Amino acid analysis, 12
Aminoacyl azide hydrobromides, 230
Aminoacyl derivatives, activated, racemization of, 336–350
Aminoacyl insertion, 95, 123
Aminoacylation of natural peptides, using active esters, 165, 166
Aminodiacylhydrazine rearrangement, 213
α-Aminoisobutyric acid, peptides of, 88
Aminolysis
  comparative rates of, 138
  electronic and steric effects on, 138
  relative rates of, in backing-off procedure, 161
3-Amino-3-methylbutanoic acid, 88
Amino-terminal proline, urethane formation at,
  circumvention of, 298
  during mixed carbonic anhydride coupling, 286
  in REMA procedure, 298
tert-Amyloxycarbonyl azide, for amine protective group introduction, 232
Analogue design
  predictive for insulin, 8
  predictive for neurohypophyseal hormones, 7, 8
Analogues with highly improved activities, tabulation of, 51
Analysis, of solution conformation of peptides, 5
Anchimeric assistance, mechanism of, 363, 364
Anderson test, for racemization, 220, 249, 251, 277, 367
Angiotensin II analogue, highly potent, [Sar^1, $Thr(Me)^8$]AII, 51

Angiotensin II synthesis, use of pentafluorophenyl esters in, 116
Anhydride method, mixed carbonic, 72, 73–75, 264–314
Anhydride(s) in peptide synthesis, 30, 71–77, 85, 243, 264–314
  classification of, 71
  coupling methods, 71
  cyclic, 73
  formation by dicyclohexylcarbodiimide, 243
  Leuchs', 85
  mixed carbonic carboxylic, 72, 73–75, 264–314
  mixed with carbonic acids, 73–75
  mixed with carboxylic acids, 72, 73
  mixed with phosphorus-derived acids, 75–77
  mixed pivalic, 72
  mixed with various acids, 77
  symmetrical, in peptide synthesis, 71, 73
Anserine, enzymatic biosynthesis of, 27
Antamanide, mixed carbonic anhydride segment condensation of, 294
Antamanide and analogue synthesis, mixed carbonic anhydride segment condensation in, 301
Applications
  of azide segment condensation, 226–228
  of mixed carbonic anhydride method, 293–304
Arginine guanidine group side reactions, in active ester method, 147, 148
Aryl carbonates, 130
Aryl esters, 112–118, 129, 135
  of benzyloxycarbonylamino acids, 129
  electron withdrawing substituent effects in, 116
  electronic and steric effects in, 115
  preparation by carbonyldiimidazole of, 135
  preparation by diphenylketene of, 135
  with sulfone activating group, 117
Asparagine dehydration, suppression by DCC–HOBt procedure, 251
Aspartame, 73
Aspartic acid, bond cleavage by dilute acid, 13, 25
Aspartic acid, cyclic anhydrides, 73
Aspartoyl formation
  in active ester method, 151, 152
  side reaction during δ-carboxyl activation of, 152
  suppression by steric hindrance of, 151, 152
Assays for chiral purity, 366–378
Azide derivatives in amine protective group introduction, tabulation of, 232
Azide method, in peptide synthesis, 30, 68–70, 179–239
  advantages, 198
  alternative procedures of hydrazide formation in, 213
  amide forming side reaction in azide formation, 214, 215
  azide coupling, 37, 217–225, 345, 360
    kinetic data on, 37
    racemization in, 219–225
    racemization by 5(4H)-oxazolone mechanism in, 345
    slow rates in use of *N*-methylmorpholine in, 360
    reaction conditions in, 217–219
    side reactions of, 225, 226
  azide formation in, 69, 213–217
    by alkyl nitrite in organic solvents, 69, 214, 215
    aqueous acid–sodium nitrite procedure of, 213, 214
    Curtius procedure, 213, 214
    by diphenyl phosphorazidate, 216, 217
    Honzl-Rudinger procedure, 69, 213–215
    illustrative procedure of Honzl-Rudinger method, 215
    reaction conditions of nitrite procedure for, 213, 214
    side reactions in nitrite procedure of, 214
    survey of procedure, 213
  azide segment condensation, 217–228
    applications to large segments, 226–228
    excess base causing racemization in, 220–225
    high molar concentration requirement in, 217, 218
    illustrative procedure of, 218
    low temperature requirement in, 217, 218
    minimal protection in, 217
    in proinsulin C-peptide synthesis, 228
    racemization in, 219–225
    reaction conditions for, 217–219

in ribonuclease A synthesis, 227, 228
in ribonuclease T_1 synthesis, 228
in secretin synthesis, 228
solvent effect on racemization, 222
solvent purity requirement in, 217
solvents useful in, 217, 218
base control, importance in, 359, 360
base recommended for minimal racemization in , 224
comparison with other methods, 232–233
direct coupling of α-carboxyls with hydrazine, 210, 212
ester hydrazinolysis in, 202–208, 288
racemization during, 206, 207
reaction conditions of, 202–204
reactivity of esters in, 205, 206
side reactions of, 207, 288
solid phase techniques of, 204, 205
stability of protective groups and amino acids during, 205
use of excess azide, 231
experimental procedure of, 199–200
history of, 198
hydrazide formation in, 200–213
incremental chain elongation by, 230
interconversion of acid, ester, hydrazide in, 201
minimal side chain protection in, 198
peptide cyclization by, 230
polyamino acid preparation by, 230
N'-protected hydrazide approach in, 69, 208–210
by solid phase procedures, 210
by solution synthesis, 69, 208–210
protective group introduction by, 230
racemization in, *see* Azide racemization
racemization in, 199, 200, 206, 207, 219–225, 359, 360
base catalysis of, 220–222
base effect, tabulation, 224
base effect, type and amount, 224
component structure influence, tabulation of, 223
carboxy-terminal influence, tabulation of, 223
danger, relative to other methods, 225
in hydrazinolysis of esters, 206, 207
5(4H)-oxazolone mechanism of, 222
recommendation for minimization of, 224
in segment condensation, 199, 200, 219–225, 359, 360
in segment condensation, tabulation of, 221, 375
solvent influence, 222
reviews of, 199
scope and limitations of, 199, 233
special applications of, 228, 230
stages of, 69, 199
stepwise chain elongation by, 230
strategic flexibility in, 69, 202
strategy and tactics of hydrazide formation in, 200–202
summary scheme of, 232
tertiary amines in excess responsible for racemization in, 359, 360
Azide racemization, *see also* Racemization
of acyl histidine azide, 69, 219, 355, 356
of acyl peptide azides, 69, 70, 77, 199, 216, 219–225, 355, 356, 359, 360, 375
influence of COOH-terminal amino acid on, 224
promotion by bicarbonate washing, 220
rational attempts at suppression of, 220
tabulation of, 221
in segment condensation, 69, 70, 77, 199, 216, 219–225, 359, 360, 375
Aziridinones, 87

B

Bacillus thermoproteolyticus, 21
Bacitracin, amino acid racemization during acid hydrolysis of, 329
Backing-off procedure, 92, 156, 160, 161, 230, 303, 304, 366
active esters of optically pure peptides, preparation by, 160, 161
mixed carbonic anhydride method in, 303, 304
for racemization suppression in peptide segment condensation, 366
Base catalysis
of acyl azide racemization, 220–222
of aminolysis, mechanism in, 70
in azide method, 70
general, intramolecular, 39, 40, 70, 93
of active ester aminolysis, 39, 40
neighboring group effect in active ester aminolysis by, 39, 40
of peptide bond formation, 39
of peptide epimerization, 329–331, 353
of serine peptides, unactivated, 330

in synthesis, 331, 353
in unactivated state, 329–331
of racemization of *N*-benzyloxycarbonyl-*S*-benzylcysteine *p*-nitrophenyl ester, 153, 154
of racemization of protected β-cyanoalanine esters, 154
of racemization of protected β-methylaspartate active esters, 154
of racemization of protected γ-methylglutamate active esters, 154
solvent effects in, 70
"Bates' reagent," 89, 90
Benzhydroxamic acid, as additive to DCC coupling, 253
Benzhydroxamic acid esters, from benzhydroamyl chloride, 132
Benzisoxazole-5-methyleneoxycarbonyl (Bic) group, 15
Benzisoxazolium salt reagents, 83
Benzoic acid, catalysis of phenyl ester aminolysis by, 141
Benzoylglycine, 72
Benzoylglycylglycine, 72
N-Benzoylhydroxylamine esters, 123
S-Benzylcysteine, side reaction in ester hydrazinolysis of, 208
Benzyloxycarbonylamino acid(s)
chlorides of, 68
5-chloro-8-hydroxy-quinoline esters, tabulation of, 186
cyclodehydration of, 87
N-hydroxyphthalimide esters, tabulation of, 183
N-hydroxysuccinimide esters, tabulation of, 184
minimal racemization of, in anhydride activation, 277
mixed carbonic anhydrides of, 296
p-nitrophenyl esters, tabulation of, 176
oxazolones derived from, 87
pentachlorophenyl esters, tabulation of, 181
preparation by reactive carbonates, 166
stepwise mixed carbonic anhydride synthesis with, 293
2,4,5-trichlorophenyl esters, tabulation of, 179
N-Benzyloxycarbonyl-*S*-benzyl-L-cysteine *p*-nitrophenyl ester
racemization by triethylamine of, 153
racemization suppression by diisopropylethylamine, 154
N-Benzyloxycarbonyl-*S*-benzyl-L-cysteine thiobenzyl ester, 153
Benzyloxycarbonylglycine, 71
symmetrical anhydride of, 71
Benzyloxycarbonyl-nitro-L-arginine pentachlorophenyl ester, 148
Benzyloxycarbonyl-L-serine 2,4-dinitrophenyl ester, 145
Benzyloxycarbonyl-L-threonine
2,4-dinitrophenyl ester of, 145
p-nitrophenyl ester of, 145
o-Benzyloxyphenyl esters, safety-catch activation by, 366
3-(*p*-Benzyloxyphenyl) - 1,1 - dimethylpropyloxy - carbonylhydrazide resin, 211
Big gastrin, human, synthesis, 49, 50
Biosynthesis
of antibiotic peptides, by multienzyme complexes, 29
of peptides and proteins, 26–29
of proteins, via ribosome cycle, 26, 27, 28
of small peptides, by enzymatic pathway, 27
Biotin
introduction into peptides by active esters, 167–169
p-nitrophenyl ester of, 167–169
reaction with *p*-nitrophenyl trifluoroacetate, 168
2-(*p*-Biphenylyl)propyl(2)oxycarbonyl azide for amine protective group introduction, 232
2-(*p*-Biphenylyl)propyl(2)oxycarbonyl (Bpoc) group, in solid phase hydrazide synthesis, 210–212
Bisalkyl pyrocarbonate, from mixed carbonic anhydride disproportionation, 288
Bisbenzyloxycarbonyl-L-cystine bis-*p*-nitrophenyl ester, as crosslinking agent, 169
N-Blocked acyl hydrazides, safety-catch activation by, 366
Blocking groups, *see* Protecting groups
Bolton-Hunter reagent, for radioactive iodine labeling, 167
"BOP reagent," 89
Bradykinin, amino acid racemization during acid hydrolysis of, 329

Brain peptides, 54
N-Bromosuccinimide, oxidation of hydrazides by, 84
Bungarotoxin synthesis, by mixed carbonic anhydride segment condensation, 301
tert-Butanol in mixed carbonic anhydride coupling, 275
tert-Butyl ester, side reaction in ester hydrazinolysis of, 208
N^{α}-*tert*-Butyloxycarbonyl group, strong racemization suppression by, 36
tert-Butyloxycarbonyl group introduction
  by *tert*-butyl *N*-hydropiperidyl carbonate, 166
  by *tert*-butyl *N*-hydroxysuccinimidyl carbonate, 166
  by *tert*-butyl *p*-nitrophenyl carbonate, 166
  by *tert*-butyloxycarbonyl azide, 232
  by *tert*-butyl pentachlorophenyl carbonate, 166
  by *tert*-butyl phenyl carbonate, 166
  by *tert*-butyl 8-quinolyl carbonate, 166
  by *tert*-butyl 2,4,5-trichlorophenyl carbonate, 166
  by reactive carbonates, 166
tert-Butyloxycarbonylamino acids
  N-hydroxysuccinimide esters, tabulation of, 185
  minimal racemization in anhydride activation, 277
  mixed carbonic anhydrides of, 296
  p-nitrophenyl esters, tabulation of, 177
  pentachlorophenyl esters, tabulation of, 182
  stepwise mixed carbonic anhydride synthesis with, 293
  2,4,5-trichlorophenyl esters, tabulation of, 180
tert-Butyloxycarbonyl-L-arginylglycine *p*-nitrophenyl ester, in insulin aminoacylation, 166
tert-Butyloxycarbonyl-*im*-benzyl - L - histidine, racemization supression by DCC–HOBt, 251, 252
tert-Butyloxycarbonyl-L-histidine azide, histidine, racemization in, 69
Bis-*tert*-butyloxycarbonyl-L-histidine *p*-nitrophenyl ester, in insulin aminoacylation, 166
Bis-*tert*-Butyloxycarbonyl-L-lysine 2,4,5-trichlorophenyl ester, in insulin aminoacylation, 166
tert-Butyloxycarbonyl-L-methionine *p*-nitrophenyl ester, in insulin aminoacylation, 166
tert-Butyloxycarbonylnitro-L-arginine *N*-hydroxysuccinimide ester, 148
2-*tert*-Butyloxycarbonyloxyimino - 2 - phenylaceto - nitrile, for *tert*-butyloxycarbonyl group introduction, 166

C

Calcitonin M synthesis, 199, 220, 222, 324, 355, 356
  acyl azide racemization in segment 11-28 preparation, 222
  diastereomer separation by countercurrent distribution in, 324
  epimerization at a COOH-terminal His during azide coupling in, 355, 356
  racemization during azide segment condensation in, 199, 220
Calcitonin (porcine) synthesis by DCC–HOSu procedure, 249
Carbodiimides, 78, 79
Carbodiimides, water soluble, 81, 242, 306
Carbodiimide method, *see* Dicyclohexylcarbodiimide method
Carbonic acid monoesters, 73–75, 264, 266, 268–271, 286, 287, 296, 304, 357, 358, 373
Carbonylbis-L-methionine *p*-nitrophenyl ester, reaction with insulin of, 169
Carbonyldiimidazole, 84, 135
Carbonyldi-*s*-triazole, 84
ω-Carboxamide group
  protection by benzyl, benzhydryl groups of, 149
  side reactions by, in active ester method, 148, 149
  side reaction by, in ester hydrazinolysis of, 207
N-Carboxyanhydrides
  in peptide synthesis, 68, 85–88, 306, 307
  formation from active esters, 153
  formation by dicyclohexylcarbodiimide, 259
  rapid peptide synthesis by, 85

Carboxyl activation
by addition reactions, 77–83
carbodiimide mechanism of, 243–252
formal classification in peptide synthesis of, 30
Carboxyl component in mixed carbonic anhydride method, 271, 272
Carboxypeptidase A and B, activation by trypsin of, 17
Carnosine, enzymatic biosynthesis of, 27
Cascade processes, of enzyme proteolysis, 16–18
Catalysis of peptide bond formation, 37–40, 93, 138–143
by acid, 141–143
in active ester reactions, 138–143
by base, intramolecular general, 93
by carboxyl acids, 37, 38
by 1-hydroxybenzotriazole, 37–39
by 3-hydroxy-4-oxo-3,4-dihydroquinazoline, 38
by imidazole, of ester aminolysis, 142
by metals, 141
Catalytic hydrogenolysis
of N^{α}-benzyloxycarbonyl groups in REMA procedure, 296, 298
of N^{α}-benzyloxycarbonyl groups, solvents for, 298
Catechol half-esters, 40, 94, 117, 363
acidolytic or hydrogenolytic activation of, 94, 117
intramolecular anchimeric assistance during aminolysis of, 40, 363
monobenzyl precursor of, 94, 117
phenacyl precursor protection of, 117, 366
Cesium salt procedure, for esterification of protected peptide acids, 201
Chain termination, in active ester method, 150, 151
Chemical cleavage, of peptide bonds, 23
Chemical synthesis of peptides
basic features in, 29–31
catalysis of, 37–40, 70, 93, 138–143, 243
kinetics in, 31–37, 139, 279, 281
racemization in, 41, 42, 336–378
Chiral integrity
definition of, 318
preservation of, 41
Chloride(s)
acyl, 67
of benzyloxycarbonylamino acids, 68
diphenylphosphoryl, 76
o-nitrophenylsulfenyl, 86
of phthaloylamino acids, 68
pivaloyl, 72
p-toluenesulfonyl, 78
of tosylamino acids, 68
of trifluoroacetylamino acids, 68
p-Chlorobenzenesulfonylhydroxamic acid, as additive to DCC coupling, 254
p-Chlorobenzhydroxamic acid, as additive to DCC coupling, 253
N-*p*-Chlorobenzoylhydroxylamine esters, 123
Chlorodimethyloxyborane, 78
5-Chloroquinoyl(8) esters, in solid phase synthesis, 173, 174
Chymotrypsin
activation of, 17
phylogenetic relation to trypsin, 17
Chymotrypsinogen, 17, 20
Cinnamycin
acid cleavage at aspartic and hydroxyaspartic acid residues in, 25
resistance to enzymatic cleavage of, 25
Cis conformation, occurrence in peptides, 8
Cleavage, of peptide bonds, 11–26
Collagen model, synthesis by tripeptide *p*-nitrophenyl ester polymerization, 165
Conformation
crystal to solution structure relationships, 10
of lysine-vasopressin in solution, 6
of *o*-nitrophenyl esters, 140
of oxytocin in solution, 5
in solution, analysis of, 5–8, 54
Constraints
of mixed carbonic anhydride procedure, 265, 308
of REMA procedure, 299
Controlled cleavage of peptide bonds, 16–26
Corticotropin
formation from β-lipotropin of, 19
lability in liquid HF of, 15
stepwise synthesis using *p*-nitrophenyl esters, 157
Countercurrent distribution, use in purification of synthetic peptides, 48
Coupling methods
classification, 66
statistics usage, comparative, 98

Coupling rate constants of Z-amino acid and dipeptide active esters, tabulation of, 36
Coupling reagents
active ester formation from, 119–123
for active ester preparation, 133–138
EEDQ, IIDQ, 289–292
EEDQ and IIDQ, intermediate mixed anhydride formation by, 289–292
polymeric, 94
selective activation by, 43
Crosslinking
active esters for, 169–171
bifunctional acylating agents for, 169
Crystal structure
of actinomycin D, 10
of glucagon, 8
of insulin, 8–10
of ribonuclease, 20
of trypsin inhibitor, bovine, 10
X-ray analysis of, 8–10
Curtius procedure, of azide formation, 213, 214
Curtius rearrangement
in azide segment condensation, 225, 226
formation of urea derivatives by, 226
monitoring by infrared spectroscopy of, 225
suppression at low temperature of, 226
Cyanides, acyl, 68, 77
β-Cyanoalanine derivatives, of asparagine, 148, 149
Cyanogen bromide
cleavage of methionyl peptide bonds by, 21, 23, 24, 26
mechanism of, 23, 24
in nisin, 21
cleavage of *S*-methylcysteryl bonds by, 24
Cyanomethyl esters
of benzyloxycarbonylamino acids, 111
preparation from chloroacetonitrile, 111, 132
p-Cyanophenyl esters, 116
Cyclic anhydrides, of aspartic or glutamic acids, 73
Cyclic dipeptides (diketopiperazines), 107, 161, 162, 332
Cyclic imidazolides, 88
Cyclization of peptides
by active ester synthesis, 92, 161–163
by the azide method, 228, 230
use of dual function groups in, 163
by mixed carbonic anhydride method, 304
by polymeric esters, 163
Cyclodehydration, of benzyloxycarbonylamino acids, 87
Cyclodimerization, 163, 165
Cysteamine derivatives, *S* → *N* acyl migration in, 112
Cysteine, racemization of derivatives of, 335, 355
Cytochrome c-(88-104) synthesis, azide segment condensation in, 218

D

Dansylglycine *N*-hydroxysuccinimide ester, for fluorescent labeling, 167
DCC–HOBt procedure, 37, 80, 110, 250–253, 256
β-lactone formation with, 110
with Boc-His(Bzl)-OH, 256
elimination of peptide trifluoroacetylation side reaction in, 251
kinetic data on, 37
racemization suppression by, 250–252
scope and limitations of, 251, 252
suppression of *N*-acylurea formation by, 250, 251
suppression of asparagine or glutamine dehydration in, 251
DCC–HOSu and DCC–HOBt procedures, racemization during segment condensation in, 360–362
[1-Deamino-2-phenylalanine, 7-(L-3,4-dehydroproline)] arginine-vasopressin, superselective analogue, 7
4-Decyloxybenzyloxycarbonyl azide, for amine protective group introduction, 232
Dehydroalanine, occurrence in subtilin of, 22
Dehydrobutyrine, occurrence in subtilin of, 22
Derivatization of peptides, active esters for, 167–169
Design of minimally racemizing activated acyl derivatives, 362–364
Diacylimide formation
in mixed carbonic anhydride coupling, 288

in solid phase mixed carbonic anhydride coupling, 295
Diagnostic amino acids, in peptide segment condensation, 45
Diastereomer separation, 323, 324
use of countercurrent distribution for, 324
efficiency of crystallization in, 323
use of high pressure liquid chromatography in, 324
use of partition chromatography for, 324
in peptides prepared by segment condensation, 323
problems of, 323, 324
in solid phase-synthetic peptides, problems of, 323
Diastereomeric content, 320–323
assay procedures for determination of, 322
isotopic dilution assay for, 323
numerical problem of, 320–323
strategies for minimization of, 322
of synthetic peptides, 320–323
5,7-Dichloro-*N*-ethylbenzoisooxazolium fluoroborate, 146
2,6-Dichloro-4-nitrophenyl esters, 115
active ester preparation by, 133, 134
complex with halophenols of, 134
complex with pentachlorophenol, 248
Dicyclohexylcarbodiimide method, 71, 74, 79–81, 119, 120, 210, 212, 241–264, 360–362, 374–375
acid catalysis of DCC activation, 243, 244
acidic reaction conditions in, 243
activation temperature in, 254
O-acylisourea intermediate of DCC activation, 243–247, 250, 252, 257
N-acylurea formation in, 242, 243, 244, 245, 246
N-acylurea formation, suppression of, 244–246
amine reaction in, 258, 259
amino acid esters, reaction with, 258
coupling conditions in, 252–258
cyclic *O*-acylisourea derivative, 243
DCC–HOBt procedure, 210, 212, 251, 252
N-hydroxy compounds as additives in, 248–254
1-hydroxybenzotriazole, 250–253, 256
N-hydroxysuccinimide, 248–250, 253
tabulation of, 253, 254
imidazole reaction in, 258, 259
kinetics of coupling, problems with, 37
mechanism of carboxyl activation, 120, 243–252
use of oxazolone in, 257
preferred reaction conditions in, 252, 254
racemization in, 242, 243, 244, 245, 360–362, 374, 375
5(4H)-oxazolone mechanism of, 244
solvent effects on DCC–HOBt, 361
suppression by 1-hydroxybenzotriazole, 250–252
suppression by *N*-hydroxysuccinimide, 248, 249
reaction conditions for peptide bond formation by, 252–258
reviews of, 243
secondary amino acid coupling by, 257, 258
side reactions in, 81, 242, 244, 254
with DCC–HOBt procedure, 251
nitrile formation, 244
α-substituted amino acid coupling by, 256, 257
symmetrical anhydride formation in, 71, 120, 246, 247, 254–256, 305, 306
trapping agents for *O*-acylurea in, 247–251
1-hydroxybenzotriazole, 250, 251, 256
N-hydroxysuccinimide, 248–250
p-nitrophenol, 247, 248
pentachlorophenol, 248
water-soluble carbodiimides in, 81, 242, 306
Diethyl chloroarsenite, 78
Diethyl chlorophosphite, 76
Diethyl phosphorocyanidate, 76
Diethylphosphoryl cyanide, 68, 77
1-(Diethyloxyphosphino)-1,2,4-triazole, 84
Diglycine ethyl ester, 107
Dihydrooxazinones, in peptide synthesis, 88
γ,δ-Dihydroxyleucine, acid cleavage at, 23,24
Diisopropylmethyloxycarbonyl azide, for amine protective group introduction, 232
Diketopiperazine(s), 107, 161, 162, 332
base catalyzed epimerization of, 332
formation by cyclization, 162
α,α-Dimethyl-3,5-dimethyloxybenzyloxycarbonyl azide, for amine protective group introduction, 232

Dimethyl malonic acid, DCC reaction with 244
Dimethyl sulfoxide
  mixed carbonic anhydride coupling in, 275
  removal by lyophilization of, 218
Dimethylacetamide, mixed carbonic anhydride formation in, 272
Dimethylamino-*tert*-butylacetylene, 78
Dimethylformamide
  active ester aminolysis in, 140
  mixed carbonic anhydride coupling in, 275
  mixed carbonic anhydride formation in, 272
  purification of, 273
2,4-Dinitrophenyl esters, 114, 132
  formation from 2,4-dinitrofluorobenzene of, 132
Di-*p*-nitrophenyl sulfite, 303
Dipeptide formation, rate constants of *p*-nitrophenyl aminolysis, 34
Diphenylketene, 78
Diphenylketene-*p*-tolylimine, 78
Diphenylmethyloxycarbonyl azide, for amine protective group introduction, 232
Diphenylphosphinic mixed anhydride procedure, 76
Diphenylphosphinyl chloride, 76
Diphenylphosphite, 76
Diphenylphosphorazidate, 76, 77, 216
  as a coupling reagent, 216
  illustrative coupling procedure of, 216
  peptide azide formation by, 213, 216
Diphenylphosphoryl
  azide, 77
  chloride, 76
Disproportionation, of mixed carbonic anhydrides, 288
  suppression by low activation temperature, 288
Duramycin
  acid cleavage at aspartic and hydroxyaspartic acid residues in, 25
  resistance to enzymatic cleavage of, 25

E

EEDQ procedure, 75, 267, 289–292, 358, 359, 374
  coupling reagent, 75, 267, 289–292
  illustrative procedure of, 292
  racemization in 289–292, 358, 359, 374
  suitable solvents for, 292
  tabulation of applications of, 290, 291
  urethane formation in, 289, 292
Elastase, activation by trypsin of, 17
Eledoisin analogue, mixed carbonic anhydride segment condensation of, 300
β-Endorphin, 19, 48, 50, 324
  diastereomer, synthetic, separation of, 324
  formation from β-lipotropin of, 19
  HPLC elution profile of, 48
  solid phase synthesis of, 50
Enkephalin, 51, 52, 53, 256
  analogue, highly potent, [DAla^2, MePhe^4, Met(O)^5-ol]EK, 51
  oral activity of analogue, 53
  sites of proteolytic cleavage in, 52
  solid phase synthesis, Met-EK, using symmetrical anhydride, 256
Enol esters, 82, 108, 109, 118, 119, 121
  derived from inamines, 121
  derived from 2,4-pentanedione, 119
Enolization mechanism, 41, 319, 326, 339
  of amino acid racemization in aqueous solution, 326
  of racemization, 41, 319
  of urethane blocked amino acid racemization during activation, 339
Enzymatic peptide biosynthesis, 27
Epimerization, *see also* Racemization
  definition of, 318
  during acidic or basic deblocking steps in synthesis, 332
  of *N*-alkylamino acid derivatives, 332, 333
  of diketopiperazines, 332
  during ester saponification, 332
  during fragment-condensation, 357–362
    of acyl azides, 359, 360
    by DCC–HOSu and DCC–HOBt procedures, 360–362
    by mixed anhydrides and EEDQ, 357–359
  during HBr–AcOH treatment, 332
  numerical problem of, 320–323
  of the penultimate COOH-terminal amino acid in, 299, 337, 348, 349
  of peptides with COOH-terminal glycine at penultimate residue, 356, 357
  of peptides with COOH-terminal proline at penultimate residue, 356, 357

of peptides and other unactivated amino acid derivatives, 329–333
during acidic or basic deblocking steps in synthesis, 332
during acidic hydrolysis or methanolysis of peptides, 329
base-catalysis in, 329–331
N^{α}-protective group influence on, 331
rate constant, definition of, 318
tabulation of conditions, 336
Ester bond formation, by mixed carbonic anhydride, 302
Ester saponification
epimerization of peptides during, 332
side products of, 201
Esterification
cesium salt–alkyl halide procedure, 201
of protected peptide acids, 201
Ethanolamide, as COOH-terminus in gramicidins A, B, C, 20
Ethyl chlorocarbonate (-chloroformate)
for mixed carbonic anhydride formation, 73, 270, 304
urethane formation by mixed anhydrides from, 280
Ethyl chloroformate, *see* Ethyl chlorocarbonate
Ethyl diisopropylamine, in mixed carbonic anhydride formation, 271, 280
N-Ethyl-*N*-dimethylaminopropyl-carbodiimide, 81
Ethyl-2-hydroximino-2-cyanoacetate, as additive to dicyclohexylcarbodiimide coupling, 254
2-Ethyl-7-hydroxybenzo-1,2-oxazolium tetrafluoroborate, 122
Ethyloxyacetylene, 71, 78, 119, 120
1-Ethyloxycarbonyl-2-ethyloxy-1,2-dihydroquinoline (EEDQ), 75
Ethyloxycarbonylglycyl-glycyl-leucine ethyl ester, 107
N-Ethyloxycarbonylhydroxylamine esters, 123
Excess active ester, in stepwise chain elongation strategy, 159
Excess acylating agent, in stepwise chain elongation strategy, 45, 296

F

Fibrinopeptide-B, synthesis by stepwise mixed carbonic anhydride procedure, 293
9-Fluorenylmethyloxycarbonyl azide, for amine protective group introduction, 232
Fluorescamine, 219, 275, 296
Fluorescamine column monitoring system, 48
Fluorescent labeling of peptides, by active esters, 167
Formic acid, catalysis of phenyl ester aminolysis by, 141
Formyl group, as NH_2-terminus in gramicidins A, B, C, 20
Formylamino acids, racemization in mixed carbonic anhydride activation, 279
Four component condensation, 45, 49, 55, 95
in peptide segment coupling, 49
Fragment-condensation
mixed anhydrides and EEDQ, 357
racemization in, 357
Free energies of amino acid and peptide formation, tabulation of, 26

G

Gastric inhibitory polypeptide (1–10) synthesis, use of diphenyl phosphorazidate, 216
General base catalysis, of ester aminolysis, 141–143
Gliadin, base-catalyzed epimerization of, 330
Glucagon, crystal structure of, 8
Glucagon synthesis
by DCC–HOSu procedure, 248, 249
use of maximum protection in, 44
7–15 segment by REMA, 297
Glutamic acid, cyclic anhydrides, 73
Glutamine dehydration, suppression by DCC–HOBt procedure, 251
Glutathione, enzymatic biosynthesis of, 27
Glycine esters, Cu^{2+} complexes, Co^{3+} complexes with, 109, 141
Glycolic acid amide esters, 111
Gonadoliberin, REMA synthesis of, 297
Gonadoliberin analogue, highly potent, des-(Gly-NH_2^{10})-[DLeu[6]]GnRH, 51
Gramicidin A, stepwise mixed carbonic anhydride synthesis of, 294

Gramicidins A, B, and C
conformation: π_{LD} helix, 20
resistance to proteolytic degradation, 20, 21
Gramicidin S
biosynthesis, 29
resin ester hydrazinolysis in synthesis of, 204
stepwise mixed carbonic anhydride synthesis of analogues, 294
synthesis by *p*-nitrophenyl ester cyclization, 162
Growth hormone, lability in liquid HF of, 15

H

Heterodetic peptides, 161
Heterodetic ring systems, in nisin and subtilin, 21
Hexachlorocyclotriphosphatriazine, active ester preparation by, 137
Hexaglycyl, cyclic, X-ray analysis of, 8
Hexamethylphosphoramide
mixed carbonic anhydride coupling in, 275
mixed carbonic anhydride formation in, 272
High performance liquid chromatography, purity test of synthetic peptides by, 48
Hindered tertiary bases, use of in active ester coupling, 157
Histidine, 81, 86, 219, 251, 252, 258, 355, 356
acylazide coupling rate of, 219
Boc-His(Bzl)-OH racemization
catalysis of, 252
suppression by DCC–HOBt procedure, 251, 252
imidazolide formation in azide coupling of, 219
racemization of derivatives of, suppression by HOBt, 355, 356
reaction with DCC, 258
reversible cyclization of His azide to imidazolide, 356
Homodetic peptides, 161
cyclic, 162, 163
Homopolyamino acid, 85
Homoserine lactone, 88, 110
intramolecular acylation by, 110
Honzl-Rudinger prodecure
of azide formation, 69, 213–215, 231
illustrative procedure of, 215
Hydrazide formation, 109, 110, 200–213
by acylation of free amino acid hydrazide, 213
chromatographic spot tests for, 204
by coupling of α-carboxyls with hydrazine, 210, 212
by ester hydrazinolysis, 109, 110, 202–208
of activated esters, 206
S-benzylcysteine side reaction in, 208
bis-acylated hydrazide product in, 207
tert-butyl ester side reaction in, 208
ω-carboxamide side reaction in, 207
of carboxyl-terminal alkyl esters, 109, 110, 200, 202–208
ornithine or aminoguanidine side product in, 207
of 5(4H)-oxazolone, 206
racemization during, 206
reaction conditios of, 202–204
reactivity of esters in, 205
selectivity of, 205, 206
side reactions in, 207
solvents for, 202, 203
stability of amino acid residues in, 205
stability of protective groups in, 205
$\alpha \rightarrow \beta$ transpeptidation in, 207
HPLC purification in, 203
hydrazide analysis, specific for, 204
hydrazinolysis of acyl peptide-resin esters for, 204
monitoring of, 203
nitrogen analysis: adequacy for, 204
from peptide acid, N_2H_2 and DCC–HOBt, 201, 210, 212
of peptides containing ω-alkyl esters, 208
precursors for, 202
via *N'*-protected hydrazide, 200, 208–210
solid phase techniques for, 204
strategy and tactics, 200–202
Hydrazides of *N*-blocked amino acid or peptides
activation by oxidation to acyldiimine, 84–85
chromatographic spot tests for, 204
oxidation and conversion to acid, 201
symmetrical bisacyl, side product in azide coupling, 226
Hydrogen bonds, different kinds of, 8
Hydrogen bromide in acetic acid, epimerization of peptides in, 332
Hydrogen fluoride, liquid, 15, 16, 44

deamination of Asn and Gln residues by, 44
decrease of bioactivity by, 15
γ-Glu acyl carbonium ion formation by, 44
methionyl peptide bond cleavage, 15, 44
N → O acyl shift caused by, 15, 44
side reactions in protecting group cleavage by, 15, 16, 44
Hydrogen isotope exchange, measurement of racemization rates by, 324
Hydrogenolysis, catalytic
N^α-benzyloxycarbonyl groups in REMA procedure, 296, 298
liquid ammonia as a solvent for, 44
Hydrolysis of peptides, 11–16
acidic
energy and entropy values in, 13, 14
Ile-Ile resistance of, 13
racemization of amino acids during, 329
alkaline, 12
for amino acid analysis, 12
leaving group departure in, 12
tabulation of heat and energy values in, 14
Hydroxamic acid(s), 108, 250–252
derivatives of, 124, 125, 250–252
2-Hydroximino-2-cyanoacetate, 136
β-Hydroxyamino acids, *N* → *O*-acyl shift in, 24
Hydroxyaspartic acid, cleavage by dilute acid of, 25
1-Hydroxybenzotriazole, 80, 89, 109, 125, 143, 155, 250–254, 256, 258
as *O*-acylisourea trapping agent, 250, 251
as additive to DCC coupling, 80, 89, 250–252, 253, 256
catalysis of active ester aminolysis by, 38, 143
catalysis of DCC reaction with histidine, 258
catalytic effects of, 38, 125
o-nitrophenylsulfenyl cleavage by, 256
racemization suppression by, 125, 155, 250–254
in active ester coupling, 155
in DCC coupling, 250–254
suppression of *N*-acylurea formation in DCC coupling, 250–254
as a trapping agent in DCC coupling, 250–252
1-Hydroxybenzotriazole esters, 37, 124, 125, 127, 250, 252
formation by DCC of, 250, 252
polymeric, 127
in situ preparation of, 252
Z-Phe-OBt reaction rates with H-Val-NHR, 37
Hydroxydihydrobenzotriazinone esters, 125
N-Hydroxyglutarimide, as additive to DCC coupling, 253
*N*Hydroxyglutarimide esters, 123
Hydroxyl group, side reactions with active esters, 145–147
Hydroxyl ion, nucleophilic addition to amide carbonyl of, 12
Hydroxylamine, derivatives, 30, 108, 123–125
N-Hydroxy-5-norbornene-2,3-dicarboximide, as additive to DCC coupling, 250, 253
3-Hydroxy-4-oxo-3,4-dihydro-1,2,3-benzotriazine, as additive to DCC coupling, 251, 253, 256
3-Hydroxy-4-oxo-3,4-dihydroquinazoline, as additive to DCC coupling, 38
o-Hydroxyphenyl ester, 94; *see also* Catechol half-esters
N-Hydroxyphthalimide
as additive to DCC coupling, 253
racemization suppression in active ester coupling, 155
N-Hydroxyphthalimide esters, 108, 123
N-Hydroxypiperidine
as additive to coupling, 253
racemization suppression in active ester coupling, 155
N-Hydroxypiperidine esters (1-Piperidyl esters), 125, 362
intramolecular anchimeric assistance in aminolysis of, 362
suppressed racemization during coupling, 362
1-Hydroxy-2(1H)-pyridone, as additive to DCC coupling, 254
8-Hydroxyquinoline esters, 124; *see also* Quinolyl(8) esters
N-Hydroxysuccinimide
as additive to DCC coupling, 80, 89, 248–250, 253
DCC reaction with, 248–250
racemization suppression by
in active ester coupling, 155
in DCC coupling, 249, 250

suppression of *N*-acylurea formation in DCC coupling, 249
as a trapping agent in DCC activation, 248, 250
N-Hydroxysuccinimide catalysis, of DCC reaction with histidine, 259
N-hydroxysuccinimide–dicyclohexylcarbo-diimide procedure, *see under* DCC–HOBt procedure
N-Hydroxysuccinimide esters, 37, 93, 123, 173, 174, 249, 302
preparation by dicyclohexylcarbodiimide of, 249
preparation by mixed carbonic anhydride method of, 302
in solid phase synthesis, 173, 174
Z-Phe-OSu reaction rates with H-Val-NHR, 37

I

IIDQ, 75, 289–292
tabulation of applications of, 290, 291
Imidazole catalysis
of methyl easter aminolysis, 142
of *p*-nitrophenyl ester aminolysis, 142
Imidazolide
cyclic, 88
formation in acyl histidine azide coupling of, 219
Iminolactone, in chemical cleavage of peptide bonds, 23
Incremental chain elongation, *see also* Stepwise chain elongation
strategy of peptide synthesis, 45
by azide method, 228, 230–231
proof of homogeneity in, 45, 46
scope and limitation of, 45, 46
strategy of peptide synthesis, 45
Industrially produced peptide pharmaceuticals, tabulation of, 53
In situ peptide synthesis
using active esters, 171, 172
procedure of, 171, 172
Inamine-derived enol esters, 121
Insoluble active esters, 109, 125-128
scope and limitations of use, 127, 128
Insulin, 8–10, 15, 18, 81, 99, 166, 169–171, 173, 226, 294, 299, 300, 361
A-chain solid phase synthesis using *p*-nitrophenyl esters, 173
aminoacylation of, 166
A1–9 azide, stability toward Curtius rearrangement, 226
B-chain-(24–30) solid phase mixed anhydride synthesis of, 294
biosynthesis of, precursor conversion into, 18
crosslinking with carbonylbis-L-methionine *p*-nitrophenyl ester, 169
crosslinking for molecular spacing of, 169, 170, 171
crystal and solution structure identity in, 10
crystal structure of, 8–10
diastereomer separation by countercurrent distribution in synthesis of, 324
monomer, dimer, hexamer association of, 8
mixed carbonic anhydride in synthesis of, 299
racemization by DCC–HOBt procedure in synthesis of, 49, 50, 361
receptor binding region of, 8
Intramolecular energy leak, for racemization suppression in segment condensation, 365
Intramolecular general base catalysis, 70, 93
Inverse addition, in mixed carbonic anhydride formation, 278
α-Inversion processes, 318
p-Iodophenylalanine 2,4,5-trichlorophenyl ester, in insulin aminoacylation, 166
Isobutyl chlorocarbonate (chloroformate), 73–75, 264, 266, 268–271, 280, 286, 287, 296, 304, 357, 358
for mixed carbonic anhydride formation, 269, 270
urethane formation by, 286
Isobutyl chloroformate, *see* Isobutyl chlorocarbonate
1-Isobutyloxycarbonyl-2-isobutyloxy-1,2-dihydroquinoline (IIDQ), 75
Isocyanate formation, in azide segment condensation, 225, 226
Isocynic acid esters, 78
Isoracemization, 324
Isothiocyanate procedure, 78
Isotopic dilution assays for racemization or epimerization, 323
Isoxazolium reagents, in peptide synthesis, 82–83, 122, 133

Izumiya test, for racemization, 222, 367

K

Kemp test (isotope dilution), for racemization, 220, 367
Kemp's reagent, 83
Ketenimines, 121, 122
Ketoketenimines, 78, 82, 122
Kinetic isotope effect, to distinguish 5(4H)-oxazolone and enolization mechanisms of racemization, 345
Kinetics
  of ester aminolysis, 31–37, 139
  of peptide bond formation, 31–37
  of peptide bond formation, tabulation of, 32, 33, 35, 36

L

Lactams and lactones, in peptide synthesis, 88–89, 110
Lactones of β-hydroxyamino acids, 110
β-Lactone
  of benzyloxycarbonyl-L-serine, 110
  of N-trityl-L-serine, 110
β-Lactones of hydroxyamino acids, side products in DCC–HOBt reaction, 146
Lactone formation
  of nitroarginine derivatives, 147
  of N^g-p-toluenesulfonylarginine derivatives, 147
Lanthionine
  occurrence in cinnamycin and duramycin of, 25
  occurrence in nisin and subtilin of, 21, 22
Lead tetraacetate oxidation, of phenylhydrazides, 84
Leuchs' anhydrides, 85
Leucine ethyl ester, 107
Limitations
  of azide method, 199
  of mixed carbonic anhydride procedure, 265, 308
  of REMA procedure, 299
β-Lipotropin
  precursor role of, 19
  solid phase synthesis of, 49, 50
Lossen rearrangement, 250, 251, 252
Luliberin analogues, REMA synthesis of, 299
Luliberin synthesis, by REMA procedure, 74
Lysinoalanine, formation from dehydroalanine of, 25
Lysozyme, lability in liquid HF of, 15

M

Maximum protection strategy, in peptide synthesis, 43, 44, 296
β-Melanotropin, formation from β-lipotropin of, 19
2-Mercaptopyridine esters, 124
8-Mercaptoquinolin esters, 114
Methyl esters, 107, 109, 111
  substituted, 107, 111
S-Methylcycteine, cyanogen bromide cleavage at, 24
O-Methyl-N,N′-dicyclohexylisourea, 259
Methyldiethylamine, influence on mixed anhydride racemization, 280
N-Methylimidazole, acyl transfer agent, 252
β-Methyllanthionine, occurrence in cinnamycin, duramycin, subtilin, 22, 25
N-Methylmorpholine
  effect on racemization of, 270, 271
  influence on mixed carbonic anhydride racemization by, 280, 281
  minimal racemization in mixed carbonic anhydride coupling, 280–283
  in mixed carbonic anhydride coupling, 274, 275
  in mixed carbonic anhydride formation, 269–271
4-Methyloxybenzyloxycarbonyl azide, for amine protective group introduction, 232
p-Methyloxybenzyloxycarbonyl group introduction, by reactive carbonates, 166
N-Methylphenylalanine, formation of urethane in mixed anhydride coupling, 286
Methylthiophenyl ester oxidation to methylsulfonyl ester, 117
Microheterogeneity of product, in solid phase peptide synthesis, 46–48
Minimum side chain protection
  in active ester method, scope and limitation of, 157
  in azide method, 198

side reaction during excess active ester acylation using, 45
strategy in peptide synthesis of, 43, 44, 45, 157, 198
Mixed anhydrides, for active ester synthesis, 129–131
Mixed carbonic anhydride method, 73–75, 129–131, 264–314, 357–359, 373, 374, *see also* REMA procedure
activation in, 269–274
active ester preparation by, 302
Anderson conditions in, for racemization suppression, 265, 268, 269, 271, 277, 279–283, 293, 301
anhydride formation, 269–274
anhydrous conditions during, 273
inverse addition in, 278
reaction conditions for, 269–270
reaction vessel for, 274
reagents for, 270
requirements for, 267–269
solvents for, 272–273
temperature during, 273
tertiary amines for, 270–271
time of, 273
applications of, 293–304
backing-off procedure by, 303, 304
carboxyl component influence, 271, 272
concentration effects in, 273
controlled reaction conditions for, 268, 269
cost of, 265
coupling conditions in, 274, 275
coupling under Schotten-Baumann conditions, 274
EEDQ and IIDQ coupling reagents, 289-292
ester bond formation by, 302
use of excess anhydride in, 273
history, 264, 265
incremental chain elongation by, 293–295
isolation of products in, 276
kinetic data on, 37
opportunities and constraints in, 308
peptide cyclization by, 304
peptide segment condensation by, 299-301
product yields in, 276
N^{α}-protective groups compatibility with, 271
racemization in, *see* mixed carbonic anhydride racemization
reaction monitoring in, 275
reaction stages, basic scheme of, 266
related methods, 304–308, *see also* Symmetrical anhydrides
resume of Anderson studies on, 282, 283
segment condensation, tabulation of applications, 300
side reactions of, 268, 269, 283, 285–288
diacylimide formation in, 288
disproportionation in, 288
reviews on, 285
steric effects on urethane formation in, 286, 287
tabulation of urethane formation, 287
urethane formation in, 285–288
urethane formation at NH_2-terminal proline, 286
in solid phase synthesis, 294, 295
special application, 301–304
steric factors in, 268, 286, 287
Mixed carbonic anhydride racemization, 268, 276–284, 357, 358, 373, 374
N^{α}-acyl substituent effect on, 277, 279
coupling time effect on, 282
minimization of, 282–284, 358
minimization by use of N^{α}-urethane-protected amino acids, 277
5(4H)-oxazolone mechanism of, 276, 345, 357
during segment condensation, 268, 276–284, 357, 358, 373, 374
solvent influence on, 281, 282
suppression by *N*-hydroxysuccinimide of, 364
suppression by short-time activation, 279, 281, 282
tabulation of, 278, 280, 284
tertiary base effect on, 280, 281
time course, during anhydride activation, 279, 281
Mixed carboxylic acid anhydrides, 307, 308
Mixed pivalic anhydride, 72, 74
Molecular organization, rhegnylogic and sychnologic types of, 54
Monitoring
of azide segment condensation, 218, 219
of Curtius rearrangemnt in azid coupling by IR, 225, 226
of hydrazide formation, 203, 204
of mixed carbonic anhydride coupling, 275

of peptide racemization, 319
of REMA coupling, 296
Motilin synthesis, use of diphenyl phosphorazidate coupling in, 216
α-MSH lability in liquid HF of, 15
β-MSH synthesis, 198
Multienzyme complex antibiotic peptide biosynthesis, 29

N

Naturally occurring peptides
in 1953, 52
aminoacylation by active esters of, 165, 166
Neighboring group 1.5 interaction, in nonenzymatic peptide bond cleavage, 23
Nisin
heterodetic ring systems in, 21, 23
structure elucidation of, 21, 24
Nicotinic acid amide, 106
Nitrile formation, of asparagine and glutamine derivatives, by DCC, 148, 149
p-Nitrophenol as a trapping agent
in DCC activation, 247, 248
in DCC-mediated *p*-nitrophenyl ester preparation, 247, 248
p-Nitrophenyl acetate,
selective amino group acylation by, 167
as substrate of hydrolytic enzymes, 167
p-Nitrophenyl chloroformate, 130
o-Nitrophenyl esters (ONo), 114, 140, 174, 178
conformation of, 140
of protected amino acids, tabulation of, 178
use in solid phase synthesis of, 174
p-Nitrophenyl esters (ONp), in peptide synthesis, 93, 108, 113, 114, 126, 129, 130, 131, 133, 134, 135, 136, 140, 143, 145, 147, 173, 174, 178, 247, 248
of acylpeptides, by backing-off procedure, 247, 248
aminolysis rate constants in DMF of, 34
of benzyloxycarbonylamino acids, tabulation of, 176
of *tert*-butoxycarbonylamino acids, tabulation of, 177
dipeptide formation by, 34
hydroxybenzotriazole catalysis of aminolysis of, 143
polymeric, 126
preparation by di-*p*-nitrophenyl carbonate of, 130
preparation by di-*p*-nitrophenyl sulfite, 145, 147
preparation by ethoxyacetylene of, 134' 135
preparation by *p*-nitrophenyl chloroformate of, 130
preparation by *p*-nitrophenyl 2-hydroximino-2-cyanoacetamide carbonate of, 136
preparation by *p*-nitrophenyl trifluoracetate, 131
preparation by *o*-phenylene *p*-nitrophenyl phosphite of, 131
preparation from phthaloylamino acid chloride of, 129
of protected amino acids, preparation by DCC, 133, 134, 247, 248
use in solid phase synthesis of, 173–174
p-Nitrophenyl 2-hydroximino-2-cyanoacetamide carbonate, *p*-nitrophenyl ester, preparation by, 136
di-*p*-Nitrophenyl sulfite, 145
o-Nitrophenylsulfenyl amino acids, racemization in mixed anhydride activation, 279
o-Nitrophenylsulfenyl chloride, 86
o-Nitro-*p*-sulfophenyl esters, 116
p-Nitrothiophenyl esters, 114
Nomenclature
of 5(4H)-oxazolone, 39
of peptide size, 198
of racemization in peptide synthesis, 317–319
Nonenzymatic cleavage
of peptide bond, 23–26
by *N*-bromosuccinimide, 23
by cyanogen bromide, 21, 23, 24, 26
1,5-interaction in, 23
neighboring group effect in, 23
scope and limitations of, 25, 26
Nonplanarity, pyramidalization, of peptide bonds, 10, 11
Nonracemizing routes to acylating agents, 364–366

O

Opioid peptides, enkephalins, endorphins, 19
Opportunities, of mixed carbonic anhydride method, 265, 308
Opportunities and constraints
in racemization control, 372, 378
of REMA procedure, 299
Optical rotation, limitations of measuring diastereomeric content of peptides by, 324
Ornithine, side product of ester hydrazinolysis, 207
Oxazolidone, from acylserine azide by Curtius rearrangement, 226
Oxazolone(s), *see under* 5(4H)-Oxazolone(s)
5(4H)-Oxazolone(s), 87, 88, 121, 153, 154, 222, 242–248, 257, 265, 268, 276, 281, 319, 339, 340, 342–350
in acyl azide racemization, 222, 345
as acylating agent, from benzoylamino acids, 121
benzyloxy, 68
catalysis of formation, 345–347
use in DCC method, 257
derived from benzyloxycarbonylamino acids, 87
evidence for, in peptide coupling, 343–345
formation in DCC method, 242, 243, 244, 245
formation in mixed carbonic anhydride method, 265, 345
formation in *p*-nitrophenyl ester preparation of acetylpeptides by DCC, 247
formation in pentachlorophenyl ester preparation of acylpeptides by DCC, 248
formation by urethane-blocked amino acids during activation, 339, 340
from neutral amine attack of *N*-methylamino acids, 345, 346
infrared absorption characteristic of, 343
kinetic isotope effect as evidence for, 345
mechanism of formation of, 41, 153, 154, 276, 339, 342–350
mechanism of racemization, 41, 153, 154, 342–350
of acylamino acids during activation, 339
nomenclature of, 39
peptide-derived
effect of hindered nucleophile on racemization of, 347, 348
racemization and aminolysis, similar rates of, 347
racemization suppression by hydroxylamine, hydrazine, 348
solvent influence on racemization, 347
trapping by HOSu or HOBt of, 348
in peptide synthesis, 87, 88, 222, 242–248, 257, 265, 268, 276, 281, 319, 339, 340, 342–350
racemic asymmetric induction in aminolysis of, 348
racemization and aminolysis of, 347–349
racemization dependence on amine basicity, 347
rate constants of ring opening and racemization of, 347
slow formation, racemization of N^{α}-urethane blocked amino acids, 339, 340
substituent effects on formation of, 349, 350
suppression in mixed carbonic anhydride coupling, 268, 281
trapping to prove intermediacy of, 343, 344
Oxidation-reduction condensation, 137, 138, 149
active ester preparation by, 137, 138
Oxytocin, 7, 49, 52, 54
analogues of,
antidiuretic to pressor ratio in, 7
superselective [4-L-threonine, 7-glycine]oxytocin, 7
conformation of
analysis by nmr in solution of, 4–8
crystal structure of 6-9 segment, 10
hydrogen bond stabilization in 7-9 segment, 4
intramolecular hydrogen bond stabilization in, 5
rational analogue design based on, 7
role of corner positions in, 7
role of peptide backbone in, 7
β-turn of 7-9 segment, 4
receptor recognition in binding of, 7
sites of proteolytic cleavage in, 52

synthesis,
25th anniversary of, 49, 52, 54
cyanomethyl esters for 1-3 segment, 111
DCC coupling in, 242
p-nitrophenyl esters in, 108
use of pentafluorophenyl esters in, 116
3-7 segment, on solid phase, by *o*-nitrophenyl esters, 174
stepwise strategy using *p*-nitrophenyl esters in, 157

P

Paleobiological dating, by racemization rates of amino acids, 325, 326
Parathyrine-(1-34)synthesis, using DCC–HOBt procedure, 258, 259
Partial rate factors of racemization, 341
Partition chromatography, use in purification of synthetic peptides, 48, 324
Pentachlorphenol, trapping agent in pentachlorophenyl ester synthesis by DCC, 248
Pentachlorophenyl esters of protected amino acids
in peptide synthesis, 93, 115
preparation by DCC, 248
Pentafluorophenyl esters, 115, 116, 139, 183
of protected amino acids, tabulation of, 183
Peptide(s) of α-aminoisobutyric acid, 88
Peptide(s), racemization of unactivated derivatives, 325
Peptide acyl derivatives, activated racemization of, 336–350
Peptide amides, formation of, 109, 110
Peptide azide hydrobromides, 230
Peptide bond, 1–64
cleavage of, 11–26
enzymatic of, 16–23
hydrolytic, 11–16
limited, by specific proteases, 16–23
by liquid HF of, 15, 16
nonenzymatic, chemical of, 23–26
conformational states of, 3
contour maps of constant electron density of, 3, 4
definition of, 2
formation of, 26–52, 55, 65–104
active esters for, 105–196
azide method for, 179–239
comparative rates of, 138
as a concentration-independent intramolecular rearrangement, 55
dicyclohexylcarbodiimide method for, 241–261
endergonic process of, 26
mixed carbonic anhydride method for, 264–314
racemization of, 34–37, 41, 42, 152–156, 206, 207, 219–225, 242–252, 276–284, 315–383
bond length and angles of, 3, 4
nature of, 2–11
nonplanarity of, 10, 11
in cyclotetraglycine, 10
in dihydrochlamidocin, 10
in glycyl-L-alanine, 11
in glycyl-L-leucine, 11
in Pro-Leu-Gly-NH_2, 11
in trypsin inhibitor, 10
solvolysis of, 11, 15, 16
Peptide chain assembly, strategy of, 42–52
Peptide hydrazide formation, *see* Hydrazide formation
Peptide hormone action, 54
Peptide isolation, 54
Peptide pharmaceuticals, tabulation of, 53
Peptide segment condensation
by active ester method, 159–161
amine capture method of, 45, 49, 55
by azide method, 217–228
by DCC–HOBt procedure, 250–252, 256, 258, 259
by dicyclohexylcarbodiimide method, 252–258
by four component condensation, 49, 95, 97
by mixed carbonic anhydride method, 299–301
overactivation of peptide acylating agents in, 362
requirements for successful coupling in, 362
Peptide size nomenclature, 198
Peptide synthesis, 29–42
acylation via tetrahedral intermediate in, 30, 31
amine component in, 42
basic scheme of, 42, 43

α-carbon chirality preservation in, 29, 30
carboxyl activation in, 42
carboxyl component in, 42
electrophylic acylation mechanism of, 30
functional group protection in, 42
in situ procedure of, 171, 172
racemization control during, 41, 351–362
by recombinant DNA techniques, 55
requirement in, 43
scope and limitations, 49–52
selection of least racemization-prone segment condensation site, 36
strategies and tactics of functional group protection in, 43, 44
Phallisine synthesis, cyclization by mixed carbonic anhydride method in, 304
Phalloidin synthesis
cyclization by mixed carbonic anhydride method in, 304
structural studies on, 23, 24
Pharmaceutical peptide drugs, tabulation of, 53
o-Phenacyloxyphenyl esters, safety-catch activation by, 117, 366
Phenyl esters, 108, 113, 117, 128, 129, 130, 141
acid catalysis of aminolysis of, 141
of acylamino acids from acyl chlorides, 128
of benzyloxycarbonylglycine, 130
halogen substituted,
nature of nucleophile on rate of aminolysis, 32–34
pK_a and reactivity correlation of, 32–34
solvent effect on reactivity of, 32–34
preparations of by mixed anhydrides, 129
substituted from isoxazolium derivatives, 122
p-Phenylazophenyl ester, 116
o-Phenylene *p*-nitrophenyl phosphite, 131
Phenylglycine active esters, racemization of, 154
Phenylhydrazides, 84
1-Phenyl-3-methyl-5-hydroxypyrazole esters (1-phenyl-3-methyl-5-pyrazolone enol esters), 117–119
1-Phenyl-3-methyl-5-pyrazolone, keto, enol forms of, 119
Phosphorus oxychloride, for carboxyl activation, 76
Phosphorus pentachloride, for carboxyl activation, 68
Phosphorus trichloride, for carboxyl activation, 76
o-Phthaldialdehyde, for fluorescent labeling, 219
Phthaloylamino acid chlorides, 68
preparation of nitrophenyl esters by, 129
Phthaloylamino acids, racemization in mixed anhydride activation, 279
1-Piperidyl esters (*N*-Hydroxypiperidine esters), 125,142
acetic acid catalysis of aminolysis of, 142
Pivalic mixed anhydride, 72; 307, 308
of N^{α}-tosylamino acids, 308
Pivaloyl chloride in mixed anhydride method, 72, 271
Pivalylbenzhydroxamic acid as additive to DCC coupling, 253
Polyamino acids, 85, 306
by *N*-carboxyanhydride polymerization, 85
preparation by the azide method, 228, 230
Polymeric active esters
derived from 4,4′-dihydroxydiphenylsulfone polymer, 126
scope and limitations in the use of, 127, 128
Polymeric coupling reagents, 94
Polymeric 1-hydroxybenzotriazole esters, 127
Polymeric *N*-hydroxysuccinimide esters, 126
Polymeric nitrophenyl esters, 126
Polymerization, by active ester synthesis, 163–165
Poly-L-serine, base-catalyzed epimerization of, 330
Predictive analogue design, 50, 54
Predictive factors
for coupling rate constants, 35, 36
of ester aminolysis, tabulation of, 35
for racemization rate constants, 36, 42
Preparation of active esters, 128–138
Presecretory proteins, 18
Pretrypsinogen, 18
Product microheterogeneity, in solid phase peptide synthesis, 46–48
Product yields, in mixed carbonic anhydride coupling, 276

Proinsulin, and preproinsulin, conversion into insulin of, 18
Proinsulin C-peptide synthesis, azide segment condensation in, 228
Prolactin-(37-55) (ovine) solid phase synthesis using preformed symmetrical anhydrides, 255
Propargyl esters, 111
Proteases, site-specific action of, 18, 19
Protected amino acid 1-hydroxypiperidine esters, tabulation of, 186
N′-Protected hydrazides, 208–210
  combinations of peptide protection with, 209
  resin supports for preparation of, 210–212
  solid phase synthesis of, 208, 210–212
  solution synthesis of, 208
  tabulation of, 209
Protecting groups
  acidolysis of, 43
  catalytic hydrogenolysis of, 43, 44
  chemical selectivity of cleavage of, 43, 44
  cleavage reagent for, 43
  combination (ideal) of, 43
  effect on epimerization of unactivated amino acids and peptides, 331
  HF cleavage of, 15
  introduction by active esters, 166, 167
  introduction by azide method, 228, 231–232
  kinetic selectivity of cleavage of, 43
  racemization during cleavage or transformation of, 352
  racemization, effected by, 340
  selective deprotection of, 43
  semipermanent protection by, 43
  temporary protection by, 43
Protection, *see under* Protecting groups
Protein biosynthesis, ribosomal pathway of, 26
Proteolysis, 16–23
  irreversible nature of limited, 17
  limited, physiological role of, 16–20
  resistance to, by antibiotic peptides, 20
  resistance to, by native protein molecules, 20
  role in structural elucidation of, 19
  single peptide bond cleavage in, 16–18
Protonation
  of amide, 12
  of peptide carbonyl, 12
Pteroyloligo-γ-glutamates, solid phase mixed anhydride synthesis of, 294, 295
Purification, of solid phase-synthetic peptides, 47, 48
Pyrazole catalysis of ester aminolysis by, 142
2-Pyridone-1-esters, 124
2-Pyridyl esters, 115, 124, 363
  intramolecular anchimeric assistance during aminolysis of, 363
3-Pyridyl esters, 115
2-Pyridylthiol esters, 40, 114, 363
  intramolecular anchimeric assistance during aminolysis of, 40, 363

Q

Quinolyl esters: 3, 6, 8, acidity of leaving groups of, 40
Quinolyl(8) esters
  no assistance of 5(4H)-oxazolone racemization by, 39
  intramolecular anchimeric assistance during aminolysis of, 39, 363
  intramolecular H-bond formation in aminolysis of, 39

R

Racemic 5(4H)-oxazolones, asymmetric induction on aminolysis of, 348
Racemization
  of activated acyl aminoacyl derivatives, 342–350
  of activated acyl derivatives of amino acids and peptides, 336–350
    amine basicity problem in, 338, 339
    catalysis by excess tertiary base of, 338
  of activated urethane-blocked amino acid derivatives, 339–342
  of amino acids, 325–328, 333–335
    in acetic acid solution, 327, 328
    during acid hydrolysis, 326, 327, 329
    N-alkylamino acid derivatives, 332, 333
    in aqueous alkaline solution, 328
    in aqueous solution: enolization mechanisms of, 326
    in dilute aqueous solution, 325, 326
    in strongly acidic aqueous solution, 326, 327

X,R,Y-model of substituent contributions in, 334, 335
in carbodiimide method, 242, 243, 244, 245
of cysteine derivatives, 355
definition of, 317
design of reagents for minimization of, 319, 320
by enolization of 5(4H)oxazolone formation, tabulation of factors in, 351
of histidine derivatives, 355, 356
mechanism of, 41, 42, 222, 276, 277, 319, 333–335, 342–351
catalysis of 5(4H)-oxazolone formation, 345–347
comparison of enolization and 5(4H)-oxazolone routes, 350, 351
enolate formation of amino acids in water, 333
enolization mechanism, 41, 42, 319
formation of 5(4H)-oxazolone, 319
5(4H)-oxazolone mechanism, 41, 42, 222, 276, 277, 319, 342–351
racemization of 5(4H)-oxazolone, 347–349
substituent effects on 5(4H)-oxazolone formation, 349–350
monitoring procedures for, 319
nomenclature of, 317–320
nonracemizing routes to acylating agents, 364–366
opportunities and constraints, 372, 378
in peptide synthesis, 315–383, *see also* Epimerization; Racemization minimization; Racemization control
by active esters, *see also* Active ester racemization, 34–36, 152–156, 376, 377
by the azide method, *see also* Azide racemization, 69, 70, 77, 199, 206, 207, 216, 219–225, 355, 356, 359, 360, 375
of benzisoxazolium salt reagents, 83
chiral integrity of peptide coupling procedures, 372
by DCC–HOSu and DCC–HOBt procedures, 360–362
by dicyclohexylcarbodiimide method, 242–245, 360–362
by mixed carbonic anhydride method, *see also* Mixed carbonic anhydride racemization, 268, 276–284, 357, 358, 373, 374
via 5(4H)-oxazolone mechanism, 343–345
promotion by triethylamine hydrochloride, 353
during protecting group transformation or removal, 352
tabulation of, 373–377
temperature coefficient, large of, 352, 353
predicting ability for, 320
rate constants of Z-amino acid and dipeptide active esters, tabulation of, 36
reviews of literature on, 320
during solid-phase synthesis, 357
tabulation of conditions, 336
tests for, 41, 366–378
by Anderson, 220, 249, 277, 367
by Bodanszky, 367
by Izumiya, 222, 283, 367
by Kemp, isotope dilution, 220, 283, 367
review of literature on, 366, 367, 368
tabulation of, 368, 370, 371
by Weygand, 367
by Young, 220, 249, 277, 367
Racemization control, 351–362
base effects in, weak, hindered tertiary amines, 353
during peptide coupling, 351–362
by amine bases and salts, 353, 354
by concentration, 354
by solvent, 354, 355
by temperature, 352, 353
solvent order in ONp ester, DCC, DCC–HOBt coupling for, 354
for special amino acids, 355–357
COOH-terminal proline or glycine, 356, 357
cysteine, 355
histidine, 355, 356
Racemization minimization, 362–366
by anchimeric assistance of aminolysis, 362–364
by use of backing-off procedure and safety-catch activation for, 366
in carbodiimide method,
by 1-hydroxybenzotriazole, 250–252
by *N*-hydroxysuccinimide, 248, 249
in concentrated solution of reactants, 354

design of activated acyl derivatives for, 362–364
by intramolecular anchimeric assistance
using catechol half-esters, 363
using 1-hydroxypiperidine esters, 362
using 2-pyridyl esters, 363
using 2-pyridylthiol esters, 363
using quinoyl(8) esters, 363
use of intramolecular energy leak for, 365
in mixed carbonic anhydride coupling, 268, 269, 277, 279, 281
by nonracemizing routes to acylating agents, 364–366
during peptide coupling, 352–355
by selection of condensation site, 36
by suppression of 5(4H)-oxazolone formation, 350
by trapping 5(4H)-oxazolone before racemization, 350, 351, 364
Racemization-free, definition of term, 324, 325
Racemization-free coupling by the azide method, 219
Racemizing tendencies of activated urethane-blocked amino acid derivatives, via enolization mechanism, 339, 340
Radiolabeling of peptides, by active esters, 167
Rate constants
of hydrolysis, 13
of *p*-nitrophenyl ester aminolysis in DMF, 34
prediction of, 34–36
Reaction temperature, racemization suppression by use of minimum temperature, 352, 353
Reactivity, comparative, of 1-hydroxybenzotriazole, *N*-hydroxysuccinimide and 2,4,5-trichlorophenyl esters, 37
Reagent K, 82
Receptor concept, 7, 54
Recombinant DNA, approach to peptide synthesis, 55
REMA procedure, 265, 295–299
concept of, 74, 295, 296
excess mixed carbonic anhydride in, 295–297
hydrolysis of excess mixed carbonic anhydride, 295, 296
illustrative example of, 296, 298
scope and limitations of, 299
tabulation of applications of, 297
urethane formation in, 286
Repetitive Excess Mixed Anhydride Procedure, *see* REMA procedure
Reverse solid-phase synthesis, 126
Review(s)
of azide method, 199
of carbodiimide method, 243
of mixed carbonic anhydride method, 265
of peptide synthesis, 66
of racemization literature, 320
of racemization tests, 366, 367, 368
of solid phase synthesis, 50
Ribonuclease A, 15, 19, 46, 85, 198, 217, 228, 307
lability in liquid HF of, 15
structure elucidation of, 19
Ribonuclease A synthesis by solid phase method, 46
Ribonuclease S-protein synthesis, 44, 49, 85, 86, 198, 217, 222, 227, 228, 307
azide racemization in segment 77-85 preparation during, 222
azide segment condensation in, 198, 227, 228
use of minimum protection in, 44, 217
use of *N*-carboxyanhydrides in, 85, 86, 307
Ribonuclease T_1 synthesis use of mimimum protection in, 44
Ribosome cycle of protein biosynthesis, 28, 29
Ring closure of aspartyl residues, 151

S

Safety-catch activation for racemization suppression in peptide segment condensation, 366
Sakakibara's reagent, 131
Salicylamide esters, substituted, 123
"Salt couplings," 92
Saponification of esters, epimerization of peptides during, 332
Schotten-Baumann reaction, 67
Scope and limitations
of active ester method, 174, 175
of azide method, 199, 233

of DCC–HOBt procedure, 251, 252
of insoluble (polymeric) active esters, 127, 128
of minimal side chain protection in active ester method, 157
of mixed carbonic anhydride method, 308
of nonenzymatic peptide bond cleavage, 25, 26
in peptide synthesis, 49–52
of peptide synthesis in solution, 49
of segment condensation strategy, 45
of solid phase peptide synthesis, 46–48, 49, 50
of stepwise chain elongation strategy, 45, 46
Scope of mixed carbonic anhydride method, 265, 308
Scope of REMA procedure, 299
Seco-ketophalloidine synthesis, cyclization by mixed carbonic anhydride method in, 304
Secondary amino acid acylation by symmetrical anhydrides, 257
Secretin synthesis
azide segment condensation in, 228
by REMA procedure, 74, 297, 299
of segment 1-6 using diphenyl phophorazidate coupling, 216
of segment 4-27 by *in situ* procedure, 172
of segment 22-27 on solid phase, using active esters, 172
stepwise strategy using *p*-nitrophenyl esters in, 157
Segment condensation
by active ester synthesis, 159–161
by mixed carbonic anhydride method, 299–301
in peptide synthesis, 45
scope and limitations of, 45
strategy of peptide synthesis, 45
Selenonaphthyl esters, 114
Selenophenyl esters, 114
Semigramicidin S cyclopentapeptide, 163
Sequential polypeptides
synthesis of, 92, 93, 165
by tripeptide *N*-hydroxysuccinimide polymerization, 165
by tripeptide *p*-nitrophenyl ester polymerization, 164, 165
by tripeptide thiophenyl ester polymerization, 164, 165
Serine-*N*-peptide bonds, acid hydrolysis sensitivity of, 13
Side chain substituent, effect on racemization, 340
Side products, of ester saponification, 201
Side reactions
in active ester method, 143–156
by *O*-acylation, 149, 150
arginine guanidine related, 147, 148
by aspartoyl formation, 151, 152
carboxamide related, 148, 149
by chain-terminating *N*-acylation, 150, 151
during coupling, 149–152
during ester preparation, 143–149
hydroxyl group related, 145–147
by racemization, 152–156
from steric effects, 143, 144
by trifluoracetylation, 131
amide formation during azide preparation, 214, 215
in azide segment condensation, 225, 226
with dicyclohexylcarbodiimide, 81, 242
in mixed carbonic anhydride methods, 268–269, 283, 285–288
in nitrite procedure of azide formation, 214
in peptide ester hydrazinolysis, 207
Silicon tetrachloride, 78
Solid phase peptide hydrazide synthesis, 204
on recycled hydroxymethyl resin, 204
tabulation of solid supports for, 211
Solid phase peptide synthesis, 46–48, 125, 126, 172–174, 246, 247, 252, 255, 256, 294, 295, 342
active esters in, 172–174
O-acylisourea intermediate in, 246, 247
asparagine, glutamine active esters in, 172–174
DCC coupling in, 246, 247
of Leu-enkephalin, 256
use of excess acylating agent in, 46
halobenzyl side chain protection in, 46, 47
hydrogen fluoride cleavage problem in, 46, 47
improvements made in, 46–48
using mixed carbonic anhydrides in, 294, 295

modified supports in, 46, 47
monitoring of, 47
product microheterogeneity in, 46
prolactin-(37-55), 255
purification problem in, 46–48
racemization expected per cycle of, 342
reverse incremental chain elongation, 230–231
reverse procedure using polymeric reagents, 126
scheme of, 47
scope and limitations of, 46–48
symmetrical anhydrides in, 46, 47, 255
Solution conformation of neurohypophyseal hormones, 50
Solution peptide synthesis, 45, 46
Solvent(s)
for azide formation, 215
of low dielectric constant, in DCC coupling, 246, 254
purity, purification, in carbonic anhydride formation, 272
purity, purification, in carbonic anhydride segment condensation, 301
Solvent effect(s)
on active ester reactions, 38, 139, 140, 354, 355
on acyl azide racemization, 222
on benzyloxycarbonyl-L-phenylalanine 2,4,5-trichlorophenyl ester aminolysis, 139
on cyanomethyl ester aminolysis, 139
of diethylformamide in active ester aminolysis, 139, 140
on dicyclohexylcarbodiimide activation, 245, 246
on 1-hydrobenzotriazole-catalysis of active ester aminolysis by, 38
on mixed carbonic anhydride formation, 268, 269
on *p*-nitrophenyl ester aminolysis, 139, 140, 354, 355
on 2-pyridyl ester aminolysis, 139, 140
on racemization in DCC and DCC–HOBt coupling, 354, 355
on racemization of mixed carbonic anhydrides, 268, 281
on racemization of urethane-protected activated amino acids, 340, 341
on solid phase coupling, 246, 247
on 2,4,5-trichlorophenyl ester aminolysis, 139
Somatostatin, sites of proteolytic cleavage in, 52
Somatostatin synthesis, 45
Somatotropin-(1-10), human, REMA synthesis of, 297
Stability of mixed carbonic anhydrides, 268
Staphylococcal nuclease, lability in liquid HF of, 15
Stepwise chain elongation
by active ester synthesis, 157–159
by the azide method, 228, 230–231
use of excess active ester in, 159
minimal side chain protection in , 157
by mixed carbonic anhydride method, 293–295
Steric factors, in mixed carbonic anhydride formation, 268
Steric factors, causing side reactions in active ester preparation, 143, 144
Steric hindrance in peptide synthesis, 87, 97
Strategy
in active ester synthesis, 156–165
of functional group protection, 43, 44
of peptide chain assembly, 42–52
of REMA procedure, 259–299
Strategy and tactics, of hydrazide formation, 200–202
Structural analysis, role of proteolysis in, 19–23
Substance P, REMA synthesis of, 297
α-Substituted amino acids, difficult acylation of, 256, 257
Subtilin
resistance to chymotryptic cleavage of, 21
structure elucidation of, 21, 22
thermolysine cleavage of, 21, 22
Subtilisin, ribonuclease S-peptide and S-protein formation by, 19, 20
Sulfide bridges, in nisin and subtilin, 21, 22
p-Sulfonamidophenyl ester, 116
Sulfur trioxide-dimethylformamide complex, 78
Symmetrical Anhydrides, 71, 73, 135, 246, 247, 252, 254–256, 288, 296, 301, 305, 306
of alkyloxycarbonylamino acids, 305, 306
application in solid phase synthesis of, 305, 306

of N^{α}-*tert*-butyloxycarbonylamino acids, 246, 247, 252, 254–256, 296
decomposition of, 255, 256
formation by DCC, 246, 247, 252, 254–256, 305, 306
formation by ethoxyacetylene of, 135
from mixed anhydride disproportionation, 288
predominant agent in solid phase synthesis, 255
preformed, in solid phase synthesis, intramolecular rearrangement of, 255, 256
preparation of, 135, 246, 247, 252, 254–256, 305, 306
in situ preparation of, 252, 257, 306
preparation by inamine of, 305
preparation in stable form of, 305, 306
Synthesis of phallotoxins, mixed carbonic anhydride segment condensation in, 301
Synthesis of proteins, problems of, 49, 50

T

Tabulation
of alkyloxycarbonyl azides used in amine protective group introduction, 232
of amine basicities in water, 338
of benzyloxycarbonylamino acid 5-chloro-8-hydroxyquinoline esters, 186
of benzyloxycarbonylamino acid *N*-hydroxyphthalimide esters, 183
of benzyloxycarbonylamino acid *N*-hydroxysuccinimide esters, 184
of benzyloxycarbonylamino acid *p*-nitrophenyl esters, 176
of benzyloxycarbonylamino acid pentachlorophenyl esters, 181
of benzyloxycarbonylamino acid, 2,4,5-trichlorophenyl esters, 179
of *tert*-butyloxycarbonylamino acid *N*-hydroxysuccinimide esters, 185
of *tert*-butyloxycarbonylamino acid *p*-nitrophenyl esters, 177
of *tert*-butyloxycarbonylamino acid pentachlorophenyl esters, 182
of *tert*-butyloxycarbonylamino acid, 2,4,5-trichlorophenyl esters, 180
of coupling and racemization rate constants of Z-amino acid and dipeptide active esters, 36
of EEDQ and IIDQ applications, 290, 291
of free energies of amino acid and peptide formation, 26
of *N*-hydroxy compounds as additives, in DCC coupling, 253, 254
of industrially produced peptide pharmaceuticals, 53
of kinetic data on peptide bond formation, 32, 33, 35, 36
of mixed carbonic anhydride segment condensations, 300
of *o*-nitrophenyl esters of protected amino acids, 178
of pentafluorophenyl esters of protected amino acids, 183
of peptide analogues with highly improved activities, 51
peptide hydrolysis heat and energy values of, 14
of predictive factors (in ester aminolysis), 35
of protected amino acid 1-hydroxypiperidine esters, 186
of *N'*-protected hydrazide, combination with peptide protection, 209
of racemization
in acyl azide coupling, 221
of acyl azides in presence of base, 224
in azide coupling, depending on component structures, 223
conditions for induction of, 336
in mixed carbonic anhydride method, 278, 280, 284
parameters of enolization or 5(4H)-oxazolone formation, 351
during peptide coupling, 373–377
of racemization suppression in mixed anhydride method, 284
of racemization tests, 368, 370, 371
of REMA applications, 297
of resin supports for solid phase synthesis of peptide hydrazides, 211
of urethane formation in mixed carbonic anhydride coupling, 287
Tactics, of functional group protection, 43, 44
extent of side chain protection, 44

mandatory, minimum protection, 44
maximum protection, 44
Temperature control
in azide segment condensation, 217
in mixed carbonic anhydride formation, 273, 275
Tetraethyl pyrophosphite, 76
Tetrahedral intermediate, 106, 108
Thermolysin, 21
action at elevated temperature of, 21
cleavage of subtilin, 20–22
p-Thiocresyl esters, catalysis of aminolysis of, 143
Thioglycolic acid *S* esters, 113
Thionyl chloride, 68
Thionyldiimidazole, 84
Thiophenyl esters, 107, 108, 112, 113, 129
preparation by mixed anhydrides, 129
Thymosin α_1-(5-10), hydrazide preparation from acid, N_2H_4 and DCC–HOBt of, 212
p-Toluenesulfonyl chloride, 78
p-Toluenesulfonylamino acid chlorides, 68
Torsion angle ω, of peptide bond, 10, 11
$\alpha \rightarrow \beta$ Transpeptidation, side reaction in ester hydrazinolysis of, 207
Trapping agents, for racemization suppression in peptide segment coupling, 247–251, 364
1,2,4-Triazole, catalysis of ester aminolysis by, 142
2,4,5-Trichlorophenyl esters, 108, 115, 143
hydroxybenzotriazole catalysis of aminolysis, 143
of Z-Phe, reaction with H-Val-NHR, 37
2,4,6-Trichlorophenyl esters, 115
Triethylamine, influence on mixed carbonic anhydride racemization by, 280
Triethylamine hydrochloride, racemization induction by, 353
Trifluoroacetyl derivatives, as side products in active ester synthesis, 131
Trifluoroacetylamino acid
chloride, 68
racemization in mixed carbonic anhydride activation, 279
2-Trifluoromethylpseudooxazolone-(5) derivatives, formation in mixed anhydride synthesis, 271
Trimethylacetic acid, catalysis of phenyl ester aminolysis by, 141
N-Trimethylacetylhydroxylamine esters, 124
Trimethylamine influence on mixed carbonic anhydride racemization by, 280
tri-*p*-Nitrophenyl phosphite, 303
Triphenylphosphite, 76, 84
Trypsin, 16, 20
action of, 16
activation of, by enterokinase, 16
conformational change during activation, 16, 20
role of NH_2-terminal isoleucine in activation, 16
phylogenetic relation to chymotrypsin, 17
specificity for Lys, Arg of, 19
Trypsin inhibitor, bovine, 10, 49, 50
crystal structure of, 10
solid phase synthesis, 49, 50
Trypsinogen, 16, 20
activation peptide of, 16
activation peptide sequence, 16
as zymogen, inactive precursor, 16
Tryptophan, nonenzymatic peptide bond cleavage at, 23
Tumor proteases, 18
Tyrocidin, biosynthesis on multienzyme complexes, 29
Tyrosine, nonenzymatic peptide bond cleavage at, 23

U

Ugi reaction, 95
Urethane amine protecting groups introduction by reactive carbonates of, 166, 167
Urethane formation
in mixed carbonic anhydride method, 285–288
in REMA procedure, 298
Urethane protected activated amino acids
tertiary amine-catalyzed racemization by enolization, 339–342
protecting group influence on, 340, 341
side chain substituent effects in, 340, 341
N^{α}-Urethane-protected amino acids, minimal racemization in mixed carbonic anhydride activation, 277

V

Vasoactive intestinal peptide-(1-6) synthesis, use of diphenyl phosphorazidate coupling in, 216
Vasopressin, 51, 53, 54, 55
Vasopressin analogues, highly potent
  [1-deamino, Phe^2, Δ^3Pro^7]AVP, 51, 55
  [1-deamino, DVal^4, DArg^8]AVP, 51
Vinyl esters, 108, 118, 120
  preparation from phenacyl esters of, 118
  preparation from vinyl acetate of, 118
  substituted, derived from ethoxyacetylene, 120

W

"Water-soluble" carbodiimides, 81, 242
Weygand test for racemization, 367
Woodward's reagent K, giving enol ester of substituted acetophenone, 109, 122
Wool, crosslinking of, 169, 171

X

X-Ray crystal structure analysis, 8, 20, 54
  multiple conformations in one molecule, 8
  of ribonuclease, 20
  of ribonuclease S, 20
  similar conformations in different molecules, 8

Y

Yields in mixed carbonic anhydride coupling, 276
Young test for racemization, 220, 249, 277, 367